Green, Energy-Efficient and Sustainable Networks

Green, Energy-Efficient and Sustainable Networks

Special Issue Editors

Josip Lorincz
Antonio Capone
Luca Chiaraviglio
Jinsong Wu

MDPI • Basel • Beijing • Wuhan • Barcelona • Belgrade

MDPI

Special Issue Editors
Josip Lorincz
University of Split
Croatia

Antonio Capone
Politecnico di Milano
Italy

Luca Chiaraviglio
University of Rome Tor Vergata
Italy

Jinsong Wu
Universidad de Chile
Chile

Editorial Office
MDPI
St. Alban-Anlage 66
4052 Basel, Switzerland

This is a reprint of articles from the Special Issue published online in the open access journal *Sensors* (ISSN 1424-8220) from 2018 to 2019 (available at: https://www.mdpi.com/journal/sensors/special_issues/green_energy_efficient_sustainable_networks).

For citation purposes, cite each article independently as indicated on the article page online and as indicated below:

LastName, A.A.; LastName, B.B.; LastName, C.C. Article Title. *Journal Name* **Year**, *Article Number*, Page Range.

ISBN 978-3-03928-038-4 (Pbk)
ISBN 978-3-03928-039-1 (PDF)

Contents

About the Special Issue Editors

Josip Lorincz received a B.Sc. (M.S. equivalent) and Ph.D. degree in telecommunications engineering and computer science from the University of Split, Croatia in 2002 and 2010, respectively. In 2003 he joined the Department of Electronics and computing at FESB – Faculty of Electrical Engineering, Mechanical Engineering and Naval Architecture, University of Split, Croatia where he currently works as an associate professor. In 2009/2010 academic year, he was a visiting researcher at the Department of electronics, informatics and bioengineering of the Politecnico di Milano, Milan, Italy. As a project leader or researcher, he participated in more than twenty scientific and professional projects funded by EU, public or private sector. He is founder and co-chair of the Symposium on Green Networking and Computing, organized in the frame of International Conference on Software, Telecommunications and Computer Networks (SoftCOM). He also serves as the technical program committee member for many international scientific conferences and reviewer for top scientific journals. He was awarded as an outstanding young researcher by the Croatian Academy of Engineering in 2013. His current research interests include energy-efficient wireless and wired networks, optimization in telecommunications, advanced design, the management and analyses of computer heterogeneous networks, and performance evolution of routing protocols. He has authored more than 40 research papers published in different scientific conferences and journals. He is a senior IEEE member, senior ACM member, and the first president of the Croatian ACM chapter. Since 2004, he has owned Cisco CCNA, CCAI, and BCMSN certificates.

Antonio Capone is Full Professor at Politecnico di Milano (Technical University of Milan), where he is the Dean of the School of Industrial and Information Engineering, Director of the Advanced Network Technologies Laboratory (ANTLab), and a member of the strategic agenda team POLIMI 2040. His expertise is on networking and his main research activities include radio resource management in wireless networks, traffic management in software-defined networks, network planning, and optimization. On these topics, he has published more than 250 peer-reviewed papers. He serves on the Technical Program Committee of major conferences in networking, he is an editor of *IEEE Trans. on Mobile Computing, and Computer Communications* (Elsevier), and he was editor of *ACM/IEEE Trans. on Networking and Computer Network*s (Elsevier). He is a fellow of the IEEE.

Luca Chiaraviglio is Associate Professor at the University of Rome Tor Vergata (Italy). He holds a Ph.D. in Telecommunication and Electronics Engineering, obtained from Polytechnic of Turin (Italy). He has spent research periods at Boston University (USA), INRIA Sophia Antipolis (France), Auckland University of Technology (New Zealand), and ETECSA S.A. (Cuba). Luca has co-authored over 130 publications in international journals, books, and conferences, and he has collaborated with more than 200 authors. He participates in the TPC of top-leading conferences, including IEEE INFOCOM, IEEE GLOBECOM, IEEE ICC, IEEE VTC, and IEEE GlobalSIP. He is also a member of the organizing committee of different conferences, such as ECOC, IEEE LANMAN, and 5G-Italy. He was the general chair of the IEEE ICCCS 2019 conference. Moreover, he is on the Editorial Board of IEEE Communications Magazine, IEEE Access and IEEE Transactions on Green Communications and Networking. He is currently a member of the H2020 LOCUS project. He has been involved in the H2020 5G-EVE, H2020 Superfluidity, FP7 Trend, FP7 EcoNet, and FP7 Bone European projects. During 2018–2019, he was the coordinator of the national project BRIGHT: Bringing 5G Connectivity in Rural and Low-Income Areas. Luca has received the Best Paper Award at the IEEE VTC and ICIN conferences. Some of his papers are listed as Best Readings on Green Communications by IEEE. Moreover, he has been recognized as an author in the top 1% most highly cited papers in the ICT field worldwide. His papers "Optimal Energy Savings in Cellular Access Networks" and "Reducing Power Consumption in Backbone Networks" are the most cited papers from all IEEE ICC conferences and IEEE ICC workshops in the period 2009–2018 (Source: Scopus). He is also an IEEE Senior Member and a founding member of the IEEE Communications Society Technical Subcommittee on Green Communications and Computing.

Jinsong Wu received a Ph.D. from Department of Electrical and Computer Engineering at Queen's University, Canada. He is the Vice-Chair, Technical Activities, IEEE Environmental Engineering Initiative, a pan-IEEE effort under IEEE Technical Activities Board (TAB). He was the Founder and Founding Chair of IEEE Technical Committee on Green Communications and Computing (TCGCC). He is also the co-founder and founding Vice-Chair of IEEE Technical Committee on Big Data (TCBD). He won both the 2017 and 2019 IEEE System Journal Best Paper Awards. His co-authored paper won the 2018 IEEE TCGCC Best Magazine Paper Award. He received IEEE Green Communications and Computing Technical Committee 2017 Excellent Services Award for Excellent Technical Leadership and Services in the Green Communications and Computing Community. He was the leading editor and co-author of the comprehensive book, entitled *Green Communications: Theoretical Fundamentals, Algorithms, and Applications*, published by CRC Press in September 2012. He is a current IEEE Senior Member.

Preface to "Green, Energy-Efficient and Sustainable Networks"

Over the last decades, information and communication technology (ICT) has radically changed many fields of living, with a significant improvement to people's lives. However, the benefits introduced by the development and the usage of ICT systems have consequences and new challenges have arisen regarding sustainability and practices that are environmentally acceptable. More specifically, ICT systems and infrastructures have constantly increased their power consumption and environmental footprint. This is primarily reflected in huge amounts of energy consumption and greenhouse gas (GHG) emissions of overall ICTs, with an additional contribution to the pollution of ICT system elements during their production and disposal phase.

Not surprisingly, such a noteworthy increase in energy consumption and GHG emissions will continue to rise due to the increase in the number of users/devices/types of ICT services, which will be coupled with the proliferation of high transmission capacity demands, mainly due to bandwidth-hungry applications and massive implementation of the Internet of Everything (IoE) technologies. In order to satisfy the detrimental economic and social demands and expectations, estimations show that the energy consumption of ICT is going to increase with an exponential trend. Although the contribution of ICT systems to global energy consumption and GHG emissions cannot be completely eliminated, these contributions should be maximally reduced, in order to limit the exponential increase of energy consumption and GHG emission trends. To face this challenge, improved or completely new algorithms, tools, platforms, methodologies, paradigms, systems, and energy models must be devised and practically implemented. Hence, greener and energy-efficient networks and ICT systems should be designed on all layers, by targeting an increase of energy efficiency, a decrease of GHG emissions, better re-use of resources, and large-scale adoption of sustainable materials and renewable energy sources. However, accomplishing this task is extremely challenging, due to the fact that it requires the combined effort of different stakeholders, e.g., from industry, academia, governments, and national and international organizations.

Additionally, sustainable networks and ICT systems refer to concepts that consider a set of programs, procedures, attitudes, and policies based on which ICT systems and corresponding elements will be implemented, used, and disposed of. Sustainable networks and ICT systems have a key role in developing the digitalized world, since technologies for the connected world have to assure the sustainability requests of new solutions and paradigms. Sustainable networks and ICT systems can be achieved only if a holistic approach in life-cycle management is targeted. This process includes structuring, developing, implementing, and disposing of ICT systems and corresponding elements, with a minimal or even without an environmental impact. Hence, sustainability is a topic of increasing importance in modern society, with a primary objective dedicated to achieving the technological, economic, social, and environmental sustainability of ICT systems and networks.

Despite such ever-growing interests in improving the energy-efficiency of ICT systems, the research on greener, energy-efficient, and sustainable networking and computing in many fields and on different levels requires improved or novel solutions and some fundamental problems are still open or are even in its infancy. Hence, green, energy-efficient, and sustainable networks are and will continue to be very relevant academic, industrial, economic, and social topics. However, recent advances in communication networks and systems have created new opportunities for the implementation of energy-efficient techniques that can be successfully built into ICT systems. This book, Green, Energy-Efficient and Sustainable Networks, focuses on all aspects of the research and development related to these areas. The book contains the outcomes of the Special Issue on "Green, Energy-Efficient and Sustainable Networks" of the Sensors journal published by MDPI (Multidisciplinary Digital Publishing Institute). Eighteen high-quality works have been collected and reproduced in this book, demonstrating significant achievements in the field. Among published scientific papers, one paper is an editorial and one paper is a review, while the remaining 16 works are research articles.

Published papers are self-contained peer-review scientific works. The editorial paper gives an introduction to the problem of ICT energy consumption and greenhouse-gas-emissions, presenting the state of the art and future trends in terms of improving the energy-efficiency of wireless networks and data centers as the major energy consumers in the ICT sector. In addition, the published articles aim to improve energy efficiency in the fields of software-defined networking (SDN), Internet of things (IoT), machine learning, authentication, energy harvesting, wireless relay systems, routing metrics, wireless sensor networks (WSNs), the device to device (D2D) communications, heterogeneous wireless networks (HetNets), and image sensing. The last paper is a review that gives a detailed overview of the energy-efficiency improvements and methods for the implementation of fifth-generation (5G) networks and beyond.

More than 80 different authors from both academia and industry backgrounds have contributed to this book. Therefore, this book can serve as a source of information for industrial, teaching, and/or research and development activities. Hence, the book gives insights and solutions for a range of problems in the field of obtaining greener, energy-efficient, and sustainable networks and it lays the foundation for solving new challenges and achieving future advances. The book editors would like to thank all authors who have submitted their articles and all reviewers for their valuable work dedicated to giving an expert review for submitted papers. Moreover, the book editors are grateful to all those involved in the publication of this book for their invaluable support, including the editors of Sensors and the team of people involved in editing the Sensors journal Special Issue on "Green, Energy-Efficient and Sustainable Networks."

We sincerely hope that this book will be a valuable source of information, presenting recent advances in different fields related to greening and improving the energy-efficiency and sustainability of those information and communication technologies particularly addressed in this book.

Josip Lorincz, Antonio Capone, Luca Chiaraviglio, Jinsong Wu
Special Issue Editors

       MDPI

Editorial

Greener, Energy-Efficient and Sustainable Networks: State-Of-The-Art and New Trends

Josip Lorincz [1,*], Antonio Capone [2] and Jinsong Wu [3]

[1] Department of electronics and computing, Faculty of electrical engineering, mechanical engineering and naval architecture (FESB), University of Split, 21000 Split, Croatia
[2] Department of electronics, informatics and bioengineering, Politecnico di Milano, 20133 Milan, Italy; antonio.capone@polimi.it
[3] Department of electrical engineering, Universidad de Chile, Santiago 8370451, Chile; wujs@ieee.org
* Correspondence: josip.lerinc@fesb.hr; Tel.: +385-21305665

Received: 2 November 2019; Accepted: 4 November 2019; Published: 8 November 2019

Abstract: Although information and communications technologies (ICTs) have the potential of enabling powerful social, economic and environmental benefits, ICT systems give a non-negligible contribution to world electricity consumption and carbon dioxide (CO_2) footprint. This contribution will sustain since the increased demand for user's connectivity and an explosion of traffic volumes necessitate continuous expansion of current ICTs services and deployment of new infrastructures and technologies which must ensure the expected user experiences and performance. In this paper, analyses of costs for the global annual energy consumption of telecommunication networks, estimation of ICT sector CO_2 footprint contribution and predictions of energy consumption of all connected user-related devices and equipment in the period 2011–2030 are presented. Since presented estimations of network energy consumption trends for main communication sectors by 2030 shows that highest contribution to global energy consumption will come from wireless access networks and data centres (DCs), the rest of the paper analyses technologies and concepts which can contribute to the energy-efficiency improvements of these two sectors. More specifically, different paradigms for wireless access networks such as millimetre-wave communications, Long-Term Evolution in unlicensed spectrum, ultra-dense heterogeneous networks, device-to-device communications and massive multiple-input multiple-output communications have been analysed as possible technologies for improvement of wireless networks energy efficiency. Additionally, approaches related to the DC resource management, DCs power management, green DC monitoring and thermal management in DCs have been discussed as promising approaches to improvement of DC power usage efficiency. For each of analysed technologies, future research challenges and open issues have been summarised and discussed. Lastly, an overview of the accepted papers in the Special Issue dedicated to the green, energy-efficient and sustainable networks is presented.

Keywords: energy-efficiency; wireless; green; sustainable; data centre; networks; ICT; 5G; power; wired access; IoT

---

1. Introduction

United Nations (UN) General Assembly have set sustainable development goals (SDGs) by the year 2030, and analyses presented in [1] show that information and communications technologies (ICTs) have the potential of enabling powerful social, economic and environmental benefits. However, a lack of exploration and innovation attempts dedicated to the search for answers on how SDGs can be achieved through the implementation of ICT, requests for more global governmental, technological, scientific and industrial attempts for accomplishing UN SDGs. The role of ICTs is twofold; while ICTs and networking currently contribute non-negligibly to the global energy consumption and carbon dioxide (CO_2) emissions, they will also contribute to the reduction of carbon dioxide (CO_2) and energy consumption of other industry sectors. This unique position of the ICT sector is confirmed in the SMARTer2030 report of the Global e-Sustainability Initiative (GeSI) [2], according to which expected carbon-dioxide equivalent (CO_{2e}) emissions of the ICT sector in 2030 can be kept at the same level as those in 2015. This means that ICTs will yield the 20% reduction of global CO_{2e} emissions by 2030 (Figure 1a). To illustrate the importance of ICTs in reducing CO_{2e} emissions, it is worth to state that contribution to CO_{2e} reduction due to the deployment of renewable energy sources by 2030 is estimated on 10.3 Gt, which is a (for 1.8 Gt) lower contribution to CO_{2e} reductions when compared with 12.1 Gt of estimated CO_{2e} reduction yield by the ICT sector (Figure 1a).

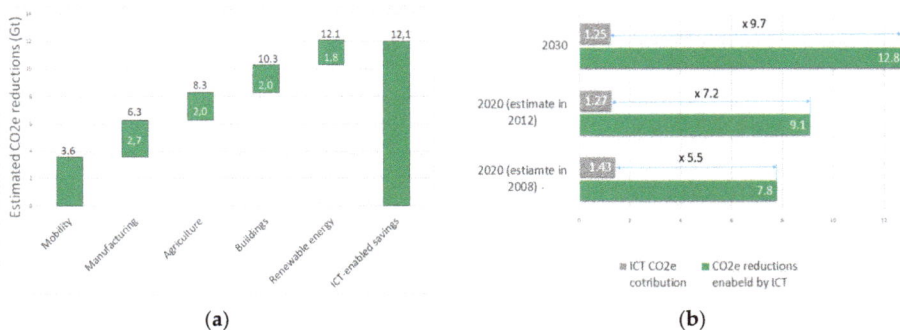

Figure 1. Estimated: (**a**) contribution of different industry sectors to global carbon-dioxide equivalent (CO_{2e}) reduction by 2030 [1], (**b**) information and communications technology (ICT) sector CO_{2e} "footprint" contribution and enabled reductions to global CO_{2e} emissions expressed in Gt [2].

According to estimates of GeSI, the ICT sector will give by 2030 significantly higher contribution to CO_{2e} emission reductions when compared with other industry sectors (e.g., mobility, manufacturing, agriculture, buildings, etc.). To achieve such CO_{2e} emission reductions until 2030, a significant decrease of ICT sector CO_{2e} emissions in global CO_{2e} emissions are envisioned by 2030 [2]. Based on results presented in SMARTer2030 report (Figure 1b), in 2020 ICT sector's CO_2 emissions "footprint" is estimated on 2.7% (1.43 Gt) of global CO_{2e} emissions, while due to expected improvements in energy efficiency of ICT systems, ICT sector will in 2030 contribute with 1.97% (1.25 Gt) to global CO_2 emissions. This means that estimated CO_{2e} emissions avoided by the use of ICT systems in 2030 (12.08 Gt) will be 9.7 times higher than the CO_{2e} emissions generated by implementing the same ICT systems (Figure 1b). Thus, an expected increase in the implementation of ICT systems in the future can potentially alleviate the need for selection among environmental protection and economic prosperity and it can pave the way to the achievement of both goals.

Despite such positive estimates, the increased demand for user's connectivity and an explosion of traffic volumes necessitate continuous expansion of current ICTs services and deployment of new infrastructures and technologies which must ensure the expected user experiences and performance.

This results in an increase in the energy consumption and energy cost of the ICTs infrastructure, which in recent years become one of the major concerns for the ICT sector. Due to the expected increase in diversity of connected objects, devices, applications and services and because of the rapid growth of the worldwide broadband subscribers, predictions related to global annual monetary costs for the energy consumption of ICT infrastructure are worrying. The energy consumption estimated for wireline (access, metro, edge, core networks and the associated data centres) and wireless access networks is presented in Figure 2a [3]. According to these forecasts, if no energy-efficiency improvements will be implemented, monetary costs for the global annual energy consumption of telecommunication networks will raise 8.6 times, more specifically form $40 billion in 2011 to $343 billion in 2025. This increase of energy consumption costs is a direct consequence of the need for satisfying explosive growth of annual global internet protocol (IP) traffic, which is estimated on 4.8 ZB/year by 2022, or 396 EB/month. In 2022 this will result in a threefold monthly increase of IP traffic since 2017 (122 EB/month), or an astonishing 14.1 times increase since 2011 (28 EB/month) [4,5].

Due to increased energy costs pushed by constantly increasing traffic volumes, current network energy costs of telecommunication service operators in developing countries already span between 40% and 50% of provider operational expenditures (OPEX), and between 7% and 15% of the OPEX for operators in developed countries [6–8]. This is confirmed by some telecom operators which start reporting energy bills of up to $1 billion, while some expect to reach these costs by 2020 [9].

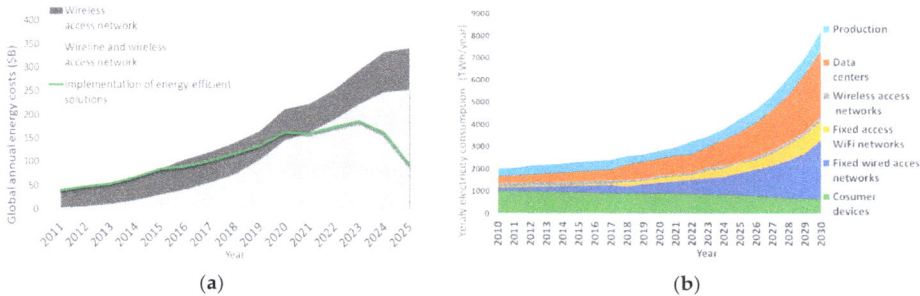

Figure 2. Estimation of (**a**) costs for the global annual energy consumption of telecommunication networks in period 2011–2025 [2], (**b**) expected total annual energy consumption per different ICT systems in period 2010–2030 [10].

High energy costs of telecommunication networks presented in Figure 2a correlate with estimations of energy consumption trends of different ICT systems presented in Figure 2b. Estimates presented for the period 2011–2030 are performed with an assumption that takes into account expected annual: future improvements in the energy efficiency of ICTs systems, trends in future IP traffic growth and future improvements in electricity usage per traffic unit [10]. According to estimations presented in Figure 2b, expected annual electricity consumption of consumer devices (including desktop, monitor, laptops, televisions (TVs) and peripherals, tablets, mobile phones, smartphones, modems, etc.) will contribute to the global electricity consumption of ICT systems by 2030 with 8.1% (670 TWh). Estimations further assume for fixed wired (core, distribution and access) networks, WiFi networks (consumer premises WiFi equipment), radio part of the wireless access network (second (2G)/third (3G)/fourth (4G)/fifth generation (5G)) and data centres (servers, power supply and cooling elements), yearly energy consumption contribution to the annual electricity footprint of ICT systems equal to 31.95% (2641 TWh), 10.75% (889 TWh), 2.35% (195 TWh) and 35.89% (2967 TWh), respectively (Figure 2b). Additionally, estimates for annual electrical energy consumed for the production of different ICT devices (user, wired and wireless network equipment, data centre devices) are anticipated at 10.92% (903 TWh) of total ICT energy consumption by 2030 (Figure 2b).

Moreover, best, expected and worst-case forecasts related to the overall yearly electricity footprint of ICT systems in 2030 equals to 2698, 8265 and 30,715 TWh, respectively, which means that energy consumption impact of ICT systems for the overall global annual energy consumption can be, in the best-case, equal to 8%, or 21% for the expected (Figure 2b) and even an astonishing 51% for the worst estimation case. To get a sense of the rapidness of ICT energy consumption increase, in 2012 it was estimated that the complete ICT sector contributes approximately 6% to global electricity consumption [11]. Hence, worst or even expected (Figure 2a) forecasts of ICT energy footprint trends in global annual energy consumption by 2030 are alarming. This dramatic increase in energy consumption of ICT systems justifies the precipice of economic unsustainability. Obviously, current technology improvements cannot cope with the increasing energy consumption of the ICT sector and it is imperative to find novel solutions that will alleviate this problem.

The rest of the paper is organised as follows. The energy consumption of user-related devices is analysed in Section 2. Sections 3 and 4 give an overview of research challenges for energy-efficiency improvements of radio access networks and data centres, respectively. A short description of all articles accepted for publication in the Special issue on green energy-efficient and sustainable networks of the Sensors journal are presented in Section 5. Finally, some concluding remarks are given in Section 6.

2. Energy Consumption of User-Related Devices

According to presented in the previous section, the energy consumption of data centres (DCs) and communication network devices is just one part of the overall ICT energy consumption, while energy consumption of user-related devices presents the other part. The energy consumption patterns of user-related devices point to different challenges and require different approaches to energy consumption reductions, than those envisioned for network and DC devices. Energy consumption estimates of user-related devices for the period 2011–2025 are presented in Figure 3 [3]. Presented estimates are performed for all connected user devices in cellular networks, internet of things (IoT) applications, public safety, intelligent buildings and generally for all consumer devices with a network connection. Estimates consider the explosive growth of user-related devices from about 50 billion in 2011 to 110 billion devices connected to the network in 2025 [3]. Forecasts for the global energy consumption of these user-related devices estimate the energy consumption raise from about 180 TWh in 2011 to 1400 TWh in 2025 (Figure 3), which represents a 7.7 time increase in the period of one and a half-decade.

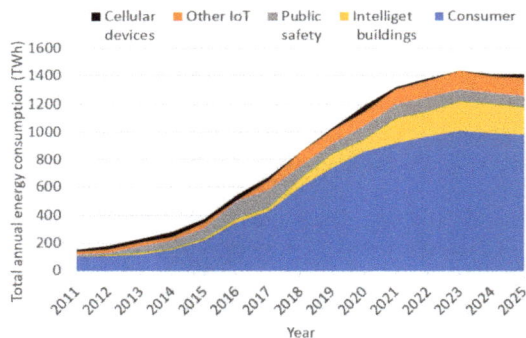

Figure 3. Estimations of energy consumption of all connected user-related devices and equipment for the period 2011–2025 [3].

Obviously, this estimated energy consumption increase is unacceptable, and attempts focused on alleviating such trends must take into account specific peculiarities of user-related devices. For example, a single sensor or IoT device, in reality, consume rather low amounts of energy in absolute values, however, it is expected that a vast number of such devices will be installed worldwide. On the other hand, battery-less user-related devices must have constant power supply while battery-powered devices must have a periodic power supply for battery recharging. This power supply can be obtained from the electricity grid, by means of renewable energy sources, by means of energy harvested from the environment or through the combination of these power sources. Hence, the problem related to the energy footprint of user devices is not solely related to their annual energy consumption trends, it is also related to the sources of energy supply and energy autonomy in the case of battery-powered devices. Some estimates show that the number of user devices powered by rechargeable, grid, network and renewable sources will increase in the period from 2011 to 2025 for 13×, 54×, 380× and 378×

times, respectively [3]. Such figures mandate a necessity for significantly higher usage of renewable energy sources.

Additionally, advances in battery storage, new solutions for lowering power consumption of user devices and relying on energy harvesting is another possibility for energy footprint improvements. This will be especially important since running power lines to a huge number of user devices or repeatedly change of batteries will not be viable from the practical or economic point of view. Hence, implementation of a fully connected world characterized by IoT and internet of everything (IoE) applications will not be possible on a large scale unless the energy supply challenges of user devices are properly solved. Future solutions must offer trade-off in the energy equation among better energy storage, more effective use of harvested and renewable energy sources and lowering power consumption of user devices.

Energy Consumption Trends

Although Figure 2a,b shows estimated trends for telecommunication networks in terms of expected total annual monetary costs and electricity consumption per different ICT systems, more detail analyses are presented in this section in order to understand the future trends in energy consumption of communication networks. In Figure 4, energy consumption is breakdown into six main network sectors, more specifically: edge and core networks, radio access, DCs, service core, fixed access and residential and businesses. Contribution to the total annual energy consumption of each network sector in 2013 and estimates for 2025 are presented in Figure 4a,b [3], respectively. Estimations are performed based on expected IP traffic growth and by assuming the potential benefits of new network architectures and technologies. According to Figure 4b, energy consumption will remain high or even increase in two sectors: the data (cloud) centres and the wireless radio access network, while in other sectors energy consumption will remain or even decrease. However, different technology improvements are required in each of these sectors to ensure that an increase in IP traffic in the future can be supported in an economically viable and sustainable way by 2025. Since wireless radio access networks and data centre sectors are the highest contributors to the overall network energy consumption, the next sections are dedicated to the presentation of main research challenges related to the improvement of energy efficiency (EE) of these sectors.

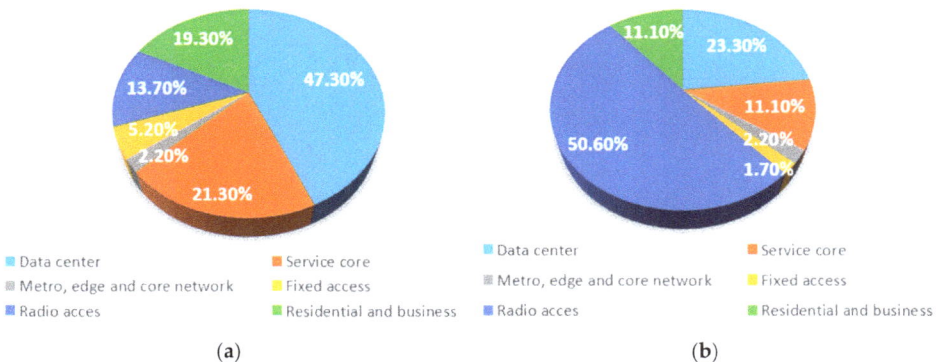

Figure 4. Estimated network energy consumption for main communication sectors in: (**a**) 2013 and (**b**) 2025 [3].

3. Research Challenges for Energy-Efficiency Improvements of Radio Access Networks

In this section, a review of the last research activities on green radio access approaches and energy harvesting for the power supply of network devices in cellular access networks is presented. Also, potential technical demands and some research topics for realizing green, energy-efficient and sustainable radio access networks are emphasized. For 5G networks, as currently the most prominent wireless network technology, tremendous performance improvements are envisioned. These improvements encompass support of: a thousand-fold increase in throughput in comparison to present networks, up to ca. 7.6 billion mobile subscribers with the connection of at least 100 billion devices worldwide, up to 10 Gb/s individual user broadband speeds, IoE communications, tactile Internet applications and the network latency of 1 ms or lower. To satisfy such demanding performance gains, different novel technologies are emerging, but performance improvements incurred by 5G networks do not come without drawbacks. One of the major consequences is the degradation of EE expressed in bits/Joule (b/J), which has been broadly accepted as the EE metric for wireless communication systems [12]. It is expressed as

$$EE = \frac{FR \times SS \times BW \times log_2(1 + SINR(D))}{P_c + P_T} \; [b/J],\qquad(1)$$

where *SS*, *BW* (Hz), *FR*, *D* (m), P_C (W) and P_T (W) represents the number of spatial streams (spatial multiplexing factor), the bandwidth of signal, frequency reuse factor, distance among communicating devices, circuit (mostly static) and transmit (mostly dynamic) power consumption of communicating devices, respectively. According to EE Equation (1), the EE of cellular networks can be increased by augmenting the signal bandwidth, the multiplexing factor, the frequency reuse factor, or by lowering the circuit and transmit power consumption. In this regard, different paradigms for 5G networks have emerged (Figure 5): Communications based on millimetre-waves (mmWave), long term evolution in unlicensed spectrum (LTE-U), ultra-dense heterogeneous networks (UDNs HetNets), device-to-device (D2D) communications and massive multiple-input multiple-output (M-MIMO) communications. The impact of each technology on EE of radio access networks is further discussed. In Table 1, an overview of technologies for EE improvements of wireless networks with future research challenges characteristic for each technology is summarised.

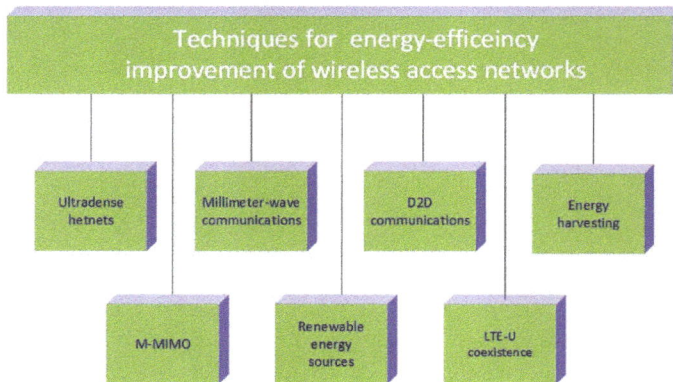

Figure 5. Techniques for energy-efficiency improvement of radio access networks.

Table 1. Technologies for energy efficiency improvements of wireless networks and future research challenges.

Technology	Energy-Efficiency Improvement Area	Future Research Challenges for EE Improvements
Ultra-dense HetNets [13–20]	Network design with decoupled data and signalling	Development of effective algorithms for the management of signalling and data decoupling
	Network design with BS on/off switching	Development of effective radio resource management algorithms for efficient BS activations and deactivations
	Network design with inter-cell interference mitigation	Development of efficient inter-cell interference management schemes
M-MIMO [12,14,21]	Design of energy-efficient antenna selection	Finding algorithms for the selection of an optimal number of antennas in M-MIMO systems
	Energy-efficient hardware design	Finding novel hardware designs for multi-antenna placement in UTs
	Energy-efficient design of pilot tones	Finding algorithms for reducing the energy consumption of pilot tome transmission
mmWave communications [12,22–24]	Energy-aware transceiver designs	Finding optimal hybrid control of RF transceiver architectures and antenna designs
	Energy-efficient analogue-to-digital converters design	Finding optimal analogy-to-digital converters in terms of sampling rate resolution
Renewable energy sources [25–27]	System design which exploits renewable energy and energy cooperation	Solutions for estimation of optimal renewable energy sources for BS sites
	System design which exploits energy cooperation	Development of systems enabling surplus power transfer among BS sites
	Design of BS site with efficient energy flows management	Development of an optimal algorithm for energy flow management on sites with renewable energy sources
D2D communications [12,28]	Network design based on the hybrid overlay and underlay communication	Development of algorithms for switching among underlay (assigned spectrum portion) and overlay (unassigned spectrum portion) communication designs
	System design which enables active users' cooperation	Development of algorithms for caching, sharing or relaying data with minimal UTs energy consumption
LTE-U coexistence with other systems [12,29]	Design of channel allocation protocols	Finding optimal protocol for RF channel scheduling among different systems in an unlicensed band
Energy harvesting [30,31]	Design of highly efficient energy harvesting systems	Development of algorithms for optimally balance between energy harvesting and data transmission
	Design of system which reduces energy conversion inefficiency	Development of systems based on energy beamforming, D2D and HetNets communications with more energy-efficient receivers
	Development of systems which exploit interference in wireless networks	Development of systems which optimally exploits interference signals for energy harvesting

3.1. Ultra-Dense Heterogeneous Networks

In essence, UDNs are heterogeneous networks based on a massive deployment of diverse types of base stations (BSs), where macro-cells (of macro BSs) ensure base signalling coverage while micro-cells (of mini/micro/pico/femto BSs) fulfils the demand for high throughput [13,14]. Such broadly accepted radio access network architecture based on decoupling data and signalling contributes to EE improvement of cellular networks and enables separation of downlink and uplink [15] communications. Due to the reduction of distance between users terminals (UTs) and BSs accomplished with densification of BSs allocation in such heterogeneous networks (HetNets), the EE improvements of the network are reflected in a significant reduction of transmit powers and consequently energy consumption for both (UT and BS) transceivers. Also, signalling and data decoupling enable replacement of the macro-cell BSs by more energy-efficient types of BSs having distant radio access unit's (RAUs) controlled from central location without impacting the small-cell BSs layer. Additionally, decoupling enables combining different radio access technologies (RATs) such as mmWave and WiFi in existing networks, which can help in achieving further EE gains. Moreover, the separation of uplink and downlink transmission enables versatile association schemes among UTs ad BSs, which also can lead to significant energy savings for both, BSs and UTs [12]. Nevertheless, UDN concept is not without drawbacks. It is expected that the realization of such HetNets requests additional equipment and BS sites that will increase telecom operators' (TOs) total network energy consumption for up to 150%–170% by 2026 [16]. Hence, novel approaches to energy control within both, the 5G network infrastructure and changes in the way TOs purchase and deliver electricity to 5G networks will become critical as they extend density, coverage and capacity over the next decades. Also, signalling and data decupling raise the complexity of HetNets management and contribute to a significant increase in signalling overheads (Table 1). This request further investigations in the development of new signalling and network designs, which will enable full exploiting of signalling and data decoupling while preserving network EE.

Another important approach to improvement of BS energy-efficiency is concept based on on/off switching of BSs (i.e., BS sleeping) to save energy [17–19]. Applicability of this concept is related to the nature of wireless traffic loads which varies in time and space. This concept enables shutting down or putting into sleep mode some BSs in periods of low traffic loads and activation of BSs when there is a need for satisfying increased traffic demands. Such dynamic management of BSs activity in the radio access networks enables tuning of BS power consumption according to real traffic variations, which eliminates the waste of energy imposed with the traditional concept based on BSs which are permanently active, even in the periods with low or without any user traffic [12]. However, ensuring full-service area coverage, signalling for smooth user handovers among BSs and elimination of overloading of those BSs that remains active is a challenging task in case of BS on/off deployments (Table 1). To ensure optimal balance between network EE and service quality, further improvements of radio resource management algorithms must be developed and implemented.

Due to the high-frequency reuse factor, inter-cell interference represents another challenge to EE implementation of ultra-dense HetNets. The Tx power increase of two neighbouring BSs initiated in process of BS radio resource management, can have a negative impact in terms of signal cancelling caused by inter-cell interference [20]. This degrades the system throughput and consequently leads to lowering of network energy-efficiency. Although complete elimination of inter-cell interference is not possible, management schemes which suppress interference, such as cooperative transmission, smart power control, interference alignment and resource scheduling and partitioning are needed for the successful proliferation of energy-efficient ultra-dense HetNets (Table 1).

3.2. Massive-MIMO Technology

Since it is based on exploiting a large number of BS antennas for serving many users with the same time-frequency resources, M-MIMO concept significantly improves multiplexing and array gain of 5G transmission systems. The drawback of M-MIMO implementation is that such concept increases significantly power consumption of individual BS sites. When compared with 4G BSs, M-MIMO strongly contributes to increase of 5G BSs power consumption due to the increase in a number of analogue-to-digital converters with corresponding digital circuitry and power amplifiers needed for M-MIMO operation [21]. More specifically, typical 4G BS contain four transmit (Tx) and receive (Rx) elements (in so-called 4×4 MIMO arrays configuration), while 5G BSs are intended to work in up to 64×64 configurations, which is the reason why it is expected that 5G BS will have three times higher power consumption than its 4G predecessor. On the other hand, M-MIMO can bring some advantages with respect to EE of wireless networks [12]. This is because the uplink Tx power of single-antenna UT can be proportionally reduced with the number of MIMO BS antennas in the case when the equivalent results as those of a related single-input single-output transceiver wants to be achieved [22]. However, only reducing the Tx power of UTs is not sufficient for significant improvements of EE in wireless networks, since power consumed by electronic circuits has linear growth with the number of MIMO signal processing circuits, which has a non-negligible impact on the overall power consumption (Table 1). Hence, determining an optimal number of antennas in M-MIMO systems arises as important research topic which generally yields assumption according to which, a larger number of antennas must be deployed in systems which Tx power dominates in the overall power consumption, and vice versa [14].

Furthermore, M-MIMO systems with a large number of antennas installed enable the implementation of simpler precoding algorithms and signal detection and transmission at the BS, which further enable significant savings in power consumption contributed by BS hardware. In comparison with the implementation of existing signal processing methods (successive cancelling of interference and dirty paper coding), implementation of advanced algorithms for signal processing such as maximum ratio transmission/combining contributes to the reduction of the dissipated energy required for signal processing computations [12]. Additionally, since M-MIMO systems demand much smaller RF Tx power (of the order of milliwatts), power amplifier losses during operations will be reduced which can bring significant power savings. Nevertheless, major challenges requesting broad investigations are currently present in the design of UTs hardware (Table 1). Major performance bottlenecks related to UT hardware are the limited physical size of UTs, lacking space for implementation of a large number of antennas and demanding requirements on battery depletion.

Above all, M-MIMO systems require accurate and timely channel state information's (CSIs) which acquisition is directly related to the Tx antenna number. This leads to the significant power consumption of pilot subcarriers and new approaches such as semi-orthogonal pilot design and pilot beamforming needs further exploiting in order to reduce the contribution of pilot transmission to the overall M-MIMO system energy consumption. Additionally, pilot interference incurred by reusing the same resources of pilots in neighbouring cells of multi-cell locations also diminishes the EE of M-MIMO systems (Table 1). Hence, designing pilot interference mitigation approaches as well as balance in the exploitation of resources in time and frequency for training of pilots in downlink and uplink, are important topics that must be slaved in order to reach high EE of M-MIMO systems.

3.3. Millimetre-Wave Communications

For transmission in the mmWave spectrum, the conventional transceiver architecture having each antenna connected to the corresponding radio-frequency (RF) chain is energy-inefficient [12]. Inefficiency is a consequence of huge power consumption which emerges from the concurrent processing of vast amounts of data burst Giga-samples/s per each RF chain. Thus, an approach to alleviate the power consumption problem is to implement both, the digital and analogue beamforming, where every RF chain can be connected to all (fully controlled architecture) or to some antennas (partially controlled architecture) of a transmission system. The signal phase of each antenna must then be scheduled by a network of digital and analogue phase shifters (PSs) [23]. The fully connected architecture demands hundreds or even thousands of PSs, which maximizes spatial degrees transmissions utilization and minimizes EE. The partially controlled architecture exploits only a limited number of PSs that improve system EE, but reduces spatial degrees freedom and consequently transmission rates. Possible solutions, which are currently a field of research, aim to find optimal hybrid control architectures (Table 1). These architectures are based on a different number of antennas and RF chains, combining and precoding approaches, Tx power allocations and antenna arrays having a lens with energy-focusing. The development of these architectures can jointly or separately optimize the system performance whit minimal impact on EE degradation [24].

Besides the high-power demands of a huge number of PSs, another power consumption problem characteristic for mmWave systems are analogue-to-digital converters (ADCs). The power dissipation of ADCs increases exponentially with the increase in the number of bits per sample and linearly with the augmentation of the sampling rate [12]. Additionally, the data circuits which connect the digital elements to the ADCs are high energy consumers and have an evident correlation with the adopted sampling rate resolution (Table 1). This motivates the search for finding optimal ADCs in terms of sampling rate resolution, which will efficiently balance between the power consumption and the data rate or ensure optimal combining of low and high-resolution ADCs in order to maximize EE.

3.4. Renewable Energy Sources

An approach based on powering BSs sites using energy harvested from renewable energy sources such as wind, solar, fuel cell or combination of these energy sources significantly contributes to the improvement of wireless network EE. Current trends in terms of integration of renewable energy into power supply systems of contemporary wireless networks are twofold. The first approach is dedicated to the replacement of an off-grid diesel-based BS power supply system with those relying solely on some renewable energy sources. The other approach is based on the so-called hybrid BSs sites which use different renewable energy sources or a mix of renewable, diesel generator and/or grid energy. In addition to EE improvements and operational expenditure reductions, such approaches significantly reduce or even completely eliminate diesel generator CO_2 emissions from BS sites [25,26]. However, the optimal selection of renewable energy sources in terms of size and power generation capacity, for the specific site remains one of the major challenges (Table 1). Hence, further investigations in the development of simulation tools that can fairly estimate the techno-economic aspect of transforming a typical BS site in green BS site must take place.

Additionally, the integration of renewable energy sources into BSs power supply systems can provide compensation for the additional circuit power consumption in case of installing more BSs on BS site [12]. Also, dense allocation of BSs employing energy harvesting from renewable sources can facilitate possible energy cooperation between BSs. This cooperation can be based on transferring through power lines superfluous energy collected on sites harvesting more energy, to BSs sites that harvest less energy (Table 1).

However, the major challenge in realization of durable BS site power supply solutions can be found in the intermittent nature of renewable energy sources, limited battery capacities installed on sites and necessity for ensuring stable and without any interruptions power supply of BSs sites. This imposes the development of resource allocation algorithms for the management of BS site power demand. Such algorithms must consider the traffic variations and wireless channel state information's, power supply impacted with the unpredictable nature of renewable energy sources and battery recharging and depletion cycles [27]. Algorithms for efficient energy flow management of BS sites are generally categorized as offline and online algorithms. The first one can be developed by exploiting optimization theory approaches and the second one assumes that some statistical data is accessible at the Tx side or they use the insights observed from the offline algorithm. Since 5G networks are characterised with very dynamic traffic variations, results of offline algorithms often serve as performance upper bounds for online algorithms. Nevertheless, the development of an optimal resource allocation algorithm for a specific hybrid BS power supply solution, continues to be an object of research interest (Table 1).

3.5. Device-To-Device Communications

This type of communication offers effective local spectrum reuse through two modes of operation: the cellular mode where UTs communicate via BSs and the D2D mode which ensure possible communication of UTs directly with each other [28]. D2D mode of communication can be realised through reuse of the spectrum portions that have not been assigned (known as overlay communication) or has already been scheduled to UTs (known as underlay communication). Overlay D2D communication does not generate co-channel interference, which results in more efficient spectral efficiency (SE) of the D2D system [12]. In periods when such interference is weak, it is possible to switch to underlay communication which offers more energy-efficient D2D communication system design (Table 1). However, for switching among underlay and overlay communication designs, effective algorithms must be envisioned, what represents a prominent research field.

Another advantage of D2D communications is the ability of proactive cooperation between users, what can bring EE improvements in 5G networks, particularly in terms of extending the mobile devices' battery lifetime. More specifically, active UTs in D2D networks can work as mobile relays or cluster heads of UT clusters and local cashing devices, and each of these working modes can bring possible EE improvements of the cellular network [12]. A mobile relay mode based on a multihop relaying of data among UTs and BSs or other UTs can reduce high energy consumption needed for direct transmission between distant UT and BS, since communication among relaying nodes can be realised with lower Tx powers. Local content caching provides a way to better exploit the UTs data storage in 5G networks and enables power consumption and backhaul loads reduction through optimal decisions related to what content to cache and at which location (Table 1). Although active user cooperation offers significant advantages in terms of improving SE and EE, further investigations must give an answer on how UTs can be managed to cache, share or relay data for other UTs at the expense of consuming their own energy.

3.6. Long-Term Evolution Coexistence with Other Systems in Unlicensed Spectrum

Implementation of LTE-U technology is constantly challenged with the need for simultaneous coexistence of different systems working in unlicensed bands, such as wireless local area network (WLAN) systems and overlay Long-Term Evolution (LTE) systems [12]. Since LTE employs scheduling-based and WLAN contention-based channel access mechanisms, lack of constraints in LTE transmissions may cause permanent interference to WLANs, where the channel is sensed as mostly unavailable [29]. This results in unending backoff times for the WLAN transmitters and poses low EE of the network due to the high energy consumption of the WLAN users lacking the possibility of transmission while waiting on backoff timer expiry. Hence, advanced modifications to resource management become critical for the coexistence of different systems in unlicensed bands, and so far, two methods have been proposed: duty cycling and the listen before talk method. The first one defines periodic turning off and on of the LTE transmitter, without checking the availability of the channel before transmitting, while the second one requires a check of channel occupancy by WLAN systems before the LTE system can start transmission. However, the first method lacks real responsibility of ensuring any transmission time window for WLAN networks since LTE carriers define on-off scheduling, while the second method has degraded performance caused by excessive transmission collisions in case of a huge number of devices contending for the channel (Table 1). Hence, currently there is no broadly accepted protocol that will ensure the harmonious coexistence among systems transmitting in the unlicensed spectrum (LTE-U, WLAN, etc.), and more advanced solutions for alleviating this coexistence issues are jet to be devised.

3.7. Energy Harvesting

Wireless power transfer (WPT) known as RF energy harvesting, allows small receivers which are expected to be massively used in 5G use cases like IoT to harvest energy from RF signals which will be received [30,31]. WPT is assumed to be a promising technology for powering a huge number of devices, since harvested energy from RF eliminates the need for powering those devices from an electric gird and also enables battery lifetime extension of mobile, sensor or actuator devices. Although WPT can be fully managed at the receiver side, in the practical implementation of RF energy harvesting, the main challenge is ensuring optimal balance among the harvested energy and the achievable transmission rates. This balance can be realized through the implementation of an approach based on exploiting simultaneous wireless information and power transfer (SWIPT), where the receiver device during reception divides the received signal into two parts, one for energy supply obtained through energy harvesting and the other for information decoding [12]. Another approach known as wireless powered communication network (WPCN) splits information transmission and energy harvesting in time, where wireless devices first harvest energy from received signals and then, by means of harvested energy perform wireless information transmission (WIT). In the case of the first approach, the development of algorithms which will minimize the power losses at the receiver in order to maximise the harvested energy and the achievable throughput must be devised. Regarding the second approach, proliferation of novel solutions which will ensure intelligent selection of WIT and WPT requests for further investigations and improvements (Table 1).

Although RF energy harvesting brings many advantages, the major implementation issue is system performance which is significantly limited by the severe RF signal path loss and consequently low energy conversion efficiency at the position of the energy harvester. One of the possible approaches to system performance improvement is in the implementation of an energy beamforming concept [28]. This concept is based on the transmission of narrow beams through multiple antennas with optimized beamforming vectors, which is fully compatible with M-MIMO and mmWave theologies. Moreover, D2D communications and ultradense networks (UDN) are technologies that contribute to performance improvements of energy harvesting systems. This is because each of these technologies ensures a reduced range among communicating pairs, what reduces the distance for energy transfer and consequently improves WPT efficiency (Table 1). Also, the substation power consumption of electronic circuits during information decoding and channel state information acquisition asks for further attempts in finding new receiver architectures which will consume less power.

Finally, the fact that energy harvesting of co-channel interference can be exploited for ensuring the power supply of receiver devices, gives completely new light on the impact of interference which can become a potential energy source. To exploit interference as an energy source, possible solutions can be based on deliberate artificial interference insertion into communication channels. This approach enables devices to harvest energy in the case of dominant co-channel interference and to decode information in case when this interference diminishes (Table 1). Obviously, more investigations related to such a paradigm shift are needed in future research.

4. Research Challenges for Improvements of Data Centres Energy-Efficiency

According to analyses presented in Figure 2b and estimation of data centres (DC) future energy consumption trends presented in Figure 4, increasing trends of DC energy consumption become a major concern. Additionally, DCs continually run at high underutilization due to fragmentation and over-provisioning of resources [32,33], with common utilization levels spanning between 5% and 25% [34–37]. Besides significant energy waste caused by such low utilisation of DCs which further worsens the energy inefficiency problem, the low DC utilization causes the energy dissipation of other DC ancillary equipment and infrastructure, such as cooling and power supply systems.

Additionally, authors in [38] analyse green issues related to the processing of the vast volume of information's characteristic for emerging big data concepts. Analyses address the green challenges related to the three phases of the big data life cycle which are characterized as data generation/acquisition/communications, storage and processing. Also, the study suggests novel green metrics for processing big data in order to accommodate the need for adopting new definitions of green metrics which will correspond to the contemporary big data concept. Although different metrics for expressing DC energy efficiency have been proposed, the widely accepted metric is power usage effectiveness (PUE) defined as [39]:

$$PUE = \frac{P_{TOT}}{P_{IT}}, \tag{2}$$

where P_{IT} is instantaneous power of the IT equipment consumed by the DCs storage, network, servers and monitoring devices (laptops or workstations), and P_{TOT} is the overall DC instantaneous power consumption which includes the aforementioned P_{IT} power and instantaneous power consumption of ancillary DC equipment (cooling system, power distribution system, uninterruptable power supply, etc.). In [39], an average value of the present DCs PUE is suggested to be 1.83 and according to the Equation (2), better DC EE means lower PUE and vice versa. Since PUE of present DCs is extremely high, different techniques and approaches for improving EE of DC arise. The Equation (2) indicates that a better PUE can be accomplished if total DC facility power will be reduced and to this end, research efforts focused on improvement of DC energy-efficiency encompass the following techniques: improvement of DC resource management, increasing DC servers efficiency through power management, developing green DC monitoring and simulations and enhanced thermal management of DC. In Table 2, each of the stated techniques for improvement of DC energy-efficiency with corresponding future research challenges is presented. Also, Figure 6 summarises techniques for energy-efficiency improvement of DCs and an overview of the latest research on green DCs is presented in the next sections.

Table 2. Technologies for EE improvements in data centres and future research challenges.

Technology	Energy-Efficiency Improvement Area	Future Research Challenges for EE Improvements
DC resource management [36–81]	Energy-aware VM/containers assignment in DCs	Finding an optimal algorithm for the implementation of energy-efficient VM/containers management
	Energy-aware DCs network traffic engineering	Development of algorithms for energy-efficient adaptation of DC traffic paths and network architectures
	Energy-efficient power distribution in DCs	Design of energy-aware solutions for intra and inter DC workload scheduling and power distribution
	Usage of renewable energy for DC power supply	Finding solutions for optimal control of DC power supply form renewable energy and implementation of stimulating energy pricing models
DC servers power management [82–98]	Energy-aware DFVS scaling of server components	Finding optimal frequency/voltage and link speed scaling solutions for minimization of the DC power consumption
	Energy-aware server/server component activity scheduling	Development of novel energy-efficient algorithms for on/off server or server components switching
	Energy-efficient hybrid (DFVS and component activity switching) solutions	Development of algorithms which combine DVFS and on/off server or server components switching
DC monitoring and simulation management [99–126]	Green DC monitoring	Development of novel DC monitoring tools which will enable analyses of green metrics
	Green DC simulators	Design of a system-oriented DC simulator for concurrent performance simulation of different DC elements
DC thermal management [127–134]	Energy-efficient cooling and workload distribution	Development of temperature-aware DC workload assignment algorithms
	DC management system which improves temperature to reliability trade-off	Design of novel temperature-resistant components for DCs with an increased average temperature

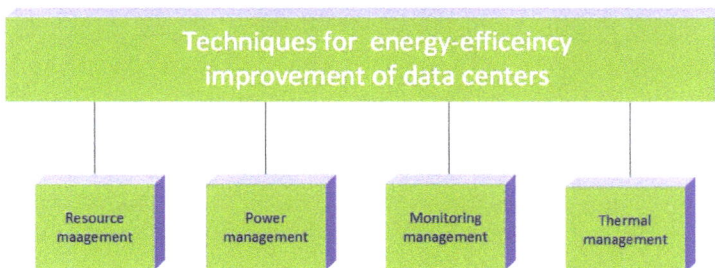

Figure 6. Techniques for energy-efficiency improvement of data centres.

4.1. DC Resource Management

To address resource underutilisation as one of the major DC problems causing an excessive energy consumption, modern servers in DC use the concept of virtualization for presenting the abstraction of many dedicated virtual machines (VMs) or containers executing separate applications (Figure 6) [40]. Hence, optimal migration, allocation and consolidation of DC server resources known as VMs/containers management is an important approach to the improvement of DC resource utilization and energy consumption reduction (Table 2). Generally, VMs/containers management is based on the efficient scheduling of VMs/containers to servers based on satisfying specific performance metrics and resource demands [39]. Although different approaches to VMs/containers management have been proposed [41–50], the main cause of why the utilization of DC resources still remains ignoble is that DC administrators and owners worry about the potential quality of service (QoS) violations caused by VMs/containers management. Additionally, in multi-tenant DCs, versatile tenants can request different levels of application performance that request heterogeneous resource management algorithms, which further increases its complexity. Hence, algorithms which will optimize DC energy-efficiency through optimal VM/container management and DC right-sizing, while preserving QoS in single and multi-tenant DCs, are at present important research issues. However, improving DC resource utilisation will consequently contribute to the DC energy-efficiency improvements and solutions which will provide efficient resource management policies that require future exploration.

Additionally, traffic engineering is a very efficient concept which enables improvement of DC energy-efficiency (Table 2). It is based on the adaptation of DC traffic paths and network architectures according to DC traffic patterns [51]. To obtain proportionality between DC traffic variations and DC power consumption, different solutions have been proposed based on traffic aggregation and VM/container assignment techniques using virtualization of network functions [52–60]. Although network function virtualization promises as an approach in providing EE improvements for deployment and management of the network services, problems such as preserving QoS and lack of accountability for the energy consumption of many implementations such as cloud networking system (for example CloudNaaS) [61] remain unsolved. Hence, finding an appropriate trade-off between network performance and EE is currently a challenging problem that solving requires further research.

Another issue related to DC energy inefficiency is the over-provisioning of DCs power distribution system, which brings high energy costs during idle periods of DCs operation. DC power distribution systems are generally over-provisioned since the deployment of such systems in terms of power capacity is based on satisfying traffic peaks and allowing DC expansions in the future. However, due to the rare occurrence of simultaneous peak power draw across all equipment in DC, power over-subscription is intentionally utilised for enhancing DC power exploitation (Table 2) [34], [61–66]. In order to more efficiently utilize the total DC power budget, proposed concepts are based on power capping, power routing and dynamic power shifting among power distribution units (PDUs) and various distributed components. These approaches are performed according to the workload variations and DC power availability. Besides dynamic power shifting, to address the peak power demand issue, a few works have introduced uninterruptible power supplies (UPSs) as an energy consumption saver [67–69]. The energy stored in batteries of UPSs is used to provide energy during periods of highest power demand, which results in DC OPEX reductions without performance degradation. Nevertheless, existing works neglect the possibility of inter-DC power scheduling were geographically distributed DCs can also offer opportunities for power distribution (Table 2). Inter-DC power scheduling enables preferment power scheduling to DCs with a larger amount of stored energy by consequence of being allocated closer to the larger sources of energy. Additionally, the low efficiency of UPSs used in DC during low UPS power demand periods, further contributes to the degradation of PUE. Some initial analyses of concept based on the simultaneous UPS and server/VMs consolidation in accordance with the DC workload variations show promising results in terms of improving DC energy-efficiency [70,71]. Still, major challenges related to achieving energy consumption reduction obtained through combining application performance, workload scheduling and power distribution in DC remains. This requests novel and more advanced solutions that can cope with DC power over-provisioning.

The use of renewable energy is another approach to improvement of DC energy efficiency (Table 2). Renewable energy sources such as solar, geothermal or wind energy are investigated for power supply of DC [72]. However, sporadic, unstable and limited nature of renewable energy production significantly determines the use of such green energy for DC power supply. Therefore, the question requesting to be addressed is how to use energy from renewable sources for the power supply of DCs and overcome the associated restrictions. To address intermittent power constraints of renewable energy, most of the previous research activities have been dedicated to the development of solutions in which the DC load had been adapted to follow the variable power supply capacities of renewable energy sources [73–75]. However, power supply solutions solely relying on unreliable renewable energy sources can experience unpredictable performance degradation and the most common approach to overcome such challenge is to have a hybrid DC power supply system combining the electrical grid and one or more types of renewable energy for DC power supply [76,77]. Such approaches use weather forecasts and historical data to estimate available renewable energy in the future, with the goal of optimal usage of renewable energy sources. Another challenge in using renewable energy for power supply of DCs is related to attenuation losses caused by the transfer of renewable energy over long distances. To avoid this, several proposed solutions suggest scheduling of server's workload to multiple DCs located in different geographical locations according to the availability of nearby renewable energy sources [78–81]. It is shown that such traffic routing based on geographical location can considerably reduce the brown energy consumption, if the energy tariff will be dynamically defined, and the degree of renewable energy usage will depend on the energy pricing model. Hence, future research must offer analyses of beneficial pricing schemes which will encourage DC operators to reduce brown energy consumption.

4.2. DC Servers Power Management

4.2.1. Dynamic Frequency and Voltage Scaling

The broadly accepted approach related to the improvements of DC servers power management is based on dynamic frequency and voltage scaling (DFVS) of server components (Table 2). The DFVS as the approach is based on lowering the frequency/voltage of components in order to achieve power savings in periods when the frequency/voltage of server components can be reduced. Due to approximate proportionality between the power consumption and the supply frequency/voltage of different hardware components, the goal is to find an optimal dynamic allocation of frequency/ voltage resources which will minimize the overall power consumption and ensure predefined performance. Vast research results related to the improvement of DC energy efficiency by implementing DFVS management according to individual servers computing and traffic load variations have been presented [82–94]. Additionally, research efforts on the level of energy-efficiency improvements of the large-scale server's warehouse through the implementation of DFVS are analysed in [37,95,96]. Since higher frequencies or voltages enable faster execution with the drawback of the increased power consumption, in [95], the optimal power allocation problem related to finding the optimal frequencies/voltages of the server components in a server farm was analysed based on server's workload. Furthermore, the implementation of an adaptive link rate (ALR) concept on the DCs network level was analysed [89,97,98]. The concept is based on an adaptive selection of speeds of links connecting DC servers in DC communication network. In order to contribute to the DC energy consumption reduction, adaptive adjustment of the Ethernet link data rate according to utilization shows that significant energy savings can be achieved since an Ethernet link can work almost 80% of the time at lower data rates [89]. However, most solutions proposed in the papers related to the DFVS concept, have been focused on power models which are assumed to be ideal. Hence, future research activities should consider models that have more similarities to real systems. More specifically, overhead which is not taken into account is mostly incurred when switching frequency or voltage speeds took place, because the central processing unit (CPU) must stop during these changes. Also, frequent changes in frequency/voltage speed can have a negative effect on CPU lifetime and a challenging issue is how to include these facts in performance analyses of practically implemented systems.

4.2.2. On/Off Server and Component Switching

Another approach related to the DC server's power management is based on the server components activity scaling, which envisions the transition of server components (such as server CPUs, memory, etc.) during idle traffic and computing periods into sleep or low-power standby mode (Table 2). The challenge is to decide when sufficiently long idle periods exist that enable component (CPU) activity state switching, while the cost for transitioning from or into the low power consumption state will not outweigh the costs incurred by this transition and will satisfy the workload demand. Analyses of the challenge of scheduling the power consumption in a two (sleep and active) states are presented in [99], and different studies extend analyses with multiple stets in [100,101]. Generally, components (such as CPU) state transition energies were assumed to be additive [100,102]. By taking into account different assumptions related to CPU state transition energy, different CPU scheduling algorithms in the case of single and multi-processor environments were proposed [104,105].

Additionally, a number of different studies proposed energy proportional computing for large hosting DCs [105–110], which are based on the concept of dynamic activation and deactivation of DC servers proportional to the DC workload demand (Table 2). Through such energy-aware provisioning, the server load is directed to the minimal active set of servers in DC and reduction of the server power consumption by 29% for characteristic web-based load is reported in [111]. Nevertheless, the novel approaches to further optimize DC energy consumption need to be devised. More specifically, in widely accepted parallel scheduling of jobs to different CPUs of servers which number is fixed, the scheduler decides about jobs that will be processed on CPUs and make a decision at any given time about the speed of each CPU. But, DCs operate on a different concept in which on-demand activation or deactivation of servers must be achieved. Hence, such DCs properties impose the development of new algorithms related to improving PUE while satisfying DC scalability and efficiency. Also, the power consumed, and latency generated during the rebooting of servers means that the effects of on/off server switching or DC networking device switching must be taken into account.

Furthermore, power-down mechanisms based on the concept of aggregating and redirecting network traffic on a few network devices which remain active are proposed in [106–109]. However, DC network architectures often ensure many communication paths between servers. This imposes the challenge of how to effectively control power consumption in DC networks and requests deeper investigation which will offer novel topologies and designs of DC networks, while satisfying demands for the network delay, congestion, loss of packets and throughput in those networks.

4.2.3. Hybrid DFVS and On/Off Server Switching

Another approach for improvement of DC server power management is based on hybrid concepts that exploit both DFVS scaling and servers or server components switching models. This hybrid approach is seen as a promising approach that can bring further improvements in DC energy-efficiency. This approach considers accelerating the processing tasks of server or server component activity, which results in longer idle periods, during which devices can be in the sleep or shut-down mode. Longer idle periods then give a higher contribution to the energy savings. The first theoretical analysis with an algorithm enabling combining system sleep mode for idle workload periods and DFVS during task processing periods are presented in [110]. In subsequent studies [111–114], improved algorithms were presented, some of which enable on-line scheduling and have low complexity orders. To enhance energy savings at the level of complete hosting DCs, in [116] a framework allowing the implementation of both approaches is introduced, while in [116], the authors considered the power consumption reduction in geographically distributed DCs. Hybrid techniques for improving the energy efficiency of network elements in DCs are used in studies [117,118]. In [117], to reduce network energy consumption, hybrid technology is implemented based on adjusting the rate of network operators to the real workload. Furthermore, in [118], the authors formulated an approach for online traffic management which reallocates the computing demands among a multitude of paths while optimizing energy consumption. Although both approaches report some energy savings, further investigations related to the implementation of the hybrid approach must take place.

Additionally, current results mostly focus on DC environments having servers with multiple homogeneous processors. Nevertheless, it is also important to take into account the DC server's heterogeneity, since the servers in DC mostly differ among themselves in terms of computing and hardware performance. Hence, results for heterogeneous environments in terms of the design of energy-efficient algorithms that can combine different DC power management methods are currently missing.

4.3. DC Simulation and Monitoring Management

Effective monitoring of DCs enable detecting the traces and tracks of thermal emission, power distribution and energy consumption for individual DCs equipment. Collected data can be used for the implementation of intelligent mechanisms based on which, the DC energy efficiency can be increased. Different DC monitoring services have been presented in [119,120]. Based on the collected data related to the VM application workloads, the resource utilization and power usage, DC online monitoring service presented in [119] enable a better understanding of the DC temperature behaviour and energy consumption. Developed monitoring solution also helps in consolidating the VM workload, which contributes to the significant energy savings. In [120], a monitoring solution based on request-tracing concept was implemented for determining energy inefficiencies in multi-tier DCs. The solution is based on collecting the resource consumption of respective requests and analyses of the characteristics of every DC tier. This further enables insights into the main causes of energy inefficiency of DCs and gives an opportunity to devise efficient power-saving methods for multitier applications.

As an example, green storage initiative (GSI) of the Storage Networking Industry Association (SNIA) works on forming a global standard for defining energy efficiency metrics of storage products working in the DC environment [121]. The proposed methodology enables the standardized and uniform method to grade the power efficiency of commercial (file, block, converged, object, etc.) storage in idle and active working states. This enables selecting the type of storage which best suits DC owner goals with the lowest power consumption contribution. This also motivates manufacturers to develop more energy-efficient storage devices since its energy efficiency can be compared among vendors.

Although monitoring of green metrics offers diagnosing of DCs energy inefficiencies, the development of monitoring tools has not been in the main research focus so far. The main obstacles in the realization of efficient DC green metrics monitoring are the availability of communication resources in DCs and a huge number of VMs and containers hosted on a large number of servers. Thus, future research activities need to be focused on solving the key research question related to the minimization of the costs incurred during collecting green DC metrics in a centralised or distributed manner, while guaranteeing monitoring accuracy.

Another approach to improvement of DC energy efficiency is based on simulation of DCs activity by means of developed simulation tools that enable understanding and identification of the design challenges that are crucial to DC energy efficiency. In this regard, different simulation platforms such as SimWare [121], GDCSim [122], GreenCloud [123] and EEFSim [124] are proposed. SimWare simulator [121] enables evaluation of the DC energy-saving policies and examination of mechanical functionalities such as management of airflow, cooling strategies and server placement. GDCSim proposed in [122] enables the iterative design of green DCs configurations for specific purposes, such as DFVS for power management, CPU sleep-state transitions and characterisation of thermal behaviour. GreenCloud [123] is a packet exchange simulator for energy-aware analyses of cloud DCs, which can be used for capturing the energy consumption of versatile DC elements such as switches, links, servers, as well as packet-based communication patterns. EEFSim [124] reproduce the behaviour of a real cloud DC and enables the possibility to examine the power consumption of different migration and scheduling policies with VMs. Nevertheless, the main drawbacks of these simulators are that each of them is very specific and dedicated only to certain functions or components of DC equipment such as CPU, VM, cooling, etc. Therefore, the task of designing a comprehensive system-oriented DC simulator that will integrate all DC components, such as the memory, CPU, cache, disks, input/output components and communication network is still an open task which requests addressing.

4.4. DC Thermal Management

Since typical DC hosts thousands of servers and communication network devices, it is reported in [125,126] that up to one-half of the total DC costs are spent on the cooling. These trends will be further contributed with previously presented servers' virtualization and consolidation techniques. These techniques increase processor utilization rates, which consequently contribute to the increase of thermal dissipation. Additionally, a combination of the abovementioned issues with trends characterised with server's concentration and high-density computing (realised through usage of many multi-core processors in single chasses), will raise the problem of thermal control as one of the most critical issues in deploying green DCs.

To cope with such a problem, in [127] the optimization of the DC cooling delivery is based on full control of the DC environment through collecting different DC attributes such as data aggregation, variable air conditioning and distributed sensing. Although this concept reports energy savings of up to 50%, another approach based on exploiting thermal energy storage (TES) tanks for the reduction of DC power is presented in [128]. In this concept, up to 28% in OPEX reduction is reported since this approach is based on TES storage of cold water or ice which are exploited as a supplement to the chillers used for cooling DCs and for heat exchange during peak power periods. Additionally, a very promising research field which can improve DC thermal management is based on DC workload migration and assignment among servers, in order to achieve thermal balance [129–131]. The general idea is dedicated to the development of algorithms for load scheduling based on temperature variations, which can reduce the energy consumption of the infrastructure dedicated to DC cooling [129]. The approach proposed is based on the dynamic transfer of server's workload from "warmed" servers and increasing the workload on remaining "colder" servers. The approach in [130] uses periodic temperature monitoring and server utilization for scheduling requests according to the DC workload weights. Also, in [131], a data-centric model dedicated to minimization of energy costs for DC cooling is developed based on dynamic file allocation in an energy and thermal-aware manner. The proposed model is developed by means of known data-semantics, cluster information and server-profile. Proposed approaches show possible energy savings between 20% and 42%. Additionally, in [132], the challenge of temperature-aware workload distribution in geo-distributed DCs is shown.

A completely different approach is based on efforts related to the reduction of the costs imposed by cooling, if the higher temperature can be sustained in DC. Basically, the concept is based on increasing temperature setting by only a few degrees, which results in an energy consumption reduction of 2%–5% [133]. However, the temperature increase of the servers and other equipment in DCs can contribute to a shortening of the DC equipment lifetime, which further contributes to the increase of capital expenditures (CAPEX) costs. Some initial studies related to the analyses of hardware (storage/memory/server) reliability and server performance presented in [134] show that in order to save energy, the DC could work at hotter temperatures than current ones, while negative effects on system reliability and performance can be partially limited. Still, better analyses of how temperature raise in DC can affect DC systems are needed and this field remains an open research topic.

Some other approaches which can offer the possibility of DCs to operate at higher temperatures are related to the development of new temperature-resistant hardware components. However, such components are still in their infancy and novel temperature-resistant chip and hardware solutions should be developed.

5. A Review of Articles for Special Issue on Green, Energy-Efficient and Sustainable Networks

The paper [135] analyses the influence of the node speed on the throughput and energy provision in an IoT network, where wireless charging stations (WCSs) are deployed to recharge IoT nodes while data transfer among nodes is limited by their abrupt links as well as the amounts of residual energy. To optimize node throughput and energy depletion of IoT nodes in such network based on wireless power transfer (WPT), authors propose a two-dimensional model based on Markov chains where the first state dimension represents the span to the closest WCS normalized with speed of nodes, while the second one represents residual energy of the node. Obtained results show that to enhance wireless charging efficiency, charging opportunity must be prioritized by WCSs based on a speed of IoT nodes, for which battery capacity can be minimized if the speed of nodes can be predicted. Also, if the same throughput must be ensured, it is shown that a lower number of WCSs per node can gain appropriate

WPT to all nodes in the area of high mobility, while a larger number of WCSs per IoT node are needed in areas of low mobility.

The next paper [136] tackles the problem of improving the energy-efficiency of software-defined networking (SDN) equipment based on the concept of traffic aggregation on links between two switches. In the paper, authors present different traffic allocation algorithms for SDN applications, which enable aggregation of the traffic flows to a few ports of the Ethernet links bundle in accordance with the traffic variations. Proposed allocation algorithms are validated in terms of packet losses, energy-efficiency and delay of packs. Obtained results show that the implementation of equipment with SDN capabilities can reduce energy consumption when Ethernet link bundles are used for up to 50%, without the necessity of changing devices firmware. Also, improvements of the two previous algorithms dedicated to offering a low-latency service for data traffic with strict requirements in terms of QoS and sustained energy consumption are proposed. According to the shown results, the algorithms can ensure the service which requests a low-delay of some orders of magnitude to time-sensitive traffic.

In [137], image compressive sensing is analysed as a potential image sensing approach that can satisfy green IoT demands in terms of finding optimal storage and data organization format suitable for sensors with limited power and bandwidth availability. The layer, patch and raster structure are proposed as three promising measurement schemes that differ in approaches related to storing and packaging of sensing measurements within an image. It is shown that each of the three proposed measurement structures restrains the image blocking artefacts and eliminate high memory requirements and huge computation complexity during image sensing and recovery. However, the layer structure shows the best results in terms of possible green IoT implementation since it has good rate and time-distortion performances and offers better visual quality than other structures.

Work [138] addresses a lack of models for energy-efficient malware detection based on gaining knowledge about devices in an IoT environment with the Android operating system (OS). In the paper, adversarial samples vulnerability of learning-based malware detection models is tackled through the development of an automated testing framework that performs security analyses for IoT devices. In order to find an appropriate fitness function that can produce the corresponding sample without impacting the characteristics of the application, authors introduce generic algorithms and specific technical enhancements built-in proposed testing framework. Obtained results show that black-box testing of the system can be done by the proposed test framework, which can create effective samples with a rate of success equal to almost 100% for the application on IoT devices with Android OS.

To eliminate drawbacks of authentication based on cipher approaches that are impacted by the large expenditures and energy constraints of smart devices, authors in [139] proposed a clustering-based physical-layer authentication scheme (CPAS) for systems with asymmetric resources in the mobile edge computing (MEC) environment. To ensure two-way authentication among edge devices and terminals, CPAS as cross-layer secure authentication merge symmetric cipher and clustering with information related to the wireless channel. Theoretical analysis of developed CPAS approach shows that CPAS can be robust to replay, spoofing and integer attacks, while experimental results show that CPAS decreases the data frame loss rate and increase the overall success rate of access authentication, without enlarging authentication latencies. Therefore, the proposed scheme reduces the complexity of resource asymmetric authentication scenarios for the edge computing systems, which contributes to the reduction of power consumption during the authentication phase.

In [140], the problem of the inter-tier interference mitigation in two-tier HetNets composed of pico-cells and underlying macro-cells has been considered. First, the near-optimal values of almost blank subframes (ABS) power reduction factor and pico-cell range expansion (CRE) bias are gained by an algorithm which uses equivalence relation between ABS and CRE for a given pico-cell base station (PBS) density. Also, by means of a linear search method, PBS density is optimized with the known factor of power reduction and constant pico-CRE bias. Lastly, to maximize network energy-efficiency of two-tier HetNets impacted with further-enhanced inter-cell interference, authors propose a heuristic algorithm for joint optimization of ABS power reduction factor, PBS density and pico-CRE bias. Results obtained through numerical simulations show that the proposed heuristic algorithm with a low complexity of computation can update the HetNets energy efficiency.

The study presented in [141] extends preceding works that use the social behaviour of the mobile users to adapt the transmission speeds of messages used for the peer discovery in D2D networks under the user equipment's (UEs) power consumption constraint. The authors introduce a three-phase energy-ratio rate decision (ERRD) algorithm, which in the first phase schedules the power budget of the network among the UEs based on their social ratios and in the second phase, based on harvested energy, allocates power quantum of each UE. Finally, in the third phase of the ERRD algorithm, the UEs beacon transmission intensities are adjusted according to their designated quantum of power. Adjusting is performed under the limitation that the overall power scheduled to the UEs cannot be above the power quanta of the network budget. Results obtained through simulations of ERRD performance show that the proposed algorithm outperforms the previously-reported algorithm by 8% and 190% on the peer discovery ratio, for a budget having the power of 20 and 1 W, respectively.

In paper [142], in order to improve the secure operation of industrial wireless sensor networks (IWSNs), a physical layer authentication based on deep learning is presented. Three different authentication methods for sensor nodes, more specifically the deep neural network (DNN), the convolutional neural network (CNN) and convolution pre-processing neural network (CPNN) have been used to deploy the PHY-layer authentication in IWSNs. According to simulation results obtained during the evaluation of algorithms performance, each algorithm can authenticate multiple nodes simultaneously trough lightweight authentication. However, the CPNN-based sensor nodes' authentication method has the best trade-off between the shortening of algorithm authentication performance and the minimal training time of the algorithm.

Paper [143] investigates the mobile directional charging vehicle (DCV) efficiency optimization in rechargeable wireless sensor networks (RWSN), through the implementation of wireless power transfer (WPT) on continuously working sensors. Authors initially design an approximation algorithm to define positions and charging orientations of the docking spots, with the constraints of maximizing the charging coverage utility and minimizing the total number of the DCVs docking spots. Then, an optimization of the DCVs energy charging is performed based on the developed moving path planning algorithm for the DCVs. Based on theoretical analyses and comprehensive simulation experiments, for the case of sparse networks, authors present that the efficiency of energy charging of the proposed DCV concept is better than those based on a model using the omnidirectional energy charging.

Authors in the next accepted work [144] have analysed the problem of green networking from the sustainability point of view. Besides energy-aware routing, authors propose pollution-aware routing with new metrics like the percentage of non-renewable energy usage and CO_2 emission factor. The proposed algorithm provides optimal control and data planes for these metric types and enables different routers scheduling and link bandwidth adaptations, while ensuring scheduling and adoption priority according to traffic demand requirements. The impact of the proposed algorithm enabling green routing was assessed for three different metrics. Obtained results show that the proposed pollution-aware routing algorithm can reduce CO_2 emissions for 20% and 36%, if compared with energy-based and shortest path routing, respectively.

A relaying system based on non-orthogonal multiple access (NOMA) in downlink transmission with the best amplify-and-forward radio-frequency energy harvesting relay was analysed in work [145]. Analyses are performed for a source node that exchanges information in parallel with multiple users and for the Rayleigh fading conditions lacking perfect RF channel state information. For such a system, authors develop expressions for the outage probability (OP), the optimal duration of energy harvesting which minimizes the OP and the ergodic capacities of each user. Based on numerical results obtained for the equal setting of parameters, the ergodic capacity of the whole system and overall performance of the proposed NOMA relaying system outperforms an equal system with the orthogonal-multiple-access (OMA) relaying.

Authors in [146] analyse the influence of using single and multiple relays on energy-efficiency and throughput of Long-Term Evolution-Advanced (LTE-A) networks, for different resource block (RB) allocation schemes. Energy-efficiency analyses for a single relay scenario is performed for the maximum throughput (MT) bisection-based power allocation (BOPA) algorithm, an alternating MT with proportional fairness (abbreviated SAMM) BOPA algorithm and SAMM equal power RB algorithm. Simulation results show that the SAMM BOPA algorithm ensures the best energy-efficiency, while SAMM equal power algorithm provides the best fairness. For a multiple relay scenario, a two-step neural network (NN) algorithm (SAMM NN) is introduced. Algorithm exploits BOPA supervised learning for power scheduling and SAMM unsupervised learning for scheduling of RBs. Results obtained for multiple relays scenario shows that SAMM NN algorithm achieves better energy-efficiency in comparison with SAMM equal power and SAMM BOPA algorithms.

Article [147] analyses the radio frequency fingerprinting identification (RFFID) approach dedicated to ensuring authentications for a high number of energy-limited user terminals working in the MEC environment. The proposed scheme combines a two-layer model with the use of non-encryption RFFID for IoT terminals. In the first layer, the MEC devices perform access authentication after signal detection and RF fingerprint features extraction with database storage of collected features. In the second layer, implementation of machine learning algorithms through collected learning features and generated decision models is done in the distant cloud, which improves the speed of authentication. Through extensive simulations performed for scenario based on IoT implementation, the gained results indicate that the approach proposed in [147] can achieve lower device energy depletion and better recognition rate than the traditional RFFID method based on wavelet features.

Paper [148] tackles the problem of scheduling consumer's requirements and the achievable electricity provision from renewable sources through the demand-response (DR) model. The proposed DR model is centralised via the data collector called the "aggregator" which schedules consumer's requirements for instantaneous power supply and supplied electricity from renewable energy sources in a home environment monitored through the implementation of IoT applications. Results of the proposed algorithm evaluation confirmed algorithm feasible costs of computation in different scenarios of consumer's behaviour and versatile communities and households. Also, it is shown that the energy reallocation costs are mostly impacted by a consumer's demand timeframe flexibility and a number of appliances.

In [149], the challenge of optimizing the power consumption of network devices in DC by means of energy-aware traffic engineering was addressed. The authors propose an optimization approach based on a mixed-integer programming algorithm, which minimizes network devices' energy consumption according to traffic load variations. The proposed approach was verified through simulations of versatile DC network topologies and obtained results demonstrate clear benefits in terms of DC power consumption reduction for different traffic volumes and DC network sizes. Furthermore, the proposed approach can be deployed as an implementation in the SDN paradigm, and therefore, it can be used in real practical implementations.

In the review paper [150], the energy efficiency of the radio access and core parts of the 5G networks are surveyed, and open issues and challenges related to the achievement of green cellular access networks are discussed. An overview of techniques for energy-efficiency improvement at the BS level encompasses next techniques for 5G networks: dynamic on/off cell switching, interference-aware energy efficiency control in UDNs, energy efficiency enhancement of BSs with radio resource control, connection management for 5G new radio, and energy-efficient cache-enabled BSs. Further analyses have been dedicated to the review of techniques for energy-efficiency enhancement at the 5G network level which includes energy-efficient: resource sharing, resource allocation in NOMA, outdoor–indoor communications and virtualization techniques. Additionally, authors perform a survey of SDN technology for improving energy-efficiency which considers energy monitoring and management in 5G with included backhaul and fronthaul and energy savings approach based on the utility of sleep mode.

Finally, the authors give an overview of techniques based on machine learning for energy-efficiency improvement of 5G networks.

6. Conclusions

Just a decade ago, the energy consumption of ICT devices and systems have been postponed in network and device design. However, a point where energy consumption optimisation of ICT devices and systems have become the new frontier for competitive differentiation and innovation is reached. Having energy-efficient ICT systems and devices is no longer a nice-to-have feature, but the mandatory requirement for the networks of the upcoming digital age. This is confirmed in papers accepted for publication in the Special Issue on the green, energy-efficient and sustainable networks, which overview in terms of addressed topics and obtained outcomes are presented in this paper. Additionally, estimations and analyses of energy costs and CO_2 emissions for different ICT systems in the period ranging up to the year 2030 are surveyed. Presented analyses confirm that ICT systems are at a critical point regarding current and future energy consumption of telecommunication networks, DCs and user-related devices. According to the presented estimations, current technology improvements of different ICT systems are not sufficient to keep up with the increasing energy costs and CO_2 emissions. This is elaborated in this paper for the case of wireless networks and DCs, which energy consumption and CO_2 emissions have the highest increase and contribution to the overall ICT energy consumption. This motivates the deeper investigation of technologies and concepts which can contribute to the improvement of energy-efficiency of these ICT sectors. As presented in this work, for the wireless networks, possible technologies that are analysed in this context include millimetre-wave communications, long term evolution in unlicensed spectrum, ultra-dense heterogeneous networks, device-to-device communications and massive multiple-input multiple-output communication. Additionally, DC resource management, DCs power management, green DC monitoring and simulation and thermal management in DCs are discussed as possible options for improvement of DCs power usage efficiency. Although each of analysed techniques and concepts can bring some energy-efficiency improvements in the corresponding area of implementation, comprehensive analyses presented in this paper shows that there is no single technology or concept which will bring energy-efficient improvements to the whole ICT sector. Hence, to achieve more power-efficient and greener ICT systems in the future, the combination of different technologies and concepts in wired, wireless and DC part of communication networks with novel solutions for energy-efficiency improvements of user-related and sensor devices must be devised. Only such an approach can result in a synergetic effect which will keep energy consumption and CO_2 emissions of ICT systems at the lowest possible levels.

Author Contributions: Writing—original draft preparation, investigation, conceptualization, methodology, data curation, visualization, J.L.; writing—review and editing, supervision, A.C. and J.W.

Funding: This research received no external funding.

Conflicts of Interest: The authors declare no conflict of interest.

References

1. Wu, J.; Guo, S.; Huang, H.; Liu, W.; Xiang, Y. Information and Communications Technologies for Sustainable Development Goals: State-of-the-Art, Needs and Perspectives. *IEEE Commun. Surv. Tutor.* **2018**, *20*, 2389–2406. [CrossRef]
2. Global e-Sustainability Initiative (GeSI). *SMARTer2030 Rapport—ICT Solutions for 21st Century Challenges*; GeSI: Brussels, Belgium, 2015; pp. 1–134.
3. Weldon, M.K. *The Future X Network: A Bell Labs Perspective*; CRC Press, Taylor and Francis Group: Boca Raton, FA, USA, 2016; pp. 1–480. ISBN 149877914X.
4. NTT Com Global Watch vol.2. Available online: https://www.ntt.com/en/resources/articles/global-watch-vol2.html (accessed on 22 October 2019).

5. *Cisco Visual Networking Index: Forecast and Trends*; 2017–2022 White Paper ©; Cisco Systems Inc.: San Jose, CA, USA, 2019.

6. Intelligent Energy. *The True Cost of Providing Energy to Telecom Towers in India, White Paper*; Intelligent Energy: Loughborough, UK, 2012.

7. GSMA. *Greening Telecoms: Pakistan and Aghanistan Market Analysis, Green Power for Mobile*; GSMA: London, UK, 2013.

8. Profitability of the Mobile Business Model … The Rise! & Inevitable Fall? Techneconomy Blog. 2014. Available online: http://techneconomyblog.com/2014/07/21/profitability-of-the-mobile-business-model (accessed on 22 October 2019).

9. Le Maistre, R. Energy Bill Shocks Orange Into Action, Light Reading. 2014. Available online: http://www.lightreading.com/energy-efficiency/energy-bill-shocks-orange-into-action/d/d-id/711038 (accessed on 22 October 2019).

10. Andrae, A.S.G.; Edler, T. On Global Electricity Usage of Communication Technology: Trends to 2030. *Chall. J.* **2015**, *6*, 117–157. [CrossRef]

11. Lambert, S.; Van Heddeghem, W.; Vereecken, W.; Lannoo, B.; Colle, D.; Pickavet, M. *Worldwide Electricity Consumption of Communication Networks*; Optical Society of America: Washington, DC, USA, 2013.

12. Wu, Q.; Li, G.Y.; Chen, W.; Ng, D.W.K.; Schober, R. An Overview of Sustainable Green 5G Networks. *IEEE Wirel. Commun.* **2017**, *24*, 72–80. [CrossRef]

13. Lorincz, J.; Matijevic, T. Energy-efficiency analyses of heterogeneous macro and micro base station sites. *Comput. Electr. Eng.* **2014**, *40*, 330–349. [CrossRef]

14. Samarakoon, S.; Bennis, M.; Saad, W.; Debbah, M.; Latva-Aho, M. Ultra Dense Small Cell Networks: Turning Density Into Energy Efficiency. *IEEE J. Sel. Areas Commun.* **2016**, *34*, 1267–1280. [CrossRef]

15. Bjornson, E.; Sanguinetti, L.; Hoydis, J.; Debbah, M. Optimal Design of Energy-Efficient Multi-User MIMO Systems: Is Massive MIMO the Answer? *IEEE Trans. Wirel. Commun.* **2015**, *14*, 3059–3075. [CrossRef]

16. The Balance between 5G Profit and Power. Available online: https://www.datacenterdynamics.com/analysis/balance-between-5g-profit-and-power/ (accessed on 20 October 2019).

17. Lorincz, J.; Capone, A.; Begusic, D. Heuristic Algorithms for Optimization of Energy Consumption in Wireless Access Networks. *KSII Trans. Internet Inf. Syst.* **2011**, *5*, 626–648. [CrossRef]

18. Zhang, S.; Gong, J.; Zhou, S.; Niu, Z. How Many Small Cells Can Be Turned off via Vertical Offloading under a Separation Architecture? *IEEE Trans. Wirel. Commun.* **2015**, *14*, 1. [CrossRef]

19. Lorincz, J.; Matijevic, T.; Petrovic, G. On interdependence among transmit and consumed power of macro base station technologies. *Comput. Commun.* **2014**, *50*, 10–28. [CrossRef]

20. Cai, S.; Che, Y.; Duan, L.; Wang, J.; Zhou, S.; Zhang, R. Green 5G Heterogeneous Networks Through Dynamic Small-Cell Operation. *IEEE J. Sel. Areas Commun.* **2016**, *34*, 1103–1115. [CrossRef]

21. Jones, D. Power Consumption: 5G Basestations Are Hungry, Hungry Hippos. Available online: https://www.lightreading.com/mobile/5g/power-consumption-5g-basestations-are-hungry-hungry-hippos/d/d-id/749979 (accessed on 20 October 2019).

22. Ngo, H.Q.; Larsson, E.G.; Marzetta, T.L. Energy and Spectral Efficiency of Very Large Multiuser MIMO Systems. *IEEE Trans. Commun.* **2013**, *61*, 1436–1449.

23. Han, S.; Chih-Lin, I.; Xu, Z.; Rowell, C. Large-scale antenna systems with hybrid analog and digital beamforming for millimeter wave 5G. *IEEE Commun. Mag.* **2015**, *53*, 186–194. [CrossRef]

24. Zeng, Y.; Zhang, R. Millimeter Wave MIMO with Lens Antenna Array: A New Path Division Multiplexing Paradigm. *IEEE Trans. Commun.* **2016**, *64*, 1557–1571. [CrossRef]

25. Lorincz, J.; Bule, I. Renewable Energy Sources for Power Supply of Base Station Sites. *Int. J. Bus. Data Commun. Netw.* **2013**, *9*, 53–74. [CrossRef]

26. Lorincz, J.; Bule, I.; Kapov, M. Performance Analyses of Renewable and Fuel Power Supply Systems for Different Base Station Sites. *Energies* **2014**, *7*, 7816–7846. [CrossRef]

27. Gunduz, D.; Stamatiou, K.; Michelusi, N.; Zorzi, M. Designing intelligent energy harvesting communication systems. *IEEE Commun. Mag.* **2014**, *52*, 210–216. [CrossRef]

28. Feng, D.; Lu, L.; Yuan-Wu, Y.; Li, G.Y.; Feng, G.; Li, S. Device-to-Device Communications Underlaying Cellular Networks. *IEEE Trans. Commun.* **2013**, *61*, 3541–3551. [CrossRef]

29. Zhang, R.; Wang, M.; Cai, L.X.; Zheng, Z.; Shen, X.; Xie, L.-L. LTE-unlicensed: The future of spectrum aggregation for cellular networks. *IEEE Wirel. Commun.* **2015**, *22*, 150–159. [CrossRef]

30. Bi, S.; Ho, C.K.; Zhang, R. Wireless powered communication: Opportunities and challenges. *IEEE Commun. Mag.* **2015**, *53*, 117–125. [CrossRef]

31. Lu, X.; Wang, P.; Niyato, D.; Kim, D.I.; Han, Z. Wireless Networks with RF Energy Harvesting: A Contemporary Survey. *IEEE Commun. Surv. Tutor.* **2015**, *17*, 757–789. [CrossRef]

32. Patel-Predd, P. Update: Energy-efficient Ethernet. *IEEE Spectr.* **2008**, *45*, 13. [CrossRef]

33. Greenberg, A.; Hamilton, J.; Maltz, D.A.; Patel, P. The cost of a cloud: Research problems in data center networks. *ACM SIGCOMM Comput Commun. Rev.* **2009**, *39*, 68–73. [CrossRef]

34. Carrega, A.; Singh, S.; Bruschi, R.; Bolla, R. Traffic merging for energy-efficient datacenter networks. In Proceedings of the SPECTS'12, Genoa, Italy, 8–11 July 2012; pp. 1–5.

35. Barroso, L.A.; Clidaras, J.; Holzle, U. *The Data Center as a Computer: An Introduction to the Design of Warehouse-Scale Machines*; Morgan & Claypool publish: San Rafael, CA, USA, 2013.

36. Benson, T.; Akella, A.; Maltz, D. Network traffic characteristics of data centers in the wild. In Proceedings of the IMC'10, Melbourne, Australia, 1–3 November 2010; pp. 267–280.

37. Abts, D.; Marty, M.; Wells, P.; Klausler, P.; Liu, H. Energy proportional datacenter networks. In Proceedings of the ISCA'10, Saint-Malo, France, 19–23 June 2010; pp. 338–347.

38. Wu, J.; Guo, S.; Li, J.; Zeng, D. Big Data Meet Green Challenges: Greening Big Data. *IEEE Syst. J.* **2016**, *10*, 873–887. [CrossRef]

39. Jin, X.; Zhang, F.; Vasilakos, A.V.; Liu, Z. Green Data Centers: A Survey, Perspectives, and Future Directions. *arXiv* **2016**, arXiv:1608.00687, pp. 1–20, 1–20.

40. Barham, P.; Dragovic, B.; Fraser, K.; Hand, S.; Harris, T.; Ho, A.; Neugebauer, R.; Pratt, I.; Warfield, A. Xen and the art of virtualization. In Proceedings of the SOSP'03, New York, NY, USA, 19–22 October 2003; pp. 164–177.

41. Stolyar, A.L.; Zhong, Y.A. Large-scale service system with packing constraints: Minimizing the number of occupied servers. In *ACM SIGMETRICS Performance Evaluation Review*; ACM: New York, NY, USA, 2013; pp. 41–52.

42. Raghavendra, R.; Ranganathan, P.; Talwar, V.; Wang, Z.; Zhu, X. No "power" struggles: Coordinated multi-level power management for the data center. In Proceedings of the ASPLOS'08, Seattle, WA, USA, 1–5 March 2008; pp. 48–59.

43. Nathuji, R.; Schwan, K. Virtual Power: Coordinated power management in virtualized enterprise systems. In Proceedings of the SOSP'07, Stevenson, WA, USA, 14–17 October 2007; pp. 265–278.

44. Bobroff, N.; Kochut, A.; Beaty, K. Dynamic placement of virtual machines for managing SLA violations. Available online: https://pdfs.semanticscholar.org/59ab/46bfd59cb43876e701389f256b93430e6273.pdf (accessed on 20 October 2019).

45. Beloglazov, A.; Buyya, R. Optimal online deterministic algorithms and adaptive heuristics for energy and performance efficient dynamic consolidation of virtual machines in Cloud data centers. *Concurr. Comput. Pract. Exp.* **2011**, *24*, 1397–1420. [CrossRef]

46. Wood, T.; Shenoy, P.; Venkataramani, A.; Yousif, M. Blackbox and gray-box strategies for virtual machine migration 2007. In Proceedings of the NSDI'07, Cambridge, MA, USA, 11–13 April 2007; pp. 229–242.

47. Van, H.N.; Tran, F.D.; Menaud, J.M. SLA-aware virtual resource management for cloud infrastructures. In Proceedings of the CIT'09, Xiamen, China, 11–14 October 2009; pp. 357–362.

48. Kusic, D.; Kephart, J.O.; Hanson, J.E.; Kandasamy, N.; Jiang, G. Power and performance management of virtualized computing environments via lookahead control. *J. Clust. Comput.* **2009**, *12*, 1–15. [CrossRef]

49. Hermenier, F.; Lorca, X.; Menaud, J.M.; Muller, G.; Lawall, J. Entropy: A consolidation manager for clusters. In Proceedings of the VEE'09, Washington, DC, USA, 11–13 March 2009; pp. 41–50.

50. Wang, M.; Meng, X.; Zhang, L. Consolidating virtual machines with dynamic bandwidth demand in data centers. In Proceedings of the INFOCOM'11, Shanghai, China, 11–15 April 2011; pp. 71–75.

51. Mahadevan, P.; Banerjee, S.; Sharma, P.; Shah, A.; Ranganathan, P. On energy efficiency for enterprise and data center networks. *IEEE Commun. Mag.* **2011**, *49*, 94–100. [CrossRef]

52. Zheng, K.; Wang, X.; Li, L.; Wang, X. Joint power optimization of data center network and servers with correlation analysis. In Proceedings of the INFOCOM'14, Toronto, Canada, 27 April–2 May 2014; pp. 2598–2606.

53. McGeer, R.; Mahadevan, P.; Banerjee, S. On the complexity of power minimization schemes in data center networks. In Proceedings of the Globecom'10, Miami, FL, USA, 6–10 December 2010; pp. 1–5.

54. Jiang, J.; Lan, T.; Ha, S.; Chen, M.; Chiang, M. Joint VM placement and routing for data center traffic engineering. In Proceedings of the INFOCOM'12, Orlando, FL, USA, 25–30 March 2012; pp. 2876–2880.

55. Xu, M.; Shang, Y.; Li, D.; Wang, X. Greening data center networks with throughput-guaranteed power-aware routing. *Comput. Netw.* **2013**, *57*, 2880–2899. [CrossRef]

56. Wang, X.; Yao, Y.; Wang, X.; Lu, K.; Cao, Q. CARPO: Correlation-aware power optimization in data center networks. In Proceedings of the INFOCOM'12, Orlando, FL, USA, 25–30 March 2012; pp. 1125–1133.

57. Wang, L.; Zhang, F.; Aroca, J.A.; Vasilakos, A.V.; Zheng, K.; Hou, C.; Li, D.; Liu, Z. GreenDCN: A general framework for achieving energy efficiency in data center networks. *IEEE J. Sel. Areas Commun.* **2014**, *32*, 4–15. [CrossRef]

58. Zhang, Y.; Ansari, N. HERO: Hierarchical energy optimization for data center networks. In Proceedings of the ICC'12, Ottawa, ON, Canada, 10–15 June 2012; pp. 2924–2928.

59. Jin, H.; Cheocherngngarn, T.; Levy, D.; Smith, A.; Pan, D.; Liu, J.; Pissinou, N. Joint host-network optimization for energy-efficient data center networking. In Proceedings of the IPDPS'13, Boston, MA, USA, 24 May 2013; pp. 623–634.

60. Vasic, N.; Bhurat, P.; Novakovic, D.; Canini, M.; Shekhar, S.; Kostic, D. Identifying and using energy-critical paths. In Proceedings of the CoNEXT'11, Tokyo, Japan, 6–9 December 2011. No. 18.

61. Benson, T.; Akella, A.; Shaikh, A.; Sahu, S. CloudNaaS: A cloud networking platform for enterprise applications. In Proceedings of the SOCC'11, Cascais, Portugal, 26–28 October 2011; No. 8. pp. 1–13.

62. Meisner, D.; Gold, B.T.; Wenisch, T.F. PowerNap: Eliminating server idle power. In Proceedings of the ASPLOS'09, Washington, DC, USA, 7–11 March 2009; pp. 205–216.

63. Pelley, S.; Meisner, D.; Zandevakili, P.; Wenisch, T.F.; Underwood, J. Power routing: Dynamic power provisioning in the dana center. In Proceedings of the ASPLOS'10, Pittsburgh, PA, USA, 13–17 March 2010; pp. 231–242.

64. Govindan, S.; Sivasubramaniam, A.; Urgaonkar, B. Benefits and limitations of trapping into stored energy for datacenters. In Proceedings of the ISCA'11, San Jose, CA, USA, 4–8 June 2011; pp. 341–352.

65. Lim, H.; Kansal, A.; Liu, J. Power budgeting for virtualized data centers. In Proceedings of the USENIX ATC'11, Portland, OR, USA, 14–17 June 2011; pp. 1–14.

66. Al-Hazemi, F.; Lorincz, J.; Mohammed, A.F.Y. Minimizing Data Center Uninterruptable Power Supply Overload by Server Power Capping. *IEEE Commun. Lett.* **2019**, *23*, 1342–1346. [CrossRef]

67. Kontorinis, V.; Zhang, L.E.; Aksanli, B.; Sampson, J.; Homayoun, H.; Pettis, E.; Tullsen, D.M.; Rosing, T. Šimunić Managing distributed ups energy for effective power capping in data centers. *ACM SIGARCH Comput. Arch. News* **2012**, *40*, 488. [CrossRef]

68. Fan, X.; Weber, W.-D.; Barroso, L.A. Power provisioning for a warehouse-sized computer. In Proceedings of the ISCA'07, San Diego, CA, USA, 9–13 June 2007; pp. 13–23.

69. Govindan, S.; Wang, D.; Sivasubramaniam, A.; Urgaonkar, B. Leveraging stored energy for handling power emergencies in aggressively provisioned datacenters. In Proceedings of the ASPLOS'12, London, UK, 3–7 March 2012; pp. 75–86.

70. Al-Hazemi, F.; Peng, Y.; Youn, C.-H.; Lorincz, J.; Li, C.; Song, G.; Boutaba, R. Dynamic allocation of power delivery paths in consolidated data centers based on adaptive UPS switching. *Comput. Netw.* **2018**, *144*, 254–270. [CrossRef]

71. AL-Hazemi, F.; Lorincz, J.; Mohammed, A.F.Y.; Salamh, F. Reducing Data Center Power Losses through UPS Serial Consolidation. In Proceedings of the SoftCOM 2019, Split, Croatia, 19–21 September 2019; pp. 1–6.

72. Gupta, P. Google to use wind energy to power data centers. *Reuters*. 2010. Available online: https://www.reuters.com/article/us-google-windpower/google-to-use-wind-energy-to-power-data-centers-idUSTRE66J3BL20100720 (accessed on 22 October 2019).

73. Sharma, N.; Barker, S.; Irwin, D.; Shenoy, P. Blink: Managing server clusters on intermittent power. In Proceedings of the ASP-LOS'11, Newport Beach, CA, USA, 5–11 March 2011; pp. 185–198.

74. Li, C.; Zhou, R.; Li, T. Enabling distributed generation powered sustainable high-performance data center. In Proceedings of the HPCA'13, Shenzhen, China, 23–27 February 2013; pp. 35–46.

75. Akoush, S.; Sohan, R.; Rice, A.; Moore, A.W.; Hopper, A. Free lunch: Exploiting renewable energy for computing. In Proceedings of the HotOS'11, Napa, CA, USA, 9–11 May 2011; pp. 1–5.

76. Goiri, I.; Le, K.; Nguyen, T.D.; Guitart, J.; Torres, J.; Bianchini, R. GreenHadoop: Leveraging green energy in data-processing frameworks. In Proceedings of the EuroSys'12, Bern, Switzerland, 10–13 April 2012; pp. 57–70.

77. Goiri, I.; Le, K.; Haque, M.E.; Beauchea, R.; Nguyen, T.D.; Guitart, J.; Torres, J.; Bianchini, R. GreenSlot: Scheduling energy consumption in green datacenters. In Proceedings of the SC'11, Seatle, WA, USA, 16 September 2011; pp. 1–11.

78. Gao, P.X.; Curtis, A.R.; Wong, B.; Keshav, S. It's not easy being green. In Proceedings of the SIGCOMM'12, Helsinki, Finland, 13–14 August 2012; pp. 211–222.

79. Gao, Y.; Zeng, Z.; Liu, X.; Kumar, P.R. The answer is blowing in the wind: Analysis of powering Internet data centers with wind energy. In Proceedings of the INFOCOM'13, Turin, Italy, 14–19 April 2013; pp. 520–524.

80. Liu, Z.; Lin, M.; Wierman, A.; Low, S.H.; Andrew, L.L.H. Greening geographical load balancing. In Proceedings of the SIGMETRICS'11, San Jose, CA, USA, 7–11 June 2011; pp. 233–244.

81. Zhang, Y.; Wang, Y.; Wang, X. GreenWare: Greening cloudscale data centers to maximize the use of renewable energy. In Proceedings of the Middleware'11, Lisbon, Portugal, 12–16 December 2011; pp. 143–164.

82. Albers, S.; Fujiwara, H. Energy-efficient algorithms for flow time minimization. *ACM Trans. Algorithms* **2007**, *3*, 1–13. [CrossRef]

83. Pruhs, K.; Stee, R.V.; Uthaisombut, P. Speed scaling of tasks with precedence constraints. *Theory Comput. Syst.* **2008**, *43*, 67–80. [CrossRef]

84. Bunde, D.P. Power-aware scheduling for makespan and flow. *J. Sched.* **2009**, *12*, 489–500. [CrossRef]

85. Bansal, N.; Kimbrel, T.; Pruhs, K. Speed scaling to manage energy and temperature. *J. ACM* **2007**, *54*, 11–39. [CrossRef]

86. Jin, X.; Zhang, F.; Song, Y.; Fan, L.; Liu, Z. Energy-efficient scheduling with time and processors eligibility restrictions. In Proceedings of the EuroPar'13, Aachen, Germany, 26–30 August 2013; pp. 66–77.

87. Greiner, G.; Nonner, T.; Souza, A. The bell is ringing in speed scaled multiprocessor scheduling. In Proceedings of the SPAA'09, Calgary, AB, Canada, 11–13 August 2009; pp. 11–18.

88. Albers, S.; Muller, F.; Schmelzer, S. Speed scaling on parallel processors. In Proceedings of the SPAA'07, San Diego, CA, USA, 9–11 June 2007; pp. 289–298.

89. Gunaratne, C.; Christensen, K.; Nordman, B.; Suen, S. Reducing the energy consumption of ethernet with adaptive link rate (ALR). *IEEE Trans. Comput.* **2008**, *57*, 448–461. [CrossRef]

90. Andrews, M.; Anta, A.F.; Zhang, L.; Zhao, W. Routing for power minimization in the speed scaling model. *IEEE/ACM Trans. Netw.* **2012**, *20*, 285–294. [CrossRef]

91. Jin, X.; Zhang, F.; Liu, Z. Discrete Min-Energy Scheduling on Restricted Parallel Processors. In Proceedings of the IPDPS'13, Cambridge, MA, USA, 20–24 May 2013; pp. 2226–2229.

92. Antoniadis, A.; Huang, C.C. Non-preemptive speed scaling. *Scheduling* **2013**, *16*, 385–394. [CrossRef]

93. Bampis, E.; Kononov, A.; Letsios, D.; Lucarelli, G.; Nemparis, I. From preemptive to non-preemptive speed-scaling scheduling. *Discret. Appl. Math.* **2015**, *181*, 11–20. [CrossRef]

94. Gandhi, A.; Harchol-Balter, M.; Das, R.; Lefurgy, C. Optimal power allocation in server farms. In Proceedings of the SIGMET- RICS'09, Seattle, WA, USA, 15–19 June 2009; pp. 157–168.

95. Wierman, A.; Andrew, L.L.; Tang, A. Power-aware speed scaling in processor sharing systems. In Proceedings of the INFOCOM'09, Rio de Janeiro, Brazil, 19–25 April 2009; pp. 2007–2015.

96. Liu, F.; Zhou, Z.; Jin, H.; Li, B.; Li, B.; Jiang, H. On Arbitrating the Power-Performance Tradeoff in SaaS Clouds. *IEEE Trans. Parallel Distrib. Syst.* **2014**, *25*, 2648–2658. [CrossRef]

97. Wang, L.; Zhang, F.; Hou, C.; Aroca, J.A.; Liu, Z. Incorporating rate adaptation into green networking for future data centers. In Proceedings of the NCA'13, Cambridge, MA, USA, 22–24 August 2013; pp. 106–109.

98. Karlin, A.R.; Manasse, M.S.; McGeoch, L.A.; Owicki, S. Competitive randomized algorithms for nonuniform problems. *Algorithmica* **1994**, *11*, 542–571. [CrossRef]

99. Irani, S.; Shukla, S.; Gupta, R. Online strategies for dynamic power management in systems with multiple power-saving states. *ACM Trans. Embed. Comput. Syst.* **2003**, *2*, 325–346. [CrossRef]

100. Chen, J.-J.; Kao, M.-J.; Lee, D.T.; Rutter, I.; Wagner, D. Online dynamic power management with hard real-time guarantees. *Theor. Comput. Sci.* **2015**, *594*, 46–64. [CrossRef]

101. Augustine, J.; Irani, S.; Swamy, C. Optimal power-down strategies. *SIAM J. Comput.* **2008**, *37*, 1499–1516. [CrossRef]

102. Angel, E.; Bampis, E.; Chau, V. Low complexity scheduling algorithms minimizing the energy for tasks with agreeable deadlines. *Discret. Appl. Math.* **2014**, *175*, 1–10. [CrossRef]
103. Demaine, E.D.; Ghodsi, M.; Hajiaghayi, M.T.; Sayedi-Roshkhar, A.S.; Zadimoghaddam, M. Scheduling to minimize gaps and power consumption. *Scheduling* **2013**, *16*, 151–160. [CrossRef]
104. Lin, M.; Wierman, A.; Andrew, L.; Thereska, E. Dynamic right-sizing for power-proportional data centers. In Proceedings of the INFOCOM'11, Shanghai, China, 10–15 April 2011; pp. 1098–1106.
105. Azar, Y.; Ben-Aroya, N.; Devanur, N.R.; Jain, N. Cloud scheduling with setup cost. In Proceedings of the SPAA'13, Montréal, QC, Canada, 23–25 July 2013; pp. 298–304.
106. Andrews, M.; Anta, A.F.; Zhang, L.; Zhao, W. Routing and scheduling for energy and delay minimization in the powerdown model. In Proceedings of the INFOCOM'10, San Diego, CA, USA, 14–19 March 2010; pp. 21–25.
107. Zhang, M.; Yi, C.; Liu, B.; Zhang, B. GreenTE: Power-aware traffic engineering. In Proceedings of the ICNP'10, Kyoto, Japan, 5–8 October 2010; pp. 21–30.
108. Heller, B.; Seetharaman, S.; Mahadevan, P.; Yiakoumis, Y.; Sharma, P.; Banerjee, S.; McKeown, N. ElasticTree: Saving energy in data center networks. In Proceedings of the NSDI'10, San Jose, CA, USA, 28–30 April 2010; pp. 249–264.
109. Bolla, R.; Bruschi, R.; Cianfrani, A.; Listanti, M. Enabling backbone networks to sleep. *IEEE Netw.* **2011**, *25*, 26–31. [CrossRef]
110. Irani, S.; Shukla, S.; Gupta, R. Algorithms for power savings. *ACM Trans. Algorithms* **2007**, *3*, 41. [CrossRef]
111. Adams, W.M. Power Consumption in Data Centers Is a Global Problem. Available online: https://www.datacenterdynamics.com/opinions/power-consumption-data-centers-global-problem/ (accessed on 21 October 2019).
112. Antoniadis, A.; Huang, C.-C.; Ott, S. A fully polynomialtime approximation scheme for speed scaling with sleep state. *arXiv* **2014**, arXiv:1407.0892.
113. Albers, S.; Antoniadis, A. Race to idle: New algorithms for speed scaling with a sleep state. *ACM Trans. Algorithms* **2014**, *10*, 1266–1285. [CrossRef]
114. Han, X.; Lam, T.-W.; Lee, L.-K.; To, I.K.K.; Wong, P.W.H. Deadline scheduling and power management for speed bounded processors. *Theor. Comput. Sci.* **2010**, *411*, 3587–3600. [CrossRef]
115. Chen, Y.; Das, A.; Qin, W.; Sivasubramaniam, A.; Wang, Q.; Gautam, N. Managing server energy and operation costs in hosting centers. In Proceedings of the SIGMETRICS'05, Banff, AB, Canada, 6–10 June 2005; pp. 303–314.
116. Yao, Y.; Huang, L.; Sharma, A.B.; Golubchik, L.; Neely, M.J. Data centers power reduction: A two time scale approach for delay tolerant workloads. In Proceedings of the INFOCOM'12, Orlando, FL, USA, 25–30 March 2012; pp. 1431–1439.
117. Nedevschi, S.; Popa, L.; Iannaccone, G.; Ratnasamy, S.; Wetherall, D. Reducing network energy consumption via sleeping and rate-adaptation. In Proceedings of the NSDI'08, San Francisco, CA, USA, 16–18 April 2008; pp. 323–336.
118. Vasic, N.; Kostic, D. Energy-aware traffic engineering. In Proceedings of the e-Energy'10, Passau, Germany, 13–15 April 2010; pp. 169–178.
119. Liu, L.; Wang, H.; Liu, X.; Jin, X.; He, W.; Wang, Q.; Chen, Y. GreenCloud: A new architecture for green data center. In Proceedings of the ICAC-INDST'09, Barcelona, Spain, 15 June 2009; pp. 29–38.
120. Lu, G.; Zhang, J.; Wang, H.; Yuan, L.; Weng, C. PowerTracer: Tracing requests in multi-tier services to diagnose energy inefficiency. In Proceedings of the ICAC'12, San Jose, CA, USA, 18–20 September 2012; pp. 97–102.
121. Yeo, S.; Li, H.-H.S. SimWare: A holistic warehouse-scale computer simulator. *Computer* **2012**, *45*, 48–55. [CrossRef]
122. Gupta, S.K.; Gilbert, R.R.; Banerjee, A.; Abbasi, Z.; Mukherjee, T.; Varsamopoulos, G. GDCSim: A tool for analyzing green data center design and resource management techniques. In Proceedings of the IGCC'11, Orlando, FL, USA, 25–28 July 2011; pp. 1–8.
123. Kliazovich, D.; Bouvry, P.; Audzevich, Y.; Khan, S.U. Green-Cloud: A packet-level simulator of energy-aware cloud computing data centers. In Proceedings of the Globecom'10, Miami, FL, USA, 6–10 December 2010; pp. 1–5.

124. Julia, F.; Roldan, J.; Nou, R.; Fito, J.O.; Vaque, A.; Goiri, I.; Berral, J.L. EEFSim: Energy Efficiency Simulator. Research Report: UPC-DAC-RR-CAP-2010-15. 2010. Available online: https://www.ac.upc.edu/app/research-reports/html/2010/19/eefsimtr.pdf (accessed on 22 October 2019).

125. Patel, C.D.; Bash, C.E.; Sharma, R.; Beitelma, M. Smart cooling of data centers. In Proceedings of the IPACK'03, Maui, HI, USA, 6–11 July 2003 ; pp. 129–137.

126. Belady, C.; Rawson, A.; Pfleuger, J.; Cader, T. The Green Grid Data Center Power Efficiency Metrics: PUE & DCiE. The Green Grid, Technical Report. 2008. Available online: https://www.academia.edu/23433359/Green_Grid_Data_Center_Power_Efficiency_Metrics_Pue_and_Dcie (accessed on 22 October 2019).

127. Zheng, W.; Ma, K.; Wang, X. Exploiting thermal energy storage to reduce data center capital and operating expenses. In Proceedings of the HPCA'14, Orlando, FL, USA, 15–19 February 2014; pp. 132–141.

128. Sharma, R.K.; Bash, C.L.; Patel, C.D.; Friedrich, R.J.; Chase, J.S. Balance of power: Dynamic thermal management for Internet Data centers. *IEEE Internet Comput.* **2005**, *9*, 42–49. [CrossRef]

129. Heath, T.; Centeno, A.P.; George, P.; Ramos, L.; Jaluria, Y.; Bianchini, R. Mercury and freon: Temperature emulation and management for server systems. In Proceedings of the ASPLOS'06, San Jose, CA, USA, 21–25 October 2006; pp. 106–116.

130. Xu, H.; Feng, C.; Li, B. Temperature aware workload management in geo-distributed datacenters. In Proceedings of the ICAC'13, Pittsburgh, PA, USA, 26–28 June 2013; pp. 303–314.

131. Moore, J.; Chase, J.; Ranganathan, P.; Sharma, R. Making scheduling "cool": Temperature-aware workload placement in data centers. In Proceedings of the ATEC'05, Anaheim, CA, USA, 10–15 April 2005; pp. 61–74.

132. Brandon, J. Going green in the data center: Practical steps for your sme to become more environmentally friendly. *Processor* **2007**, *29*, 1–30.

133. Kaushik, R.T.; Nahrstedt, K. *T*∗: A data-centric cooling energy costs reduction approach for big data analytics cloud. In Proceedings of the SC'12, Salt Lake City, UT, USA, 10–16 November 2012; pp. 1–11.

134. El-Sayed, N.; Stefanovici, I.A.; Amvrosiadis, G.; Hwang, A.A.; Schroeder, B. Temperature management in data centers: Why some (might) like it hot. In Proceedings of the SIGMETRICS'12, London, UK, 11–15 June 2012; pp. 163–174.

135. Ko, S.-W.; Kim, S.-L. Impact of Node Speed on Energy-Constrained Opportunistic Internet-of-Things with Wireless Power Transfer. *Sensors* **2018**, *18*, 2398. [CrossRef]

136. Fondo-Ferreiro, P.; Rodríguez-Pérez, M.; Fernández-Veiga, M.; Herrería-Alonso, S. Matching SDN and Legacy Networking Hardware for Energy Efficiency and Bounded Delay. *Sensors* **2018**, *18*, 3915. [CrossRef]

137. Li, R.; Duan, X.; Li, Y. Measurement Structures of Image Compressive Sensing for Green Internet of Things (IoT). *Sensors* **2019**, *19*, 102. [CrossRef]

138. Liu, X.; Du, X.; Zhang, X.; Zhu, Q.; Wang, H.; Guizani, M. Adversarial Samples on Android Malware Detection Systems for IoT Systems. *Sensors* **2019**, *19*, 974. [CrossRef]

139. Chen, Y.; Wen, H.; Wu, J.; Song, H.; Xu, A.; Jiang, Y.; Zhang, T.; Wang, Z. Clustering Based Physical-Layer Authentication in Edge Computing Systems with Asymmetric Resources. *Sensors* **2019**, *19*, 1926. [CrossRef]

140. Sun, Y.; Xu, H.; Zhang, S.; Wu, Y.; Wang, T.; Fang, Y.; Xu, S. Joint Optimization of Interference Coordination Parameters and Base-Station Density for Energy-Efficient Heterogeneous Networks. *Sensors* **2019**, *19*, 2154. [CrossRef] [PubMed]

141. Hailemariam, Z.L.; Lai, Y.-C.; Chen, Y.-H.; Wu, Y.-H.; Chang, A.; Chang, A.A. Social-Aware Peer Discovery for Energy Harvesting-Based Device-to-Device Communications. *Sensors* **2019**, *19*, 2304. [CrossRef] [PubMed]

142. Liao, R.-F.; Wen, H.; Wu, J.; Pan, F.; Xu, A.; Jiang, Y.; Xie, F.; Cao, M. Deep-Learning-Based Physical Layer Authentication for Industrial Wireless Sensor Networks. *Sensors* **2019**, *19*, 2440. [CrossRef] [PubMed]

143. Xu, X.; Chen, L.; Cheng, Z. Optimizing Charging Efficiency and Maintaining Sensor Network Perpetually in Mobile Directional Charging. *Sensors* **2019**, *19*, 2657. [CrossRef]

144. Hossain, M.; Georges, J.P.; Rondeau, E.; Divoux, T. Energy, Carbon and Renewable Energy: Candidate Metrics for Green-aware Routing? *Sensors* **2019**, *19*, 2901. [CrossRef]

145. Hoang, T.M.; Le Van, N.; Nguyen, B.C.; Dung, L.T.; Van, N. On the Performance of Energy Harvesting Non-Orthogonal Multiple Access Relaying System with Imperfect Channel State Information over Rayleigh Fading Channels. *Sensors* **2019**, *19*, 3327. [CrossRef]

146. Hassan, H.; Ahmed, I.; Ahmad, R.; Khammari, H.; Bhatti, G.; Ahmed, W.; Alam, M.M. A Machine Learning Approach to Achieving Energy Efficiency in Relay-Assisted LTE-A Downlink System. *Sensors* **2019**, *19*, 3461. [CrossRef]

147. Chen, S.; Wen, H.; Wu, J.; Xu, A.; Jiang, Y.; Song, H.; Chen, Y. Radio Frequency Fingerprint-Based Intelligent Mobile Edge Computing for Internet of Things Authentication. *Sensors* **2019**, *19*, 3610. [CrossRef]
148. Cruz, C.; Palomar, E.; Bravo, I.; Gardel, A. Towards Sustainable Energy-Efficient Communities Based on a Scheduling Algorithm. *Sensors* **2019**, *19*, 3973. [CrossRef]
149. Charalampou, P.; Sykas, E.D. An SDN Focused Approach for Energy Aware Traffic Engineering in Data Centers. *Sensors* **2019**, *19*, 3980. [CrossRef]
150. Usama, M.; Erol-Kantarci, M. A Survey on Recent Trends and Open Issues in Energy Efficiency of 5G. *Sensors* **2019**, *19*, 3126. [CrossRef] [PubMed]

sensors

MDPI

Article

An SDN Focused Approach for Energy Aware Traffic Engineering in Data Centers

Paris Charalampou * and Efstathios D. Sykas *

School of Electrical & Computer Engineering, National Technical University of Athens, 9 Iroon Polytechniou Street, 15773 Athens, Greece
* Correspondence: pchara@cn.ntua.gr (P.C.); sykas@cn.ntua.gr (E.D.S.); Tel.: +30-210-772-1512 (P.C.)

Received: 22 July 2019; Accepted: 12 September 2019; Published: 14 September 2019

Abstract: There is a lot of effort to limit the impact of CO_2 emissions from the information communication technologies (ICT) industry by reducing the energy consumption on all aspects of networking technologies. In a service provider network, data centers (DCs) are the major power consumer and considerable gains are expected by regulating the operation network devices. In this context, we developed a mixed integer programming (MIP) algorithm to optimize the power consumption of network devices via energy aware traffic engineering. We verified our approach by simulating DC network topologies and demonstrated that clear benefits can be achieved for various network sizes and traffic volumes. Our algorithm can be easily implemented as an application in the software-defined networking (SDN) paradigm, making quite feasible its deployment in a production environment.

Keywords: software defined networking (SDN); data center; optimization; traffic engineering; energy awareness

1. Introduction

The problem of climate change due to global warming is already identified and the research community as well as industry are working on methods to limit its impact. The following areas have been identified as the main contributors of CO_2 emissions: (a) energy production (29%), (b) transportation (27%), (c) industry (21%), (d) agriculture (9%) and (e) information communication technologies (ICT) (9%) [1]. In the area of ICT, a large increase that can reach a total of 15% is expected by the end of 2020 due to the deployment of 5G networks, mass introduction of IoT (Internet of Things) devices, IP traffic generated by video streaming and augmented reality applications. Data centers (DCs) in internet service providers (ISPs) account for more than 45% of the power consumption [2]. Although a marginal 10% of it is caused by network equipment [3], data center networking is responsible for 0.36% of the total power consumption [4]. This figure is expected to rise even more because of NFV (Network Functions Virtualization) and software defined networking (SDN) based service deployment [5]. In 5G networks, the number of DCs is expected to rise in order to support mobile edge computing (MEC) and user plane intensive applications [6]. Also, the majority of legacy applications will be migrated to cloud infrastructure increasing the workloads served by DCs.

Data center networks are characterized by traffic patterns and volumes that significantly vary from typical IP based networks that reside in ISP's premises. Typically, DC networks are overprovisioned with a large number or redundant devices and links. Depending on the deployed topology, access and aggregation layer links rarely exceed 10% of utilization due to the high number of redundant links on these layers [7]. Links and devices on higher layers of the topology (closer to root or data center routers) tend to have higher levels of utilization and thus consume more energy. Accordingly, only 60% of installed links could potentially serve traffic even during a busy hour [8].

If we consider the utilization of available resources during a whole day and not only during the busy hour, DC network infrastructure stays in idle state serving zero to minimal traffic (mostly for management purposes and maintaining the routing algorithms) for almost 70% of the day. Even on higher traffic volumes, the average link utilization is not surpassed for the remaining 25% of the day. Since DC networks are mainly constructed to provide a high level of resilience during a busy hour, we identify the opportunity to minimize the power consumption of network devices and links between them for the largest part of the day.

In this paper we address the problem of minimizing the power consumption within DCs. The main obstacle in developing a practical optimization method for DCs is the requirement for global knowledge of network topology and the flows matrix between the hosts. DC networking presents high variability mostly from virtual machines (VMs) instantiation that have ephemeral life span in cloud environments. Thus, there is a need to continually monitor for new traffic flows and the optimization algorithm should act accordingly. Building a distributed algorithm for this purpose is not appropriate since it would require the introduction of new or updated protocols. Nevertheless, SDN architectures are already deployed in production environments and the centralized implementation of SDN controllers (SDN-C) now allows practical implementations of such optimization.

Taking for granted that all modern DCs follow the SDN architecture, we show that, via energy aware traffic engineering, an SDN [9] application can successfully address the problem. Depending on hardware properties and traffic conditions, links can be put on a lower power state or completely shut off via traffic steering. We formulate the optimization process as a mixed integer programming (MIP) problem [10] that models power consumption taking into account all relevant constraints. Using standard optimization tools and heuristics for its solution we show that significant power savings result. Since the solution is optimal only for a short time frame, as long as traffic loads do not change considerably, periodical repetitions and fast execution times are required and fine tuning of the solver is mandatory. We took great care to ensure that the optimization algorithm can be executed in a timely manner and confirmed the viability of our approach on a variety of topologies and network sizes. Finally, we implemented an application following the SDN paradigm to demonstrate the applicability of our approach and verified it against emulated topologies for performance evaluation and experimental ones for software verification purposes.

2. Related Work

For the ICT domain, the main method to limit CO_2 emissions is by decreasing the power consumption of network topologies. A lot of work is performed in this area on various layers of the network architecture. Existing methods can be classified into two main categories (a) evolving hardware to support power consumption proportional to traffic served [11] and (b) putting unused links and devices into sleep mode after applying traffic engineering techniques to reroute traffic [12].

Techniques to allow unused links to get into sleep states can be applied to 'access' part of the network (both wired and wireless), to 'transport' (covering wired aggregation and mobile backhauling based on optical links) and to 'core' [13]. Based on work performed in core networks, a number of mechanisms have been proposed and that can be separated into four main categories (a) new routing protocols [14], (b) traffic engineering techniques [15], (c) new network architectures [16], and (d) modifications to existing routing protocols [17,18] so that routing decisions will be based on energy consumption criteria. The main drawback of existing approaches is the requirement for full knowledge of network topology and link utilizations. Even though getting a full layer-3 topology is rather easy even on large scales networks, it is not feasible to gather per flow statistics even when utilizing modern monitoring systems with telemetry [19] and big data analytics [20].

On the DC domain, a more holistic approach is taken into consideration that addresses the power consumption not only on networking and computing devices but on supporting equipment such as cooling devices. On the networking layer, similar approaches have been proposed that try to put unused links in sleep state. Furthermore, they only focus on legacy devices with sleep states, neglecting other possible power states [21]. In contrast, modern switches and especially fiber optics links in DCs support a certain degree of energy proportionality with a step-wise approach [22]. A typical modular network device consumes power only in two concrete states (a) maximum power when operating in normal mode and (b) minimum power when operating in sleep state. This mode of operation is applicable on chassis layer, line cards and routing processors. There is no direct correlation between traffic volumes and power consumption, but only on environmental aspects like DC room temperature and the number of routing table entries. Interfaces on the other hand operate on energy aware states according to the amount of traffic they serve. The number of states and the maximum power consumption of an interface varies among manufacturers and depends on the link type (multimode optics, single mode optics, DAC cables, CAT-6 copper cables) [23].

Until recently, all proposed mechanisms have been applied only to experimental topologies and lack adoption from major networking equipment manufacturers. In addition, there is a lack in real life deployments of such solutions due to a number of reasons: (a) protocol expansions are not eagerly adopted from manufactures, (b) network statistics (on per flow basis) and topology monitoring cannot scale well, and (c) advances in the area of SDN are not taken into account.

The SDN architecture [24] of separated data and control plane has been successfully used to build networks inside a DC regardless of the purpose served and the workloads hosted, from 5G and NFV services to generic IT applications and data storage. Furthermore, SDN already provides the mechanisms to efficiently collect measurements and apply forwarding rules to existing equipment without modifications to devices or routing protocols. Finally, SDN design allows to easily integrate external applications via well-defined software APIs in SDN controllers (SDN-C). Therefore, the use of an SND-C application seems to be the only viable method for network policies in any aspect of DC networking.

In this paper we pursue the SDN paradigm as applied for traffic engineering purposes, so as to mitigate the aforementioned gaps. First, we define an optimization problem that takes into account new hardware capabilities by modeling the interface power consumption in concrete power states. Then we show that the optimization problem can be solved in a timely manner. Finally, by harvesting SDN capabilities we get the opportunity to develop an optimization application using a real, production ready SDN-C and to apply the optimal solution in a scalable and practical manner. Our optimal solution can be applied without traffic loss and is independent of the traffic patterns (UDP or TCP based).

Relevant approaches [8] consider only selected topologies, are optimized for these and assume that there is a-priori knowledge of traffic patterns. Even though they make use of SDN as a method to apply the optimal solution, they do not fully exploit SDN capabilities. In particular, they resort on legacy methods to collect statistics via SNMP. As a result, they cannot lead to practical implementations in production environments. Finally, since their solution is developed as a part of the SDN-C, not as an application following SDN paradigm, it is coupled only to the specific implementation and cannot be scaled to other controllers and topologies.

3. Optimization Problem

We formulate the following optimization problem to determine the optimal topology and state of links (either disabled or in an optimal state for served traffic). We consider only fixed sized network devices since this is the trend in currently deployed DC network topologies. Furthermore, since we envisage our approach to be deployed in real-life scenarios, we exclude the option to put into sleep mode a whole device.

Definitions:

V: set of network devices in a given topology. Device $v \in V$ has a base power consumption of B_v

E: set of links between the devices. Links are assumed to be bi-directional thus link $l = (v, v')$ originates from device v and terminates at device v' where $v, v' \in V$

F: set of flows from Access only devices. Flow $(s,t) \in F$ originates from device s and terminates at device t where $s,t \in V$

ST: set of possible power states n for a link

$P_l^{(n)}$: power consumption of link l when operating in state n

C_l^s: capacity of link $l \in E$ operating on a state $n \in ST$

$x_l^{(s,t)}$: flow from s to t, where $s,t \in V$, passing through link $l \in E$

$\Phi_l(r)$: power consumed at link $l \in E$ when carrying traffic r

$\tau_{(s,t)}$: flow from device s to t where $s,t \in V$

Decision Variables:

$S_l^{(n)}$: a binary variable vector to describe the state of a link. Value $S_l^{(n)}$ equals to 1 if link l is in state n

y_v: a binary variable describing the state of device v and equals to 0 when a device is in sleep state

z_l: a binary variable defining if link l is in sleep mode and equals to 1 when is serving any volume of traffic

Our objective is to minimize the following gain function that calculates the total power consumed from network devices

$$\text{Minimize}: \sum_{v \in V} B_v \cdot y_v + \sum_{l \in E} \sum_{n \in ST} S_l^{(n)} \cdot P_l^{(n)} \tag{1}$$

While maintaining the following constraints:

$$\sum_{(v,\,v') \in E} x_{(v,v')}^{(s,t)} - \sum_{(v,\,v') \in E} x_{(v',v)}^{(s,t)} = \begin{cases} \tau_{s,t} & \text{if } v = s \\ -\tau_{s,t} & \text{if } v = t \\ 0 & \text{otherwise} \end{cases}, \forall\ (s,t) \in F \text{ and } v \in V \tag{2}$$

$$\sum_{(s,\,t) \in F} x_l^{(s,t)} \le C_l^{(n)}, \forall\ l \in E,\ n \in ST \tag{3}$$

$$z_l \cdot C_l^{(n)} \ge \sum_{(s,t) \in F} x_l^{(s,t)}, \forall\ l \in E,\ n \in ST \tag{4}$$

$$z_{(v',\,v)} = z_{(v',\,v)}, \forall\ l = (v,v') \in E \tag{5}$$

$$y_v \cdot \left(\sum_{(v,v') \in E} C_{(v,v')}^{(n)} + \sum_{(v',v) \in E} C_{(v',v)}^{(n)} \right) \ge \sum_{(s,t) \in F} \left(\sum_{(v,v') \in E} x_{(v,v')}^{(s,t)} + \sum_{(v',v) \in E} x_{(v',v)}^{(s,t)} \right) \tag{6}$$

$$C_l^{(n-1)} \le \sum_{(s,t) \in F} x_l^{(s,t)} \le C_l^{(n)} \tag{7}$$

$$\sum_{l \in E, n \in ST} S_l^{(n)} = 1 \tag{8}$$

$$\Phi_l(r) = \sum_{n \in ST} S_l^{(n)} \cdot P_l^{(n)} \tag{9}$$

Equation (2) guarantees flow preservation, i.e., flow that enters a device is also coming out of it without packet loss. Equation (3) guarantees that capacity constrains are not violated and the sum of all the flows coming from an interface cannot be greater than the capacity of the interface. The fact that a link must be on a specific state and not on sleep mode and in parallel be on the same state on both interconnected devices is defined in Equations (4) and (5). Equation (6) guarantees that a device (and the interconnected one) are kept online if they serve any amount of traffic. Capacity constraint in Equation (7) ensures that the traffic through an interface is not grater then the capacity for a given state.

Sensors **2019**, *19*, 3980

Each interface can only be on one state and this is guaranteed by Equation (8). Equation (9) expresses the consumption of a link based on the power state defined by the traffic crossing each interface.

This is a typical MIP (mixed integer programming) problem and the solution is relatively fast in case of small topologies with a limited number of devices and interconnecting links. Network topologies in DCs are much larger with a high number of nodes (hundreds of switches and thousands of links) and solving the above optimization problem becomes time consuming. Thus, CPLEX solver [25], a well-known solver for optimization problems and constrained programming for linear and integer programming problems, was used to apply a number of heuristics and produce a suboptimal solution in a timely manner. In addition, flows and capacity values are expressed as integers (bytes per second) to accelerate CPLEX execution times.

CPLEX MIP solver applies a number of algorithms to automatically select the best method to solve complex problems. Our problem is by definition feasible since traffic loads can be served in the initial network state. That is, if links are operated on higher energy state, all constrains are fulfilled. In our work, CPLEX as MIP solver uses pre-processing and probing by setting all binary variables to either 0 or 1 and checking the logical implications. It automatically selects an appropriate Branch and Cut algorithm to solve the optimization problem. If heuristics are required, a neighborhood exploration search starts which is called solution polishing after the time limit is reached. Polishing is based on the integration of an evolutionary algorithm within an MIP branch and bound framework.

Since we aim at practical implementations, we applied a hard time limit to the execution of the algorithm. A number of parameters were evaluated in order to get a solution in a timely manner. Initially we obtained a sub-optimal solution (that reduces the total power consumption) and then explored for a better one without violating the timing restriction. The CPLEX parameters evaluated are shown in the following Table 1.

Table 1. Description of CPLEX software parameters that were considered during the execution of optimization algorithm.

Parameter	Description
tilim	Duration in seconds that CPLEX looks for a solution to the optimization problem
threads	Manages the number of parallel threads used during the calculations (maximum value depends on the available CPUs)
parallelmode	Sets the parallel optimization mode. Possible modes are automatic, deterministic, and opportunistic
mipemphasis	Controls trade-offs between speed, feasibility, optimality, and moving bounds in MIP.
probe	Sets the amount of probing on variables to be performed before MIP branching. Higher settings perform more probing
varsel	Sets the rule for selecting the branching variable at the node which has been selected for branching
lbheur	Controls whether CPLEX applies a local branching heuristic to try to improve new incumbents found during a MIP search
fpheur	Turns on or off the feasibility pump heuristic for mixed integer programming (MIP) models

The first parameters were selected so as to fit the available hardware resources where CPLEX was executing. In more detail, *threads* value was set to 8 to fully utilize available vCPU capacity of hosting VM. *Tilim* was set to a value that can produce a viable solution even on large data set. *Mipephasis* was set to feasibility mode in order to produce an initial suboptimal value given the timeframe and later search for the optimal value. *Probe* was set to the highest possible value since the initial time spend looking for a suboptimal solution guaranteed that we could have power savings for all scenarios. The other parameter (*varsel*, *lbheur* and *fpheur*) values were selected to fit the nature of the problem and the available input data. Multiple executions of the algorithm on the same data set were performed to find an optimal set considering the time limitations imposed in a production grade network environment.

4. Emulation Results

We evaluated the results of our optimization algorithm on various network topologies, sizes (number of nodes) and traffic volumes. We emulated three different network topologies applicable to DC networking (a) classical three tier topology [26], (b) fat tree [27], and (c) leaf and spine [28]. We focused only to inter-switch communications and did not consider the impact of multihomed servers where further benefits can be achieved. Physical topologies and switches were emulated using Mininet [29] that implements OpenvSwitch (OVS) [30] as an Openflow switch, because of its ability to create and maintain large network topologies [31]. Mininet is responsible for instantiating a number of Openflow enabled switches, connecting them to an external SDN-C, creating a number of physical hosts and interconnecting them with virtual links. Mininet allows the development of scripts to deploy large scale topologies based on pre-defined parameters. In this context, we developed a series of scripts that automatically generate the topologies under examination with the number of devices described in Table 2.

Table 2. Number of devices on evaluated scenarios.

Size	Access (Number of Switches)	Aggregation (Number of Switches)	Core (Number of Switches)
Classical 3-Layer			
Size 1 (XS)	8	2	2
Size 2 (S)	16	4	2
Size 3 (M)	32	4	2
Size 4 (L)	64	4	2
Size 5 (XL)	64	8	2
Fat Tree			
Size 1 (XS)	8	2	2
Size 2 (S)	16	4	2
Size 3 (M)	32	4	2
Size 4 (L)	64	4	2
Size 5 (XL)	64	8	2
Leaf and Spine			
Size 1 (XS)	8	2	n/a
Size 2 (S)	16	4	n/a
Size 3 (M)	32	4	n/a
Size 4 (L)	32	8	n/a
Size 5 (XL)	64	8	n/a

We have chosen Opendaylight (ODL) [32] as the SDN controller and we executed our experiments employing ODL's RESTful API. The experimentation procedure is depicted in Figure 1. Initially we spin up the desired topology on a VM instance executing Mininet software. The OVS instance inside Mininet is connected to an external VM that runs ODL. A python module collects the topology information and the configured link capacities. The CPLEX solver runs on a separate dedicated VM in order not to interfere with ODL. Our software module is fed with the device power consumption model and randomized flows (depending on the scenario) so as to calculate the initial power consumption and link utilization according to current, non-optimized state where all links operate on the highest possible state. The same module includes the routing functionality and generates traffic flows between all devices. Based on the scenario, the appropriate input configuration file for CPLEX is generated and the MIP solver is invoked. According to the results, new optimal consumption and link utilizations are calculated, and the new topology is stored in an external file for review.

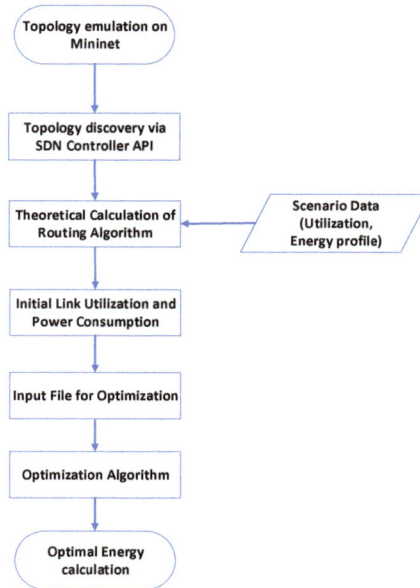

Figure 1. Process followed during the simulation.

The power consumption model of an interface consists of a baseline value due to the transceiver and a traffic proportional part due to the electronics parts on the Linecards. Typical values for transceiver consumption start from 1.5 W per transceiver for a 10 Gbps multimode optical module till 3.5 W/module for 100 Gbps single mode fiber [33]. On the other hand, the nominal value on the switch side is 3.5 W for a 10 Gbps interface and 14 W for a 40 Gbps one. In our emulation, a non-linear power model is assumed where the initial states tend to consume higher volumes of power. We considered four different states regarding the link power consumption, where the last state corresponds to the maximum power consumption of a link. The link is considered to consume no power when in sleep state. The first step includes the power requirements to maintain the link state (thus consumes relatively more than intermediate steps). The next steps are according to power consumption data sheet from switch manufacturers [34].

Network sizes varied from 12 switches (4 server racks equivalent in real deployments) to 74 switches (32 server racks) were emulated. For each topology the exact figures for switches per network layer are listed in Table 2. Traffic profile (we evaluated only east–west communications) and total traffic volume were considered constant for each execution of the experiment.

Figure 2 demonstrates the results for various network sizes and different network topologies. Even though there is a different level of redundancy, starting from limited resiliency in a 3-tier architecture up to the highest level of resiliency in leaf and spine topology, we recorded a power reduction of at least 65% in the worst case. By adding redundant paths, especially on the spine layer, the power savings increased up to 90%. Leaf and spine topology by design includes the higher number of redundant links compared to the other two topologies. Adding switches on the spine layer increases the number of redundant paths without serving more hosts and thus increases the power savings potential. There is no similar improvement in the other topologies although savings are around 70%. The classic 3-tier architecture has less redundant links. Energy savings derive from partially loaded link power states and are independent of the number of switches in each layer. In the fat tree architecture, where the number of alternate paths depends on the switches at the aggregation level, power consumption is affected only by the number of devices at this layer.

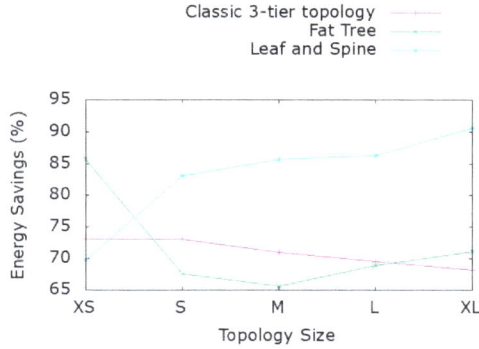

Figure 2. Energy savings for various network topologies and sizes.

On the leaf and spine topology, the most promising in terms of power savings potential, we examined the impact of various workloads. To compare different traffic volumes we introduce the concept of traffic amplifier, being a percentage of the same maximum traffic across all topologies. The maximum traffic is calculated assuming that hosts load the network as much as possible without violating the link capacity on a given topology. Traffic amplifier is the factor used to multiply the reference flows between the hosts (the ones resulting in full core link utilization without traffic engineering algorithm in place). Thus, a traffic factor of 10% corresponds to a random generation of flows between the hosts that lead to a maximum 10% utilization on the core links. Figure 3 depicts the impact of increasing traffic volumes on the energy savings in conjunction with the number of interfaces that operate on specific power states. The power state is derived from the traffic level at each interface, whereas on the 1st state the load is 25% of the nominal and the 4th state corresponds to the maximum capacity.

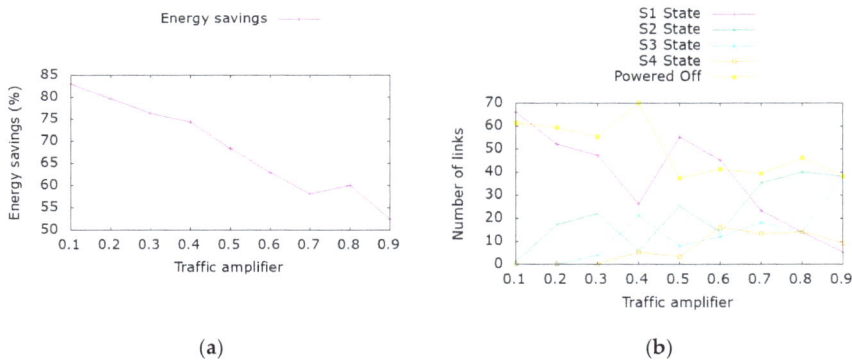

(a)   (b)

Figure 3. Analysis for average size leaf and spine topology under various traffic loads. (**a**) Energy savings from optimization algorithm. (**b**) Number of interfaces per state after the execution of the algorithm.

As depicted in Figure 3, even for high load utilization, where a number of links have to operate on their maximum capacity, we still recorded energy savings up to 52%. Notice that, since the maximum gains are achieved when a link is at sleep state, the MIP solver tends to reroute traffic on selected links causing them to operate at higher rate, instead of distributing the load to multiple links which is the current mode of operation in existing DC deployments. As traffic increases, the margins for energy savings via traffic engineering are narrower as shown in Figure 3a where energy savings decrease as link utilization increases from 85% savings to nearly 50%.

Next, we evaluate the gains from considering enhanced power scaling capabilities of interfaces compared to only disabling unused interfaces and devices. In Figure 4a, we plot the energy savings gain versus traffic volume achieved in a leaf and spine topology. Even for high traffic volumes, there is at least 28% benefit when taking into account the interface's power state behavior. Similar results can be achieved regardless of the selected topology as demonstrated in Figure 4b. The benefits of interface power scaling decrease as traffic increases since more interfaces operate closer to their maximum capacity consuming the maximum power. Power scaling is far more beneficial in the classic 3-tier topology due to the absence of redundant paths on core links and the fact that only a small portion of the links can be disabled. In the fat tree and leaf and spine topology, benefits are still significant compared to the benefits of disabling only unused interfaces even though a large number of redundant links can be switched off. Figure 4 demonstrates that a combined strategy of traffic engineering (to disable the highest number of interfaces) and exploiting interface power states is required so as to achieve the maximum benefits in power savings since greater values of power savings can be achieved regardless of the traffic volumes and topology size.

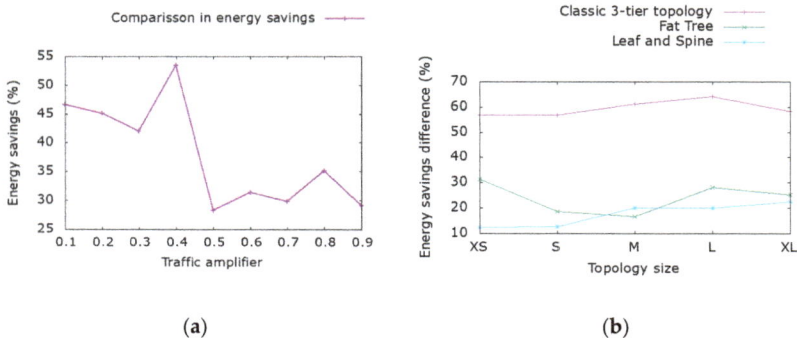

Figure 4. Comparison between approaches based on disabling interfaces and our interface power scale proposal. Difference under (**a**) various workloads on leaf and spine topology, and (**b**) various topologies and number of devices for the same workload.

We further compare our results to existing studies that do not consider hardware capabilities and only try to suspend unused interfaces. We selected the studies based on the relevance to our study and the methodology followed. Existing literature in this domain can be classified into two main categories: (a) studies that can be applied to all types of networks [35], (b) studies that focus only on DC topologies. In both cases full knowledge of the topology and the relevant traffic matrix is required. The optimal solution is applied either via SDN methods or following legacy approaches, i.e., by modifying the routing information. Since there is no direct comparison on topology level with the studies in the first category, we can only examine power savings achieved for equivalent average link utilizations of 10%, 50% and 90%. Compared to the results in Figure 6 of [35], our method can provide five times more energy savings in the worst-case scenario of 90% link utilization, achieving 53% power savings instead of only 10%. Considering 50% link utilization, we achieve more than triple savings namely 68% in our case compared to 22%. For low link utilizations of 10% the savings we get are more than double, 85% savings in our study compared to 35%.

Compared to studies in the second category, we can consider the same traffic volumes and network topology. In particular, we compared our method towards the fat-tree network topology examined in [8] under mid-traffic profile (50% on near nodes, 50% on far nodes) for 1024 nodes and maximum link utilization of 20%. In this case, our approach can produce around 45% more savings, i.e., 65% instead of 45% power savings. For lower number of hosts and similar link utilization, we observe 55% more savings when our approach is applied. Based on the aforementioned comparison with DC focused methods, the energy savings potential of our method is far greater compared to approaches that try to optimize power consumption in DCs without considering hardware characteristics.

5. SDN Application

In SDN architecture, control and forwarding planes are clearly separated defining a discrete device for control plane functions, the SDN controller (SDN-C), and keeping the forwarding process on the physical switches. This architecture, as defined by ONF (open networking foundation), allows the development of SDN applications that harvest controller APIs to collect statistics or routing information, to modify port configuration and to reroute traffic. To demonstrate the applicability of our approach in a DC environment, we developed an application that can be easily integrated to any SDN controller.

The flow chart in Figure 5 depicts a high-level description of the internal activity of such an SDN application. First, it discovers all relevant switches and hosts that reside in the SDN-C database. Then, a full mesh list of flows between hosts is generated for the given topology and their values are stored internally in application's configuration. In accordance with the Layer-2 topology as created by the SDN-C using standard STP (Spanning Tree Protocol) algorithms, a set of link flows results. The SDN application provisions these flows without modifying the existing routing information. On a configurable time interval, the application collects flow statistics with the Openflow build-in mechanism. Based on flow statistics and power profile for each device type (power consumption per state), the CPLEX module calculates the new optimal state for all device and interfaces. The new topology is then provisioned to devices directly or via the SDN-C. After a sleeping period, the whole process is executed again removing all host specific flows generated on the initial execution. All ports and devices are re-provisioned on the initial state where they consume the maximum amount of energy and can serve nominal traffic.

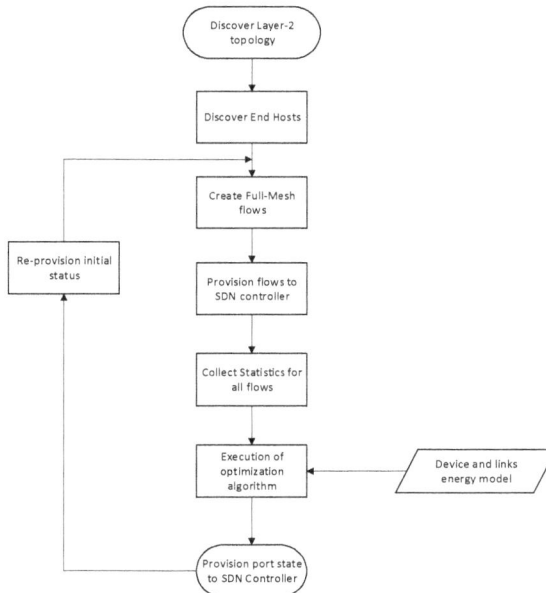

Figure 5. Software defined networking (SDN) application internal sequence of activities.

Note that redundant links remain in service only if there is enough traffic in the topology. In general, as expected from similar approaches in core networks, a certain stretch on the path length between end devices is expected accompanied by a minor degradation in service quality. Since communication inside DCs involves a small number of hops, especially in leaf-and-spine topology, this stretch is not expected to affect applications. The SDN application guarantees that connectivity of end hosts is not disrupted at any case.

The key point is that the optimization problem can be solved by an external system irrespective of the software architecture and programming language of the controller. Nevertheless, a number of additional software components are needed. In order to verify our proposal, an SDN application coded in python, was deployed at NTUA's Computer Networks Laboratory using a testbed consting of Opendaylight version 7.3 SDN controller, an HPE switch running firmware WB.16.05.0003 and several software implementations based on OpenVSwitch. In particular, we have developed the following generic modules:

Topology_discovery: This module uses the RestAPI of an SDN-C to automatically discover all openflow enabled switches, interconnection links between the devices and end-hosts for any topology. The outcome of the discovery process is stored in a single file, The SDN-C is scanned periodically for topology changes.

Flow_generator: Based on the topology and hosts discovered, the flow_generator module creates a full table of traffic flows between end hosts. Communication between Openflow [36] switches is omitted as it is expected to consist of management traffic, marginal compared to the volume of production traffic. These flows are provisioned via Rest API of the SDN-C to all devices.

Stats_Collector: It runs periodically to collect the statistics and aggregates the results according to the operator needs. Stats_Collector uses the build-in mechanism of Openflow protocol and gathers the values of flow counters based on "Counters" field for the provisioned flows of the previous step.

Green_Topology_Optimizer: This is the CPLEX module performing the optimization and some python modules to control its execution. Based on a preconfigured link power consumption model and the configuration files as created from Topology_discovery and Stats_collector modules, this module generates the optimal power state for all devices and interfaces.

Port_Modifier: According to the solution generated, this module provisions the new state either via the SND-C programmable API, via OVSDB or via ovs-vsctl.

The verification of our application was performed into two steps. The first step, the functional verification, consisted of the successful integration of the above compents in the physical and virtual lab environment. The second step, the performance evaluation, was carried out on the same Mininet emulated topologies as in Section 4. Thus, the optimal solution in terms of power consumption savings for each topology does not change. Topologies emulated by Mininet are depicted in Figure 6.

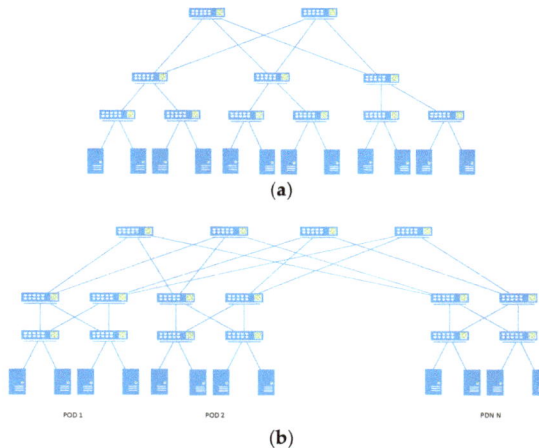

(a)

(b)

Figure 6. *Cont.*

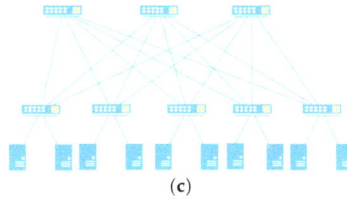

(c)

Figure 6. Network topologies under evaluation: (**a**) classical-3-layer (**b**) fat tree (**c**) leaf and spin.

The network size of emulated topologies affects the time required for our SDN application to calculate and provision the optimal solution. Thus, we evaluated its total execution time and not only the optimization (CPLEX) part. We performed measurements for the concrete phases: (a) time required to discover network topology and compute link power consumption based on the device's energy model and link capacity; (b) flow generation for full mesh communication among hosts and their provisioning via Openflow commands; and (c) collection of flow statistics and initialization of optimization problem CPLEX solution. As shown in Table 3, the topology discovery part is the faster phase even for large topologies. The times for the flow generation and the collection of flow statistics depend only on the number of flows and end hosts and are independent of the complexity of the topology. The two initial phases have to be executed sequentially in less than 100 s in the worst case scenario. After the bootstrap, the statistics collection can forked to different processes. Measured values appearing in table must be regarded as the upper limit for this phase.

Table 3. Performance evaluation of SDN application.

Size	Topology Discovery (s)	Flow Provisioning (s)	Statistics Collection (s)
	Classical 3-layer		
Size 1 (XS)	0.45	0.942	0.808
Size 2 (S)	0.825	3.649	3.313
Size 3 (M)	1.495	17.059	13.203
Size 4 (L)	2.495	76.116	71.189
Size 5 (XL)	3.739	71.941	72.298
	Fat Tree		
Size 1 (XS)	0.608	0.995	0.879
Size 2 (S)	1.198	3.916	3.258
Size 3 (M)	5.068	35.099	23.061
Size 4 (L)	5.024	89.483	72.062
Size 5 (XL)	4.539	79.729	70.433
	Leaf and Spine		
Size 1 (XS)	0.599	0.952	0.856
Size 2 (S)	1.72	4.316	3.277
Size 3 (M)	3.342	17.562	13.587
Size 4 (L)	6.5	18.853	13.815
Size 5 (XL)	13.556	78.15	71.851

Timely execution of the SDN application is of high importance since we aim at a practical implementation. As demonstrated in Table 3, execution is fast even when the MIP optimization is applied to large network topologies. Next, we compare the performance of our solution to a similar approach in existing literature [37] that requires full knowledge of the flow matrix and applies heuristic algorithms (referred to as QRTP and RQRTP) for traffic engineering. The method in [37] is selected for comparison since it is developed as an SDN application like ours, it is focused on DC network topologies and requires equivalent input for the optimization problem (flow matrix). Applying traffic engineering decisions is based on QoS and network performance metrics, not on power saving criteria. Since the comparison cannot be direct for the full scope (energy savings and complexity) of our study, we compare only the execution times of the optimization algorithm, i.e., the time to calculate the optimal solution for the same number of flows and the same network topology. Our optimization algorithm can be solved significantly faster for small number of flows. Namely, for 100 flows, we need 0.08 s to calculate the optimal solution in our algorithm compared to 3.16 and 2.66 s, respectively, for the QRTP and RQRTP algorithms, using equivalent hardware resources (i.e., number of CPUs) while emulating a classical 3-layer network topology. The number of hosts and traffic volumes do not impact the complexity of the algorithms thus are not mentioned in detail. For larger topologies and 500 flows, our solution generates an optimal solution in 1.57 s whereas QRTP in 178.13 and RQRTP in 41.45 s. In the extreme scenario of 1000 flows, our optimization problem can be solved in 3.08 s compared to 1227.26 and 46.07 s respectively, for QRTP and RQRTP, and the same type of topology.

6. Conclusions

Due to the nature of DC network topologies, deployments tend to be overprovisioned with sparse utilization even in peak hour. We showed that DC networking can be largely optimized regarding power consumption regardless of the topology selected. The benefits in power consumption range from 65% to 90% in all typical scenarios depending on the total load. Since power benefits are coupled with traffic volumes, harvesting hardware capabilities for traffic steering can guarantee these savings even for high workloads reaching 50% for fully utilized leaf and spine topology which is the benchmark topology for DCs. Furthermore, we demonstrated that our proposal is a viable solution for DCs where SDN is deployed. It can be implemented as an SDN application regardless of network equipment manufacturer and SDN controller user and therefore easily applied to real life deployments.

Author Contributions: Authors contributed equally to the preparation of the manuscript and the concept of the research. Both authors contributed to the design of the study and interpreted the results. P.C. performed the software development, optimization problem definition and experiment execution. E.D.S. performed critical revisions to the manuscript and to the data analysis.

Funding: This research received no external funding.

Conflicts of Interest: The authors declare no conflicts of interest.

References

1. Belkhir, L.; Elmeligi, A. Assessing ICT global emissions footprint: Trends to 2040 & recommendations. *J. Clean. Prod.* **2018**, *177*, 448–463.
2. Cisco. Cisco Visual Networking Index: Forecast and Trends, 2017–2022. 2018, pp. 2–3. Available online: https://www.cisco.com/c/en/us/solutions/collateral/service-provider/visual-networking-index-vni/white-paper-c11-741490.html (accessed on 12 July 2019).
3. Kolias, C. SDN & NFV for energy-efficient clouds. In Proceedings of the E2DC Workshop—ACM e-Energy 2013, Berkeley, CA, USA, 21 May 2013; p. 29.
4. Dayarathna, M.; Wen, Y.; Fan, R. Data Center Energy Consumption Modeling: A Survey. *IEEE Commun. Surv. Tutor.* **2016**, *18*, 732–794. [CrossRef]
5. Han, B.; Gopalakrishnan, V.; Ji, L.; Lee, S. Network function virtualization: Challenges and opportunities for innovations. *IEEE Commun. Mag.* **2015**, *53*, 90–97. [CrossRef]
6. Chen, S.; Zhao, J. The requirements, challenges, and technologies for 5G of terrestrial mobile telecommunication. *IEEE Commun. Mag.* **2014**, *52*, 36–43. [CrossRef]
7. Benson, T.; Anand, A.; Akella, A.; Zhang, M. Understanding Data Center Traffic Characteristics. *ACM SIGCOMM Comput. Commun. Rev. Arch.* **2010**, *40*, 65–72. [CrossRef]
8. Heller, B.; Seetharaman, S.; Mahadevan, P.; Yiakoumis, Y.; Sharma, P.; Banerjee, S.; McKeown, N. ElasticTree: Saving Energy in Data Center Networks. In Proceedings of the NSDI'10 7th USENIX Conference on Networked Systems Design and Implementation, San Hose, CA, USA, 28–30 April 2010.

9.   Haleplidis, E.; Pentikousis, K.; Denazis, S.; Salim, J.H.; Meyer, D.; Koufopavlou, O. Software-Defined Networking (SDN): Layers and Architecture Terminology. *RFC 7426 Internet Res. Task Force (IRTF)* **2015**, 7–12. [CrossRef]

10. Bertsimas, D.; Weismantel, R. *Optimization Over Integers*; Dynamic Ideas: Charlestown, MA, USA, 2005.

11. Vishwanath, A.; Hinton, K.; Ayre, R.W.; Tucker, R.S. Modeling Energy Consumption in High-Capacity Routers and Switches. *IEEE J. Sel. Areas Commun.* **2014**, *32*, 1524–1532. [CrossRef]

12. Charalampou, P.; Zafeiropoulos, A.; Vassilakis, C.; Tziouvaras, C.; Giannikopoulou, V.; Laoutaris, N. Empirical evaluation of energy saving margins in backbone networks. In Proceedings of the 2013 19th IEEE Workshop on Local & Metropolitan Area Networks (LANMAN), Brussels, Belgium, 10–12 April 2013; pp. 1–6.

13. Bolla, R.; Bruschi, R.; Davoli, F.; Cucchietti, F. Energy Efficiency in the Future Internet: A Survey of Existing Approaches and Trends in Energy-Aware Fixed Network Infrastructures. *IEEE Commun. Surv. Tutor.* **2011**, *13*, 223–244. [CrossRef]

14. Coudert, D.; Koster, A.M.; Phan, T.K.; Tieves, M. Robust Redundancy Elimination for Energy-Aware Routing. In Proceedings of the 2013 IEEE International Conference on Green Computing and Communications and IEEE Internet of Things and IEEE Cyber, Physical and Social Computing, Beijing, China, 20–23 August 2013; pp. 179–186.

15. Addis, B.; Capone, A.; Carello, G.; Gianoli, L.G.; Sanso, B. Energy-aware multiperiod traffic engineering with flow-based routing. In Proceedings of the 2012 IEEE International Conference on Communications (ICC), Ottawa, ON, Canada, 10–15 June 2012; pp. 5957–5961.

16. Chiaraviglio, L.; Mellia, M.; Neri, F. Minimizing ISP Network Energy Cost: Formulation and Solutions. *IEEE/ACM Trans. Netw.* **2012**, *20*, 463–476. [CrossRef]

17. Capone, A.; Cascone, C.; Gianoli, L.G.; Sanso, B. OSPF optimization via dynamic network management for green IP networks. In Proceedings of the 2013 Sustainable Internet and ICT for Sustainability (SustainIT), Palermo, Italy, 30–31 October 2013; pp. 1–9.

18. Bianzino, A.P.; Chiaraviglio, L.; Mellia, M. GRiDA: A green distributed algorithm for backbone networks. In *Proceedings of the 2011 IEEE Online Conference on Green Communications*; IEEE: Piscataway, NJ, USA, 2011; pp. 113–119.

19. Doorn, V.; Rock, C.; Kiok, J.; Chester, W. Monitoring and Using Telemetry Data. U.S. Patent 2018/0293618 A1, 11 October 2018.

20. Bär, A.; Finamore, A.; Casas, P.; Golab, L.; Mellia, M. Large-scale network traffic monitoring with DBStream, a system for rolling big data analysis. In Proceedings of the 2014 IEEE International Conference on Big Data (Big Data), Washington, DC, USA, 27–30 October 2014; pp. 165–170.

21. Xia, W.; Zhao, P.; Wen, Y.; Xie, H. A Survey on Data Center Networking (DCN): Infrastructure and Operations. *IEEE Commun. Surv. Tutor.* **2017**, *19*, 640–656. [CrossRef]

22. European Telecommunications Standards Institute (ETSI); Green Abstraction Layer (GAL). *Power Management Capabilities of the Future Energy Telecommunication Fixed Network Nodes*; ETSI: Sophia Antipolis, France, 2014.

23. Niewiadomska-Szynkiewicz, E.; Sikora, A.; Arabas, P.; Kołodziej, J. Control system for reducing energy consumption in backbone computer network: Control system for reducing energy consumption in backbone network. *Concurr. Comput. Pract. Exp.* **2013**, *25*, 1738–1754. [CrossRef]

24. Mendiola, A.; Astorga, J.; Jacob, E.; Higuero, M. A Survey on the Contributions of Software-Defined Networking to Traffic Engineering. *IEEE Commun. Surv. Tutor.* **2017**, *19*, 918–953. [CrossRef]

25. CPLEX Optimizer. Available online: https://www.ibm.com/analytics/cplex-optimizer (accessed on 12 July 2019).

26. Bilal, K.; Manzano, M.; Khan, S.U.; Calle, E.; Li, K.; Zomaya, A.Y. On the Characterization of the Structural Robustness of Data Center Networks. *IEEE Trans. Cloud Comput.* **2013**, *1*, 1–14. [CrossRef]

27. Al-Fares, M.; Loukissas, A.; Vahdat, A. A Scalable, Commodity Data Center Network Architecture. In Proceedings of the ACM SIGCOMM 2008 Conference on Applications, Technologies, Architectures, and Protocols for Computer Communications, Seattle, WA, USA, 17–22 August 2008; p. 12.

28. Alizadeh, M.; Edsall, T. On the Data Path Performance of Leaf-Spine Datacenter Fabrics. In Proceedings of the 2013 IEEE 21st Annual Symposium on High-Performance Interconnects, San Jose, CA, USA, 21–23 August 2013; pp. 71–74.

29. Lantz, B.; Heller, B.; McKeown, N. A network in a laptop: Rapid prototyping for software-defined networks. In Proceedings of the Ninth ACM SIGCOMM Workshop on Hot Topics in Networks—Hotnets '10, Monterey, CA, USA, 20–21 October 2010; pp. 1–6.

30. Pfaff, B.; Pettit, J.; Koponen, T.; Jackson, E.; Zhou, A.; Rajahalme, J.; Gross, J.; Wang, A.; Stringer, J.; Shelar, P.; et al. The Design and Implementation of Open vSwitch. In Proceedings of the 12th USENIX Symposium on Networked Systems Design and Implementation (NSDI '15), Oakland, CA, USA, 4–6 May 2015; p. 15.

31. De Oliveira, R.L.S.; Schweitzer, C.M.; Shinoda, A.A.; Rodrigues Prete, L. Using Mininet for emulation and prototyping Software-Defined Networks. In Proceedings of the 2014 IEEE Colombian Conference on Communications and Computing (COLCOM), Bogota, Colombia, 4–6 June 2014; pp. 1–6.

32. Medved, J.; Varga, R.; Tkacik, A.; Gray, K. OpenDaylight: Towards a Model-Driven SDN Controller architecture. In Proceedings of the IEEE International Symposium on a World of Wireless, Mobile and Multimedia Networks 2014, Sydney, Australia, 16–19 June 2014; pp. 1–6.

33. Optical Transceivers. Available online: https://www.finisar.com/optical-transceivers (accessed on 1 July 2019).

34. Cisco. Cisco Nexus 9500 Platform Switches Datasheet. 2014. Available online: https://www.cisco.com/c/en/us/products/collateral/switches/nexus-9000-series-switches/datasheet-c78-729404.html (accessed on 12 July 2019).

35. Vasić, N.; Bhurat, P.; Novaković, D.; Canini, M.; Shekhar, S.; Kostić, D. Identifying and using energy-critical paths. In Proceedings of the Seventh Conference on Emerging Networking EXperiments and Technologies on—CoNEXT '11, Tokyo, Japan, 6–9 December 2011; pp. 1–12.

36. Braun, W.; Menth, M. Software-Defined Networking Using OpenFlow: Protocols, Applications and Architectural Design Choices. *Future Internet* **2014**, *6*, 302–336. [CrossRef]

37. Tajiki, M.; Akbari, B.; Shojafar, M.; Mokari, N. Joint QoS and Congestion Control Based on Traffic Prediction in SDN. *Appl. Sci.* **2017**, *7*, 1265. [CrossRef]

    MDPI

Article

Towards Sustainable Energy-Efficient Communities Based on a Scheduling Algorithm

Carlos Cruz *, Esther Palomar, Ignacio Bravo and Alfredo Gardel

Department of Electronics, University of Alcala, Alcala de Henares, 28871 Madrid, Spain;
esther.palomar@uah.es (E.P.); ignacio.bravo@uah.es (I.B.); alfredo.gardel@uah.es (A.G.)
* Correspondence: carlos.cruzt@uah.es; Tel.: +34-91-8856-589; Fax: +34-91-885-6591

Received: 11 August 2019; Accepted: 10 September 2019; Published: 14 September 2019

Abstract: The Internet of Things (IoT) and Demand Response (DR) combined have transformed the way Information and Communication Technologies (ICT) contribute to saving energy and reducing costs, while also giving consumers more control over their energy footprint. Unlike current price and incentive based DR strategies, we propose a DR model that promotes consumers reaching coordinated behaviour towards more sustainable (and green) communities. A cooperative DR system is designed not only to bolster energy efficiency management at both home and district levels, but also to integrate the renewable energy resource information into the community's energy management. Initially conceived in a centralised way, a data collector called the "aggregator" will handle the operation scheduling requirements given the consumers' time preferences and the available electricity supply from renewables. Evaluation on the algorithm implementation shows feasible computational cost (CC) in different scenarios of households, communities and consumer behaviour. Number of appliances and timeframe flexibility have the greatest impact on the reallocation cost. A discussion on the communication, security and hardware platforms is included prior to future pilot deployment.

Keywords: cooperative smart community; scheduling algorithm; consumer preferences; renewables

1. Introduction

There exists a global aim to conceive novel sustainable services and energy infrastructures to balance supply and demand. Over the last decade, many sustainable development initiatives across the globe have been promoting regulatory campaigns, such as pricing or optional/mandatory thermal retrofit policies, looking at the engagement of cost-effective social behaviour and/or a social pro-environmental morality [1]. To this regard, the Internet of Things (IoT) and Demand Response (DR) combined have transformed the way Information and Communication Technologies (ICT) contribute to saving energy and reducing costs, while also giving consumers more control over their energy footprint [2,3]. Connected devices (e.g, household items, machines, vehicles or gadgets) can automatically influence each other in order to increase the overall potential for energy efficiency and the range of management systems' involvement.

DR programmes, designed to stimulate changes in consumers' electric usage patterns, thus appear to bolster not only energy efficiency, but also renewable energy resource management initiatives. Current DR strategies are based on providing end-users with individualised tailored advice about their particular habits with incentive payments for load reductions when needed to ensure reliability [4]. For instance, as control and communication technologies become more widely accessible, electricity prices and information are delivered more effectively to consumers. This allows consumers to identify and more easily target discretionary loads that can be curtailed or shifted. On one hand, we can find new challenges to the analysis of these loads and the extraction of consumer/community patterns that produce more automatic and user-friendly DR systems as well as driven by congestion management instead of being price-based. On the other hand, this automation should be enabled by on-site energy controls fed by near-real-time pricing information without significant customer effort or intervention. Furthermore, the real exploitation of renewable sources for energy supply presents multiple challenges not only to utilities, grid and system operators, but also to the consumer that knows very little about its availability or potential from microproduction [5] and energy harvesting processes [6,7]. For instance, according to the Eurostat survey (https://ec.europa.eu/eurostat) (Figure 1), only 19% of the final energy consumption in residential sector comes from renewable resources.

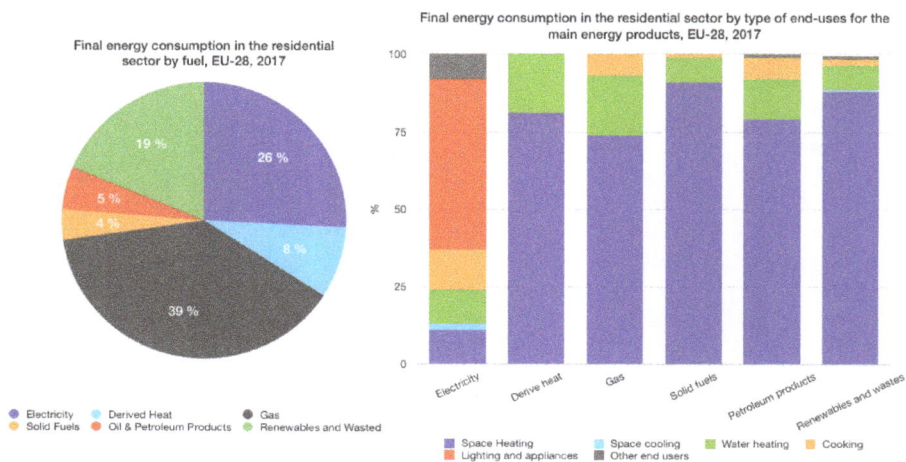

Figure 1. Final energy consumption in the European residential sector from Eurostat survey 2017.

Our proposal intends to bridge the aforementioned gap between utilities and consumers by leveraging consumer cooperation towards a joint daily schedule of their household appliances operation using supply generated from renewable energy sources. In this work, we assume the existence of an Utility entity (a set of energy providers or substations) generating, accumulating, storing and ultimately serving electricity to the consumers. This role, the Utility, is therefore in charge of allocating the available supply from the different energy sources at disposal of the community; it is not, however, dealing with the final destination of the supply (whether to power low-energy electronics or bigger appliances). As an application scenario, imagine a smart community of electricity consumers who, empowered by a better access to their consumption controls and appliance interconnection, are provided with sufficient incentives to coordinate and adjust their energy demands for a certain purpose. These consumption controls are coordinated by a Home Energy Management System, which enables energy management at homes. By doing this, consumers are able to visualise the energy data and make optimum use of energy by controlling their electrical appliances. They autonomously adapt their energy consumption by means of sharing nearly real-time electricity demand information. An aggregator device, capable of shifting the consumers' use of the resource, will be able to make the overall consumption pursue common goals such as being sustainable, ecofriendly or cheaper. On the other hand, utilities, allowed to perform real-time billing, profiling and fault detection, are also creating incentives for users consuming renewable sources (e.g., guaranteeing the lowest price if the load demand does not exceed a certain threshold). They produce, store, distribute and serve the supply to the consumers who will now benefit from additional information about the supply availability. It has to be observed that neither utilities nor consumers are considering microgenerated energy in the current model. Hence, it is responsibility of the aggregator the computation and rescheduling of the total daily load of the community to avoid overloading the utility supply from renewables. This scheduling represents an optimisation problem whose main factors are the 24 h-vector of the next day's supply from renewable sources and the duration and activation time preferences of every consumer's appliance.

In this work, we present a cooperative DR system designed to promote behavioural changes in small or large communities with common interests. The involved entities will reach binding agreements and coordinated behaviour through the aggregator, a device that collects 24-h vectors with the consumers' demands and the expected supply from renewables. It also centralises the supply allocation algorithm that optimises the distribution of the available green supply between the consumers taking into consideration their time preferences of appliance activation. Experimentation on the algorithm implementation is conducted using estimated values and benchmarks. We include the analysis of the power consumption in watts for most commonly used appliances taking average measures informed by manufacturers; for each appliance type, we show the efficiency label (according to EU normative), the estimated cost while in operation mode and standby mode as well as the average consumption in 24 h time. In addition, we evaluate the algorithm over different strategies of player order selection as well as over the application of four heuristics that optimise the objective search. Evaluation results throw feasible computational cost in all these different programming configurations as well as considering a series of scenarios for household and community settings, and consumer behaviours. Finally, from the empirical results we can discuss on the hardware and networking requirements for an efficient pilot deployment.

The paper is organised as follows. We discuss the related work in Section 2. Section 3 states the system model and design decisions. We describe the simulation of the implemented scheduling algorithm and estimate the performance cost in Section 4. Technical considerations in terms of communication, network protocols, security and hardware platforms are drawn in Section 5. Finally, Section 6 concludes and establishes future research directions.

2. Related Work

The starting point of our research can be found in the works by the authors of [8,9], where an adaptive model for DR is envisioned over the deployment of smart meter networks. Special focus is taken on the software design in order to facilitate the integration and scalability of the community system future development. An example of a DR aggregator model is designed in the work by the authors of [10] to facilitate renewable energy integration, where end consumers play a key role.

One of the major challenges in the energy efficiency context is the way to involve end-users in energy markets. This fact can be exemplified in the works by the authors of [11,12], where systems are designed to facilitate DR for residential prosumers. For instance, the work by the authors of [12] shows a system based on an aggregator of residential prosumers that participate in the day-ahead energy market to minimise operation costs by controlling appliances. The performance of an optimisation-based residential energy management scheme is presented in the work by the authors of [13]. This work applies a constrained swarm intelligence model to minimise the total cost of household electricity consumption. As it has been stated by the authors of [14], models based on DR, smart technologies and intelligent controllers can lead to a considerable energy consumption reduction.

The vast majority of the related work addresses energy-efficient solutions and optimisation algorithms from a single consumer/home viewpoint. The appliance scheduling optimisation solution in the work by the authors of [15] considers time ranges and consumer preferences along with different types of appliance consumption profiles. Their solution is based on the Mixed Integer Linear Programming (MILP) technique under "Gurobi" solver, which is addressed to minimise both the total energy cost and the peak load of all the home appliances used per day. This model is unscalable though. Similarly, household load scheduling is also approached in the work by the authors of [16] by the MILP optimisation model. The MILP model and a heuristic algorithm accounting for a typical household user are simulated taking into account overall costs, climatic comfort level and timeliness.

MILP is also the technique applied to load shifting by the authors of [17] to optimise the interaction between an aggregator and smart consumers' operation. The specific DR program incentives and the consumers' needs are the main parameters that a Smart Home (SH) controller considers to reshape the consumers' demand profile through shifting the operation of flexible loads. Focused on users' individual preferences, the work by the authors of [18] sets priorities and preferred time intervals for load scheduling, along with making efforts to optimise the consumption curves of household, commercial and industrial consumers.

A remarkable interaction between the utility and its consumers is modelled through a two-step centralised game in [19], where consumers reduce the peak-to-average power ratio by optimising their energy schedules. The utility supplier pulls consumers in a round-robin (RR) fashion and provides them with energy price parameter and current consumption summary vector. Each user, then, optimises its own schedule and reports it back to the supplier, which, in turn, updates its energy price parameters before pulling the next consumers. Also centralised but considering renewable energy technologies to improve energy efficiency and reduce costs through optimisation algorithms, approaches in the work by the authors of [20] focus on the context of microgrids and storage at residential and commercial building environments. In addition, heuristics based on genetics algorithms [21] and neural networks [22] work on the scheduling of the consumer consumption to save the peak formation. Their simulation results show that the proposed algorithms reduce the peak-to-average ratio and help users minimise their energy expenses without compromising comfort.

Applying a distributed and an autonomous Demand Side Management (DSM) within a neighbourhood, the consumers' schedulers in the work by the authors of [23] are assumed to be built inside smart meters and connected to the power grid and a local area network. In order to reduce the total energy cost, these schedulers interact automatically by running a distributed algorithm to find the optimal energy consumption schedule. Subscribers also receive incentives to use the schedulers via a novel pricing model derived from a game-theoretic analysis. The authors of [24] formulate a power allocation game, where multiple companies, leaders and their consumers are the followers to reach a unique pure-strategy Nash equilibrium via a distributed algorithm. Authors find that the multi-period scheme, compared with the single-period one, provide more incentives for energy consumers to participate in DR. For a comprehensive description of the many algorithms that can be used to solve the resource allocation problem, see the work by the authors of [25].

In summary, several optimisation Pareto-efficient approaches to the load and/or consumption adaptive scheduling have been the focus of much attention in demand side management, SHs, wireless sensor networks, broadband networks, and smart grids [26,27].

3. System Model

Our proposal embraces the use of renewable resources aiming three main actors: *Consumer*, *Aggregator* and *Utility*. Figure 2 illustrates the main roles and processes within the adopted cooperative DR framework.

The first actor, *Consumer*, provides the home energy usage to be managed and automatically controlled by Home Energy Management System (HEMS) that performs three main functions: (1) schedule demand, (2) appliance control and (3) information provider. It selects the daily scheduling preferences, managing a profile for collaboration in a DR system and viewing its account and consume information. The consumer can manage them from a portable device (i.e., an app installed on a mobile phone or tablet) that is connected to a communication network for preferences scheduling. A community will comprise a set of consumers sharing electricity supplier or substation. HEMS pulls scheduling information and generates processed data to the *Aggregator*. HEMS is also responsible for collecting information from the *Aggregator* and controlling a variety of home appliances.

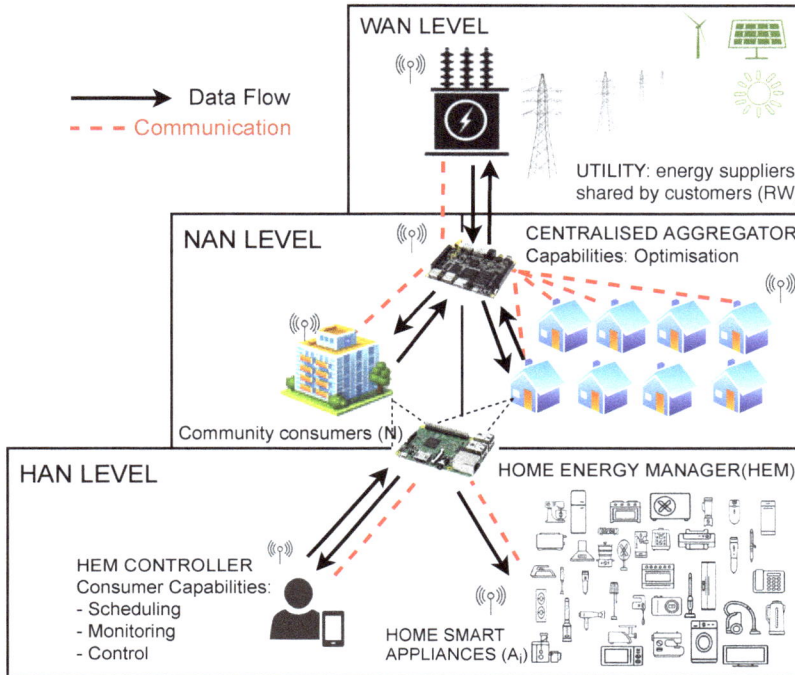

Figure 2. Smart cooperative system divided into Home Area Network (HAN), Neighbour Area Network (NAN) and Wide Area Network (WAN).

In our proposal, *Consumers* adapt their energy consumption cooperatively on a centralised way, that is, sharing their demand schedule with a data collector, which facilitates the integration of energy consumption information into a common view. This integration is performed over the so-called *Aggregator*, the second actor, which implements an optimised resource allocation algorithm as a response to supply conditions, in particular, targeting renewable sources. The *Aggregator* is defined as the optimal system providing energy management services in order to efficiently manage demand in SH [28]. HEMS acts as a central node and receives the demand scheduled information from the *Aggregator*. Then it loads the power consumption preferences to each appliance and establishes communication for managing the appliances. The *Aggregator* allows the local distribution of the energy provided, according to the availability of renewable resources. This energy management system will be connected to the *Utility*, the third actor, which is a set of energy suppliers shared by customers. We presume utilities implement a distributed generation that allows to gather energy from mainly renewable sources addressed to give lower environmental impacts and improve supply security.

3.1. Consumer System Design

Let $\mathcal{N}$ denote an ordered set of *Consumers* that are willing to cooperate in the pursuit of global community targets (i.e., becoming greener) by sending their data to the *Aggregator*. Each *consumer* $i \in \mathcal{N}$ has a set of household appliances labeled as $\mathcal{A}_i$. Fixed energy load is identified by factors such as the consumers' habits, their behaviours and their use of appliances, as well as a variable load resulting from the use of such appliances and other equipments. Formulae and benchmarks can be used to estimate appliance and home electronic energy use in kilowatt hours (kWh) as well as household local records.

Bearing in mind a discrete time slot system, and without loss of generality, we assume that time granularity is one hour per day. Regarding the appliances, each *Consumer* is supposed to preallocate a certain amount of fixed demand and variable consumption planned for the next 24 h [29]. For each appliance, $a_{ij} \in \mathcal{A}_i$, we assume both daily fixed and variable energy consumption scheduling vectors at each time slot $t \in \{0, \dots, 23\}$ to control its non-shiftable and shiftable consumption respectively.

We define $f\mathcal{D}^t_{i,a_{ij}}$ and $v\mathcal{D}^t_{i,a_{ij}}$ by denoting the corresponding one-hour fixed and variable energy consumptions respectively. Variable energy demand is characterised by its flexibility, as the *Consumer* preference for an appliance starting in a specific period of time is taken into consideration. For each appliance, there is an execution window (i.e., a closed interval) that selects a minimal starting time, and a maximal ending time labelled by t^i_{beg} and t^i_{end}. t^i_{sched} is defined as the working time of appliance "i" and matches the range of operation start time $t^i_{sched} \in [t^i_{beg}, t^i_{end}]$. $\mathcal{L}$ is defined as the duration of the planned operation of appliance a_{ij} in the next day. Load needs to be switched on for a time between two predefined moments: $\forall_{ij} \in \mathcal{A}_i, t^i_{sched} \geq t^i_{beg}$. In this line, load also needs to be switched off: $\forall_{ij} \in \mathcal{A}_i, t^i_{sched} \leq t^i_{end}$. In other words, *Consumer i* will set the following data for its appliance $a_{ij} \in \mathcal{A}_i$ (see Table 1).

Table 1. Appliance configuration.

Appliance Configuration				
Consumption (kWh)	Fixed consumption (kWh)	Duration (hours)	Time ON	Time OFF
$v\mathcal{D}^t_i$	$f\mathcal{D}^t_i$	$\mathcal{L}$	t_{beg}	t_{end}

- Fixed consumption (kWh) when appliance a_{ij} is in standby status
- Consumption (kWh) when a_{ij} is on
- Duration (hours/minutes) of the planned operation of appliance a_{ij} in the next day
- Point in time (hour, e.g., 8am) of preferred start of appliance a_{ij} activation
- Point in time (hour, e.g., 12pm) of preferred end of appliance a_{ij} operation

A centralised home controller provides access to all the appliances and devices at home via wireless networks; it will receive and apply the 24-hour reallocated vector from the *Aggregator* to systematically activate/deactivate every appliance without human intervention.

Moreover, we have developed an energy consumption scheduling app based on the Adobe XD template [30], which provides the consumer with an interface to control, monitor, visualise and program the functioning of appliances. More specifically, it allows the configuration and setting of the aforementioned data for each appliance a_{ij}. Figure 3 depicts a usage sequence to explain how the application works. The app allows users to check the resources used in the previous 24 h as well as to select the appliance in relation to the dwelling zones such as the kitchen or the bathroom, among others. At this stage, consumer will be able to indicate the time range and the duration of activation for each appliance. The last window summarises the introduced demand information. It also provides an estimated power cost in operation and standby for each appliance (according to benchmark analysis in Table 2). Consumers have to give consent by sending these data to the home controller. Finally, a vector is sent to the *Aggregator* with the data structure shown in Table 1.

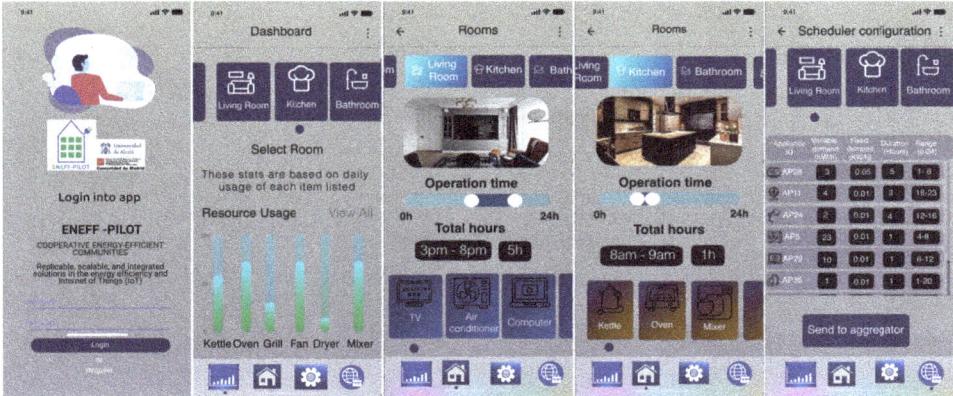

Figure 3. Consumer scheduling app.

Data aggregation is defined as a centralised system with aggregation tasks that communicates with the *Utility* and the *Consumers* as shown in Figure 4. An algorithm is originally designed to optimise the allocation of the expected electricity supply from renewables among the community's *Consumers* related to their chosen preferences.

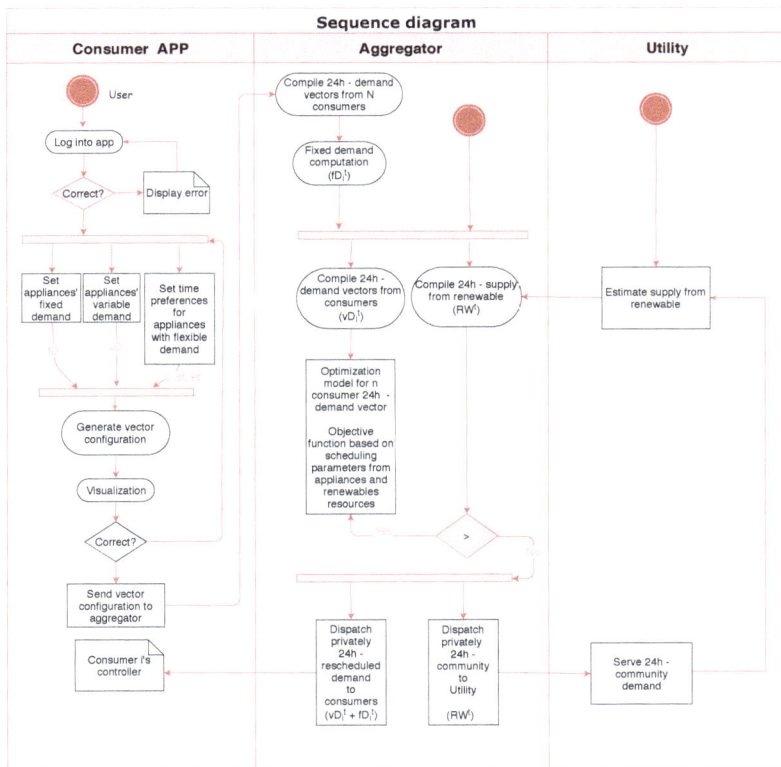

Figure 4. Sequence stating main processes and message exchange among the system's players.

3.2. Aggregator System Design

As an example of a renewable source efficient use, the *Utility* provides essential information on reliable renewable source and fossil energy programmed for the next 24 h by replacing carbon-intensive energy. The energy supply generated from a set of renewable sources at a time slot $t \in \{0, \ldots, 23\}$ is denoted by $\mathcal{RW}^t$. $f\mathcal{U}^t$ represents the energy supply at time t generated from a set of fossil sources. The *Utility* centralises the distribution of the energy, the notification to the *Aggregator*, and the billing process. The renewable supply vector $\mathcal{RW}$ is essential for the *Aggregator* in the optimisation of a fair allocation of such supply between the *Consumers'* fixed (non-shiftable) and variable (shiftable) energy demands.

The daily fixed demand for consumer $i \in \mathcal{N}$ is denoted by $f\mathcal{D}_i = \sum_{t=0}^{23} \sum_{a_{ij} \in \mathcal{A}_i} f\mathcal{D}_{i,a_{ij}}^t$ as the aggregated load of non-shiftable local consumption of the appliances and frequent behaviours. The *Aggregator* can then easily compute the daily fixed demand for the whole community of consumers at a time t as $f\mathcal{D}^t = \sum_i^N f\mathcal{D}_i^t$. On a daily basis, the *Aggregator* verifies that the total energy consumed by all appliances in the system fulfils the daily utility service provided by the *Utility*. It is critical that the community does not reach the worst case such as $\sum_i^N \sum_{t=0}^{23} f\mathcal{D}_i^t \gg \sum_{t=0}^{23} \mathcal{RW}^t$. On the contrary, aggregation of the variable energy is more complex given the consumers' time preferences. The *Aggregator* will execute a fair-share rescheduling of the community's requested variable demand per hour $v\mathcal{D}_i^t$ aiming at $\forall t \in \{0, \ldots, 23\}, \sum_i^N (f\mathcal{D}_i^t + v\mathcal{D}_i^t) \leq \mathcal{RW}^t$. We will show refinements of the proposed scheduler algorithm looking at the max–min fairness, Pareto-efficiency, envy-freeness, and truthfulness while serving Consumers' preferences. Perhaps the simplest way to give each Consumer equal chance against all other is to recursively apply a "round-robin" strategy in the allocation of each Consumer's needs. Fair random assignment is one of the refinement methods to be compared. A global centralised optimisation problem is faced here, where only a "Nash bargaining solution" is possible such as $\forall i \in \{1 \ldots \mathcal{N}\}, \mu_i^t = f\mathcal{D}_i^t + min\{\mathcal{DFC}(v\mathcal{D}_i^t)\} \leq \mathcal{RW}^t$.

Therefore, solutions to the optimisation problem should satisfy t_{sched}^i and $\mathcal{L}$ while avoiding overconsumption at $\mathcal{RW}^t$. The formulation is explained in Algorithm 1 (Demand Calculation Function, $\mathcal{DCF}$) and it will be shaped as its minimum, i.e., $min.\mathcal{DFC}(\cdot)$ upon request of Algorithm 2. In $\mathcal{DFC}$ function, a search for the optimum time slot for every appliance activation takes place given its activation time, its preference interval and the available supply in kW from the renewable utility. In particular, taking into account $\mathcal{A}_i$, t_{sched}^i, t_{beg}^i and t_{end}^i variables, the optimisation will determine how appropriate an adjustment is by minimising the total overconsumption (in hours) of the community appliances against the available renewable supply at a certain time slot.

Finally, upon reaching the optimisation objective, the *Aggregator* will notify the community that an agreement has been reached and privately release the reallocated demand vector $\overrightarrow{\mu}_i \forall i \in \mathcal{N}$ to each Consumer.

Algorithm 1 Demand Calculation Function ($\mathcal{DCF}$)

1: RW^t(renewable vector) = $\{\sigma_1, \ldots, \sigma_{24}\}$
2: $\mathcal{N} = size(\mathcal{A}_i)$
3: Defining variables t^i_{beg}, t^i_{end}
4: **for** ihour time to the total number of hours **do**
5:     **if** ihour doesn't belongs to interval $[t^i_{beg}, t^i_{end}]$ **then**
6:         $f\mathcal{D}^t_i$ computation
7:     **end if**
8: **end for**
9: **for** iappliance 1 to size of appliances configuration ($\mathcal{A}_i$) **do**
10:     $\mathcal{RW}^t = \mathcal{RW}^t - -\mathcal{A}_i(f\mathcal{D}^t_i)$
11: **end for**
12: $\mathcal{A}_i(v\mathcal{D}^t_i) = \mathcal{A}_i(v\mathcal{D}^t_i) - -\mathcal{A}_i(f\mathcal{D}^t_i)$
13: $\mathcal{A}_i(\mathcal{D}^t_i(\mathcal{D}^t_i < 0)) = 0$
14: $\mathcal{A}_i(f\mathcal{D}^t_i) = \mathcal{A}_i(f\mathcal{D}^t_i) - -\mathcal{A}_i(f\mathcal{D}^t_i)$
    Objective Function $\mathcal{F}(\mathcal{A}_i, \mathcal{RW}^t, t^i_{sched})$
Require: $\mathcal{A}_i$ configuration: $v\mathcal{D}^t_i, f\mathcal{D}^t_i, \mathcal{L}_i, t^i_{beg}, t^i_{end}$
Ensure: $t^i_{beg} < t^i_{end}$
15: $\mathcal{HC}$ initialisation (consume Hourly Energy)
16: **for** iappliance 1 to size of appliance configuration **do**
17:     Set t^i_{beg}
18:     Set t^i_{end} based on $\mathcal{L}_i$ and t^i_{beg}
19:     **for** ihour time to the total number of hours **do**
20:         **if** ihour belongs to interval $[t^i_{beg}, t^i_{end}]$ **then**
21:             $\mathcal{HC}(ihour) \leftarrow \mathcal{HC}(ihour) + \mathcal{A}_i(v\mathcal{D}^t_i)$
22:         **else**
23:             $\mathcal{HC}(ihour) \leftarrow \mathcal{HC}(ihour) + \mathcal{A}_i(f\mathcal{D}^t_i)$
24:         **end if**
25:     **end for**
26: **end for**
27: $\mathcal{RW}^ts = \mathcal{RW}^t - -\mathcal{HC}^t$
28: $\mathcal{RW}^ts(\mathcal{RW}^ts < 0) = 0$
29: $Demanded\_\mathcal{RW}^t \leftarrow min(\mathcal{RW}^t, \mathcal{HC}^t)$
30: $\mathcal{R}1 = sum(\mathcal{RW}^ts); \mathcal{R}2 = max(\mathcal{HC}^t)$
31: $Result = sum(\mathcal{R}1 + \mathcal{R}2)$
32: **return** $Result, Demanded\_\mathcal{RW}^t, \mathcal{HC}^t(t^i_{sched})$

Algorithm 2 RR strategy

1: Generate parameters for consumer allocation
2: Define global variable $\mathcal{RW}$
3: **while** (user < $\mathcal{N}$) and (min($\mathcal{RW}$) >= 0) **do**
4:     **if** Optimisation needs **then**
5:         Load consumer preferences. $\mathcal{A}_i$ size from preference array
6:         Call Optimisation Function under variables preferences: $\mathcal{RW}, \mathcal{A}_i, \mathcal{N}$
7:         Number of user ++
8:         **if** ($\mathcal{RW}$ equals to 0) **then**
9:             Break
10:         **end if**
11:     **end if**(No consumer to optimise)
12: **end while**

3.3. Proposed Algorithm: A Fair Division Game

The *Aggregator* can apply different approaches to optimisation search within the aggregated load vector vD. It contains all consumer appliances' scheduling that could be shifted within their preferred activation time frame. This scheduling problem at the *Aggregator* can be seen as a division game given a set of players (either the consumers in $\mathcal{N}$ or all its appliances) and a set of assets (the supply from renewables in $\mathcal{RW}$). In our algorithm we have opted for a turn-based sequential game played by Appliance instead of Consumer for optimisation purposes.

The scheduling problem has to produce a fair division of $\mathcal{RW}$, i.e., a set of rules that, when properly used by the players, guarantees at the end of the game each player will have received a fair share of the assets. In our view, a fair share means that consumers' preferences on the appliance activation are considered by the *Aggregator* with equity and privacy. As in turn-based sequential games, defining the order under which players start within a turn could be approached in terms of (A) Round-Robin (RR) start: the first player selection policy is RR; (B) random RR; (C) ranking: the first player being the same every time; and (D) randomness: the first player is randomly selected (likewise the sequence order), as follows.

(A) The RR principle, known from other fields such as network scheduling and processor queuing, is based on a process/game/technique, where each task/person/device takes an equal share of something in turn. The RR scheduling can allocate the available electricity from renewables both simple and fairly among the *Consumers/Appliances*, because (1) the consumers' number is known and fixed and (2) the reallocation process is centralised by the *Aggregator* which, starting on its own, will satisfy the demand of the *Consumers/Appliances* in a periodically repeated order. We include pseudocode of our algorithm's main function in the round robin strategy, being the rest pseudocodes similar with exception of the player turn selection on Algorithm 2-line 3. RR results in max–min fairness if the *Consumers/Appliances'* demands are equally sized; otherwise, fair queuing that establishes a fair share size would be desirable.

(B) A random RR scheduling: A similar process as in A), though the election of first *Consumer* is random.

(C) A picking-sequence has several merits as a fair division protocol [31]. Assuming that each agent has a (private) ranking over the set of objects, the allocator must find a policy (i.e., a sequence of agents that maximises the expected value of some social welfare function). Moreover, picking sequences are a natural way of allocating (indivisible) items to agents in a decentralised manner: at each stage, a designated agent chooses an item among those that remain available. The goal of the method is to identify the fairest sequence.

(D) A random process could, or could not, introduce efficiency (no other "random" assignment dominates) in the aforementioned methods while keeping them Pareto-efficient, envy-free and giving good approximation to the social welfare. Efficiency in terms of computational time is also at stake.

4. System Validation

In this section, we measure the Computational Cost (CC) of the implemented scheduling algorithm evaluating the suitability of a number of four heuristics applied to the optimisation search and on a series of different case scenarios of consumer communities. In particular, the evaluation of the heuristics and their behaviour on our algorithm under the same input parameters will assist in the selection of the hardware platform for an efficient deployment.

4.1. Optimisation Algorithms Used

We adopt heuristic techniques to perform a partial random search of optimal solutions to our objective, i.e., either when the reallocation demand is met or when the number of predefined iterations is reached. We have identified and implemented the following four optimisation methods to guide the search of a workable solution, i.e., the nearest local minimum standard strategy in our Algorithm 1 as follows.

(i) Simulated Annealing (SA) [32] finds a local minimum solution for our Algorithm 1 ($\mathcal{DCF}$) starting at an initial operation time t^i_{sched}. As explained in Algorithm 3, SA starts generating trial point based on current estimates and evaluates the function by accepting a new value generated after $\mathcal{T}$ parameter is set. The solution must consider the $[t^i_{beg}, t^i_{end}]$ time constraints. t^i_{sched} can randomly generate and filter by $\mathcal{L}$. In case of better $\mathcal{D}$, the original one $\mathcal{D}'$, $t^i_{sched'}$ could be accepted as better solution if $\mathcal{D}'$ is worst than $\mathcal{D}$. After the internal counter reaches its threshold, $\mathcal{T}$ is cooled down and re-select the best solution again with the reset counter.

Algorithm 3 Optimisation based on SA algorithm

1: Let $\mathcal{T} > 0$ as initial parameter
2: Let $\mathcal{N}(\mathcal{T})$ as maximum number of iterations
3: **while** stop criterion has not been met **do**
4:      Randomly generate a fasible solution t_{sched}
5:      Evaluate t^i_{sched}, $\mathcal{D} = f(t^i_{sched})$
6:      n = 1
7:      **while** while n $<= \mathcal{N}(\mathcal{T})$ **do**
8:          Generate solution $t^i_{sched'}$ based on t^i_{sched}
9:          Evaluation of $t^i_{sched'}$; $\mathcal{D}' = f(t^i_{sched'})$; $\delta = f(t^i_{sched'}) - f(t^i_{sched})$
10:         **if** $f(t'_{osi}) < f(t^i_{sched})$ **then**
11:              $t^i_{sched} = t^i_{sched'}$
12:         **else**
13:              **if** $\delta >= 0$ and u$< \exp((f(t^i_{sched'}) - f(t^i_{sched}))/\mathcal{T})$ **then**
14:                  $t^i_{sched} = t^i_{sched'}$
15:              **end if**
16:         **end if**
17:         n = n+1
18:      **end while**
19:      $\mathcal{T}$ reduction and update t^i_{sched} at each reduction
20: **end while**

(ii) Genetic algorithm (GA) [33] is identified as a method mainly used to solve optimisation problems based on a natural selection process similar to biological evolution. GA finds an optimal operative time from our Algorithm 1 ($\mathcal{DCF}$) for the $\mathcal{A}_i$ variables. As explained in Algorithm 4, GA can find a solution beginning with random population of points. GA repeatedly modifies a population of individual solutions. At each step, GA produces a next generation population based on a randomly selection of individuals from the current population. After that, the population turns into an optimal solution. The evaluation number is increased when the method finishes by calculating one generation $\mathcal{P}$. Each generation is a feasible solution for the appliance scheduling (t^i_{sched} per appliance). In the evaluation stage, the best solution $t^i_{sched'}$, which has the lowest demand, is inserted to the best solution set. Mutation and crossover operators are selected to generate the next evaluation from the current generation. The mutation operator randomly shifts the scheduled start times of some appliances in order to generate newly solutions that may have a better result in demand efficiency. They are screened with the constraints to filter out the infeasible ones. The crossover driver swaps scheduled t^i_{sched} under feasible solutions.

Algorithm 4 Optimisation based on GA algorithm

1: Generate Solutions. Build a set of PopSize $\mathcal{P}$ solution
2: Reformulation of solutions. Selection of a local search method to each solution in $\mathcal{P}$
3: **while** number of evaluations < MaxEval **do**

4:     t^i_{sched} introduction to P. Evaluation of solution in $\mathcal{P}$ and update
5:     Probability of survival based on the quality of the solution
6:     $\mathcal{P}$ solution is partially selected to apply the mutation and crossover operation
7:     Number of evaluation ++
8:     Constraint validate $\mathcal{P}$ for each t^i_{sched}. Discard solutions which are disqualified
9: **end while**

(iii) Pattern Search (PS) [34] polls the values around the current point and determines the direction that will minimise our Algorithm 1 ($\mathcal{DCF}$) starting at an initial operation time t^i_{sched}. For each possible direction, an all linear combination of the current position is created, and each pattern is multiplied by the size of the mesh to obtain a new one. As presented in Algorithm 5, PS investigates nearest neighbourhood of a possible solution always in the range of lower and upper bounds $[t^i_{beg}, t^i_{end}]$ for each appliance. This solution seeks to find a better one. A failure improvement generation by neighbours ($\mathcal{L}$ and $\mathcal{D}$) would reduce the search step (Δ). Search finishes when the step gets sufficiently short, ensuring the convergence to a local minimal overconsumption.

Algorithm 5 Optimisation based on PS algorithm

1: Initialise predefine default search step Δ_0; t^i_{sched} and $\Delta=\Delta_0$
2: **while** Termination condition not reached **do**

3:     init current solution $\mathcal{D}= (t^i_{sched}+\mathcal{L}*\Delta)$
4:     Evaluate nearest neighbours in $\mathcal{D}$
5:     **if** betters in $\mathcal{D}$ **then**

6:         Update the current solution to the best neighbour in $\mathcal{D}$; $\Delta=\Delta_0$
7:     **else**

8:         Search step reduction $\Delta=\Delta_0/2$
9:     **end if**
10: **end while**

(iv) Particle Swarm Optimisation (PSO) [35] is a stochastic search method and simulates the social behaviour of particles used to find parameters that minimise a given objective. The optimisation determines the minimum value and the best location evaluating our Algorithm 1 ($\mathcal{DCF}$) through iterations.

Algorithm 6 illustrates this search procedure, which is initialised with the generation of particles assigning initial velocities and positions. The operative appliances time is defined as a set of lower and upper bounds $[\overrightarrow{t_{beg}}, \overrightarrow{t_{end}}]$, where the solution is found in operation time range $\overrightarrow{t_{beg}} = (t^{i1}_{beg}, ..., t^{ij}_{beg})$, and $\overrightarrow{t_{et}} = (t^{i1}_{end}, ..., t^{ij}_{end})$. The vectors $\overrightarrow{x} = (x^{i1}, ..., x^{ij})$ and $\overrightarrow{v} = (v^{i1}, ..., v^{ij})$ are the current position and velocity, respectively. Each individual adjusts its position according to a linear combination of its inertia ω, the best location of individual particle $\overrightarrow{p} = (p^{i1}, ..., p^{ij})$ and the best location of particle swarm $\overrightarrow{g} = (g^{i1}, ..., g^{ij})$. The confidence degree is determined by the random operators ϕ_p and ϕ_g in the range [0,1] together with the confidence coefficients c_p and c_g. They are responsible for moving in the direction of the best position of a particle and the global best position. The new displacements are no more than one way of trying to imitate other individuals. It then iteratively updates the solution positions (the new location is the old one plus the velocity, modified to keep particles within $[\overrightarrow{t_{beg}}, \overrightarrow{t_{end}}]$, velocities and neighbours). The solution, above $\mathcal{A}_i$, tries to find the optimal ones. After several iterations, particles converge to the best solution.

Algorithm 6 Optimisation based on PSO algorithm

1: Initialise population of particles with random values positions in the search space $\overrightarrow{x} \sim U[\overrightarrow{t_{beg}}, \overrightarrow{t_{end}}]$
2: Set each particle best known position to its initial position $\overrightarrow{p} \leftarrow \overrightarrow{x}$
3: Initialise each particle velocity to random values $\overrightarrow{v} \sim U[-\overrightarrow{d}, \overrightarrow{d}]$ where $\overrightarrow{d} = \overrightarrow{beg} - -\overrightarrow{end}$
4: Initialise the best known position $\overrightarrow{g}$ to the $\overrightarrow{x}$ where $f(\overrightarrow{x})$ is lowest
5: **while** Termination condition not reached **do**
6:    **for** Each particle i **do**
7:      **if** $i > 1$ **then**
8:         Choose two random numbers ϕ_p, ϕ_g
9:         Adapt velocity $\overrightarrow{v} \leftarrow \omega \overrightarrow{v} + c_p \phi_p (\overrightarrow{p} - \overrightarrow{x}) + c_g \phi_g (\overrightarrow{g} - \overrightarrow{x})$
10:         Bound $\overrightarrow{v}$ for all dimensions i ($\overrightarrow{v}$, - $\overrightarrow{d}$, $\overrightarrow{d}$)
11:         Update the position of the particle $\overrightarrow{x} \leftarrow \overrightarrow{x} + \overrightarrow{v}$
12:         Bound population x_i for all dimensions i ($\overrightarrow{x}$,$\overrightarrow{t_{beg}}$, $\overrightarrow{t_{end}}$)
13:      **end if**
14:      **if** $f(\overrightarrow{x}) < f(\overrightarrow{p})$ **then**
15:         update the particle's best known position $\overrightarrow{p} \leftarrow \overrightarrow{x}$
16:      **end if**
17:      **if** $f(\overrightarrow{x}) < f(\overrightarrow{g})$ **then**
18:         update the particle's best known position $\overrightarrow{g} \leftarrow \overrightarrow{x}$
19:      **end if**
20:    **end for**
21:    $\overrightarrow{g}$ holds the best found position in search space
22: **end while**

4.2. Performance Analysis

Simulation runs on a computer with the following specifications: CPU: 2.3 GHz Intel Core i5; Memory: 8 GB 2133 MHz LPDDR3 and MATLAB R2018b [36]. We evaluate the computational cost of the proposed Algorithm 1 on a series of experiments that represent a variety of possible scenarios of community sizes, consumption patterns or consumer behaviour as depicted in Table 3. Experimentation will help us to identify the most influential factor/s in the computation of the community scheduling.

We have conducted an analysis on the most common appliances' real consumption estimation from manufacturers and data sources from the authors of [37], the U.S. Department of Energy (http://www.energy.gov/), the National Grid report (http://www.nationalgrid.com), the authors of [38] and the reports (https://standby.lbl.gov/docs) as well as the manufacturer data to set the scenarios. Our benchmark is depicted in Table 2. Scenarios were envisioned from the design of a residential building as in Figure 5. In particular, we have generated eight scenarios as illustrated in Table 4 and conducted hundreds of experiments for the different factor values to obtain results on a boxplot shape. On the one hand, we can denote as altruistic or flexible a consumer whose time preferences range is big (e.g., from 0h to 23 h); such types of communities are represented by Cases 1, 3, 5 and 7. On the other hand, Cases 2, 4, 6 and 8 illustrate communities on a more selfish setting. Duration is set equally in both situations.

Table 2. Common household appliance energy use.

ID	Appliance	Model	Watts (W)	Efficiency Ranges European Union A, A^+, A^{++}, A^{+++}	Estimated Average Power in 24 h (kWh)	Estimated Standby Power in 24 h (kWh)	Estimated Operative Time in 24 h (h)
AP1	Water Heater	Wesen ECO30	2000	A++	10–14.73	0.010	1–15
AP2	Clothes Dryer	Balay 3SB285B	4350	A	1–2.22	0.015	1–10
AP3	Clothes Washer	Eutrotech 1106	1800	A++	1–2.67	0.015	0.5–10
AP4	Iron	Rowenta DX1411	2100	A	0.1–3	0.002	1–3
AP5	Air conditioner	Fujitsu STG34KMTA	9400	-	3.9–24.3	0.015	0.3–15
AP6	Room air conditioner	Rinnai RPC26WA	2600	-	8–24.3	0.015	3–18
AP7	Heater	DeLonghi HSX3324FTS	2400	-	1–7	0.08	0.1–10
AP8	Fan heater	Dyson AM09	2000	-	1–6.7	0.015	0.1–10
AP9	Dehumidifier	DeLonghi DEX	210	A++	4–24.3	0.005	1.1–9
AP10	Electric blanket	Medisana HDW	120	-	1–3	0.08	1.2–9
AP11	Ceiling Fans	Westinghouse Bendan	80	B	0.5–9	0.01	0.5–5
AP12	Attic Fans	Remigton	500	-	4.73–6	0.01	0.1–18
AP13	Tower Fan	Sunbeam FA7250	40	-	1.4–3	0.03	0.1–18
AP14	Hoover	BGLS4TURBO	750	-	3–6	0.02	0.3–18
AP15	Boiler	Greenstar Ri	9000	B	8–22	0.05	0.1–3
AP16	Coffee maker	DeLonghi ECOV	1100	A	9–12	0.05	0.1–3
AP17	Refrigerator	Bosch KDN46VI20	500	A	8.77–10	0.05	4.77–24
AP18	Dishwasher	Bosch SMS88TI36E	1500	A+++	0.5–1.5	0.015	0.3–4
AP19	Food processor	Becken BFP-400	110	A	0.5–2	0.015	0.1–5
AP20	Freezer	Bosch GSN36BI3P	350	A++	6–8	0.009	0.1–24
AP21	Microwave	Balay 3CG5172N0	1700	A	0.9–3	0.01	0.1–4
AP22	Oven	Bosch VBD5780S0	5000	A	10.96–12	0.01	0.1–8
AP23	Toaster	Russell Hobbs 21973	1100	A	0.2–1	0.01	0.1–1
AP32	Lighting	Osram	100	-	0.7–3	0.01	0.1–24
AP25	Vaporizer	Philips GC362/80	400	A+++	0.3–2	0.07	0.1–8
AP26	Printer	HP Officejet 3833	100	-	0.8–1	0.05	0.1–4
AP27	Computer	Samsung ls24a450	350	A++	0.7–15.3	0.05	0.1–24
AP26	TV	Panasonic TX43E302B	54	A++	0.1–100	0.05	0.1–24
AP29	Kettle	Philips HD4644/00	3000	A	6–19	0.01	0.1–1
AP30	Security Alarm	Vbestlife	20	-	0.6–1	0.02	0.1–24
AP31	Auto Cook	MUC88B68ES	1200	A++	1–3	0.09	0.1–3
AP32	Air Cleaner	Balay 3BC598GN	150	A	1.1–6	0.01	0.1–6
AP33	Vacuum Cleaner	Hoover TH31HO01	1000	A++	0.9–3	0.06	0.2–4
AP34	Electric Fryer	DeLonghi F26237	1800	-	13–16	0.05	0.2–3
AP35	LedTV	LG 49LJ515V	250	A++	1.9–5	0.05	0.2–24
AP36	Electric Store	Dura Heat EUH4000	4000		2.4–4	0.05	0.3–23
AP37	Speaker	Logitech Z120	180	A+++	0.3–4	0.01	0.2–20
AP38	Hair Dryer	Rowenta CV3812F0	2100	A+++	0.99–4	0.01	0.2–6
AP39	Smart Camera	Yi Home	4	-	0.99–2	0.01	0.2–24
AP40	Monitor Sensor	iHome	5	-	0.99–10	0.01	0.1–24

Table 3. List of factors for the different case scenarios.

Factor	Type	Value
Community Size	High, Low	30, 5
N. of Appliances	High, Low	1200, 40
Distribution of Appliances	Same, Different	S, D
Fixed Demand	High, Low	Not influenced by optimisation
Variable Demand	High, Low	Up to 18 kWh[6], Up to 9 kWh[6]
Consumer Flexibility	High, Low	24 h, $\mathcal{A}$ duration: $\mathcal{L}$
Vector of $\mathcal{RW}$	Even, Uneven	10 kWh, [10 kWh–20 kWh] 50% SD

Table 4. Possible load-shape situations.

	Community Size $\mathcal{N}$	N. of Appliances $\mathcal{A}$	Distribution of Appliances	Fixed Demand $f\mathcal{D}$ (kWh)	Variable Demand $v\mathcal{D}$ (kWh)	Consumer Flexibility $\mathcal{CF}$	RW Vector per Hour (kWh)
Case 1	From 5 to 30	From 40 to 1200	S	Up to 0.43	Up to 9	24 h	10
Case 2	From 5 to 30	From 40 to 1200	S	Up to 0.43	Up to 9	$\mathcal{L}$	10
Case 3	From 5 to 30	From 40 to 1200	D	Up to 0.43	Up to 9	24 h	10
Case 4	From 5 to 30	From 40 to 1200	D	Up to 0.43	Up to 9	$\mathcal{L}$	10
Case 5	From 5 to 30	From 40 to 1200	S	Up to 0.43	Up to 18	24 h	10
Case 6	From 5 to 30	From 40 to 1200	S	Up to 0.43	Up to 18	$\mathcal{L}$	10
Case 7	From 5 to 30	From 40 to 1200	D	Up to 0.43	Up to 18	24 h	10
Case 8	From 5 to 30	From 40 to 1200	D	Up to 0.43	Up to 18	$\mathcal{L}$	10

We will compare our method's performance with the four different heuristics mentioned in Section 4.1, i.e., SA, PSO, GA and PS, and evaluate the efficiency of the different strategies presented in Section 3.3 on the search for the optimisation objective.

Figures 6–9 depict the scheduling cost for the different case scenarios of consumers using strategy C picking-sequence. These scenarios represent extreme conditions either considering high number of appliances and/or an uneven distribution of them, and also the flexibility of the consumers' time preferences. Communities with selfish settings or fixed consumption display the best results over all different optimisation procedures (cases 2, 4, 6 and 8).

Figure 5. Distribution of the appliances, consumers and aggregator in the community.

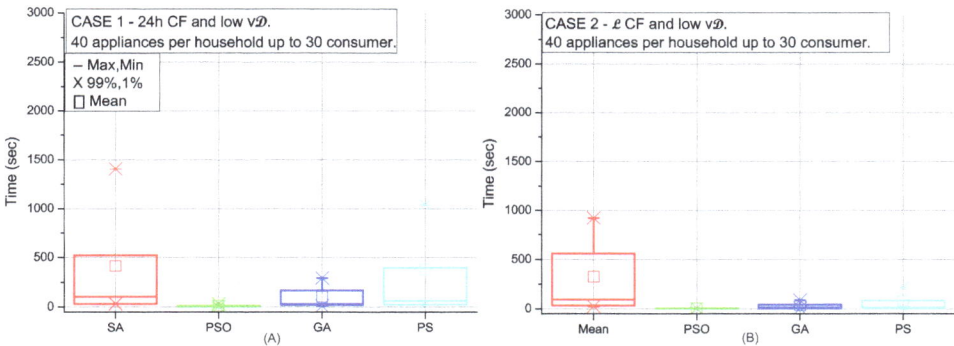

Figure 6. Comparison of the SA, PSO, GA and PS methods for low variable and high fixed consumption in 24 h CF (**A**) and $\mathcal{L}$ CF (**B**).

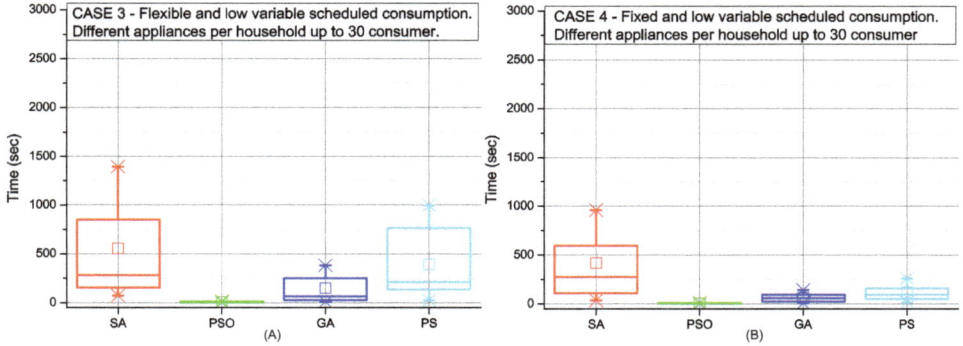

Figure 7. Comparison of the SA, PSO, GA and PS methods for low variable and high fixed consumption in 24 h CF (**A**) and $\mathcal{L}$ CF (**B**).

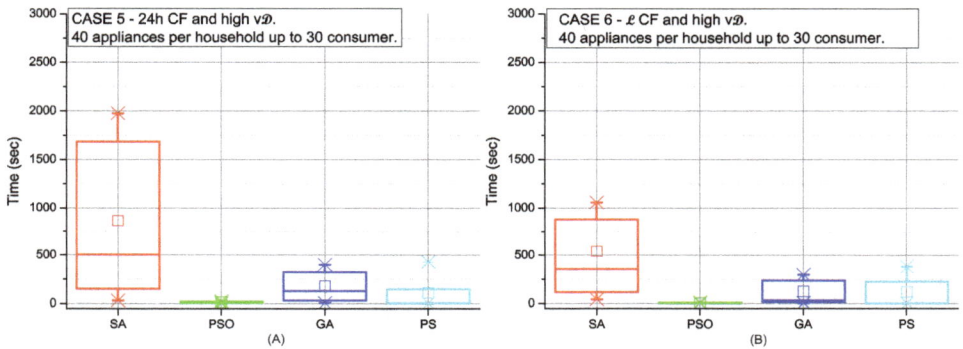

Figure 8. Comparison of the SA, PSO, GA and PS methods for both high variable and fixed consumption in 24 h CF (**A**) and $\mathcal{L}$ CF (**B**).

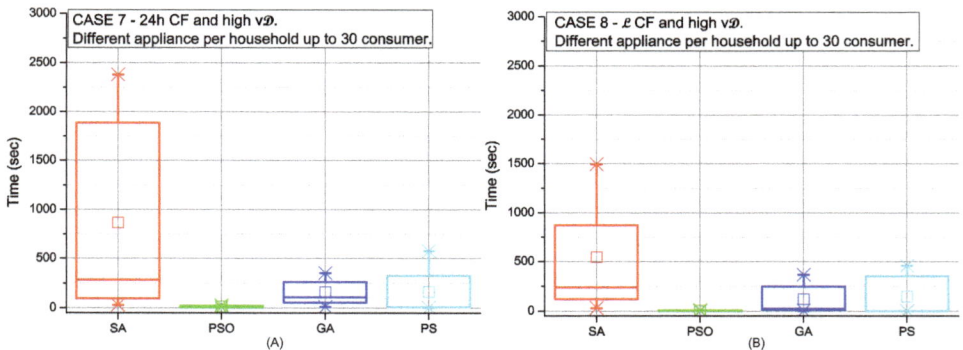

Figure 9. Comparison of the SA, PSO, GA and PS methods for both high variable and fixed consumption in 24 h CF (**A**) and $\mathcal{L}$ CF (**B**).

The CC is higher when consumers have an uneven number of appliances. This effect can be observed in Figure 7 and in comparison with Figure 6. The same occurs in scenarios with high variable demand (see Figure 9 and Figure 8). In addition, a high variable demand (Figures 8 and 9) penalises the CC when compared with low settings (Figures 6 and 7). We find the worst case situation on altruistic communities with high variable demand when applying strategy C under SA optimisation. As Figure 9 (red colour) shows, it takes 30 minutes. Both PSO and GA work with sets of solutions that interact between themselves. Both perform better due to the number of solutions managed at the same time. We can also conclude that strategy C under all possible scenarios can be solved within the next 24 h, being PSO the most computationally efficient for scheduling (28sec).

Additional simulation measures the performance of communities of 20 appliances per consumer in Case 1. Figure 10 (left) compares all the algorithms and shows that PSO achieves a global optimum solution quickly. GA obtains similar outcomes. Applying SA, Figure 10 (right) shows that the cost needed for the scheduling increases linearly with the number of appliances.

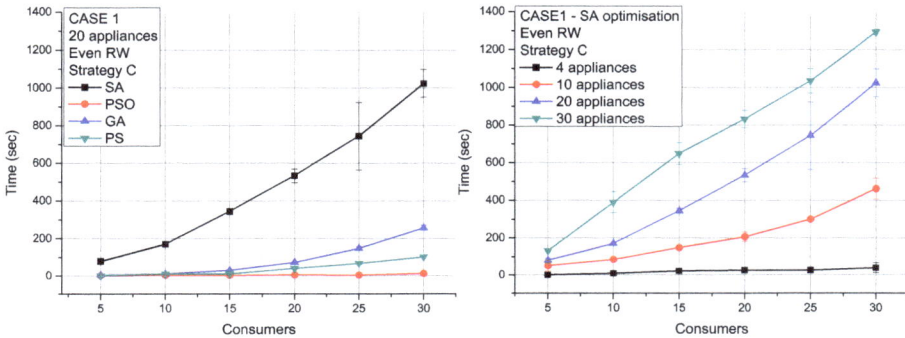

Figure 10. Different approaches (**left**) and appliances number results (**right**).

So far, experiments have been mainly focused on strategy C. Figure 11 depicts the optimised cost obtained after applying all strategies, and taking into account the different factors (see Table 4). These factors are differentiated by branches and data are expressed as a percentage of the required CC.

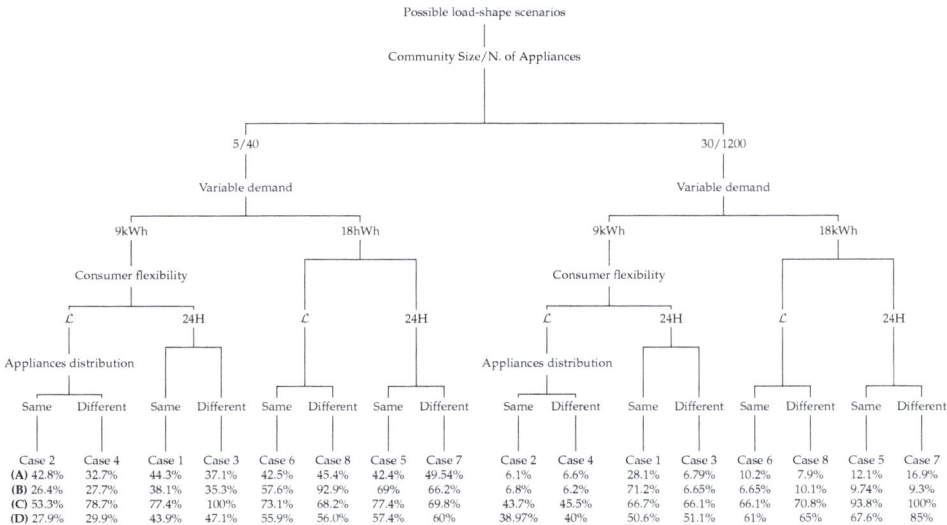

Figure 11. CC results (in %) after applying different strategies (Section 3.3) performed under SA: Round-Robin (**A**), Randomly Round-Robin (**B**), "having the first consumer always the same" (**C**) and randomly (**D**).

Strategy C performs badly, with higher CC in all circumstances. This is mainly because the *Aggregator* needs more resources when it has to optimise all consumers and their appliances all together. The simulation performed with consumers by adding their preferences randomly (strategy D) shows similar cost when compared to strategy C. By contrast, the *Aggregator* under strategy A optimises consumers preferences consecutively in an individual way. A new variant of RR is to perform this strategy when the first consumer starts randomly (strategy B). Both strategies act equally, though dispersed. Strategy A appears as the most appropriate strategy on our system.

Further conclusions can be extracted globally for all the strategies. The CC is higher in four possible situations: when users set a very flexible demand need, when the community is large, when the number of appliances is also large and when a high demand is needed (Figure 11, last branches). A better performance is achieved under strategy B when consumers demand low variable load, in a selfish and small community (Figure 11, first branch). The distribution of appliances also impacts the CC, being higher in large communities with uneven number of appliances per neighbour. The highest CC, which exceeds the half an hour of computation, is obtained in strategy C scenario under a high demand flexibility for an optimisation of 30 neighbours with different appliances distribution per dwelling (Figure 11, last branch).

Finally, Figure 12 compares the CC considering two different $\mathcal{RW}$ vector structure provided from the available sources at the *Utility*: uniform $\mathcal{RW}$ vector and the 50% standard deviation of $\mathcal{RW}$ values. For the eight different cases, and using strategy B and SA, testing is performed for communities of 5–30 consumers. In terms of the chosen strategy, both situations display similar behaviours.

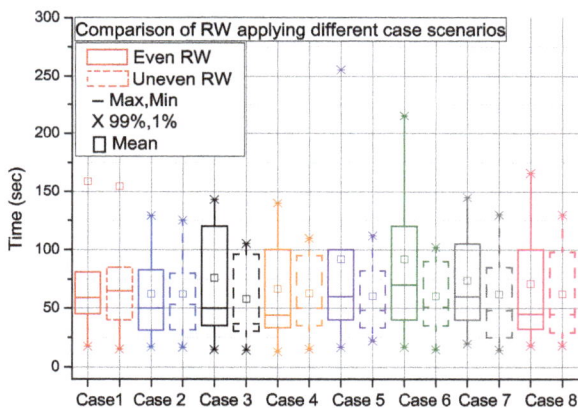

Figure 12. Comparison of RW factor from cases 1 to 8 by applying strategy A under SA.

5. Technical Considerations: Communication, Security and Hardware

Discussion on the development of a pilot testbed for our system over the existent smart home technologies, their security properties and more feasible communication protocols are included in this section. Extensive work on networking infrastructures has been proposed for smart metering data transmission [14,39]. Some approaches focus on fiber-optics for a high-speed data exchange transmission [40], whose deployment cost would only worth when high data transmission rates are required. Power Line Carrier (PLC) is generally applied to computer networks, wired smart meters among other purposes such as remote monitoring and direct control applications offered by utility companies to consumers [41,42]. Note that regulation should be taken into account to allow the use PLC technologies in outdoor deployments as discussed in the work by the authors of [43].

Infrastructures in a Home Area Network (HAN) comprise the communication technologies for deploying HEMS integrating the household appliances. Communication protocols for data transmission between appliances can be provided with a variety of unwired techniques [28] such as (1) *ZigBee*, which offers an adequate range communication with low data rate and power consumption; (2) *Z-Wave*, which has been used for short range communication due to the low latency communication of small data packets in scalable environments; (3) *6LoWPAN*, which can be applied to building automation designs [44] and to home automation architectures [45]; (4) *Bluetooth*, which is widely used to exchange data over short distances in low energy usage and fast data exchange [46]; and (5) *GSM* networks and *WLAN*, which provide low latency robust communications [47].

Neighbourhood Area Network (NAN) connects customers' HEMS on a two-way communication infrastructure responsible for transmitting their demands and time preferences to the *Aggregator*, as well as the traditional control messages and power grid sensing data. Wireless cellular is widely used in this type of scenarios as described in the work by the authors of [14].

A Wide Area Network (WAN) establishes communication between the *Aggregator* and the Utility substations. Distance to cover is in a radius of a thousand meters comprising of two interconnected networks [48]. Protocols *LoRaWan* and *5G* demonstrate high speed, bandwidth and responsiveness while operating on various licensed and unlicensed frequency bands. Moreover, *LPWAN (LoRa)* will fulfil most of the IoT challenges and applications. By contrast, the introduction of *5G* into the IoT world is still slow and other technologies sound more promising at present time. Table 5 summarises the main features of the discussed technologies and includes recommendations on more appropriate application areas.

In terms of security and privacy, HEMS involve the deployment of physical controls, cyber-security countermeasures as well as privacy leakage prevention [49]. In addition, a gateway architecture for high system availability is proposed in the work by the authors of [50]. Anonymous authentication applying zero-knowledge proof of knowledge could be the solution to provide anonymous authentication between consumers and *Aggregator*. The latter needs to guarantee compliance with the General Data Protection Regulation (GDPR). Furthermore, a methodology to assess the security risks within the HAN domain should be developed as in the work by the authors of [51]. Further details can be found in the work by the authors of [52], where the authors explain the different IoT security threats scenarios (e.g., personal information leak) and provide an evaluation method within a situational smart home framework.

Table 6 identifies the most promising hardware platforms to build our HEMS emphasising low-cost, compatibility, scalability, easy programming and lightweight properties [48,53]. Raspberry Pi 3 [54] is a single-board computer with integrated Bluetooth and WiFi module and enough resources to control the smart appliances and send/receive our system messages. The emergence of cheap microcontrollers like the Arduino has enabled the implementation of low-cost HEMSs mainly devoted to obtain the consumption data as to generate demand/load profiles [55]. For example, the work by the authors of [56] designs and implements a remotely controlled, energy-efficient and highly scalable HEMS using Zigbee in Arduino Mega board as a central controller. In [57] it is discussed and evaluated the performance of *BeagleBone blue* for HEMS developments, an open-source hardware platform with similar principles of Raspberry Pi. Similarly, the proposal by the authors of [58] develops a remote monitoring system using "Libelium Waspmot", a modular device that allows the integration of different sensors and radio transceivers. Additionally, deep learning implementations on Field Programmable Gate Array (FPGA) performs fast due to the exploration of parallel computing [59]. Particularly in the work by the authors of [60], Zedboard implementation (Zynq-ARM Cortex-A9 processor) allows the control of unpredictable loads in a deterministic demand management model. In the work by the authors of [61], the algorithm is modelled in Verilog language on a FPGA allowing dynamic reconfiguration of the HAN. A HEMS prototype is developed on a Cubietruck board (Linux based cortex A7 processor) using a WiFi module [62]. It transmits real-time sensing data using TCP/IP communication protocols.

Table 5. Wireless networks.

Technology	Standard	Data Rate	Frequency Band	Power Consumption	Complexity Transmission Range	Strengths	Application Areas	Encryption/Authentication
Bluetooth	IEEE802.15.1	24 Mbps (v3.0)	2.4 GHz	Low	10 m typical	Small networks Security, speed Easy access Flexibility	HAN	Challenge response scheme/CRC32
WiFi	EEE802.11x	11,54 to 300 Mbps outdoor	2.4 GHz 5 GHz	Very high	Up to100 m	Popular in HAN Speed, flexibility	HAN	4-Way handshake/CRC32
Z-Wave	802.11	100Kbps	2.4GHz 868.42 MHz (EU)	Low	30 m indoor; 100 m outdoor	No interferences	HAN, NAN	AES128/32bit home I.D
Zigbee	IEEE802.15.4	256 Kbps	2.4 GHz	Very low	10–100 m	Low cost Low consume Flexible topology	HAN,NAN	ENC-MIC-128 Encrypted key/CRC16
LPWAN	SigFox LoRaWAN NB-IoT	0.3 to 50 kbit/s per channel	915 MHz	Low	10 km in rural settings	Low power Low cost	NAN,WAN	Symmetric key cryptography/AES 128b
6LoWPAN	IEEE802.15.4	250 Kbps	2.4 GHz	Low	Up to 200 m	Low energy use	HAN, NAN	Symmetric key cryptography/AES 128b
GSM/GPRS	ETSI GSM EN 301349 EN 301347	14.4 Kbps (GSM) 114 Kbps (GPRS)	935 MHz Europe 1800 MHz	Low	Several Km	Low cost Signal quality	HAN, NAN WAN	64 bit A5/1 encryption/Session key generation
WLAN	IEEE 802.11	150 Mbps	2.4 GHz Europe	Low	250m	Robustness	HAN, WAN	WEP, WPA, WPA2/Open, Shared EAP
5G	5G Tech Tracker	Up to 20 Gbps	3400-3800 MHz awarding trial licenses (EU)	Very Low	46 m indoor; 92m outdoor	High speed Low latency	HAN, WAN	Symmetric key encryption/Mobility management entity
3G	UMTS	Up to 14.4 Mbps	450,800 MHz 1.9 GHz	Low	Up to 100 m	Fast Data Transfer	HAN,WAN	CDMA2000/Authentication and Key Agreement

Sensors **2019**, *19*, 3973

Table 6. Hardware platforms.

Hardware	Features	Communication Transceivers	Operating System	Power Consumption	Strengths/Weakness
Raspberry Pi 3	1.2 GHz Quad Core BCM2837 64bit CPU 1GB	4 USB, Wi-Fi, Bluetooth, optional ZigBee and Z-Wave	Raspbian Ubuntu Windows 10	1.8 W	Open source platform; Use Python or C++; Cost: 50€
Arduino	32 MHz Micro controller based on ATmega2560 32 kB	WiFi, Bluetooth, ZigBee, GSM	Processing-based	0.2W	Open source platform hardware/software; High flexibility. Cost: 30€; Appliances compatibility
BeagleBone	720 MHz MR Cortex-A8 processor 512 MB	1 USB port, PLC, Bluetooth, Ethernet	Angstrom Linux	1 W	Open source platform similar to Raspberry; Easy setting up; Cost: 90€
RADXA	ROCK Pi 4 is a Rockchip RK3399 based SBC six core ARM processor, 1GB	WiFi, Bluetooth 5.0, USB Port, GbE LAN	Linux	2.3 W	Open source platform; High flexibility; Cost:50€
Libelium Waspmote	14.7 MHz ATmega1281 28 kB	1USB, 802.15.4/ZigBee LoRaWAN,WiFi PRO GSM/GPRS,4G modules	Linux	2 W	High flexibility; Starter kit:200€; ZigBee,WiFi and LoRaWAN support
Xilinx Spartan FPGA	16 Mb SPI flash memory, 100 MHz	Ethernet, USB port	Linux	2 W	SH, Deep Learning, Autonomous System
PYNQ	Embedded systems Xilinx Zynq Systems on Chips (SoCs)	Bluetooth, Ethernet, USB port	Linux	2.3 W	IoT hardware development in Python
Control4Home Automation	Control4Home owners enjoy personalised smart living experiences	Bluetooth, WiFi Z-Wave and ZigBee	Licensed	-	Operation with internet connection; Not user installation
Nexia	Smart home automation system	Z-Wave	Licensed	-	No knowledge of installation required/ Only Z-Wave support; Low compatibility
LG smart appliance	Control key features on LG smart appliances from your smartphone	WiFi	Licensed	-	No knowledge of installation required/ Only for LG appliances; Closed source

In terms of our DR system, and from preliminary design decisions, our prototype will consist of a Raspberry testbed as the main processor of the HEMS, as it offers good computing performance at a very low price. Its interoperability will provide the performance of alternatives protocols such as ZigBee, WiFi or Z-wave. In the proposed architecture, the WiFi wireless communication between the Aggregator, HEMS and appliances can transfer the data at around a hundredth millisecond, a suitable speed for our current application. The proposed centralised DR system (*Aggregator*) will also operate in an open-source HW platform.

6. Conclusions

Globally, smart communities are envisioned more efficient as residents gain autonomy and self-organisation for reducing and shifting any resource consumption. Strategies for energy demand response applied to smart residential communities can lead to improved scenarios of energy efficiency. Consumers have the opportunity to reduce their electricity cost and/or peak-to-average ratio through scheduling their power consumption. In this article, we have described a DR model that integrates the electricity supply available from renewable energy sources into the scheduling process, which is centralised via the community *Aggregator*. We have showed details of community scheduling algorithm implementation and evaluated it in terms of its computational cost. Empirical comparison of our algorithm design on different implementation strategies for player turn selection and optimisation heuristics as well as on a series of case scenarios of community's consumption patterns showed feasible results in all cases (less than 1 minute to compute the rescheduled community vector). Simulations are conducted with data from our own benchmark of appliance power cost. We have also illustrated development decisions of a mobile app for consumers introducing their demands and time preferences. Finally, we included the discussion of the preliminary decisions on the hardware requirements and communication protocols for a pilot deployment. Immediate future work includes the pilot deployment comprising the algorithm/Aggregator running on the most suitable HW platform as well as the home controllers that autonomously activate/deactivate the smart appliances. We also plan to refine the scheduling algorithm as to consider the usage or purpose of the consumption along with the device type. Furthermore, Utilities and Aggregators in possession of the real-time data from microgeneration and other energy harvesting generators would enhance the conceptual demand response model. Finally, a study of the community patterns will be conducted through game theory and evolutionary computation methods.

Author Contributions: Conceptualisation, C.C. and E.P.; methodology, C.C. and E.P.; editing and visualisation, C.C.; Software, C.C.; validation, C.C.; Resources C.C. and E.P.; formal analysis, C.C. and E.P.; investigation, C.C., E.P., I.B. and A.G.; writing—original draft preparation, C.C.; writing—review and editing, C.C., E.P., I.B. and A.G.; supervision, E.P., I.B. and A.G.; project administration, E.P.; funding acquisition, E.P., I.B. and A.G.

Funding: This research was funded by Comunidad de Madrid (Spain) under grants talent attraction 2017. Reference 2017-T1/TIC- 5184.

Acknowledgments: This research was supported by Comunidad de Madrid (Spain) under Grant Talent Attraction 2017 (ENEFF-PILOT http://www3.uah.es/eneffpilot; Reference 2017-T1/TIC- 5184). The authors would like to express our gratitude to Manuel Aleixandre (from Tokyo Institute of Technology) for his cooperation and suggestions during this research.

Conflicts of Interest: The authors declare no conflicts of interest.

Abbreviations

The following abbreviations are used in this manuscript.

$\mathcal{N}$	Consumer number
$\mathcal{A}_i$	Appliance number
i	Consumer identifier
j	Appliance index
a_{ij}	Consumer i's appliance identifier
t	Certain time
$\mathcal{RW}$	24-hour supply vector from renewables
t_{beg}	Earliest start time appliance
t_{end}	Latest final time appliance
t_{sched}	Scheduled start time of appliance
$\mathcal{D}$	Consumer demand
$v\mathcal{D}$	Variable demand
$f\mathcal{D}$	Fixed demand
CF	Consumer Flexibility
$\mathcal{L}$	Duration of the planned operation of appliance a_{ij} in the next day
SH	Smart Home
$HEMS$	Home Energy Manager System
HAN	Home Arena Network
NAN	Neighbour Area Network
WAN	Wide Area Network
IoT	Internet Of Things
ICT	Information and Communication Technologies
SG	Smart Grid
DSM	Demand System Manager
$MILP$	Mixed Integer Linear Programming
SA	Simulates Annealing
PSO	Particle Search Optimisation
GA	Genetics Algorithm
PS	Pattern Search
RR	Round-Robin
PLC	Power Line Carries
CC	Computational Cost

References

1. Steg, L. Promoting household energy conservation. *Energy Policy* **2008**, *36*, 4449–4453. [CrossRef]
2. Lui, T.J.; Stirling, W.; Marcy, H.O. Get Smart. *IEEE Power Energy Mag.* **2010**, *8*, 66–78. [CrossRef]
3. Gubbi, J.; Buyya, R.; Marusic, S.; Palaniswami, M. Internet of Things (IoT): A Vision, Architectural Elements, and Future Directions. *Future Gener. Comput. Syst.* **2013**, *29*, 1645–1660. [CrossRef]
4. Schaffers, H.; Komninos, N.; Pallot, M.; Trousse, B.; Nilsson, M.; Oliveira, A. Smart Cities and the Future Internet: Towards Cooperation Frameworks for Open Innovation. In *The Future Internet*; Lecture Notes in Computer Science; Springer: Berlin, Germany 2011; Volume 6656, pp. 431–446.
5. Palensky, P.; Dietrich, D. Demand Side Management: Demand Response, Intelligent Energy Systems, and Smart Loads. *IEEE Trans. Ind. Inform.* **2011**, *7*, 381–388. [CrossRef]
6. Babayo, A.A.; Anisi, M.H.; Ali, I. A Review on energy management schemes in energy harvesting wireless sensor networks. *Renew. Sustain. Energy Rev.* **2017**, *76*, 1176–1184. [CrossRef]
7. Yang, Z.; Zhou, S.; Zu, J.; Inman, D. High-Performance Piezoelectric Energy Harvesters and Their Applications. *Joule* **2018**, *2*, 642 – 697. [CrossRef]
8. Palomar, E.; Liu, Z.; Bowen, J.P.; Zhang, Y.; Maharjan, S. Component-based modelling for sustainable and scalable smart meter networks. In Proceedings of IEEE International Symposium on a World of Wireless, Mobile and Multimedia Networks, Sydney, Australia, 19 June 2014; pp. 1–6. [CrossRef]

9.   Palomar, E.; Chen, X.; Liu, Z.; Maharjan, S.; Bowen, J. Component-Based Modelling for Scalable Smart City Systems Interoperability: A Case Study on Integrating Energy Demand Response Systems. *Sensors* **2016**, *16*, 1810. [CrossRef] [PubMed]

10.   Ponds, K.T.; Arefi, A.; Sayigh, A.; Ledwich, G. Aggregator of Demand Response for Renewable Integration and Customer Engagement: Strengths, Weaknesses, Opportunities, and Threats. *Energies* **2018**, *11*, 2391. [CrossRef]

11.   Jacobsen, R.H.; Gabioud, D.; Basso, G.; Alet, P.; Azar, A.G.; Ebeid, E.S.M. SEMIAH: An Aggregator Framework for European Demand Response Programs. In Proceedings of the 2015 Euromicro Conference on Digital System Design, Madeira, Portugal 26–28 August 2015; pp. 470–477. [CrossRef]

12.   Correa-Florez, C.A.; Michiorri, A.; Kariniotakis, G. Comparative Analysis of Adjustable Robust Optimization Alternatives for the Participation of Aggregated Residential Prosumers in Electricity Markets. *Energies* **2019**, *12*, 1019. [CrossRef]

13.   Lin, Y.H.; Hu, Y.C. Residential Consumer-Centric Demand-Side Management Based on Energy Disaggregation-Piloting Constrained Swarm Intelligence: Towards Edge Computing. *Sensors* **2018**, *18*, 1365. [CrossRef]

14.   Shareef, H.; Ahmed, M.; Mohamed, A.; Al Hassan, E. Review on Home Energy Management System Considering Demand Responses, Smart Technologies, and Intelligent Controllers. *IEEE Access* **2018**, *6*, 24498–24509. [CrossRef]

15.   Qayyum, F.; Naeem, M.; Khwaja, A.S.; Anpalagan, A.; Guan, L.; Venkatesh, B. Appliance scheduling optimization in smart home networks. *IEEE Access* **2015**, *3*, 2176–2190. [CrossRef]

16.   Agnetis, A.; De Pascale, G.; Detti, P.; Vicino, A. Load scheduling for household energy consumption optimization. *IEEE Trans. Smart Grid* **2013**, *4*, 2364–2373. [CrossRef]

17.   Sarris, T.; Messini, G.; Hatziargyriou, N. Residential demand response with low cost smart load controllers. In Proceedings of the Mediterranean Conference on Power Generation, Transmission, Distribution and Energy Conversion (MedPower 2016), Belgrade, Serbia, 6–9 November 2016; pp. 1–8. [CrossRef]

18.   Logenthiran, T.; Srinivasan, D.; Phyu, E. Particle swarm optimization for demand side management in smart grid. In Proceedings of the 2015 IEEE Innovative Smart Grid Technologies–Asia (ISGT ASIA), Bangkok, Thailand, 3–6 November 2015; pp. 1–6. [CrossRef]

19.   Fadlullah, Z.M.; Quan, D.M.; Kato, N.; Stojmenovic, I. GTES: An optimized game-theoretic demand-side management scheme for smart grid. *IEEE Syst. J.* **2014**, *8*, 588–597. [CrossRef]

20.   Poolla, C.; Ishihara, A.K.; Milito, R. Designing near-optimal policies for energy management in a stochastic environment. *Appl. Energy* **2019**, *242*, 1725–1737. [CrossRef]

21.   Iqbal, Z.; Javaid, N.; Mohsin, S.; Akber, S.; Afzal, M.; Ishmanov, F. Performance analysis of hybridization of heuristic techniques for residential load scheduling. *Energies* **2018**, *11*, 2861. [CrossRef]

22.   Mahapatra, C.; Moharana, A.K.; Leung, V.C.M. Energy Management in Smart Cities Based on Internet of Things: Peak Demand Reduction and Energy Savings. *Sensors* **2017**, *17*, 2812. [CrossRef] [PubMed]

23.   Mohsenian-Rad, A.; Wong, V.W.S.; Jatskevich, J.; Schober, R. Optimal and autonomous incentive-based energy consumption scheduling algorithm for smart grid. In Proceedings of the 2010 Innovative Smart Grid Technologies (ISGT), Gaithersburg, MA, USA, 19–21 January 2010; pp. 1–6. [CrossRef]

24.   Alshehri, K.; Liu, J.; Chen, X.; Bacsar, T. Privacy-Preserving Multi-Period Demand Response: A Game Theoretic Approach. *arXiv* **2017**, arXiv:1710.00145 .

25.   Kelly, F.P.; Maulloo, A.K.; Tan, D.K.H. Rate Control for Communication Networks: Shadow Prices, Proportional Fairness and Stability. *J. Oper. Res. Soc.* **1998**, *49*, 237–252. [CrossRef]

26.   Yaïche, H.; Mazumdar, R.R.; Rosenberg, C. A Game Theoretic Framework for Bandwidth Allocation and Pricing in Broadband Networks. *IEEE/ACM Trans. Netw.* **2000**, *8*, 667–678. [CrossRef]

27.   Fan, Z. A Distributed Demand Response Algorithm and its Application to PHEV Charging in Smart Grids. *IEEE Trans. Smart Grid* **2012**, *3*, 1280–1290. [CrossRef]

28.   Zhou, B.; Li, W.; Chan, K.W.; Cao, Y.; Kuang, Y.; Liu, X.; Wang, X. Smart home energy management systems: Concept, configurations, and scheduling strategies. *Renew. Sustain. Energy Rev.* **2016**, *61*, 30–40. [CrossRef]

29.   Myerson, R.B. Conference structures and fair allocation rules. *Int. J. Game Theory* **1980**, *9*, 169–182. [CrossRef]

30.   Schwarz, D. *Jump Start Adobe XD*, 1 ed.; Sitepoint: Melbourne, Australia, 2017.

31.   Bouveret, S.; Lang, J. Manipulating picking sequences. *ECAI* **2014**, *14*, 141–146.

32. Nahar, S.; Sahni, S.; Shragowitz, E. Simulated Annealing and Combinatorial Optimization. In Proceedings of the 23rd ACM/IEEE Design Automation Conference, Las Vegas, NV, USA, 29 June–2 July 1986; pp. 293– 299. [CrossRef]

33. Goldberg, D.E. *Genetic Algorithms in Search, Optimization and Machine Learning*, 1st ed.; Addison-Wesley Longman Publishing Co., Inc.: Boston, MA, USA, 1989.

34. Abramson, M. Pattern Search Algorithms for Mixed Variable General Constrained Optimization Problems. Ph.D. Thesis, Rice University, Houston, TX, USA, 2003.

35. Kennedy, J.; Eberhart, R. Particle swarm optimization. In Proceedings of the ICNN'95–International Conference on Neural Networks, Perth, WA, Australia, 27 November–1 December 1995; Volume 4, pp. 1942–1948. [CrossRef]

36. *MATLAB*. version 9.5.0 (R2018b); The MathWorks Inc.: Natick, MA, USA, 2010.

37. Laicane, I.; Blumberga, D.; Blumberga, A.; Rosa, M. Evaluation of Household Electricity Savings. Analysis of Household Electricity Demand Profile and User Activities. *Energy Procedia* **2015**, *72*, 285–292. [CrossRef]

38. Ross, J.; Meier, A.K. Whole-House Measurements of Standby Power Consumption. In *Energy Efficiency in Household Appliances and Lighting*; Technical Report; Springer: Berlin, Germany, 2000; pp. 278–285.

39. Andreadou, N.; Guardiola, M.O.; Fulli, G. Telecommunication Technologies for Smart Grid Projects with Focus on Smart Metering Applications. *Energies* **2016**, *9*, 375. [CrossRef]

40. Shakerighadi, B.; Anvari-Moghaddam, A.; Vasquez, J.C.; Guerrero, J.M. Internet of Things for Modern Energy Systems: State-of-the-Art, Challenges, and Open Issues. *Energies* **2018**, *11*, 1252. [CrossRef]

41. Pitì, A.; Verticale, G.; Rottondi, C.; Capone, A.; Lo Schiavo, L. The Role of Smart Meters in Enabling Real-Time Energy Services for Households: The Italian Case. *Energies* **2017**, *10*, 199. [CrossRef]

42. Fang, X.; Wang, N.; Gulliver, T.A. A PLC channel model for home area networks. *Energies* **2018**, *11*, 3344. [CrossRef]

43. Galli, S.; Scaglione, A.; Wang, Z. Power Line Communications and the Smart Grid. In Proceedings of the Power Line Communications and the Smart Grid, Gaithersburg, MD, USA, 4–6 October 2010; pp. 303–308. [CrossRef]

44. Han, S.; H. Cao, Q.; Alinia, B.; Crespi, N. Design, Implementation, and Evaluation of 6LoWPAN for Home and Building Automation in the Internet of Things. In Proceedings of the IEEE/ACS 12th International Conference of Computer Systems and Applications (AICCSA), Marrakech, Morocco, 17–20 November 2015. [CrossRef]

45. Aradindh, J.; Srevarshan, V.B.; Kishore, R.; Amirthavalli, R. Home automation in IOT using 6LOWPAN. *Int. J. Adv. Comput. Eng. Netw.* **2017**, *5*, 26–28.

46. Collotta, M.; Pau, G. A Solution Based on Bluetooth Low Energy for Smart Home Energy Management. *Energies* **2015**, *8*, 11916–11938. [CrossRef]

47. Wen, M.H.; Leung, K.C.; Li, V.O.; He, X.; Kuo, C.C.J. A survey on smart grid communication system. *APSIPA Trans. Signal Inf. Process.* **2015**, *4*, e5. [CrossRef]

48. Saleem, Y.; Crespi, N.; Rehmani, M.H.; Copeland, R. Internet of things-aided smart grid: technologies, architectures, applications, prototypes, and future research directions. *IEEE Access* **2019**, *7*, 62962–63003. [CrossRef]

49. El-hajj, M.; Fadlallah, A.; Maroun, C.; Serhrouchni, A. A Survey of Internet of Things (IoT) Authentication Schemes. *Sensors* **2019**, *19*, 1141. [CrossRef] [PubMed]

50. Lin, H.; Bergmann, N.W. IoT Privacy and Security Challenges for Smart Home Environments. *Information* **2016**, *7*, 44. [CrossRef]

51. Ali, B.; Awad, A. Cyber and Physical Security Vulnerability Assessment for IoT-Based Smart Homes. *Sensors* **2018**, *18*, 817. [CrossRef] [PubMed]

52. Park, M.; Oh, H.; Lee, K. Security Risk Measurement for Information Leakage in IoT-Based Smart Homes from a Situational Awareness Perspective. *Sensors* **2019**, *19*, 2148. [CrossRef] [PubMed]

53. Froiz-Míguez, I.; Fernández-Caramés, T.; Fraga-Lamas, P.; Castedo, L. Design, Implementation and Practical Evaluation of an IoT Home Automation System for Fog Computing Applications Based on MQTT and ZigBee-WiFi Sensor Nodes. *Sensors* **2018**, *18*, 2660. [CrossRef] [PubMed]

54. Qureshi, M.U.; Girault, A.; Mauger, M.; Grijalva, S. Implementation of home energy management system with optimal load scheduling based on real-time electricity pricing models. In Proceedings of the IEEE 7th International Conference on Consumer Electronics—Berlin (ICCE-Berlin), Berlin, Germany, 3–6 September 2017; pp. 134–139. [CrossRef]
55. Amer, M.; M El-Zonkoly, A.; Aziz, N.; M'Sirdi, N. Smart Home Energy Management System for Peak Average Ratio Reduction. *Ann. Univ. Craiova.* **2014**, *38*, 180–188.
56. Baraka, K.; Ghobril, M.; Malek, S.; Kanj, R.; Kayssi, A. Low Cost Arduino/Android-Based Energy-Efficient Home Automation System with Smart Task Scheduling. In Proceedings of the 2013 Fifth International Conference on Computational Intelligence, Communication Systems and Networks, IEEE Computer Society, Washington, DC, USA, 5–7 June 2013; pp. 296–301. [CrossRef]
57. Nayyar, A.; Puri, V. A Review of Beaglebone Smart Board's-A Linux/Android Powered Low Cost Development Platform Based on ARM Technology. In Proceedings of the 9th International Conference on Future Generation Communication and Networking (FGCN), Jeju Island, Korea, 25–28 November 2015; pp. 55–63.
58. Quintana-Suárez, M.A.; Sánchez-Rodríguez, D.; Alonso-González, I.; Alonso-Hernández, J.B. A Low Cost Wireless Acoustic Sensor for Ambient Assisted Living Systems. *Appl. Sci.* **2017**, *7*, 877. [CrossRef]
59. Alhafidh, B.M.H.; Daood, A.I.; Alawad, M.M.; Allen, W. FPGA Hardware Implementation of Smart Home Autonomous System Based on Deep Learning. In Proceedings of the International Conference on Internet of Things, Santa Barbara, CA, USA, 15–18 October 2018; Springer: Berlin, Germany 2018; pp. 121–133.
60. Khoury, J.; Mbayed, R.; Salloum, G.; Monmasson, E. Design and implementation of a real time demand side management under intermittent primary energy source conditions with a PV-battery backup system. *Energy Build.* **2016**, *133*, 122–130. [CrossRef]
61. Bazydło, G.; Wermiński, S. Demand side management through home area network systems. *Int. J. Electr. Power Energy Syst.* **2018**, *97*, 174–185. [CrossRef]
62. Vivek, G.V.; Sunil, M.P. Enabling IOT services using WIFI–ZigBee gateway for a home automation system. In Proceedings of the IEEE International Conference on Research in Computational Intelligence and Communication Networks (ICRCICN), Kolkata, India, 20–22 November 2015, pp. 77–80. [CrossRef]

sensors

MDPI

Article

Radio Frequency Fingerprint-Based Intelligent Mobile Edge Computing for Internet of Things Authentication †

Songlin Chen [1], Hong Wen [2,*], Jinsong Wu [3,4,*], Aidong Xu [5], Yixin Jiang [5], Huanhuan Song [1] and Yi Chen [1]

[1] National Key Laboratory of Science and Technology on Communications, University of Electronic Science and Technology of China, Chengdu 611731, China
[2] School of Aeronautics and Astronautics, University of Electronic Science and Technology of China, Chengdu 611731, China
[3] School of Artificial Intelligence, Guilin University of Electronic Technology, Guilin 541004, China
[4] Department of Electrical Engineering, Universidad de Chile, Av Tupper 2007, Santiago 8370451, Chile
[5] EPRI, China Southern Power Grid Co. Ltd., Guangzhou 510080, China
* Correspondence: sunlike@uestc.edu.cn (H.W.); wujs@ieee.org (J.W.)
† This paper is the expanded version of "A Novel Terminal Security Access Method Based on Edge Computing for IoT" published in the Proceedings of the 2018 International Conference on Networking and Network Applications (NaNA), Xi'an, China, 12–15 October 2018.

Received: 10 June 2019; Accepted: 16 August 2019; Published: 19 August 2019

Abstract: In this paper, a light-weight radio frequency fingerprinting identification (RFFID) scheme that combines with a two-layer model is proposed to realize authentications for a large number of resource-constrained terminals under the mobile edge computing (MEC) scenario without relying on encryption-based methods. In the first layer, signal collection, extraction of RF fingerprint features, dynamic feature database storage, and access authentication decision are carried out by the MEC devices. In the second layer, learning features, generating decision models, and implementing machine learning algorithms for recognition are performed by the remote cloud. By this means, the authentication rate can be improved by taking advantage of the machine-learning training methods and computing resource support of the cloud. Extensive simulations are performed under the IoT application scenario. The results show that the novel method can achieve higher recognition rate than that of traditional RFFID method by using wavelet feature effectively, which demonstrates the efficiency of our proposed method.

Keywords: mobile edge computing; IoT; RF Fingerprinting; authentication

---

1. Introduction

In recent years we have seen an innovative Internet-of-Things (IoT) paradigm, which combines mobile edge computing (MEC) with traditional IoT architecture [1–3]. MEC is used as a bridge between IoT devices and remote cloud devices to provide edge intelligent services to meet the critical needs of industry digitization in terms of agile connectivity, real-time services, data optimization, application intelligence, security and privacy protection, which are key issues for the industry control applications [4–6]. However, such new architecture has aroused many security protection requirements, including security access authentication, security transmission, and data privacy etc., in which the most important one is security access authentication [7]. Due to constraints of terminals under the IoT system, and resource constraints of the existing authentication methods that rely on encryption, some lightweight and effective security access authentication measurements [8,9] are necessary. Recently, many researchers have turned to using the physical (PHY) layer information to enhance wireless security [10–12]. MEC operates on the wireless media. The innovative PHY-layer security designs can cope with the unique PHY layer weakness of the MEC in which physical characteristics, such as the channel responses between communication peers, the hardware property of the wireless transmitter, have been explored as a form of fingerprint in the scenario of wireless security.

Many scholars have made contributions to the development of physical layer security [13–15]. RFFID is of vital importance to physical layer security technology. In fact, radiofrequency fingerprints (RFF), which embody the hardware property of the wireless transmitter to be identified and have the characteristics difficult to be cloned, are a good candidate to be used to enhance device identification [16–21]. Additionally, RFFID is a lightweight authentication method for the transmitters, because the authentication algorithm is mainly performed on receivers and transmitters that do almost nothing. Therefore, it is especially suitable for the source-constrained terminals of IoT to perform access identification.

The RFFID method is different from the device signal authentication method proposed in [22]. The stochastic features of dynamic watermarking signal are used as identity information. Hall et al. [23] first proposed RF fingerprinting technology in Bluetooth wireless network device identification research in 2003. After that, studies have found that the transmitter can also be identified by transmitting the steady-state portion of the signal. Hu et al. [24] utilized RF signals to identify mobile phones in a mobile cellular network. Hall et al. [23] and Ureten et al. [25,26] used RF fingerprinting technology to achieve wireless positioning and access control for wireless network. In order to further improve the authentication rate of RFFID, a machine learning algorithm has been introduced in extensive research as the classification algorithm of RFFID [27–29]. However, a machine learning algorithm needs a certain amount of computing resources to ensure a higher recognition and authentication rate. Especially in offline training, the number of offline training samples will affect the effect of machine learning. With available computing resources, MEC can perform limited tasks in offline training of machine learning. When a large number of samples need to be trained, while uploading these computing cost tasks to cloud computing platform, the authentication rate can be expected to be further improved. In this article, we propose an efficient and flexible RFFID-MEC authentication method, in which RFFID is combined with MEC and the cloud, making full use of the characteristics of MEC-IoT framework to establish the two-layer model. The first layer provides data collection, extraction of RF fingerprint features, dynamic data storage and access authentication decision, which consumes less limited computing resources running at the MEC platform. The second-layer provides powerful computing for more complex and resource-consuming tasks in the remote cloud, such as feature learning, generating decision model, and establishing a machine learning algorithm. Since the authentication algorithm is mainly performed on the MEC, the terminals do almost nothing. The novel model, the edge computing, and cloud computing work collaboratively to ensure that the method has strong computing resources and improve the authentication rate. Compared with the conventional physical authentication method [10–12,30–33], our method makes efficient use of the characteristics of edge computing to collect transmitting signals and computing support of the cloud, and performs the fast identity authentication of terminals in the IoT scenario with asymmetric computing resources. Therefore, our proposed novel authentication scheme is light-weight to the IoT terminals. Our contributions can be summarized as follows:

(1) To the best of our knowledge, we are the first to propose the radio frequency fingerprint-based authentication, that combines physical characteristics of wireless device radio frequency and machine learning algorithms under the collaborative work of edge computing and cloud computing to achieve fast and efficient authentication.

(2) We present the typical scenario that uses an RFFID-MEC method for IoT devices authentication applications and demonstrate the effectiveness of the algorithm.

The rest of this article is organized as follows. In Section 2, we introduce the related work about the background information of MEC-IoT architecture and RFFID. The secure access authentication method based on RFFID- MEC is proposed in Section 3. Section 4 includes the application of the novel method to typical scenario and the evaluation of the proposed methods via experiments. Finally, the conclusions are given in Section 5.

2. Background of RFFID-MEC

2.1. MEC Architecture in IoT

The emergence of MEC brings a high-performance computing platform that provides data preprocessing, storage, and edge intelligent services [2]. As shown in Figure 1, the MEC-IoT architecture encompasses three different layers, the IoT devices, MEC, and the remote cloud platform. Each layer is characterized by different constraints on computation ability, memory, and energy availability. Among them, the computation ability of IoT terminals is the weakest. The MEC layer can perform limited tasks with available computing resources and the cloud layer provides strong computing support to the other two layers.

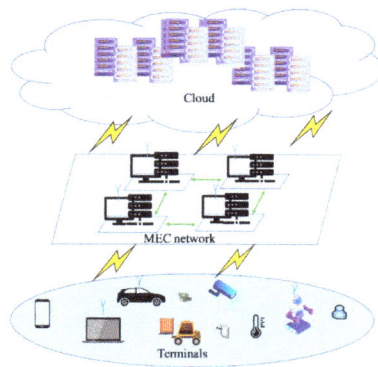

Figure 1. Mobile edge computing-Internet-of-Things (MEC-IoT) architecture.

A key transformation is to perform information processing based on servers at network edge, applying the concepts of cloud computing. MEC can be seen as a cloud server running at the edge of a mobile network and performing specific tasks that are necessary to some scenarios, such as agile connectivity, real-time services, security and privacy protection, which could not be achieved with traditional IoT network infrastructure [1]. In addition, IoT devices are connected with MEC that provide (when needed) the computational resources for more complex and resource-demanding application or processing tasks. MEC devices are interconnected through MEC networking and linked to the remote cloud depending on application needs. As the traditional cloud-centered IoT architecture, MEC-IoT architecture is also confronted with security problems. Meanwhile, access authentication is envisioned as the primary problem to be solved. As a physical layer security technology, RFFID is regarded as a lightweight access authentication method, and applicable to the MEC-IoT architecture.

2.2. Radio Frequency Fingerprinting Identification (RFFID)

The RFFID method refers to identification based on the radio frequency signal fingerprint of the wireless devices to confirm the access of the legal wireless devices, thereby realizing the identity authentication of the wireless devices. The RFFID, which embodies the hardware property of the wireless transmitters, is difficult to be cloned and can be used for non-cryptographic authentication for the wireless transmitters. Cobb et al. [30] made a further explanation of the mechanism of RFF and introduced electronic component tolerances due to differences in hardware devices, such as printed circuit board traces, integrated circuit internal components, and RF front-end circuits. The electronic component tolerance effect of wireless transmitters is the main reason for generating RFF. Since the hardware of any two wireless devices is different and hard to be faked, it is feasible to uniquely identify electronic components by RF signal fingerprinting.

As shown in Figure 2, the RFFID method consists of six steps: Signal collection, signal analysis and process, feature extraction and classification, fingerprint database, and identification. RFFID mainly includes two processes. The first one is offline to establish a fingerprint database for legitimate wireless devices by implementing, analyzing and processing the radiation signals after collecting the signals of legitimate devices. The second process is an online authentication process. The signals of the wireless devices to be identified are collected and the fingerprint features are extracted through signal analysis and processing. Then, matching and recognition are carried out in the existing legitimate fingerprint database.

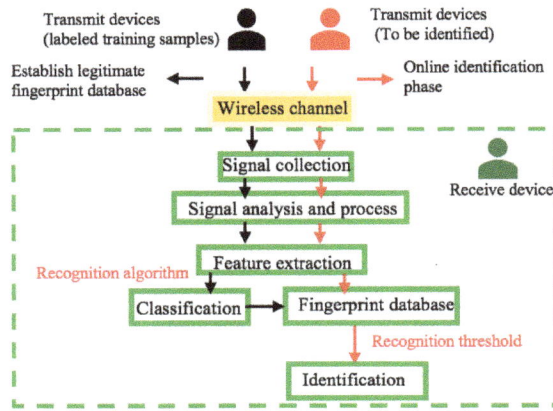

Figure 2. Radio frequency fingerprinting identification (RFFID) authentication method.

Recently, in order to further improve the authentication rate of RFFID, machine learning method is taken as a recognition algorithm [27–29]. However, this method consumes resources in offline samples learning due to a large number of samples training required to ensure the authentication effect. Otherwise, the authentication rate will be compromised if only limited training samples are used. Therefore, abundant computing resources are necessary to guarantee the authentication rate of RFFID. When the computing power of MEC devices is inadequate, it can upload tasks to the cloud platform. By this means, MEC can rationally get computational resources to support and fully ensure the accuracy of tasks. Therefore, a combination of the RFFID method, MEC and cloud can strengthen hardware resource guarantee.

3. Security Access Authentication Method Based on RFFID-MEC

In this section, we propose a lightweight algorithm for resource-constrained terminals to accomplish access authentication with a satisfied authentication rate. This algorithm that combines the mobile edge computing with the cloud may improve the accuracy of the authentication-based RFFID. The architecture of the proposed RFFID-MEC authentication consists of two layers: The first layer provides signal collection, signal analysis, and process, feature extraction and classification, and establish fingerprint feature database, which will be performed on the MEC layer. The second layer includes learning features, generating decision models, and implementing machine learning algorithms for recognition, which need the powerful computing support for much more complex and resource-consuming tasks, will be implemented on the remote cloud due to the limited computing resources of MEC. Figure 3 shows a detailed logical flow of this authentication process.

The authentication process includes two processes that are an offline training process and an online decision-making process.

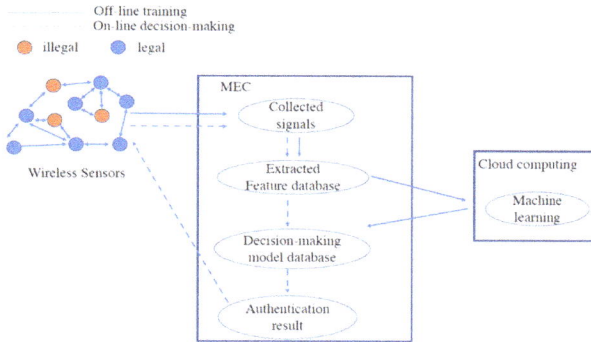

Figure 3. Detailed authentication process.

In the offline training authentication process, a terminal initiates an access request to the MEC platform. After that, the MEC platform collects the signal of the terminal with the identity information, and then performs feature extraction and establishes dynamic fingerprint feature database. The feature information is transmitted to the cloud computing platform. The cloud computing platform makes use of the machine learning algorithm to generate the authentication decision-making model, and transmits the resulted decision model that meets the target authentication rate back to the MEC platform. At this time, the offline training authentication process ends. In the online decision-making authentication process, a terminal initiates an access request to the MEC platform, and the MEC platform collects signals, extracts features, and performs fast identity authentication through the trained authentication model that was established from the previous step.

The RFFID-MEC algorithm is illustrated in Figure 4. Notations of frequently-used variables are described in Table 1 for steps. Steps 1–3, 5, 6 are carried out by the first layer, meanwhile Step 4, including Steps 4.1–4.6, are carried out by the second layer.

Table 1. Notations of frequently-used variables.

Symbol	Description
i	The i-th terminal
N	Discrete points of signal acquisition
$x_i^{<l>T}$	The l-th collection of the i-th terminal's vector
$X_i^{<l>T}$	The total l-th times collection of the i-th terminal's set
$x_i^{<m>T}$	The vector after remove the outline from the $x_i^{<l>T}$
$X_i^{<m>T}$	The set after remove the outline from the set $X_i^{<l>T}$
$x_i^{<m>T}$	The data normalization of vector $x_i^{<m>T}$
$X_i^{<m>T}$	The data normalization of set $X_i^{<m>T}$
$\overline{x_i^{<m>T}}$	The vector $\overline{x_i^{<m>T}}$ generated after DTWT
$\overline{X_i^{<m>T}}$	The set $\overline{X_i^{<m>T}}$ generated after DTWT
T	The training data set
y_i	The category of the instance

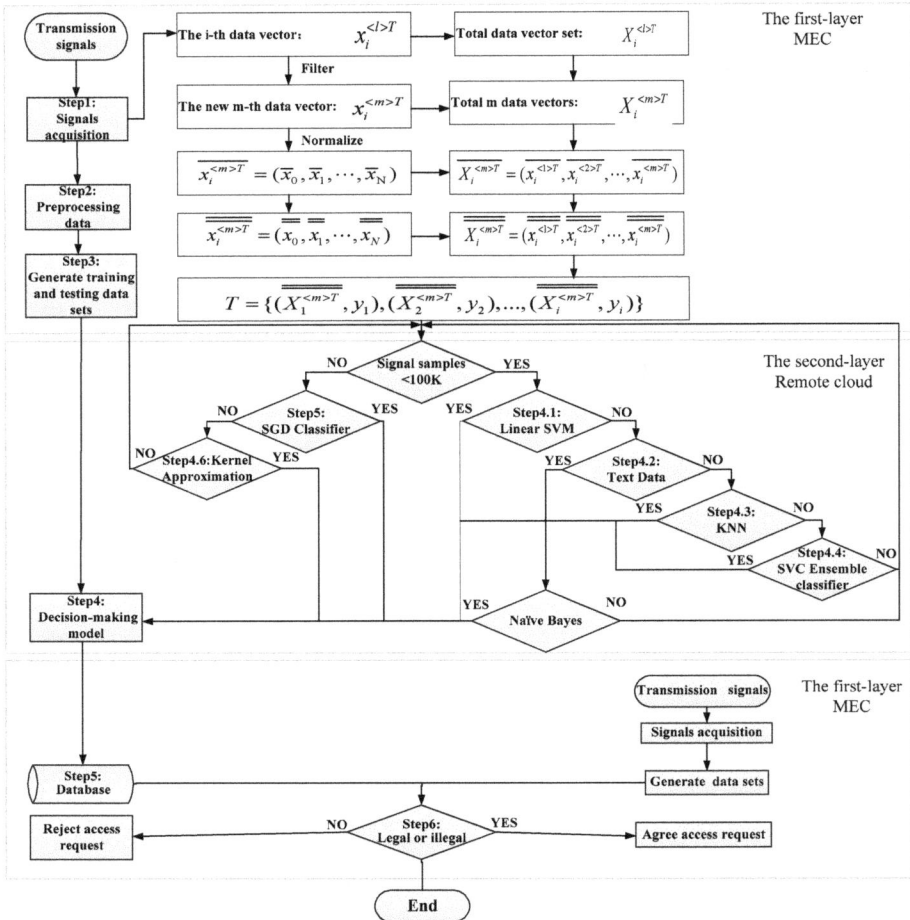

Figure 4. Flow chart of RFFID-MEC method algorithm.

Step 1: The MEC platform continuously acquires signals:

The MEC platform collects the RF signals of the IoT devices with identity tags:

1. The vector of the $l-th$ collection of the $i-th$ terminal device is: $x_i^{<l>T} = (x_0, x_1, \dots, x_N)$, (N represents the discrete sample points of the collected signals)

2. The data set of the total L acquisitions of the $l-th$ terminal devices is: $x_i^{<l>T} = (x_i^{<1>T}, x_i^{<2>T}, \dots, x_i^{<L>T}), l = (1, 2, \cdots, L)$

Step 2: Data preprocessing in MEC platform:

The MEC platform preprocesses data sets for filtering and normalizing.

1. According to the data set, we obtain the mean $E(X_i^{<l>T})$ and, standard deviation $\sigma_{X^{<l>T}}$, and remove the outliers from the data set $X_i^{<l>T}$. Then $x_i^{<l>T}$ and, $X_i^{<l>T}$ were changed to: $x_i^{<m>T} = (x_0, x_1, \cdots, x_N)$ and, $X_i^{<m>T} = (x_i^{<1>T}, x_i^{<2>T}, \cdots, x_i^{<m>T}), m = (1, 2, \cdots, M), M < L.$

2. $x_i^{<m>T} = (x_0, x_1, \cdots, x_N)$ was normalized to new value:

$$\bar{x} = \frac{1}{N} \sum_{i=1}^{N} x_i, i = 1, 2, 3, \cdots, N \tag{1}$$

$$\sigma^2 = \frac{1}{N-1} \sum_{i=1}^{N} (x_i - \bar{x})^2 \tag{2}$$

$$\overline{x_i} = \frac{x_i - \bar{x}}{\sigma} \tag{3}$$

$x_i^{<m>T}$ and, $X_i^{<m>T}$ were changed to:

$$\overline{x_i^{<m>T}} = (\overline{x_0}, \overline{x_1}, \cdots, \overline{x_N})$$

and

$$\overline{X_i^{<m>T}} = (\overline{x_i^{<1>T}}, \overline{x_i^{<2>T}}, \ldots, \overline{x_i^{<m>T}})$$

where $\overline{x_i^{<m>T}}$ has a standard normal distribution with mean zero and unit variance.

Step 3: The MEC platform generates training and testing data sets:

The normalized data sets $\overline{X_i^{<m>T}}$ are used by MEC platform to generate the feature vector as the training and testing data sets T as follows:

$\overline{x_i^{<m>T}}$ are changed to $\overline{\overline{x_i^{<m>T}}}$

$$\varphi_{ik(n)} = f_i(n - 2^{i+1}k), (i = 0, 1, \cdots, J-2)$$
$$\overline{\overline{x_i^{<m>T}}} = \sum_i \sum_k \overline{x_i^{<m>T}} \varphi_{ik}(n) \tag{4}$$

$\overline{X_i^{<m>T}}$ are changed to $\overline{\overline{X_i^{<m>T}}}$.

$$\overline{\overline{x_i^{<m>T}}} = (\overline{\overline{x_0}}, \overline{\overline{x_1}}, \cdots, \overline{\overline{x_N}})$$

$$\overline{\overline{X_i^{<m>T}}} = (\overline{\overline{x_i^{<1>T}}}, \overline{\overline{x_i^{<2>T}}}, \cdots, \overline{\overline{x_i^{<m>T}}})$$

T is the final generated training data sets given by:

$$T = \{(\overline{\overline{X_i^{<m>T}}}, y_1), (\overline{\overline{X_2^{<m>T}}}, y_2), \ldots, (\overline{\overline{X_i^{<m>T}}}, y_i)\}$$

$$m = (1, 2, \ldots, M), y_i \in Y = \{+1, -1\}$$

(+1 is represented as a legal terminal device, −1 is an illegal terminal device.)

Step 4: The cloud platform generates a decision-making model.

Step 4.1: When the number of sample data <100 K, we will choose a support vector machine (SVM) classification algorithm to generate a decision-making model:

$$\min \frac{1}{2} \sum_{i=1}^{N} \sum_{j=1}^{N} \alpha_i \alpha_j y_i y_j K(x_i, x_j) - \sum_{i=1}^{N} \alpha_i$$
$$s.t \sum_{i=1}^{N} \alpha_i y_i = 0, C \geq \alpha_i \geq 0, i = 1, 2, \ldots, N \tag{5}$$

The decision-making model using the linear kernel function $K(\vec{x}, \vec{z}) = \vec{x} \cdot \vec{z}$, is applied to the linear classification of large data sets to find the optimal solution: $\vec{a}^* = (a_1^*, a_2^*, \cdots, a_N^*)^T$, $\vec{w}^* = \sum\limits_{i=1}^{N} a_i^* y_i \vec{x}_i$, choosing $C > a_j^* > 0$, $b^* = y_j - \sum\limits_{i=1}^{N} a_i^* y_i K(\overline{\overline{X_i^{<m>T}}}, \overline{\overline{X_j^{<m>T}}})$.

The decision-making model is defined by:

$$f(\vec{x}) = sign(\sum_{i=1}^{N} a_i^* y_i K(\overline{\overline{X_i^{<m>T}}}, \overline{\overline{X_j^{<m>T}}}) + b^*) \tag{6}$$

According to the training data sets testing model, if it can satisfy the correct target recognition rate, the current model is the decision-making model and transmitted to the MEC platform database, otherwise the algorithm will jump into Step 4.2.

Step 4.2: The cloud platform determines whether the data sets are text data. (a) If they are text data, using Naive Bayes, which can achieve the correct target recognition rate, then the current model is the decision-making model and transmitted to the MEC platform database, otherwise the algorithm will jump to Step 4. (b) If it is not text data, the algorithm will jump to Step 4.2.

Step 4.3: The cloud platform uses the (k-nearest neighbor) KNN classification algorithm to determine whether the correct recognition rate is greater than the preset one.

(a) Input the training data set T:

$$T = \{(\overline{\overline{X_1^{<m>T}}}, y_1), (\overline{\overline{X_2^{<m>T}}}, y_2), \cdots, (\overline{\overline{X_i^{<m>T}}}, y_i)\}$$

$\overline{\overline{X_i^{<l>T}}} \in \chi \subset R^n$ is the feature of the instance, and $y_i \in Y = \{+1 - 1\}$ is the category of the instance.

(b) Calculate the Euclidean distance:

$$L_p(\overline{\overline{X_i^{<m>T}}}, \overline{\overline{X_j^{<m>T}}}) = (\sum_{m=1}^{M} \left| \overline{x_i^{<m>T}} - \overline{x_j^{<m>T}} \right|^2)^{\frac{1}{2}} \tag{7}$$

(c) Find the k samples closest to $\overline{x_i^{<m>T}}$ in the training data sets T, let the neighborhood of this t point be $N_t(\overline{x_i^{<m>T}})$.

(d) Determine the category of $\overline{x_i^{<m>T}}$ in $N_t(\overline{x_i^{<m>T}})$, according to the classification decision is y_i:

$$y = \underset{C_j}{argmax} \sum_{\overline{x_i^{<m>T}} \in N} I(y_s = c_j) \tag{8}$$
$$(s = 1, 2, \cdots, M * i; j = 1, 2)$$

where I is the indicator function. If $y_s = c_j$, then $I(y_s = c_j)$ is 1. By a similar argument, if $y_s \neq c_j$, then $I(y_s = c_j)$ is 0.

According to the training data sets testing model, if it can satisfy the correct recognition of the target, the current model is the decision model and transmitted to the MEC platform database, otherwise the algorithm will jump to Step 4.4.

Step 4.4: The cloud platform uses the integrated classifier to determine whether the correct recognition rate is greater than the preset. Integrated classifier, using a variety of existing learning algorithms from the training data to generate individual learners and based on Adaboost binary classification algorithm process, is as follows:

(a) Input the training data set T:

$$T = \{(\overline{\overline{X_1^{<m>T}}}, y_1), (\overline{\overline{X_2^{<m>T}}}, y_2), \cdots, (\overline{\overline{X_i^{<m>T}}}, y_i)\}$$

(b) Initialize the weight distribution of training data

$$D_1 = (w_{11}w_{12}\cdots w_{1q}\cdots, w_{1i}), w_{1q} = \frac{1}{i}q = 1, 2, \cdots, i$$

(c) Use the $D_h(h = 1, 2, \cdots, H)$ training data set with weights to learn to get the basic classifier

$$G_h(\overline{\overline{x_i^{<m>T}}}) : X \rightarrow \{-1, +1\}$$

Calculate the classification error rate of $G_h(\overline{\overline{x_i^{<m>T}}})$ on the training data set given by:

$$\begin{aligned} e_h &= P(G_h(\overline{\overline{x_i^{<m>T}}}) \neq y_i) \\ &= \sum_{q=1}^{i} w_{hq}I(G_h(\overline{\overline{x_i^{<m>T}}}) \neq y_i) \end{aligned}$$

Calculate the coefficients of $G_h(\overline{\overline{x_i^{<m>T}}})$

$$\alpha_h = \frac{1}{2}\log\frac{1-e_h}{e_h} \tag{9}$$

Update the weight distribution of the training data sets:

$$\begin{aligned} D_{h+1} &= (w_{h+1,1}\cdots w_{h+1,q}\cdots, w_{m+1,i}) \\ w_{h+1,q} &= \frac{w_{hq}}{z_h}\exp(-\alpha_h y_q G_h(\overline{\overline{x_i^{<m>T}}})) \\ (q &= 1, 2, \cdots, i) \end{aligned} \tag{10}$$

z_h is a normalization factor:

$$z_h = \sum_{q=1}^{i} w_{hq}\exp(-\alpha_h y_q G_h(\overline{\overline{x_i^{<m>T}}})) \tag{11}$$

It makes D_{h+1} a probability distribution.

(d) Build a linear combination of basic classifiers:

$$f(x) = \sum_{h=1}^{H}\alpha_h G_h(\overline{\overline{x^{<m>T}}}) \tag{12}$$

(e) Get the final classifier:

$$\begin{aligned} G(\overline{\overline{x^{<m>T}}}) &= \text{sign}(f(x)) \\ &= \text{sign}(\sum_{h=1}^{H}\alpha_h G_h(\overline{\overline{x^{<m>T}}})) \end{aligned} \tag{13}$$

According to the testing set of test models, if it can satisfy the correct target recognition rate, the current model is the decision-making model and transmitted to output the MEC platform database, otherwise the algorithm will jump to Step 4.

Step 4.5: When the number of sample data is greater than 100 K, the SVM classification algorithm based on stochastic gradient descent is selected and the cost model is optimized by stochastic gradient descent method given by:

$$
\begin{aligned}
J(\theta) &= \frac{1}{s}\sum_{i=1}^{s}\frac{1}{2}(y^i - h_\theta(x^i))^2 \\
&= \frac{1}{s}\sum_{i=1}^{s}\cos t(\theta,(x^i,y^i))
\end{aligned}
\tag{14}
$$

The final decision-making model is an SVM algorithm based on the multi-class linear kernel. According to the testing data sets testing model, if it satisfies the correct target recognition rate, the current model is a decision-making model and transmitted to the MEC platform database, otherwise the algorithm will jump to Step 4.6 and continue to select the classification algorithm.

Step 4.6: Kernel approximation is a nonlinear classification model. Nonlinear SVM is a classification model based on linear SVM. Different kernel functions are used to realize the transformation of high-dimensional space map to low-dimensional space. The optional kernel functions are:

$$
\begin{aligned}
k(\vec{x},\vec{z}) &= (\gamma(\vec{x}\vec{z}+1)+r)^p \\
k(\vec{x},\vec{z}) &= \exp(-\gamma\|\vec{x}-\vec{z}\|^2) \\
k(\vec{x},\vec{z}) &= \tanh(\gamma(\vec{x}\vec{z})+r) \\
k(x,y) &= \sum_i \frac{2x_iy_i}{x_i+y_i} \\
k(x,y) &= \prod_i \frac{2\sqrt{x_i+c}\sqrt{y_i+c}}{x_i+y_i+2c}
\end{aligned}
\tag{15}
$$

According to the testing set test model, if it satisfies the correct target recognition rate, the current model is the decision-making model and transmitted to the MEC platform database, otherwise the algorithm will jump to Step 4.

Step 5: The MEC platform stores the decision-making model and data set in the database.

Step 6: The MEC platform implements access authentication to determine whether it is legal.

Step 7: The MEC platform continuously collects the RF signals of the IoT devices with identity tags: The MEC platform collects the signal and preprocesses the data, and then passes the processed data and the training data set in the database through the decision model to judge whether the terminal identity is legal. If it is the legal device, MEC platform consents the access request. If it is not legal, the MEC platform refuses to access the request.

The main features of the RFFID-MEC architecture are:

1. Low-complexity: There is no need for encryption algorithm at the terminal node, and all the identification algorithms are completed by MEC. Therefore, the novel authentication method is especially beneficial to the terminals that are resource-constrained.

2. Low-latency: As the decision-making model has been generated by cloud computing and transmitted to MEC platform, it considerably reduces decision latency. This becomes particularly important for IoT scenarios, for example, when dealing with a large number of legitimate users' access requests that need low latency and real-time access authentication such as a driverless scenario.

3. Universality: This method is suitable for interconnection of resource-constrained IoT devices in 5G networks. Meanwhile, it has the characteristics of low computational complexity and high authentication accuracy.

4. RFFID-MEC Authentication Method Evaluation

We demonstrate a typical application scenario of RFFID-MEC Authentication method, as illustrated in Figure 5. IoT terminals are many NRF24Les nodes. More specifically, NRF24LE is a single RF transceiver chip and the operating frequency range is from 2.4 to 2.525 GHz. Its internal components include frequency synthesizer, power amplifier, crystal oscillator, GFSK modulator, and filters. NRF24LE chip is characterized by small power consumption, monolithic and small size. It is widely used in home automation and factory control [34]. MEC platform was composed of Universal Radio Software Peripheral (USRP). USRP is an open-source software-defined radio platform, which is consisted of a mother-board equipped with a dual 14-bit analog to digital converter (ADC) operating at 100 MHz and dual 16-dit digital to analog converter (DAC) operating at 400 MHz, and two UBX160 daughter boards and vert2450 antennas. UBX160 transceiver daughter boards that act as a front end and have a frequency range from 10 MHz to 6 GHz, which allows transmitting and receiving in the 2.4 GHz industrial, scientific, and medical radio band (ISM band) [35]. Cloud server is taken as a cloud platform.

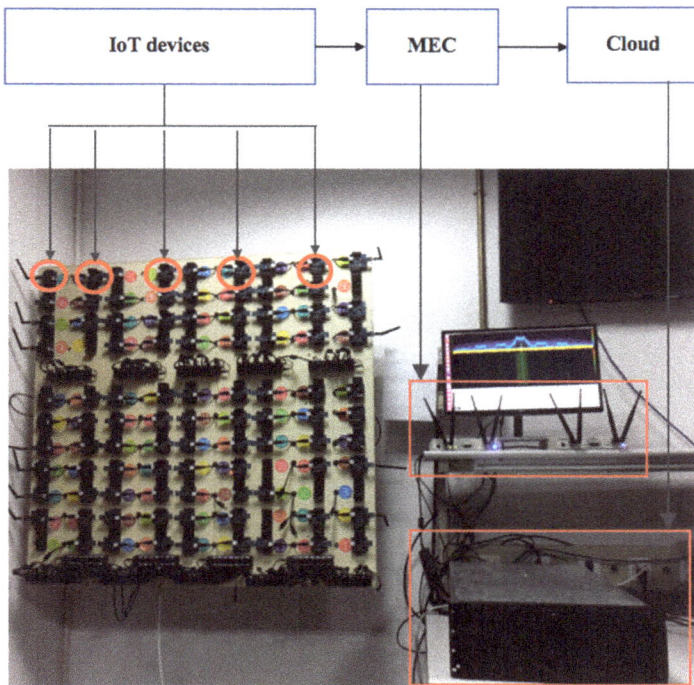

Figure 5. Typical application scenarios of RFFID-MEC authentication method.

There are several steps in our experiment:

The steps of the offline training authentication are as follows.

Step 1: Two terminal nodes with identity tags (illegal, legal) send signals to MEC platform, then the MEC platform collects the signals;

Step 2: The MEC platform preprocesses the collected signals to establish a fingerprint feature database;

Step 3: The MEC platform generates training and testing data sets, which are transmitted to the cloud platform;

Step 4: The cloud platform performs the training processing and generates a decision-making model, which is transmitted back to the MEC platform;

Step 5: The MEC platform stores the decision-making model. Online decision-making authentication includes one step as follows:

Step 6: Terminals (illegal, legal) send signals to the MEC platform. The MEC platform generates feature data sets and determines whether it is legal or not via a previously trained model.

In addition, we have compared RFFID-MEC with traditional RFFID methods. Our simulation utilizes each of four kinds of RF fingerprint features to verify the authentication effect under different SNR. Compared with the traditional RFFID method, the RFFID-MEC method takes advantage of the cloud computing platform to increase the number of offline training samples in a machine learning algorithm. As shown in Figure 6, from the four simulation results, it can be seen that whether using envelope, phase, STFT, or wavelet feature, the correct identification probability of RFFID-MEC method is higher than that of RFFID method at different SNR. Besides, the simulation indicates that the correct identification probability of wavelet feature clearly outperforms the other ones and achieves a higher correct identification rate at low SNR, because the fingerprint of wavelet transform possesses strong anti-noise characteristics [36]. Therefore, we demonstrate the effectiveness of the proposed RFFID-MEC method choosing wavelet RF feature, which can be applied to security authentication.

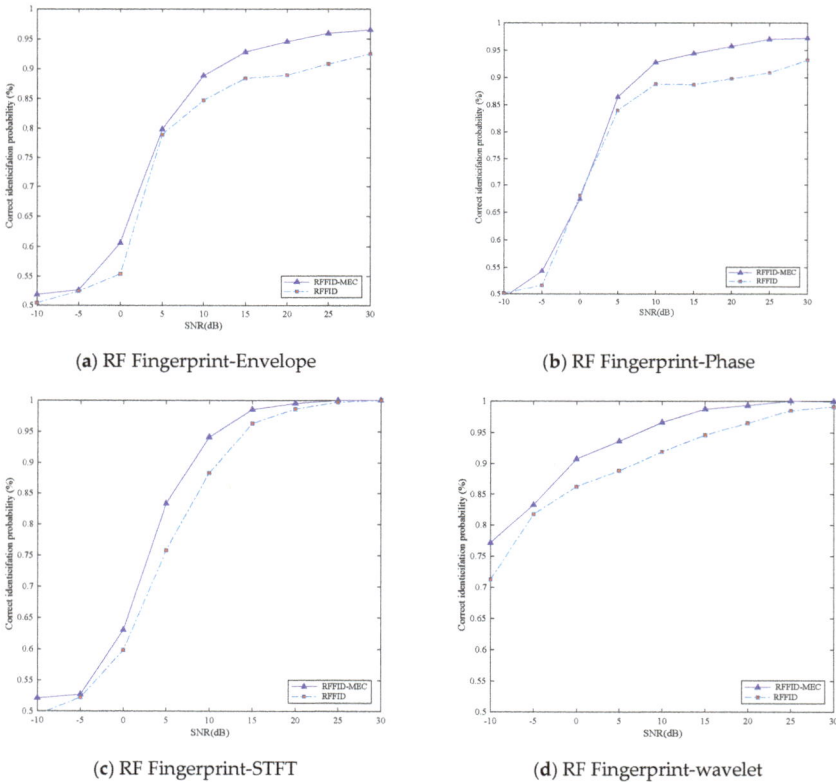

(**a**) RF Fingerprint-Envelope

(**b**) RF Fingerprint-Phase

(**c**) RF Fingerprint-STFT

(**d**) RF Fingerprint-wavelet

Figure 6. Correct identification probability versus SNR for RFFID-MEC and RFFID using four different RF fingerprint features including: Envelope, phase, STFT, and wavelet feature.

Sensors **2019**, *19*, 3610

5. Conclusions

The paper has developed a lightweight RFFID-MEC authentication method by taking advantage of attributes-based MEC, cloud, and non-encryption RFFID for IoT terminals. The presented two-layer model is extremely suitable for the MEC-based IoT paradigms. Compared with the traditional RFFID security access authentication, our light-weight RFFID-MEC authentication method has achieved higher authentication accuracy and improved the work efficiency of IoT terminals, in which all the computing burdens are taken by the edge devices and the cloud. Subsequently, we put this method into an application scenario. Our simulations have demonstrated the effectiveness of this method in actual IoT environments.

Author Contributions: S.C. and H.S. contributed to the main results and code implementation. H.W. and J.W. organized the work, and revised the draft of the paper. S.C. and H.W. designed the experiments. S.C., H.S. and Y.C. performed the experiments. S.C. and H.W. analyzed the experimental results. S.C., H.W., A.X., Y.J., H.S., and Y.C. discussed the results. S.C. wrote the original manuscript.

Funding: This research was supported by National Key R&D Program of China (2018YFB0904900, 2018YFB0904905). This research was also supported in part by Chile CONICYT FONDECYT Regular Project 1181809.

Conflicts of Interest: The authors declare no conflicts of interests.

References

1. Chiang, M.; Zhang, T. Fog and IoT: An overview of research opportunities. *IEEE Internet Things J.* **2016**, *3*, 854–864. [CrossRef]
2. Mao, Y.; You, C.; Zhang, J.; Huang, K.; Letaief, K.B. A survey on mobile edge computing: The communication perspective. *IEEE Commun. Surv. Tuts.* **2017**, *19*, 2322–2358. [CrossRef]
3. Xie, Y.; Wen, H.; Wu, B.; Jiang, Y.; Meng, J. A Modified Hierarchical Attribute-Based Encryption Access Control Method for Mobile Cloud Computing. *IEEE Trans.Cloud Comput.* **2015**, *7*, 383–391. [CrossRef]
4. Zhang, K.; Leng, S.; He, Y.; Maharjan, S.; Zhang, Y. Mobile edge computing and networking for green and low-latency internet of things. *IEEE Commun. Mag.* **2018**, *56*, 39–45. [CrossRef]
5. Chen, S.; Wen, H.; Wu, J.; Lei, W.; Hou, W.; Liu, W.; Xu, A.; Jiang, X. Internet of Things Based Smart Grids Supported by Intelligent Edge Computing. *IEEE Access* **2019**, *7*, 74089–74102. [CrossRef]
6. Pan, F.; Pang, Z.; Wen, H.; Luvisotto, M.; Xiao, M.; Liao, R.F.; Chen, J. Threshold-Free Physical Layer Authentication Based on Machine Learning for Industrial Wireless CPS. *IEEE Trans. Indus. Inf.* **2019**. [CrossRef]
7. Pan, F.; Pang, Z.; Luvisotto, M.; Jiang, X.; Jansson, R.N.; Xiao, M.; Wen, H. Authentication Based on Channel State Information for Industrial Wireless Communications. In Proceedings of the IECON the 44th Annual Conference of the IEEE Industrial Electronics Society, Washington, DC, USA, 21–23 October 2018.
8. Jing, Q.; Vasilakos, A.V.; Wan, J.; Lu, J.; Qiu, D. Security of the internet of things: Perspectives and challenges. *Wirel. Netw.* **2014**, *20*, 2481–2501. [CrossRef]
9. Zhang, K.; Liang, X.; Lu, R.; Shen, X. Sybil attacks and their defenses in the internet of things. *IEEE Internet Things J.* **2014**, *1*, 372–383. [CrossRef]
10. Chen, Y.; Wen, H.; Song, H.; Chen, S.; Xie, F.; Yang, Q.; Hu, L. Lightweight one-time password authentication scheme based on radiofrequency fingerprinting. *IET Commun.* **2018**, *12*, 1477–1484. [CrossRef]
11. Wen, H.; Li, S.; Zhu, X.; Zhou, L. A framework of the PHY-layer approach to defense against security threats in cognitive radio networks. *IEEE Netw.* **2013**, *27*, 34–39.
12. Wen, H.; Wang, Y.; Zhu, X. Physical layer assist authentication technique for smart meter system. *IET Commun.* **2013**, *3*, 189–197. [CrossRef]
13. Wen, H.; Ho, P.H.; Qi, C.; Gong, G. Physical layer assisted authentication for distributed *ad hoc* wireless sensor networks. *IET Inf. Secur.* **2010**, *4*, 390–396. [CrossRef]

14. Rehman, S.U.; Sowerby, K.; Coghill, C. RF fingerprint extraction from the energy envelope of an instantaneous transient signal. In Proceedings of the Australian Communications Theory Workshop (AusCTW), Wellington, New Zealand, 30 January–2 February 2012.

15. Dubendorfer, C.K.; Ramsey, B.W.; Temple, M.A. An RF-DNA verification process for ZigBee networks. In Proceedings of the IEEE Military Communications Conference, Orlando, FL, USA, 29 October–1 November 2012.

16. Patel, H. Non-parametric feature generation for RF-fingerprinting on ZigBee devices. In Proceedings of the IEEE Symposium on Computational Intelligence for Security and Defense Applications (CISDA), Verona, NY, USA, 26–28 May 2015.

17. Baldini, G.; Giuliani, R.; Steri, G.; Neisse, R. Physical layer authentication of internet of things wireless devices through permutation and dispersion entropy. In Proceedings of the Global Internet of Things Summit (GIoTS), Geneva, Switzerland, 6–9 June 2017.

18. Chen, S.; Wen, H.; Wu, J.; Chen, J.; Liu, W.; Hu, L.; Chen, Y. Physical layer channel authentication for 5G via machine learning algorithm. *Wirel. Commun. Mob. Comput.* **2018**. [CrossRef]

19. Xie, F.; Wen, H.; Li, Y.; Chen, S.; Hu, L.; Chen, Y.; Song, H. Optimized coherent integration-based radio frequency fingerprinting in internet of things. *IEEE Internet Things J.* **2018**, *5*, 3967–3977. [CrossRef]

20. Ferdowsi, A.; Saad, W. Deep Learning for Signal Authentication and Security in Massive Internet-of-Things Systems. *IEEE Trans. Commun.* **2019**, *67*, 1371–1387. [CrossRef]

21. Hu, L.; Wen, H.; Wu, B.; Pan, F.; Liao, R.F.; Song, H.; Tang, J.; Wang, X. Cooperative jamming for physical layer security enhancement in internet of things. *IEEE Internet Things J.* **2018**, *5*, 219–228. [CrossRef]

22. Hu, L.; Wen, H.; Wu, B.; Tang, J.; Pan, F.; Liao, R.F. Cooperative jamming-aided secrecy enhancement in wireless networks with passive eavesdroppers. *IEEE Trans. Veh. Technol.* **2018**, *67*, 2108–2117. [CrossRef]

23. Hall, J.; Barbeau, M.; Kranakis, E. Detection of transient in radio frequency fingerprinting using signal phase. *Wirel. Opt. Commun.* **2003**, 13–18. [CrossRef]

24. Hu, N.; Yao, Y. Identification of legacy radios in a cognitive radio network using a radio frequency fingerprinting based method. In Proceedings of the IEEE International Conference on Communications (ICC), Ottawa, ON, Canada, 10–15 June 2012.

25. Ureten, O.; Serinken, N. Bayesian detection of wi-fi transmitter RF fingerprints. *Electron. Lett.* **2005**, *41*, 373–374. [CrossRef]

26. Ureten, O.; Serinken, N. Wireless security through RF fingerprinting. *Can. J. Electr. Comput. Eng.* **2007**, *32*, 27–33. [CrossRef]

27. Merchant, K.; Revay, S.; Stantchev, G.; Nousain, B. Deep Learning for RF Device Fingerprinting in Cognitive Communication Networks. *IEEE J. Sel. Top. Signal Process.* **2018**, *12*, 160–167. [CrossRef]

28. Youssef, K.; Bouchard, L.; Haigh, K.; Silovsky, J.; Thapa, B.; Valk, C.V. Machine Learning Approach to RF Transmitter Identification. *IEEE J. Radio Freq. Identif.* **2018**, *2*, 197–205. [CrossRef]

29. McGinthy, J.M.; Wong, L.J.; Michaels, A.J. Groundwork for Neural Network-Based Specific Emitter Identification Authentication for IoT. *IEEE Internet Things J.* **2019**, *4*, 6429–6440. [CrossRef]

30. Weiner, M.; Jorgovanovic, M.; Sahai, A.; Nikolié, B. Design of a low-latency, high-reliability wireless communication system for control applications. In Proceedings of the IEEE International Conference on Communications (ICC), Sydney, Australia, 10–14 June 2014.

31. Cobb, W.E.; Laspe, E.D.; Baldwin, R.O.; Temple, M.A.; Kim, Y.C. Intrinsic physical-layer authentication of integrated circuits. *IEEE Trans. Inf. Forensics Secur.* **2012**, *7*, 14–24. [CrossRef]

32. Suski II, W.C.; Temple, M.A.; Mendenhall, M.J.; Mills, R.F. Using spectral fingerprints to improve wireless network security. In Proceedings of the IEEE Global Telecommunications Conference, New Orleans, LO, USA, 30 November–4 December 2008.

33. Pan, F.; Pang, Z.; Xiao, M.; Wen, H.; Liao, R.F. Clone detection based on physical layer reputation for proximity service. *IEEE Access* **2019**, *7*, 3948–3957. [CrossRef]

34. Nordic, Nordic Semiconductor-nrf24 Series. Available online: https://www.nordicsemi.com/Products/Low-power-short-range-wireless/nRF24-series (accessed on 29 January 2019).

35. Ettus Research, Universal Software Radio Peripheral. Available online: https://www.ettus.com/ (accessed on 29 January 2019).
36. Klein, R.W.; Temple, M.A.; Mendenhall, M.J. Application of wavelet-based RF fingerprinting to enhance wireless network security. *J. Commun. Netw.* **2009**, *11*, 544–555. [CrossRef]

sensors

MDPI

Article

A Machine Learning Approach to Achieving Energy Efficiency in Relay-Assisted LTE-A Downlink System [†]

Hammad Hassan [1], Irfan Ahmed [2,*], Rizwan Ahmad [1], Hedi Khammari [3], Ghulam Bhatti [3], Waqas Ahmed [4] and Muhammad Mahtab Alam [5]

[1] School of Electrical Engineering and Computer Science, National University of Sciences & Technology (NUST), Islamabad 44000, Pakistan
[2] Department of Electrical Engineering, Higher Colleges of Technology, Ruwais Campus 12389, UAE
[3] College of Computers and Information Technology, Taif University, Taif 21974, Saudi Arabia
[4] Pakistan Institute of Engineering and Applied Sciences (PIEAS), Islamabad 45650, Pakistan
[5] Thomas Johann Seebeck Department of Electronics, Tallinn University of Technology, Tallinn 19086, Estonia
* Correspondence: iahmed3@hct.ac.ae
† This paper is an extended version of our paper published in 16th Biennial Baltic Conference on Electronics and Embedded Systems, Tallinn, Estonia, 8–10 October 2018.

Received: 7 May 2019; Accepted: 22 July 2019; Published: 8 August 2019

Abstract: In recent years, Energy Efficiency (EE) has become a critical design metric for cellular systems. In order to achieve EE, a fine balance between throughput and fairness must also be ensured. To this end, in this paper we have presented various resource block (RB) allocation schemes in relay-assisted Long Term Evolution-Advanced (LTE-A) networks. Driven by equal power and Bisection-based Power Allocation (BOPA) algorithm, the Maximum Throughput (MT) and an alternating MT and proportional fairness (PF)-based SAMM (abbreviated with Authors' names) RB allocation scheme is presented for a single relay. In the case of multiple relays, the dependency of RB and power allocation on relay deployment and users' association is first addressed through a k-mean clustering approach. Secondly, to reduce the computational cost of RB and power allocation, a two-step neural network (NN) process (SAMM NN) is presented that uses SAMM-based unsupervised learning for RB allocation and BOPA-based supervised learning for power allocation. The results for all the schemes are compared in terms of EE and user throughput. For a single relay, SAMM BOPA offers the best EE, whereas SAMM equal power provides the best fairness. In the case of multiple relays, the results indicate SAMM NN achieves better EE compared to SAMM equal power and BOPA, and it also achieves better throughput fairness compared to MT equal power and MT BOPA.

Keywords: machine learning; LTE-A; energy efficiency; resource block allocation; bisection based optimal power allocation; water filling algorithm; proportional rate constraint

---

1. Introduction

Green Radio communication has received a lot of attention in the past few years with an aim to decrease the carbon foot print of wireless networks. It has been estimated that nearly 70% of the energy being used by cellular operators is on the radio part [1] and around 9% of the global CO_2 emission is from the communication systems [2]. In addition, one of the main concerns is the User Equipment (UE) battery, which has not shown progression at par with the Radio Access Technology (RAT). This phenomena is highly visible for the cell edge users that despite spending higher energy (due to high pathloss shadow fading and adjacent cell interference) are unable to achieve fair share of the radio resources. In this context, Green communications employing cooperative and fair resource allocation techniques can help in reducing the carbon footprint and increasing Energy Efficiency (EE).

Most of the existing wireless systems use Orthogonal Frequency Division Multiple Access (OFDMA) to distribute radio resources among UEs. One of the existing RAT to use OFDMA is Long Term Evolution-Advanced (LTE-A), which has a similar structure to its predecessor LTE. In LTE, each Resource Block (RB) is a time frequency grid element. The basic RB structure contains 15 subcarriers of 12 KHz each and a 10 ms frame. Each frame is subdivided into 10 subframes of 1 ms and each subframe is further divided into 2 slots of 0.5 ms each. Each slot may contain 6 or 7 OFDM symbols depending on a normal or an extended cyclic prefix. The RB allocation can be changed after every Transmission Time Interval (TTI) based upon channel conditions or RB allocation algorithm. OFDMA offers flexibility of RBs Allocation to tailor user and network requirement, such as throughput, fairness, Energy EfficiencyEE, and Spectral Efficiency (SE). For example, in order to support higher peak data rates Carrier Aggregation (CA) is introduced to obtain wider bandwidth. Compared to LTE, CA in LTE-A can support maximum 5 adjacent/non-adjacent component carriers of maximum 20 MHz to achieve 100 MHz bandwidth.

In addition, LTE-A allows Layer 3 (L3) relays to be incorporated in the network that can decode and forward the data to a UE [3]. This cooperative communication addresses EE and throughput of the cell edge users by providing channel diversity. As the network can only accommodate finite relays, their placement is, therefore, crucial to manage the overall throughput. The RB allocation between direct link that is Base Station (BS)–UE and two hop link that is (BS)–Relay Node (RN)–UE can be done independently or in a shared manner. In [4], the authors have presented a thorough comparison of basic RB allocation schemes, which are Round Robin (RR), Proportional Fairness (PF), Maximum Throughput (MT), and Maximum Minimum (MM); they presented an alternating MT and PF based resource allocation scheme SAMM (abbreviated with Authors' names) without considering any relay. The paper considered LTE system with a basic RB structure [5], 5 MHz bandwidth with fixed 10 users that are uniformly placed from the BS. These schemes have been compared in terms of sum throughput, individual user throughput and fairness based on JFI (Jains fairness index). The SAMM scheme provides a better tradeoff between throughput and fairness. Cell edge users show some throughput gains due to proportional fairness, however, this scheme fails to address EE. Authors in [6] considered EE for generic OFDMA based downlink system. They presented Bisection based Optimal Power Allocation (BOPA) algorithm for a given RB assignment. The BOPA works as an iterative approach based on water filling principle. This work uses equal power allocation among users for initial users' rate calculation, whereas, a modified algorithm in [7] uses equal power per resource block.

In [8], the authors proposed a quality of service (QoS) aware optimization problem for relay-based multi-user cooperative OFDMA uplink system. The main goal is to find optimal solutions for relay selection, power allocation and subcarrier assignment that maximize the system throughput. Aiming to support and attain the green wireless LTE network, an energy-efficient resource allocation scheduler with QoS aware support for LTE network is proposed in [9]. The authors of [10] proposed a two-stage method to solve the inter-cell interference problem. In the first stage, the subcarrier allocation and time scheduling are jointly conducted with sequential users' selection and without considering the interference. The power control optimization is left to the second stage, using a geometric programming method.

In [11], energy efficient resource block and power allocation optimal and low complexity suboptimal schemes are presented for OFDMA relay-assisted downlink. Authors use fractional programming to make the non-linear mixed integer problem to convex subtractive problem. In order to reduce the computational complexity of the optimal solution, they present two-stage RB allocation and transmission power control algorithms. The system model of this paper is similar to our model but they use relays (small eNB in that paper) with frequency reuse factor of one, and the users employ maximal ratio combining to maximize the received signal-to-noise ratio (SNR). In our case, we use multiplexing gain instead of the diversity gain at users' end by exploiting the knowledge of their location in relay selection and users association algorithm. We have compared our proposed schemes with the low complexity energy-efficient resource block and power allocation (LERPA) algorithm 3 and 4 of [11].

Artificial intelligence techniques can be used in highly dynamic and stringent constraint Next-Generation networks. Since machine learning is a most promising technique of artificial intelligence, it can be directly/indirectly employed to achieve the goals of 5 G in cognitive radios, massive multiple-input multiple-output (MIMO), hybrid beamforming, femto/small cells, smart grid, wireless power transfer, device-to-device communications, non-orthogonal multiple access (NOMA) etc. [12]. This paper [12] gives an overview of the applications of machine learning in Next-Generation wireless networks. Specifically, supervised learning techniques are suitable for massive MIMO channel estimations and spectrum sensing, unsupervised learning could be helpful in users grouping and clustering; and reinforcement learning can be applied in resource allocation problems.

A detailed review on existing techniques and methods have been provided in [13]. For example, in [14], a cooperative Q-learning approach was applied as an efficient approach to solve the resource allocation problem in a multi-agent network. The quality of service QoS for each user and fairness in the network are taken into account and more than a four-fold increase in the number of supported small cells. The authors in [15], proposed a machine learning framework for resource allocation to determine the optimal or near-optimal solutions based on the learning of the most similar historical scenario.

In paper [16], the authors proposed an approximated solution to a wireless network capacity problem using flow allocation, link scheduling, and power control. The Support Vector Machine (SVM) was used to classify each link to be assigned maximal transmit power or be turned off, whereas, the deep belief networks (DBNs) computes an approximation of the optimal power allocation. Both learning approaches have been trained on offline computed optimal solutions. A novel resource allocation method using deep learning to squeeze the benefits of resource utilization was developed in [17]. It was reported that when the channel environment is changing fast, the deep learning method outperforms traditional resource optimization methods. The resource allocation is to be optimized by a convolutional neural network using channel information. A similar problem has been explored in [18] that use Upper Confidence Bound learning for Greedy Maximal Matching (GMM) when the channel statistics are unknown. Since the subchannel and power allocation problem is a non-convex combinatorial problem, the optimal solution of the subchannel and power allocation problem requires an exhaustive search over all possible combinations of subchannels and power levels. In order to train the deep neural network (DNN) for an optimal solution, Ref. [19] utilizes the genetic algorithm to get the training data for DNN. It shows that the prediction accuracy increases with the size of dataset and the number of hidden layers. A four-step reinforcement learning based intercell interference coordination (ICIC) scheme is presented in [20]. The users selection, resource allocation, power allocation, and retransmit packet identification are handled by reinforcement learning to reduce the intercell interference.

However, to the best of our knowledge no available literature discusses LTE-A with L3 relays for SE and EE consideration. In this work,

- We present an energy efficient algorithm based on SAMM and BOPA for LTE-A system with a L3 relay. Performance evaluation in terms of throughput, fairness, power consumption, SE and EE is shown between two best performing schemes i.e., MT and SAMM considering equal power and BOPA.
- Considering the practical deployment, where there may be more than one relay supporting the cell edge users, we devise a clustering strategy to obtain near optimal placement of L3 relays and users' association.
- In a multiple relay scenario, to optimize EE and reduce computational complexity of running algorithm every TTI, we present a two step machine learning process that uses both the SAMM and BOPA approach for resource and power allocation of the cell users. The proposed approach is compared to MT equal power, MT BOPA and SAMM equal power in terms of users' throughput and EE.

A complete list of notations used in this paper is given in Table 1.

Table 1. List of notations.

Notation	Definition
$\mu_{k,n}$	the RB assignment indicator
R_k	capacity of user k is given by the Shannon Capacity
$SNR_{k,n}$	the signal-to-noise ratio for user k on RB n
B	System bandwidth
N	Number of RB
$W = \frac{B}{N}$	the RB bandwidth
$P_{total}^{RN}, P_{total}^{BS}$	the total power at which RN and BS transmit
$g_{k,n}^{Relay\_link}, g_{k,n}^{Direct\_link}$	channel gains for user k on RB n for RN and BS
$h_{k,n}^{Relay\_link}, h_{k,n}^{Direct\_link}$	random channel coefficients for user k on RB n for RN and BS
$SNR_{k,n}^{Direct\_link}$	signal-to-noise ratio for user k via Direct Link
$SNR_{k,n}^{Relay\_link}$	signal-to-noise ratio for user k via Relay Link
$SINR_{k,n}^{Direct\_link}$	signal-to-Interference-and-noise ratio for user k via Direct Link
α_k	proportional rate constraint for user k
λ_k	rate parameter for user k
D_k	allocated RB set for user k
R_k	rate matrix for user k
$\hat{p}_{k,n}$	optimal power allocation
P_T	total transmit power
θ	Lagrangian multiplier

Rest of the paper is organized as follows: system model is described in Section 2, algorithms and performance for MT, SAMM and BOPA with single relay network are given in Section 3. Multiple relay users' association and deployment with machine learning based power and RB allocation for SAMM is presented in Section 4. Complexity analysis is given in Section 5, followed by the conclusions in Section 6.

2. System Model

We consider a two-tier LTE-A system with a BS supported by L3 relays as shown in Figure 1. The relays are assumed to be In-band type 1b [3] and full duplex, placed in the center of BS to the most distant user. A total of K users and N RBs are considered with users placed at a uniform distance from BS. The total powers of BS and RN are denoted by P_{total}^{BS} and P_{total}^{RN}, respectively. The LTE-A system uses OFDMA transmission in the downlink. Let the system bandwidth is B with N number of RB, then, $W = \frac{B}{N}$ is the bandwidth of one RB. We express the channel gains $g_{k,n}^{Direct\_link}$ and $g_{k,n}^{Relay\_link}$ for user k where $k \in \mathcal{K} = \{1..., K\}$ on RB n where $n \in \mathcal{N} = \{1..., N\}$ for BS and RN respectively. Practically, the channel gain depends upon various factors, including thermal noise at receiver, receiver noise figure, antenna gains, distance between transmitter and receiver, path loss exponent, log normal shadowing and fading. Therefore, for all the links, we can write

$$g_{k,n} = -\varrho - \phi 10 \log_{10} d_k - \zeta_{k,n} + 10 \log_{10} h_{k,n} \tag{1}$$

In the above equation, ϱ (83.46 dB) is a constant depending upon thermal noise at receiver, receiver noise figure, and antenna gains, ϕ is path loss exponent, d_k is the distance in Km from UE k to the BS/relay, $\zeta_{k,n}$ (10.5 dB) is shadowing parameter modeled by a normally distributed random variable with standard deviation 8 dB, and $h_{k,n}$ corresponds to the Rayleigh fading channel coefficient of user k in subchannel n [21].

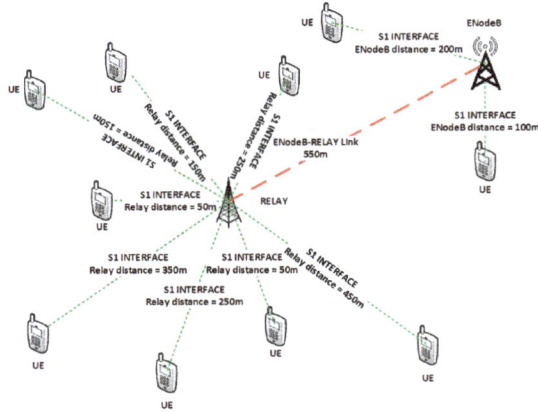

Figure 1. Topology.

The throughput of user k is given by,

$$
R_k = \begin{cases} \frac{1}{2}\frac{B}{N}\sum\limits_{n=1}^{N}\mu_{k,n}\log_2(1+SNR_{k,n}), & \text{Access link users} \\ \frac{B}{N}\sum\limits_{n=1}^{N}\mu_{k,n}\log_2(1+SNR_{k,n}), & \text{Direct link users} \end{cases}
\tag{2}
$$

where the factor $1/2$ in access link shows the two time-slots transmission from BS-RN and RN-UE, and $\mu_{k,n}$ is the binary variable such that $\mu_{k,n} = 1$ when RB n is allocated to the user k, $SNR_{k,n}$ is the maximum average signal-to-noise ratio for user k between direct and relay links. Let $SNR_{k,n}^{Direct\_link}$ be the signal-to-noise ratio for user k via Direct Link, and $SNR_{k,n}^{Relay\_link}$ be the signal-to-noise ratio for user k via Relay Link, then, the $SNR_{k,n}$ is given as

$$
SNR_{k,n} = \max(\mathbb{E}\{SNR_{k,n}^{Direct\_link}\}, \mathbb{E}\{SNR_{k,n}^{Relay\_link}\}),
\tag{3}
$$

$$
SNR_{k,n}^{Direct\_link} = \frac{p_{k,n}^{BS}g_{k,n}^{Direct\_link}}{N_0\frac{B}{N}},
\tag{4}
$$

$$
SNR_{k,n}^{Relay\_link} = \min(SNR_{k,n}^{backhaul\_link}, SNR_{k,n}^{access\_link}),
\tag{5}
$$

where, $SNR_{k,n}^{backhaul\_link/access\_link}$ is

$$
SNR_{k,n}^{backhaul\_link/access\_link} = \frac{p_{k,n}^{BS/RN}g_{k,n}^{backhaul\_link/access\_link}}{N_0\frac{B}{N}}
\tag{6}
$$

The Energy Efficiency EE in terms of $bits/s/Watts$ can be expressed as

$$
EE = \frac{\sum\limits_{k=1}^{K}\sum\limits_{n=1}^{N}\mu_{k,n}\log_2(1+SNR_{k,n})}{\sum\limits_{k=1}^{K}\sum\limits_{n=1}^{N}p_{k,n}}.
\tag{7}
$$

The EE optimization problem for the above scenario can be written as

$$\begin{aligned}
\text{maximize} \quad & EE \\
\text{subject to} \quad & \sum_{k=1}^{K} \sum_{n=1}^{N} p_{k,n} \leq P_{total} \\
& p_{k,n} \geq 0, \forall\, k, n \\
& \mu_{k,n} = \{0, 1\}, \forall\, k, n \\
& \sum_{k=1}^{K} \mu_{k,n} = 1, \forall\, n \\
& R_1 : R_2 : \ldots\ldots R_K = \alpha_1 : \alpha_2 : \ldots\ldots \alpha_K
\end{aligned} \tag{8}$$

where α_k is the proportional rate constraint [22]. We assume that channel state information (CSI) of all the users is known to the BS. Also, it is assumed that the RB allocation decision and assignment is done in less than channel coherence time so that CSI information can be used. This further puts constraints on the RB allocation algorithm complexity. The two-hop transmission to the RN users will be carried out in two TTI's. In the first TTI, the BS will only send data to the RN users that are in close proximity of RN or have better RN-UE channel conditions than the direct link BS-UE. In the second TTI RN-UE data will be sent. BS will choose the path to the user (direct or via RN) with best channel coefficient in each TTI. The centralized scheduling minimizes the possibility of interference for In-band type of RNs. Frequency division duplexing ensures that the RN may handle backhaul data simultaneously with the access link data so that from the second TTI onwards backhaul BS-RN transmission is carried out simultaneously with the access link RN-UE transmission.

The LTE-A downlink is an OFDM based system which supports M-ary quadrature amplitude modulation (MQAM). We can use Equation (2) to calculate the throughput of user k on RB n for both direct and relay-link paths. The two paths provide channel diversity to increase the users and system level throughput. We use MT and SAMM criteria for RB allocation with equal power allocation to all RBs or BOPA as explained below.

3. Fairness-Aware Power and Resource Block Allocation with Single Relay LTE-A Network

There are several well-known resource allocation schemes for cellular systems, namely, round robin RR, maximize throughput MT, maximize the minimum throughput (max-min), and proportional fairness PF. An improved hybrid MT and PF scheme, SAMM is presented in [4]. We briefly summarize MT, PF, and SAMM, and then present our fairness-aware power and resource allocation algorithm.

3.1. Maximum Throughput

In a Maximize Throughput MT scheme, the aim is to maximize the sum throughput of the network. It assigns more RBs to the user which has better channel conditions on direct link or two hop link thereby adding more throughput to the system but its drawback is that users with the worst channel conditions are essentially ignored. The maximum throughput criterion in mathematical form is given as,

$$D_k = \arg\max_{k}(R_k) \tag{9}$$

where D_k is RB allocation matrix and R_k is rate matrix.

3.2. Proportional Fairness

The proportional fairness based resource allocation schemes are widely used in practical wireless communication systems. In this scheme, the system allocates the resource to a user who has the maximum PF metric. The PF criterion in mathematical form is given as,

$$D_k = \arg\max_k \ \frac{R_k(t)}{\bar{R}_k(t)} \tag{10}$$

where $R_k(t)$ is the throughput of user k at scheduling time t, and $\bar{R}_k(t)$ is the average user throughput (moving average) over a past window of length $T_w = 1/\alpha$ [23], as

$$\bar{R}_k(t) = \alpha R_k(t-1) + (1-\alpha)\bar{R}_k(t-1), \tag{11}$$

3.3. SAMM

In SAMM [4] PF and MT are run one after the other, i.e., in first TTI PF run for K users and in second TTI MT run for $K-1$ users ignoring the user with highest throughput in previous TTI. This results in maximizing fairness and throughput alternatively in each TTI.

3.4. BOPA Algorithm

Bisection based optimal power allocation BOPA Algorithm 1 allocates the power to the RBs assigned to a particular user. Given the RB allocation from MT or SAMM and throughput of each user at equal power allocated to all RBs we can calculate λ "rate parameter" as given below:

$$\lambda_k = \frac{R_1}{\alpha_1} = \frac{R_2}{\alpha_2} = ... = \frac{R_K}{\alpha_K} \tag{12}$$

where R_k is the rate of each user and α_k is proportional rate constraint set for fairness [6]. Optimal power allocation is water filling operation and obtained for single user as

$$\hat{p}_{k,n} = \max\left\{\frac{1}{\theta_L \ln 2} - \frac{1}{g_{k,n}}, 0\right\} \tag{13}$$

where θ_L is Lagrangian multiplier and its value is chosen such that R_k is satisfied. Hence, the user power can be expressed as $P_k(\lambda \alpha_k | D_k)$ and the total transmit power $P_T(\lambda)$ can be rewritten as

$$P_T(\lambda) = \sum_{k \in K} P_k(\lambda \alpha_k | D_k) \tag{14}$$

EE can be given as user rate divided by power consumed to achieve that rate.

$$EE(\lambda) = \frac{\lambda \sum\limits_{k \in K} \alpha_k}{P_T}, \tag{15}$$

and total transmit power is also limited by

$$\lambda \leq \lambda_{\max}. \tag{16}$$

According to [24] if transmit power $P_T(\lambda)$ is strictly convex in rate then $EE(\lambda)$ is quasi-concave, global optimal solution proof is given in the appendix of paper [22]

$$f(\lambda) = P_T(\lambda) - \lambda \ln 2 \sum_{k \in K} \min_{n \in D_k} \left\{\frac{1 + \hat{p}_{k,n} g_{k,n}}{g_{k,n}}\right\} \alpha_k \tag{17}$$

Bisection method is a simple and robust. Since the method brackets the root, it is guaranteed to converge. We apply BOPA on the RB allocation scheme SAMM, an alternating MT and PF scheme for the relay-assisted LTE-A for the optimal power allocation with the objective of maximizing the EE. In addition, we trained neural network with the dataset generated by the BOPA. Since power is a monotonically increasing function of the rate parameter λ, we apply bisection method on the following equation to find the root,

$$P(\lambda) = \sum_{k} \sum_{n} \frac{2^{\lambda \alpha_k N} - 1}{g_{k,n}} - P_{total} = 0 \qquad (18)$$

Algorithm 1 BOPA Algorithm

1: **Require:** $\hat{p}_{k,n}$ is the optimal power allocation matrix.
2: **Ensure**: Prior RB allocation through any algorithm and given as D_k.
3: Getting all the λ then calculate λ_{max} which gives the max energy Efficiency by substitution in
   Equation (6).
4: Using λ_{max} set user rate as $\alpha_k \, \lambda_{max}$, do water filling using Equation (13) and calculate f(λ_{max})
   based on Equation (17).
5: **If** f (λ_{max}) $\geq$ 0
6: **Return** ; $\hat{p}_{k,n}$
7: **Else** Go to Step 9;
8: **End if**
9: Set $\lambda_{high} = \lambda_{max}$, $\lambda_{low} = 0$, $\lambda_{current} = \lambda_{max}$ /2
10: **Repeat:** Set user rate according to $\alpha_k \, \lambda_{max}$, do water filling using Equation (13) and calculate
    f(λ_{max}) based on Equation (17).
11: **If** f ($\lambda_{current}$) $>$ 0
12: Set $\lambda_{low} = \lambda_{current}$
13: **Else** Set $\lambda_{high} = \lambda_{current}$
14: **End if**
15: Set $\lambda_{current} = \lambda_{high} + \lambda_{low}$ / 2
16: **Return** $\hat{p}_{k,n}$
17: **End if**

3.5. Performance Evaluation

A single cell is considered for generating simulations results. The cell consists of a BS, RN and UEs equipped with Omni-directional antennas. The throughput, energy and spectral efficiency is averaged over 1000 TTIs, with the duration of a TTI being 0.5 ms. The channel involves Raleigh fading and distance based path loss as shown in Figure 1. BS is located in the center of the cell coverage and most distant user is 1 Km distant from BS with RN in between at 0.55 Km. RN are In-band full duplex relays and bit error rate (BER) considered for MQAM modulation is 10^{-3}. Table 2 below summarizes all simulation parameters used to derive results shown next.

Figure 2 shows the result of average throughput for MT and SAMM with equal power and BOPA based power allocation. It can be seen that SAMM curves remain on top of MT curves for most of the users due to inherent fairness which ensures all users get due share of RBs. However as evident from Figure 3 Sum throughput of MT is higher as compared to SAMM for overall averaged throughput of sum users due to channel exploitation of users with good channel conditions. This makes MT better than SAMM as BOPA has proportional rate constraint set for assigning user priorities.

Table 2. Simulation Parameters.

Parameter	Value
Cell Radius	1 Km
Noise Density (σ)	-171 dBm/Hz
No of users (K)	10
Bandwidth (B)	5 MHz
Number of resource blocks (RB) (N)	25
No of subcarriers per RB	12
Subcarrier bandwidth	15 KHz
BS Transmitter Power	46 dBm
Relay Power	34 dBm
TTI duration	0.5 ms
Relay distance from BS	0.55 Km
Bit Error Rate (BER)	10^{-3}
OFDM symbols per TTI	7
Relay Type (In-band / Out-band Type 1 / Type 2)	In-band with Type 1b (full duplex)

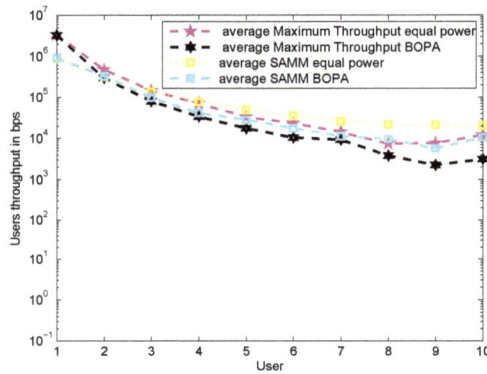

Figure 2. User vs. Throughput averaged over thousand times for SAMM and MT with equal power and BOPA.

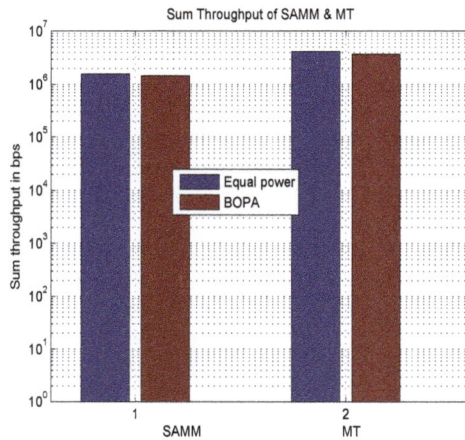

Figure 3. Sum throughput of all users.

Figure 4 shows energy efficiency per user in bits per seconds per watts. SAMM BOPA outperforms for initial users and remains considerably lower for rest of the users. Whereas MT BOPA compared to all other schemes performs better for every user of the system with consistency due to convergence of BOPA to maximize throughput and minimize energy.

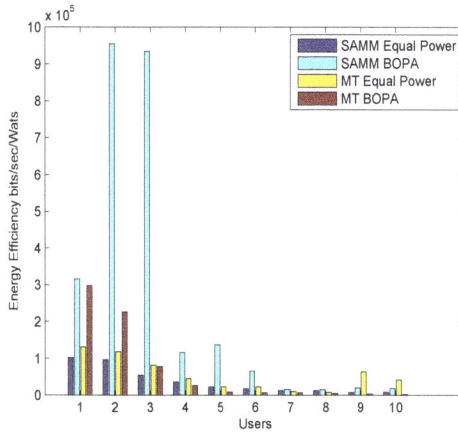

Figure 4. Users vs. Energy Efficiency (bits per second per watts) for all four schemes.

Figure 5 shows fairness Index using Jains fairness Index [25] using below equation

$$FI = \frac{\left(\sum\limits_{k=1}^{K} r_k\right)^2}{K \sum\limits_{k=1}^{K} r^2{}_k},$$

(19)

where r_k can be throughput or EE. Figure 5 shows SAMM has better fairness in terms of throughput due to PF in its algorithm. Figure 6 depicts the system's energy efficiency EE with and without power allocation. The BOPA-based power allocation algorithm allocates the available power to the RB to maximize the energy efficiency EE, therefore, both MT-BOPA and SAMM-BOPA outperforms their corresponding MT and SAMM schemes with equal power allocation.

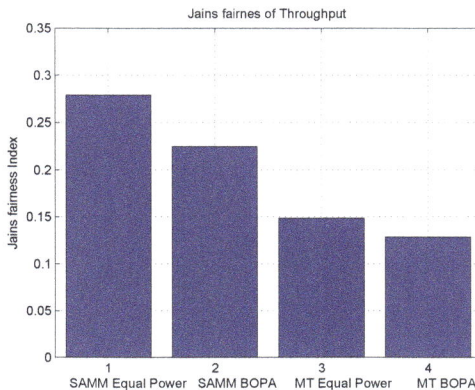

Figure 5. Jains fairness Index for Throughput.

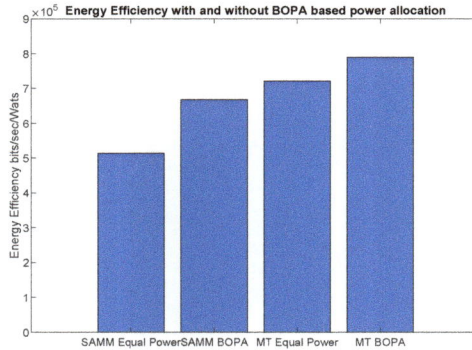

Figure 6. The system energy efficiency with and without BOPA based power allocation.

4. Fairness-Aware Machine Learning Based Power and RB Allocation with Multiple Relays

In practical scenarios, multiple relays are deployed to facilitate the cell-edge users as shown in the Figure 7. The multiple relay deployment causes inter-relay interference. This interference can be minimized by the careful deployment of relays, transmit power control, and the scheduling of time/frequency resources. Though, L3 relays incur more processing delay as compared to the L1 and L2 relays but they provide robust transmission in the presence of interference [26]. Assume there are Q relays in a cell, such that relay $q \in \mathcal{Q} = \{1, ..., Q\}$. The signal-to-interference-and-noise ratio (SINR) at UE k in direct link is given as

$$SINR_{k,n}^{Direct\_link} = \frac{p_{k,n}^{BS} g_{k,n}^{Direct\_link}}{\sum_{q \in \mathcal{Q}} p_{k',n}^{q} g_{k,n}^{q} + N_0 \frac{B}{N}} \tag{20}$$

where $p_{k',n}^{q}$ is the transmit power of relay q assigned to its associated user k' and $g_{k,n}^{q}$ is the channel gain between relay q and the UE k. Similarly, the SINR at UE k in relay q link is given as

$$SINR_{k,n}^{q} = \frac{p_{k,n}^{q} g_{k,n}^{q}}{\sum_{q' \in \mathcal{Q}-\{q\}} p_{k',n}^{q'} g_{k,n}^{q'} + p_{k,n}^{BS} g_{k,n}^{Direct\_link} + N_0 \frac{B}{N}} \tag{21}$$

As seen from the above equation, the interference and fairness causes a significant increase in the computational cost when deploying multiple relays. Therefore, we present a machine learning based approach that utilizes relay deployment and users' association data to develop RB allocation and Power allocation strategy that maximizes the sum EE. Once trained, the proposed approach can save cost of scheduling in every TTI. This is shown in Figure 8, the machine learning model takes the inputs: number of relays, relays' coordinates, CSI, SNR, and total transmit power and produces the outputs: optimal relays' coordinates with associated users, set of RBs assigned to each user k, and the optimal power allocation ($p_{k,n}^*$) to each user k in the RB n. Based on single relay performance, the RB allocation block is trained using SAMM and power allocation block is trained using BOPA. Since the relay deployment can significantly alter the RB and power allocation, a clustering approach is presented that determines relay positioning and corresponding users' association based on a pre-defined metric.

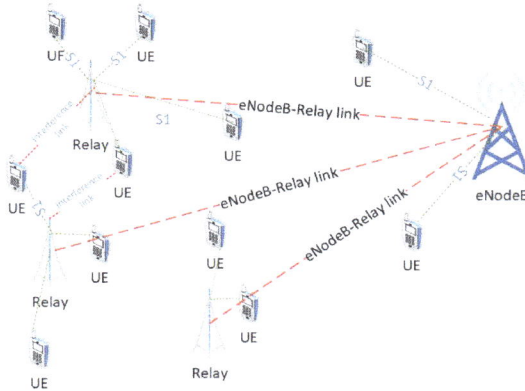

Figure 7. LTE-A Network with Multiple L3 Relays.

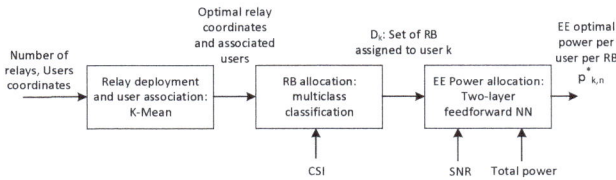

Figure 8. Block diagram of machine learning based resource and power allocation.

4.1. Relays Deployment and Users Association

In this section, we present an autonomous unsupervised machine learning scheme that provides users association with optimally deployed relay nodes in the cell-edge area. Machine learning algorithms can broadly be divided into two main categories, namely supervised learning and unsupervised learning algorithms. The former class of algorithms learn by training on the input labeled examples, called training dataset, $\{(x^{(1)}, y^{(1)}), (x^{(2)}, y^{(2)}), (x^{(3)}, y^{(3)}), ..., (x^{(m)}, y^{(m)})\}$, where the ith example $(x^{(i)}, y^{(i)})$ consists of the ith instance of feature vector $x^{(i)}$ and the corresponding label $y^{(i)}$. Given a labeled training dataset, these algorithms try to find the decision boundary that separates the positive and negative labeled examples by fitting a hypothesis to the input dataset. Unsupervised machine learning algorithms, on the other hand, are given an unlabeled input dataset. These algorithms are used for extracting information or features from the dataset. These features might be related, but not confined, to the underlying structures or patterns in the input data, relationships in data items, grouping/clustering of data items, etc. Discovered features are meant to provide a deeper insight into the input dataset that can subsequently be exploited for achieving specific goals. Clustering algorithms make an important part of unsupervised learning where the input examples are grouped into two or more separate clusters based on some features. The K-Means (KM) algorithm, is probably the most popular clustering algorithm. It is an iterative algorithm that starts with a set of initial centroids given to it as input. During each iteration, it performs the following two steps.

1. **Assign Cluster:** For every user, the algorithm computes the distance between the user and every centroid. The user is then associated to the cluster with the closest centroid. During this step, a user might change its association from one cluster to another one.
2. **Recompute centroids:** Once all users have been associated to their respective cluster, the new position of centroid for every cluster is then calculated.

Let us define the following notations to be used later in this section.

K = Total number of clusters being formed.

$x^{(i)}$ = Location coordinates of user $u^{(i)}$. In our case, $x^{(i)} \in \mathbb{R}^2$

$c^{(i)}$ = Cluster to which the user $u^{(i)}$ is currently associated.

μ_k = Centroid of kth cluster, $\mu_k \in \mathbb{R}^2$

$\mu_{c^{(i)}}$ = Centroid of the cluster to which the user $u^{(i)}$ is currently associated.

Now the cost function J can be defined as

$$J(c^{(1)}, c^{(2)}, ..., c^{(m)}, \mu_1, \mu_2, ..., \mu_K) = \frac{1}{m}\sum_{i=1}^{m} ||x^{(i)} - u_{(c^i)}||^2 \tag{22}$$

with the following optimization objective function.

$$\min_{c^{(1)}, ..., c^{(m)}, \mu_1, ..., \mu_K} J(c^{(1)}, c^{(2)}, ..., c^{(m)}, \mu_1, \mu_2, ..., \mu_K)$$

It may be pointed out that Equation (22) allows us to compare multiple clustering layouts based on their cost and select the one with the lowest cost.

In this section, we use the KM algorithm for optimal clustering of m users competing for resources in a particular cell. The clustering is performed based on their geographic location, thus our input dataset $\{u^{(1)}, u^{(2)}, u^{(3)}, ..., u^{(m)}\}$ has m vectors $u^{(i)}$, $1 \leq i \leq m$, consisting of location coordinates, of ith user. For the sake of simplicity, we assume these users are deployed in a two dimensional area, i.e., a plane and so $u^{(i)} = (x_1^{(i)}, x_2^{(i)})$, i.e., an ordered pair of location coordinates. Our clustering algorithm is summarized in Algorithm 2.

The proposed algorithm takes the location coordinates of m users as input. It also takes two numbers min_k and max_k as additional inputs. The algorithm outputs the best number of clusters, k, such that $min_k \leq k \leq max_k$, and corresponding members of each cluster. It starts with $k = min_k$ and randomly selects k user locations as the initial centroids (line 6). It assigns the closest centroid to each user (line 8) and then computes new centroids by calculating the center/average location of all nodes in each cluster (line 11). So, in effect, the location of centroids keeps moving in successive iterations. It repeats the above two steps until the change in centroids' positions is zero or negligible. We repeat the test max_t times with a new set of randomly chosen initial centroids every time. During every test, the discovered centroids, corresponding centroid assignment to users, and the cost are saved (lines 14–16) for later comparison. After running the loop for max_t times, we select and store the best k centroids resulting from the test with the lowest cost while discarding the remaining (lines 19–21). The same is repeated for the next value of k, i.e., $k = k + 1$, until $k > max_k$. At the end we have $cnt = max_k - max_k$ vectors μ_k, one for each value of k, the corresponding assignment vector a_k and cost c_k. Finally, we choose the vector μ having the lowest cost and corresponding assignment vector **a** among cnt stored cases. That is the best number of clusters and corresponding centroids that the algorithm found. A snapshot of the relay deployment and users's association algorithm output is shown in Figure 9.

Algorithm 2 Users association clustering algorithm

1: $cnt = 0$

2: **for** $k = min_k : max_k$ **do**

3:     $cnt = cnt + 1$

4:     **for** $t = 1 : max_t$ **do**

5:         **repeat**

6:             Randomly choose initial k centroids $\mu_1, \mu_2, \mu_3, ..., \mu_k$

7:             **for** $i = 1 : m$ **do**

8:                 $a^{(i)} = j, \quad 1 \leq j \leq k$, such that μ_j is the centroid closest to $u^{(i)}$

9:             **end for**

10:           **for** $l = 1 : k$ **do**

11:               μ_l = mean of all users/points $u^{(i)}$ assigned to lth centroid

12:           **end for**

13:         **until** converges

14:         $\boldsymbol{\mu}^{(t)} = (\mu_1, \mu_2, \mu_3, ..., \mu_k)$

15:         $\mathbf{a}^{(t)} = (a^{(1)}, a^{(2)}, a^{(3)}, ..., a^{(m)})$

16:         $c^{(t)} = cost(\mu_1, \mu_2, \mu_3, ..., \mu_k)$

17:     **end for**

18:     $idx = \arg_{min}\{c(t), 1 \leq t \leq max_t\}$

19:     $\boldsymbol{\mu}_k^{(k)} = \boldsymbol{\mu}^{(idx)}, 1 \leq idx \leq max_t$

20:     $\mathbf{a}_k^{(k)} = \mathbf{a}^{(idx)}, 1 \leq idx \leq max_t$

21:     $c_k^{(k)} = c^{(idx)}, 1 \leq idx \leq max_t$

22: **end for**

23: $index = \arg_{min}\{c_k^{(k)}, 1 \leq k \leq cnt\}$

24: $\boldsymbol{\mu} = \boldsymbol{\mu}_k^{(index)}, 1 \leq index \leq cnt$

25: $\mathbf{a} = \mathbf{a}_k^{(index)}, 1 \leq index \leq cnt$

26: $n = index$

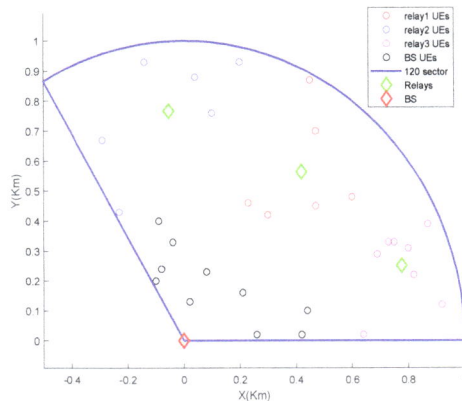

Figure 9. A snapshot of the relay deployment and users' association algorithm output in a 120 degree sector.

4.2. Resource Allocation by Multiclass Classification

The resource block allocation problem has multiple discrete outputs, i.e., the users, therefore, we use the multiclass classification to classify one out of K users. The multiclass classification is an extension of One-Vs-All classification. The input of the training network comprises of channel state information in terms of the SNR and the output consists of a particular user that maximizes the utility function (throughput for MT and PF metric for the proportional fairness). The training data is obtained from the implementation of SAMM algorithm of [4] as 25,000 K-dimensional samples of received SNR and the corresponding selected users. The dataset is partitioned into three parts, the training dataset, the validation dataset, and the test dataset. These are divided in 70%, 15%, and 15% ratio, respectively. The Matlab Neural Network Pattern Recognition Apps is used to train and deploy the neural network. It uses Scaled Conjugate Gradient algorithm [27] for training. Our application requires $K = 10$ neurons in input layer and 10 neurons in output layer. A hit and trial choice of eight neurons in hidden layer gave the best result. The neural network architecture is shown in Figure 10.

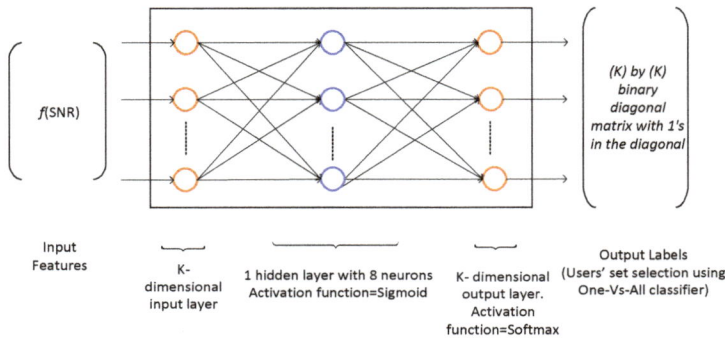

Figure 10. Neural network architecture for RB allocation, $K = 10$.

The neural network loss function is a generalization of the logistic regression's loss function. In logistic regression classification problem, we try to find the weighted parameter θ, such that the mean square error between the predicted output and the actual output is minimized. This is called loss function (LF) or the cost function and is given by

$$LF(\theta) = \frac{1}{m} \sum_{i=1}^{m} (h_\theta(x^{(i)}) - y^{(i)})^2 \tag{23}$$

where the prediction or hypothesis function $h_\theta(x)$ is a sigmoid function, i.e., $h_\theta(x) = \frac{1}{1+e^{-\theta^T x}}$. In the above equation, $(x^{(i)}, y^{(i)})$ is a training dataset with $1, ..., m$ input-output pairs. However, loss function with sigmoid function leads to a non-convex function, therefore, a cross entropy based loss function is used to make it convex function as,

$$LF(\theta) = -\frac{1}{m} \sum_{i=1}^{m} [y^{(i)} \log(h_\theta(x^{(i)})) + (1 - y^{(i)}) \log(1 - h_\theta(x^{(i)}))] + \frac{\lambda_R}{2m} \sum_{j=1}^{n} \theta_j^2 \tag{24}$$

where the second summation is for the regularization of weight or bias units θ_j and λ_R is a regularization parameter.

In case of neural networks with multiclass classification, the prediction variable becomes K-dimension, $h_\Theta(x) \in \mathbb{R}^K$, therefore, the loss function is given as

$$LF(\Theta) = -\frac{1}{m} \sum_{i=1}^{m} \sum_{k=1}^{K} [y_k^{(i)} \log(h_\Theta(x^{(i)}))_k + (1 - y^{(i)}) \log(1 - (h_\Theta(x^{(i)}))_k)] + \frac{\lambda_R}{2m} \sum_{l=1}^{L-1} \sum_{i=1}^{s_l} \sum_{j=1}^{s_{l+1}} (\Theta_{j,i}^{(l)})^2 \tag{25}$$

where L is the number of layers in neural network, s_l is the number of neurons in layer l, and $\lambda_R = 5 \times 10^{-4}$ is a regularization parameter to control the tradeoff between fitting the training dataset and keeping the parameter Θ small. The neural network is trained using the stochastic gradient descent algorithm. The gradient or partial derivative is calculated by the backpropagation algorithm and weights (θ) are updated. The amount at which the weights are updated is called learning rate. It our case, we set learning rate to 0.01. Batch size is a matrix of input (or output) vectors applied to the network simultaneously to produce the update on network weights and biases. In our work, batch size of 128 (MATLAB default), 10×1 input vectors is used.

We use MATLAB 2019a App, Neur al Network Pattern Recognition (nprtool) which is a two-layer (one for hidden layer activation functions and other for output layer activation functions) feedforward network.

Lower the cross entropy higher the classification accuracy, zero cross entropy means no error. Figure 11 shows that cross entropy reaches 0.0078318 at iteration 136. Figure 12 shows variation in gradient coefficient with respect to number of epochs. The final value of gradient coefficient at epoch number 142 is 0.001787 which is approximately near to zero. Minimum the value of gradient coefficient better will be training and testing of networks. From the figure, it can be seen that the gradient value is decreasing with the increase in number of epochs. Large number of validation fails indicate the overtraining. In Figure 12 validation fails are the iterations when validation mean square error (MSE) increased its value. A lot of fails means overtraining. MATLAB automatically stops training after 6 fails in a row.

Figure 13 shows the error histogram of the trained neural network for the training, validation and testing parts. In this figure we can see that the data fitting errors are minimum and they are distributed within a closed range around zero. The confusion matrix Figure 14 visualizes the performance of supervised learning. The rows correspond to the predicted user (Output Class) and the columns correspond to the true user (Target Class). The diagonal cells correspond to observations that are correctly assigned the user-RB pairs. The off-diagonal cells correspond to incorrectly assigned user-RB pairs. The trained neural network provides 97.5% classification accuracy. The Figure 15 represent the receiver operating characteristics (ROC) curves. The ROC curve plot shows the true positive rate versus the false positive rate as the threshold is varied. A perfect test would show points in the upper-left corner, with 100% sensitivity and 100% specificity [28]. In the RB allocation module, it worked very well.

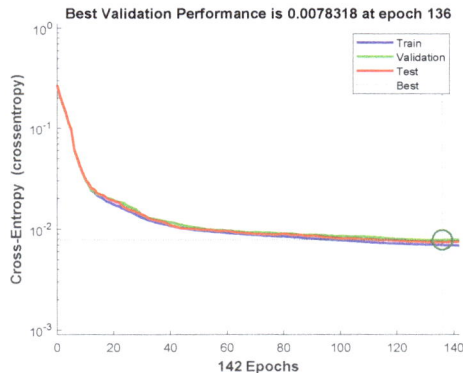

Figure 11. The mean-squared-error for the training and testing of the RB allocation module.

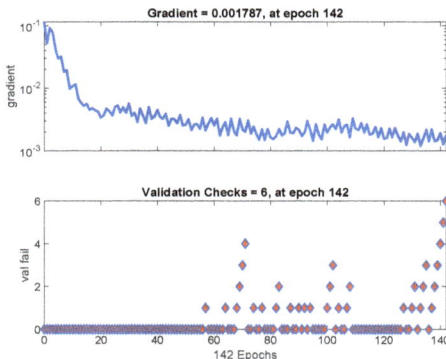

Figure 12. The neural network training states with gradient and validation fail statistics as a function of number of epochs.

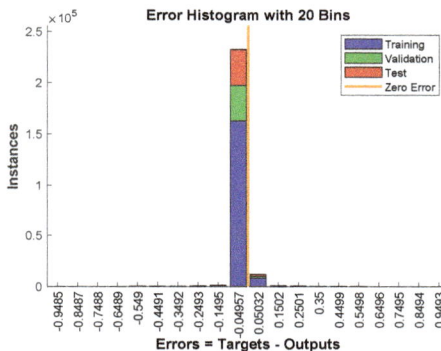

Figure 13. The error histogram of the trained neural network for the training, validation and testing phases.

Figure 14. The confusion matrix for test dataset.

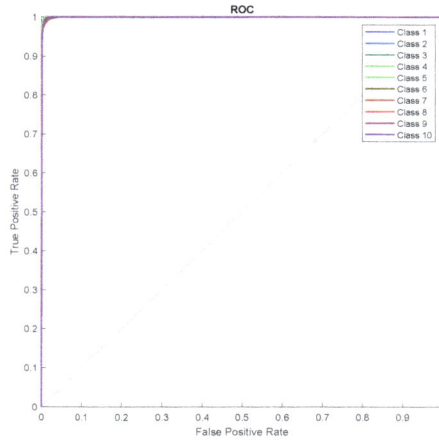

Figure 15. Receiver Operating Characteristic (ROC curve).

4.3. Power Allocation through Two-Layer Feedforward Neural Network

In the power allocation problem, we have to map the numeric input dataset (SNR) to the numeric output dataset (allocated power) per user per RB. Therefore, we use neural network curve fitting technique. The training dataset is generated by the Algorithm 2 as input received SNR and output allocated transmit power. Given the resource blocks allocation set $\mathcal{D}_k \quad \forall k \in \mathcal{K}$, the power allocation problem has been solved using two-layer feedforward neural network. The hidden layer neurons use sigmoid function as activation function and output neurons implement linear function as shown in Figure 16. We use Bayesian Regularization method to train the neural network. This method typically requires more training time but gives good results for difficult and noisy dataset. The Bayesian Regularization method uses Levenberg-Marquardt optimization to update the weight and bias values. It minimizes a combination of squared errors and weights, and then determines the correct combination for better generalization. In this method, the training does not stops after six consecutive validation (improve) fails and by default max_fail = inf. The training continues until an optimal combination of errors and weights is reached. More detail on the use of Bayesian regularization, along with Levenberg-Marquardt training, can be found in [29].

We use MATLAB 2019a App, Neural Net Fitting (nftool) which is a two-layer (one for hidden layer activation functions and other for output layer activation functions) feedforward network.

The mean-squared-error graph for the training and testing is shown in Figure 17. It shows that the MSE reaches to 0.087358 in 498 epochs. Our input/output samples to training network were channel gain/allocated power. Since, the total transmit power is a sum of linear functions of the channel gain, therefore, the neural network is got trained in a single epoch. An epoch is a full pass through the entire dataset and the calculation of new weights and biases. Figure 18 shows that the gradient coefficient reaches to 0.00076591 in 499 epochs. The lower value of gradient ensures the training and testing of the network. Other parameters such as Mu, Num Parameters, and Sum Squared Param are the stop criteria defined in Bayesian regularization backpropagation function 'trainbr' [30]. Error histogram in Figure 19 visualizes the errors between target values and the predicted values after training a feedforward neural network. In this figure we can see that the data fitting errors are minimum and they are distributed within a closed range around zero. Around 88.1% errors fall between −0.3 and 0.33.

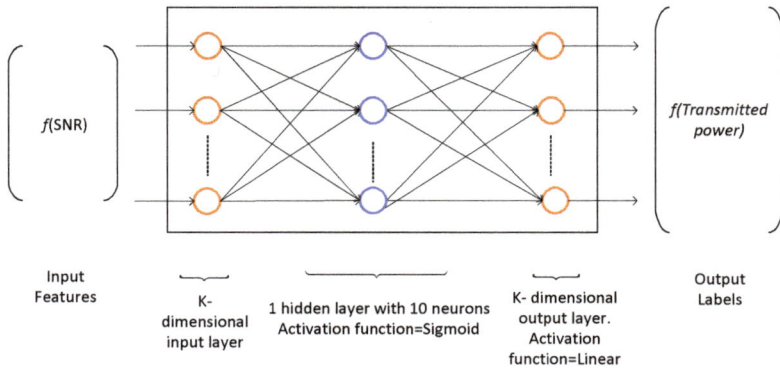

Figure 16. Neural network architecture for power allocation, $K = 10$.

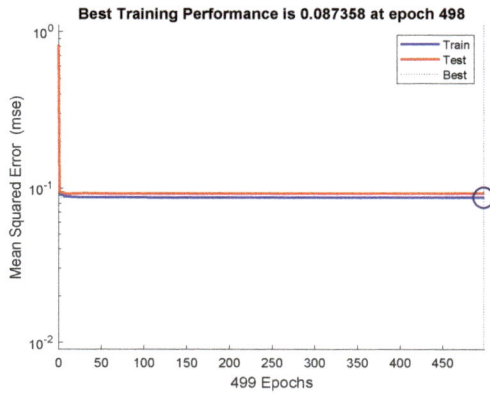

Figure 17. The mean-squared-error for the training and testing of the power allocation module.

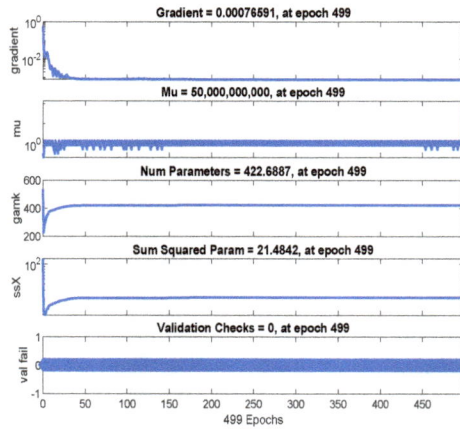

Figure 18. The neural network training states.

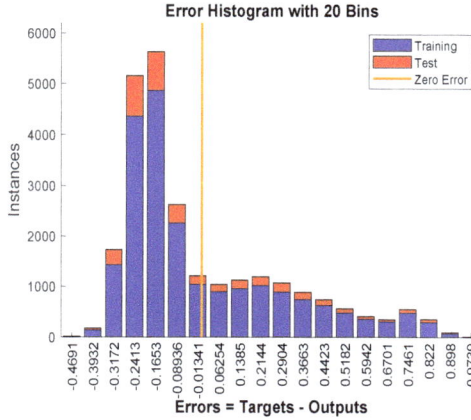

Figure 19. The error histogram.

4.4. Performance Evaluation with Machine Learning Techniques

First, we apply the neural network for the RB and power allocation modules with a single relay. For the SAMM scheme in Figure 20 shows 30.25% increase in the EE. This is because of the limitations of the BOPA method which sometimes returns no result, whereas, the neural network is trained on diverse dataset and always gives the output result. We also compare our proposed schemes with LERPA of [11]. LERPA uses max–min criteria for RB allocation and fractional programming based transmission power control. In case of LTE network with multiple relays as shown in Figure 7 or Figure 9, the users associated with relay q experience interference due to the neighboring relays q_{neigh}. This interference decreases the users' throughput as shown in Figure 21. However, the EE maximization based NN power allocation continues to dominate in the multiple relay scenario. Since the transmission is orthogonal between BS and RNs, only the relay's associated users are affected by the other relays transmissions. The equal power MT throughput does not affect because almost all the users are associated with BS. This further reduces the required transmit power of the relay, hence a net increase in EE has been observed in Figure 22. Addition of multiple relays slightly affect the SAMM NN and SAMM equal power in positive and negative way, respectively. The PF component of the SAMM forces the association of low throughput users to increase the fairness. This association goes in positive way for the SAMM NN due to the EE based power allocation, but goes in negative way for the SAMM equal power because of no compensation of the interference power. The increased fairness of SAMM NN is evident from the Figure 21, where, even the farthest users 9 and 10 have higher throughput. It can be seen that in LERPA, closer users get lower throughput but fairly large throughput is given to the farther users. This is because it uses max-min criterion for the RB allocation, which assigns the RB to the users who have lowest received SNR.

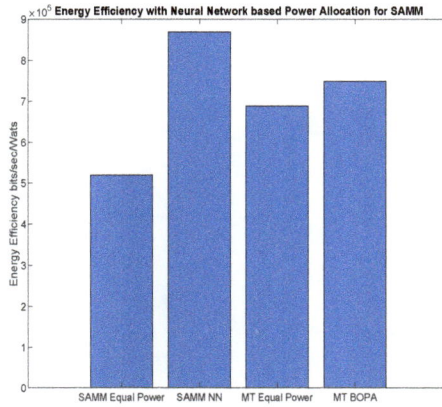

Figure 20. The system energy efficiency with neural network for SAMM which is trained on waterfilling based power allocation among users and BOPA based power allocation among subchannels in a single relay scenario.

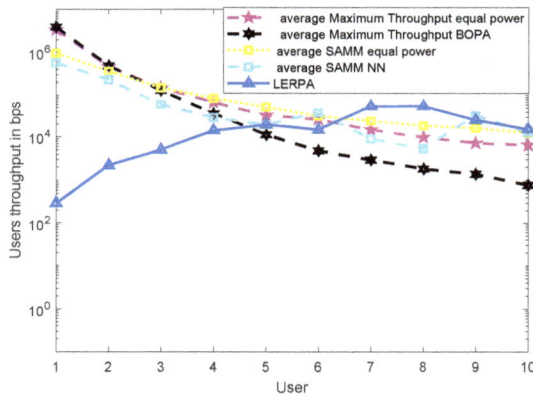

Figure 21. The users' throughput with neural network for SAMM along with LERPA of [11] in multiple relays scenario.

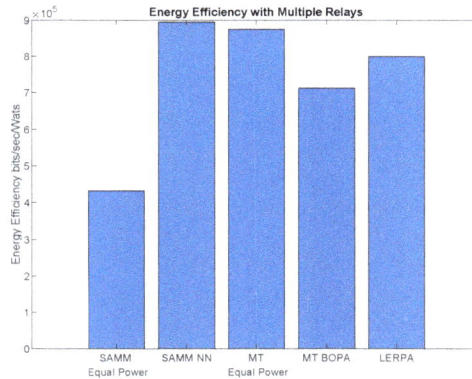

Figure 22. The system energy efficiency with neural network for SAMM which is trained on waterfilling based power allocation among users and BOPA based power allocation among subchannels along with LERPA of [11] in a multiple relays scenario.

Table 3 summarizes simulation results.

Table 3. Simulation Results.

KPI	SAMM Equal Power	SAMM BOPA	SAMM NN	MT Equal Power	MT BOPA	LERPA [11]
Energy Efficiency (Mbps/Watts)	0.5128	0.5481	1.0630	0.6660	0.5202	0.8862
System average throughput (Mbps)	2.1471	1.1845	1.1037	4.0778	4.5775	0.4137
Throughput fairness index	0.3155	0.2337	0.2234	0.1453	0.1366	0.5797

It can be seen that SAMM with BOPA and NN compete well in fairness with best EE. Tradeoff has to be done on system throughput. LERPA has better fairness performance but is less efficient in EE and system throughput, whereas, the hypothetical MT performs better in average system throughput. We say hypothetical because it only allocates the RB and power to the users with the highest SNR which can not be applicable on practical scenarios.

5. Complexity Analysis

The RB allocation scheme SAMM uses alternate MT and PF metrics to assign the N RBs to K users. MT assigns $\frac{N}{2}$ RBs to K users and PF assigns $\frac{N}{2}$ RBs to $K-1$ users in alternate TTI. Therefore, the computational complexity of SAMM is $\mathcal{O}\left(N(K-\frac{1}{2})\right)$. The BOPA Algorithm 1, first requires λ_{min} and λ_{max} in line 4 using water-filling algorithm for which the worse-case complexity is $\mathcal{O}(2NK)$. After that, BOPA uses binary search method to estimate the roots of Equation (17). In the worse case, with N_p points in the search space, binary search requires $\log_2(N_p)$ iterations to find the roots of polynomial. In our case, $N_p = \frac{\lambda_{max}-\lambda_{min}}{\epsilon}$, where ϵ is the error tolerance. Therefore, the overall complexity of the Algorithm 1 is $\mathcal{O}(2NK^2\log_2(N_p))$. In case of the optimal exhaustive search (K^N) RB allocation combining with the BOPA; the complexity is $\mathcal{O}(2NK^{N+2}\log_2(N_p))$, whereas, the complexity of SAMM-BOPA is $\mathcal{O}((NK)^2(2K+1)\log_2(N_p))$.

The running-time complexity of the K-mean algorithm is $\mathcal{O}(kmdi)$ [31], where k is the number of clusters, m is the number of objects to be clustered, d is the dimension of objects, and i is the number of iterations. In our application of K-mean Algorithm 2, we use $min_k < k < max_k$ and two-dimensional geographical location of the users. Therefore, the worse-case computational complexity is given as $\mathcal{O}(max_k Ki)$.

6. Conclusions

In this paper, we have investigated the impact of using single and multiple L3 relays in terms of EE and throughput. For a single relay scenario, equal power and BOPA are used in conjunction with the SAMM and MT RB allocation algorithms. Simulation results show that SAMM BOPA has 26% power saving when compared with MT BOPA. Whereas, when comparing SAMM with equal power allocation to all RBs, our proposed scheme gives 77% increase in EE. For a multiple relay scenario, a clustering scheme is proposed that addresses relay placement and users' association. This information acts as an input to a machine learning process (SAMM NN) that cognizes both the SAMM and BOPA approaches using One-Vs-All classification and feedforward neural networks, respectively. The SAMM NN approach when compared with the SAMM Equal Power, gives a 2.07 times increase in EE at the cost of 0.72 times decrease in throughput. A SAMM BOPA approach adopted in the case of single relay still provided the best tradeoff in terms of energy efficiency EE, throughput and fairness in the case of multiple relays.

Author Contributions: H.H., I.A., W.A. and R.A. contributed the key idea and defined the main system model. W.A. and M.M.A. assisted with the system model and the mathematical analysis. I.A. developed the machine learning framework. I.A., G.B. and H.K. developed the model for multiple relay scenario, relay selection and user association. I.A., W.A. and R.A. analysed all the results and added the relevant discussions. All authors contributed to the paper write up.

Funding: This research was supported in part by King Abdul Aziz City for Science and Technology Project under Grant PC-37-66. This research was also partially supported by European Union's Horizon 2020 Research and Innovation Program under Grant 668995, European Union Regional Development Fund in the framework of the Tallinn University of Technology Development Program 2016-2022, and NATO-SPS funding grant agreement No. G5482.

Conflicts of Interest: The authors declare no conflict of interest.

Abbreviations

The following abbreviations are used in this manuscript:

BER	Bit Error Rate
BOPA	Bisection based Optimal Power Allocation
BS	Base Station
CA	Carrier Aggregation
CoMP	Coordinated Multipoint
CSI	Channel State Information
EE	Energy Efficiency
JFI	Jains Fairness Index
L3	Layer 3
LTE-A	Long Term Evolution Advanced
MIMO	Multiple-input multiple-output
MQAM	M-ary Quadrature Amplitude Modulation
MT	Maximum Throughput
OFDM	Orthogonal Frequency Division Multiplexing
PF	Proportional Fairness
QoS	Quality of service
RB	Resource Block
RN	Relay Node
RR	Round Robin
SAMM	hybrid proportional fairness scheme

SINR	signal-to-interference-and-noise ratio
SNR	signal-to-noise ratio
TTI	Transmission Time Interval
UE	User Equipment

References

1. Edler, T.; Lundberg, S. Energy efficiency enhancements in radio access networks. *Ericsson Rev.* **2004**, *81*, 42–51.
2. Kumar, R.; Mieritz, L. Conceptualizing green IT and data center power and cooling issues (ID: G00150322), 7 September 2007. *Retrieved from Gartner Database*; Available online: https://www.gartner.com/en/documents/519717/conceptualizing-green-it-and-data-center-power-and-coolin (accessed on 20 July 2019).
3. 3GPP; Evolved Universal Terrestrial Radio Access (E-UTRA); Study on mobile relay. Technical Report (TR) 36.836, 3rd Generation Partnership Project (3GPP); 2014, Version 12.0.0. Available online: https://portal.3gpp.org/desktopmodules/Specifications/SpecificationDetails.aspx?specificationId=2427 (accessed on 21 May 2019).
4. Al-Amri, A.; Al-Zahrani, S.; Al-Harthi, M.; Al-Qarni, M.; Ahmed, I. Hybrid frequency-time domain proportional fair resource allocation scheme for LTE downlink. In Proceedings of the 2012 18th IEEE International Conference on Networks (ICON), Singapore, 12–14 December 2012; pp. 286–290. [CrossRef]
5. 3GPP; Long Term Evolution - Evolved Packet System RAN part (LTE). Release 8, 3rd Generation Partnership Project (3GPP); 2014, Version 0.3.3. Available online: https://www.3gpp.org/specifications/releases/72-release-8 (accessed on 21 May 2019).
6. Ren, Z.; Chen, S.; Hu, B.; Ma, W. Energy-efficient resource allocation in downlink OFDM wireless systems with proportional rate constraints. *IEEE Trans. Veh. Technol.* **2014**, *63*, 2139–2150. [CrossRef]
7. Hassan, H.; Ahmed, I.; Ahmad, R.; Ahmed, W.; Alam, M.M. Energy efficiency for bisection based power allocation with proportional fairness in relay-assisted LTE-A downlink system. In Proceedings of the Biennial Baltic Electronics Conference, Tallinn, Estonia, 8–10 October 2018; pp. 1–4. [CrossRef]
8. Alam, M.S.; Mark, J.W.; Shen, X.S. Relay Selection and Resource Allocation for Multi-User Cooperative OFDMA Networks. *IEEE Trans. Wirel. Commun.* **2013**, *12*, 2193–2205. [CrossRef]
9. Yusoff, R.; Dani Baba, M.; Ali, D. Energy-efficient resource allocation scheduler with QoS aware supports for green LTE network. In Proceedings of the 2015 6th IEEE Control and System Graduate Research Colloquium, Shah Alam, Malaysia, 10–11 August 2015; pp. 109–111. [CrossRef]
10. Hua, Y.; Zhang, Q.; Niu, Z. Resource Allocation in Multi-cell OFDMA-based Relay Networks. In Proceedings of the 2010 Proceedings IEEE INFOCOM, San Diego, CA, USA, 14–19 March, 2010, pp. 1–9. [CrossRef]
11. Yang, K.; Martin, S.; Quadri, D.; Wu, J.; Feng, G. Energy-Efficient Downlink Resource Allocation in Heterogeneous OFDMA Networks. *IEEE Trans. Veh. Technol.* **2017**, *66*, 5086–5098. [CrossRef]
12. Jiang, C.; Zhang, H.; Ren, Y.; Han, Z.; Chen, K.C.; Hanzo, L. Machine Learning Paradigms for Next-Generation Wireless Networks. *IEEE Wirel. Commun.* **2017**, *24*, 98–105. [CrossRef]
13. Sun, Y.; Peng, M.; Zhou, Y.; Huang, Y.; Mao, S. Application of Machine Learning in Wireless Networks: Key Techniques and Open Issues. Available online: https://arxiv.org/abs/1809.08707 (accessed on 2 May 2019).
14. Amiri, R.; Mehrpouyan, H.; Fridman, L.; Mallik, R.K.; Nallanathan, A.; Matolak, D. A Machine Learning Approach for Power Allocation in HetNets Considering QoS. In Proceedings of the IEEE International Conference on Communications, Kansas City, MO, USA, 20–24 May 2018; pp. 1–7. [CrossRef]
15. Wang, J.B.; Wang, J.; Wu, Y.; Wang, J.Y.; Zhu, H.; Lin, M.; Wang, J. A Machine Learning Framework for Resource Allocation Assisted by Cloud Computing. *IEEE Netw.* **2018**, *32*, 144–151. [CrossRef]
16. Sun, C.; Cheng, Y.; Shi, H.; Liu, L.; Ma, R.; Cao, X. A Machine Learning-Based Algorithm for Joint Scheduling and Power Control in Wireless Networks. *IEEE Internet Things J.* **2018**, *5*, 4308–4318. [CrossRef]
17. Huang, D.; Gao, Y.; Li, Y.; Hou, M.; Tang, W.; Cheng, S.; Li, X.; Sun, Y. Deep Learning Based Cooperative Resource Allocation in 5G Wireless Networks. *Mob. Netw. Appl.* **2018**, *23* 1–8.[CrossRef]
18. Stahlbuhk, T.; Shrader, B.; Modiano, E. Learning Algorithms for Scheduling in Wireless Networks with Unknown Channel Statistics. In Proceedings of the Eighteenth ACM International Symposium on Mobile Ad Hoc Networking and Computing, Mobihoc '18, Los Angeles, CA, USA, 26–29 June 2018; ACM: New York, NY, USA, 2018; pp. 31–40. [CrossRef]

Sensors **2019**, *19*, 3461

19. Ahmed, K.I.; Tabassum, H.; Hossain, E. Deep Learning for Radio Resource Allocation in Multi-Cell Networks. *IEEE Netw.* **2019**, 1–8.[CrossRef]

20. Témoa, D.; Förster, A.; Kolyang.; Yamigno, S.D. A reinforcement learning based intercell interference coordination in LTE networks. *Future Internet* **2019**, *11*, 1–23. [CrossRef]

21. Ahmed, I.; Mohamed, A.; Shakeel, I. On the group proportional fairness of frequency domain resource allocation in L-SC-FDMA based LTE uplink. In Proceedings of the 2010 IEEE Globecom Workshops, GC'10, Miami, FL, USA, 6–10 December 2010. [CrossRef]

22. Shen, Z.; Andrews, J.G.; Evans, B.L. Adaptive resource allocation in multiuser OFDM systems with proportional rate constraints. *IEEE Trans. Wirel. Commun.* **2005**, *4*, 2726–2737. [CrossRef]

23. Escudero-Garzas, J.J.; Devillers, B.; Garcia-Armada, A. Fairness-Adaptive Goodput-Based Resource Allocation in OFDMA Downlink with ARQ. *IEEE Trans. Veh. Technol.* **2014**, *63*, 1178–1192. [CrossRef]

24. Miao, G.; Himayat, N.; Li, G.Y. Energy-efficient link adaptation in frequency-selective channels. *IEEE Trans. Commun.* **2010**, *58*, 545–554. [CrossRef]

25. Jain, R.; Chiu, D.M.; Hawe, W.R. *A Quantitative Measure of Fairness and Discrimination for Resource Allocation in Shared Computer System*; DEC Research Report TR-301; Eastern Research Laboratory, Digital Equipment Corporation: Hudson, MA, USA 1984; Volume 38.

26. Mikio, I.; Hideakii, T.; Satoshi, N. LTE-Advanced Relay Technology Self-backhauling. *NTT DOCOMO Tech. J.* **2011**, *12*, 29–36.

27. Møller, M.F. A scaled conjugate gradient algorithm for fast supervised learning. *Neural Netw.* **1993**, *6*, 525–533. [CrossRef]

28. Mathworks. Classify Patterns with a Shallow Neural Network—MATLAB & Simulink. Available online: https://www.mathworks.com/help/deeplearning/gs/classify-patterns-with-a-neural-network.html (accessed on 20 May 2019).

29. Dan Foresee, F.; Hagan, M. Gauss-Newton approximation to Bayesian learning. In Proceedings of International Conference on Neural Networks (ICNN'97), Lausanne, Switzerland, 8–10 October 1997; Volume 3, pp. 1930–1935. [CrossRef]

30. Mathworks. Bayesian Regularization Backpropagation—MATLAB Trainbr. Available online: https://www.mathworks.com/help/deeplearning/ref/trainbr.html (accessed on 22 May 2019).

31. Hartigan, J.A.; Wong, M.A. A K-Means Clustering Algorithm. *Appl. Stat.* **1979**, *28*, 100. [CrossRef]

![sensors logo]

Article

On the Performance of Energy Harvesting Non-Orthogonal Multiple Access Relaying System with Imperfect Channel State Information over Rayleigh Fading Channels

Tran Manh Hoang [1,2], **Nguyen Le Van** [1], **Ba Cao Nguyen** [1] **and Le The Dung** [3,4,*,†]

1   Faculty of Radio Electronics, Le Quy Don Technical University, Hanoi 11917, Vietnam
2   Faculty of Telecommunications Services, Telecommunications University, Khanh Hoa 650000, Vietnam
3   Division of Computational Physics, Institute for Computational Science, Ton Duc Thang University, Ho Chi Minh City 758307, Vietnam
4   Faculty of Electrical and Electronics Engineering, Ton Duc Thang University, Ho Chi Minh City 758307, Vietnam
*   Correspondence: lethedung@tdtu.edu.vn
†   Current address: 19 Nguyen Huu Tho Street, Tan Phong Ward, District 7, Ho Chi Minh City 758307, Vietnam.

Received: 31 May 2019; Accepted: 19 July 2019; Published: 29 July 2019

Abstract: In this paper, we propose a non-orthogonal multiple access (NOMA) relaying system, where a source node communicates simultaneously with multiple users via the assistance of the best amplify-and-forward (AF) relay. The best relay is selected among N relays which are capable of harvesting the energy from radio frequency (RF) signals. We analyze the performance of the proposed NOMA relaying system in the conditions of imperfect channel state information (CSI) and Rayleigh fading by deriving the exact expressions of the outage probability (OP) and the approximate expression of the ergodic capacities of each user and the whole system. We also determine the optimal energy harvesting duration which minimizes the OP. Numerical results show that, for the same parameter settings, the performance of the proposed NOMA relaying system, especially the ergodic capacity of the whole system, outperforms that of the orthogonal-multiple-access (OMA) relaying system. Monte-Carlo simulations are used to validate the correctness of the analytical results.

Keywords: NOMA; energy harvesting; amplify-and-forward; imperfect CSI; successive interference cancellation (SIC)

---

1. Introduction

Nowadays, the Internet of Things (IoT) has received increasing attention from both industry and academia. It is considered an important mean for wireless connections in the fourth industrial revolution. IoT is also being used in the fourth generation (4G) mobile communications and will be applied to the fifth generation (5G). In order to support a large multiuser system such as IoT, the non-orthogonal multiple access (NOMA) is a very potential technique due to its high bandwidth efficiency [1,2]. Moreover, compared with conventional orthogonal multiple access (OMA) systems, such as time division multiple access (TDMA), code division multiple access (CDMA), orthogonal frequency division multiple access (OFDMA), the NOMA systems offer better fairness among users, even for users with weak channel conditions such as the cell-edge users. The fundamentals of the NOMA system can be found in Reference [3] while a study of NOMA system in cellular communication with machine-to-machine in IoT is given in Reference [4].

Recently, the power supply for terminal devices in wireless networks has become an important matter and has attracted much interest from researchers. Besides using the optimal power allocation for the fifth generation (5G) and sixth generation (6G) networks (6G network will start to enter the market by 2026 [5]) to reduce the power consumption [6], another promising method to improve the lifetime of communication devices is to generate electric power from some external energy sources such as solar, wind, and radio frequency (RF) signal to charge the batteries. Unfortunately, natural energy sources are not suitable for small-size mobile devices and in some cases they cannot be used in the healthcare monitoring networks and the sensor networks with real-time requirements. In contrast, the RF energy is often available due to its increased power density and availability, and is independent on environmental conditions, including weather, climate, and temperature. As the result, the RF energy harvesting (EH), also called simultaneous wireless information and power transfer (SWIPT), has been widely used compared with other kinds of energy harvesting techniques [7–9]. SWIPT has been applied not only in the point-to-point systems but also in relaying systems because deploying relays can improve the amount of harvested energy and the coverage area of wireless networks. The authors of References [10,11] investigated information and energy receiver architecture for SWIPT networks. Reference [11] especially considered a non-linear energy harvesting model which described the practical system well.

To prolong network lifetime and improve the spectral utilizing efficiency, NOMA is combined with SWIPT [12]. In Reference [13], the authors investigated the tradeoff among the energy efficiency, fairness, harvested energy, and system sum rate of NOMA systems in power domain. Investigation of an integrated wireless communication system including NOMA, full-duplex relaying, and energy harvesting techniques was conducted in Reference [14]. The authors of References [15,16] studied the system performance of cooperative NOMA systems and derived the expressions of outage probability in the conditions of perfect successive interference cancellation (SIC) and perfect channel state information (CSI). In Reference [17], the near users which are close to the base station will harvest the RF energy and forward signals to far users. The analysis results showed that if the time switching ratio in NOMA system with SWIPT is appropriately chosen, the diversity gain will not be impaired. The authors of Reference [18] proposed a NOMA system where source node communicates with two users via the assistance of the best relay with the RF energy harvesting capability. The exact expressions of the outage probability and throughput were used as the criteria to evaluate the system performance. The effects of power allocation and time switching ratio on the performance of multi-user NOMA system were investigated in Reference [19]. Specifically, the authors derived the outage probability expression and determined the optimal power allocation coefficient for two NOMA power allocation policies, namely NOMA with fixed power allocation (F-NOMA) and cognitive radio inspired NOMA (CR-NOMA). It was shown that when a reasonable power allocation coefficient is selected, higher system performance can be achieved in comparison with the conventional multi-user system.

We observe that all previous works only mentioned the case of perfect CSI and used only one relay to forward signals to multiple users. Moreover, although the partial relay selection has been widely studied in conventional wireless systems, it has not been analyzed in NOMA systems. Another observation is that the NOMA systems perform superimposing signals in power domain, thus they always require CSI to allocate power for all users. However, due to variation in the communication quality of wireless environment, the imperfect CSI may happen [20,21]. Perfect CSI exists if and only if the amount of feedback CSI from users to the base station is large and the length of the pilot sequences which are used to estimate channel is very long. Unfortunately, these conditions rarely happen in practice. Therefore, investigation of the impact of imperfect CSI on the relay selection and power allocation is vitally important to the design of practical NOMA systems.

Motivated by the above issues, in this paper we propose a downlink NOMA relaying system with partial relaying selection. In this system, source node transmits superposition modulated signals to multiple users via the assistance of the best relay. The best relay is chosen from a set of relays which are capable of harvesting RF energy and grouped by their locations. Based on the feedback CSI from all users, the source node performs power allocation and chooses the best communication link. The main contributions of this paper can be summarized as follows:

- We overcome the limitation of current multiple access techniques and the energy demand of wireless networks by proposing the downlink NOMA relaying system where the best relay is selected from a set of multiple RF energy harvesting relays.
- We study the system performance in terms of the outage probability and the ergodic capacity of each user and the whole system in the condition of imperfect CSI and Rayleigh fading. The imperfection of the CSI is modeled by the correlation coefficient and its impact on the system performance is investigated by using both analysis and simulation approaches. We also compare the outage performance and the ergodic capacity of the proposed NOMA relaying system with those of OMA relaying system.
- We determine the optimal time switching ratio to balance between the energy harvesting and the signal processing so that the outage probability can be minimized. All analysis results are validated by simulation results.

The rest of this paper is organized as follows. Section 2 describes the proposed downlink NOMA relaying system with partial relaying selection and time switching (TS) protocol. The analysis of the outage probability and ergodic capacity of the proposed system are presented in Sections 3 and 4, respectively. Section 5 shows numerical results to evaluate the system performance. Finally, the conclusions are given in Section 6.

For the sake of clarity, the frequently used mathematical notations together with their descriptions are summarized in Table 1.

Table 1. The mathematical notations used in this paper.

Notation	Description
$F_U(u)$	Cumulative distribution function (CDF)
$f_U(u)$	Probability density function (PDF)
$\mathcal{CN}(\mu, \sigma^2)$	Circularly symmetric complex Gaussian distribution X with mean μ and variance σ^2
γ_{th}	Predefined outage threshold
$\mathbb{E}\{\cdot\}$	Expectation operator
$\mathcal{K}_n(\cdot)$	Second order Bessel function n [22]
$I_0(\cdot)$	Modified zero order Bessel function of first kind [22]
α	Time switching ratio
η	Energy conversion efficiency
ρ	Channel correlation coefficient
T	Transmission period

2. System Model

Figure 1 illustrates the proposed downlink NOMA relaying system. In this system, source node S transmits the signals which are coded and superposed in power domain to multiple users D_m, $m \in \{1, \cdots, M\}$, via the assistance of the best relay which is selected from a set of relays R_n, $n \in \{1, \cdots, N\}$. The direct link S-D_m is assumed not available because the distance between S and D_m is larger than the coverage area of S or due to deep shadow fading.

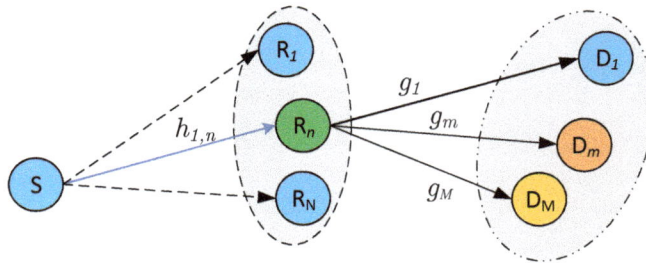

Figure 1. System model of downlink non-orthogonal multiple access (NOMA) relaying system with simultaneous wireless information and power transfer (SWIPT).

We consider that all nodes are equipped with single antenna and operate in half-duplex mode. All channels between S and R_n and between R_n and D_m are influenced by block fading, that is, the symbol rate is larger than channel varying rate so that it can be considered as constant over each symbol duration. The communication links from S to R_n and from R_n to D_m are respectively modeled as complex Gaussian distributions with zero mean and variances $\Omega_{1,n}$ and $\Omega_{R_nD_m}$, that is, $h_{1,n} \sim \mathcal{CN}(0, \Omega_{1,n})$ and $g_m \sim \mathcal{CN}(0, \Omega_{R_nD_m})$. The Additive White Gaussian Noise (AWGN) at the relays and users are $w_{R_n} \sim \mathcal{CN}(0, \sigma_{R_n}^2)$ and $w_{D_m} \sim \mathcal{CN}(0, \sigma_{D_m}^2)$, respectively. Because of the time varying characteristic of wireless channel, its coherent time may be altered when the feedback delay is larger than the transmission block period of a symbol. Thus, the received CSI is always outdated at the transmitter, which often happens in practice [23,24].

We denote ρ_i, $0 \leq \rho_i \leq 1$, $i \in \{1,2\}$, as the correlation coefficients between the past channel $h_{1,n}$ and the current channel $\hat{h}_{1,n}$, similarly for $\hat{g}_m$ and g_m. These coefficients can be considered as the measurements of the fluctuation rate of wireless channels and are related solely to the time delay. Based on the Markov chain, the relationship between $\hat{h}_{1,n}$ and $h_{1,n}$ and between $\hat{g}_m$ and g_m can be presented as [25]

$$\hat{h}_{1,n} = \rho_1 h_{1,n} + \sqrt{1 - \rho_1^2}\varepsilon_{1,n}, \tag{1}$$

$$\hat{g}_m = \rho_2 g_m + \sqrt{1 - \rho_2^2}\varepsilon_m, \tag{2}$$

where $\varepsilon_{1,n}$ and ε_m are the circular symmetric complex Gaussian random variables which can be modeled as $\varepsilon_{1,n} \sim \mathcal{CN}(0, \sigma^2)$ and $\varepsilon_m \sim \mathcal{CN}(0, \sigma^2)$, respectively.

In this paper, a partial relay selection (PRS) scheme [26] is used to select the best relay from a set of relays. According to the PRS scheme, S continuously monitors the gain of S–R_n channels by using the feedback signal and selects the communication link that gives the largest instantaneous channel gain, that is,

$$\gamma_b = \arg \max_{n=1,2\cdots,N} \gamma_{1,n}, \tag{3}$$

where $\gamma_{1,n}$ is the instantaneous SINR of S-R_n link.

The advantage of using PRS scheme is that the system configuration is simpler and easier than using the full relay selection (FRS) scheme [27]. In addition, the results in Reference [28] show that both PRS and FRS schemes have the same average channel capacity in a high SNR regime and the outage probability of PRS is higher than that of FRS when the number of relays is more than 2. On the other hand, FRS scheme may not applicable for multiple-user systems because the distances from the best relay to users are different, thus the calculation complexity of FRS scheme is significantly increased with the number of users.

After a link from the source node to the best relay is established, the transmission period T for communication process is spitted into two parts (in this system, we use the time-division multiple access (TDMA) scheme). According to the time switching (TS) protocol [29], a time duration αT is used for EH. The remaining time duration $(1 - \alpha)T$ is divided into two equal time sub-slots, which are used for the information transmission. The first half $(1 - \alpha)T/2$ is used for the information transmission from source node to the relay and the remaining half $(1 - \alpha)T/2$ is used for the information transmission from the relay to the user. It should be noted that the case $\alpha = 1$ is not considered in this paper because when the energy harvesting time duration takes the whole transmission period T, i.e., the relay does not process any signals, the basic role in signal forwarding of the relay is eliminated [30]. Hence, we only consider the time switching ratio α in the range $0 \leq \alpha < 1$. Then, the harvested energy of the selected relay in time duration αT is expressed as [29]

$$E_h = \alpha T \eta P_S \max_{n=1,\cdots,N} |\hat{h}_{1,n}|^2, \tag{4}$$

where η is the energy conversion efficiency coefficient which varies from 0 to 1 and closely depends on the quality of energy harvesting electric circuitry, P_S is the transmission power of S.

In our proposed relaying system, since the harvest-use (HU) architecture is used, the relay does not need an energy buffer to store the harvested energy. Since all amounts of harvested energy during EH phase is consumed by R_n for signal transmission from R to D_m, from (4), the transmission power of the best relay is given by

$$P_R = \frac{E_h}{(1 - \alpha)T/2} = \frac{2\alpha\eta P_S}{1 - \alpha} \max_{n=1,\cdots,N} |\hat{h}_{1,n}|^2. \tag{5}$$

According to the NOMA technique in power domain, during the first time sub-slot $\frac{1-\alpha}{2}$, source node transmits the superimposed signal $x_S = \sum_{m=1}^{M} \sqrt{P_S a_m} x_m$, where x_m and a_m are the signal and power allocation coefficient of mth user, respectively. At the end of this time sub-slot, the received signal at R_n is

$$y_R^n = \hat{h}_{1,b} \sum_{m=1}^{M} \sqrt{a_m P_S} x_m + w_{R_n}, \tag{6}$$

where $\hat{h}_{1,b} = \max_{n=1,\cdots,N} |\hat{h}_{1,n}|^2$.

In the remaining second time slot $\frac{1-\alpha}{2}$, the relay employs the AF protocol to broadcast y_R^n to all users after multiplying it with an amplifying factor G. To keep the output power constraint at relay, it is required that $E\{\|Gy_R^n\|^2\} = P_R$, where P_R is given in (5), thus the amplifying factor G is given by

$$G = \sqrt{\frac{2\alpha\eta P_S |\hat{h}_{1,b}|^2}{(1 - \alpha)(P_S |\hat{h}_{1,b}|^2 + \sigma_R^2)}} \approx \sqrt{\frac{2\alpha\eta}{(1 - \alpha)}}. \tag{7}$$

Therefore, the received signal at D_m in the case of perfect SIC is expressed as

$$y_{D_m} = G\hat{h}_{1,b}\hat{g}_m \sqrt{a_m P_S} x_m + \underbrace{G\hat{h}_{1,b}\hat{g}_m \sum_{j=m+1}^{M} \sqrt{a_j P_S} x_j}_{\text{signals of other users}} + \underbrace{G\hat{g}_m w_{R_n} + w_{D_m}}_{\text{noise}}, \tag{8}$$

where $\hat{g}_m$ denotes the channel coefficient between R_n and D_m.

The received signals at the best relay and each user is comprised of the desired signal and the signals of other users, which are treated as the interferences. Hence, to mitigate the negative effect of the inter-user interference, successive interference cancellation (SIC) method is applied.

For the downlink communication considered in this paper, the optimal SIC algorithm performs decoding signals in an order of increasing channel gain [31] ($|g_{D_1}|^2 \leq |g_{D_2}|^2 \leq \cdots \leq |g_{D_m}|^2 \leq |g_{D_M}|^2$). To ensure the fairness among all users, the power allocation coefficients are assumed to be

$a_1 \geq a_2 \geq \cdots a_m \geq a_M$, with $\sum_{m=1}^{M} a_m = 1$. Hence, at the D_j, the signal of D_m, $j < m$, will be detected and then be removed from the received signal by SIC method. Specifically, D_j first decodes symbol x_m while treating x_j as noise.

Then, the SINR of symbol x_m at D_j is given by

$$\gamma_{m,j}^{D} = \frac{G^2 a_m P_S |\hat{h}_{1,b}|^2 |\hat{g}_j|^2}{G^2 \sum_{j=m+1}^{M} a_j P_S |\hat{h}_{1,b}|^2 |\hat{g}_j|^2 + G^2 |\hat{g}_j|^2 \sigma_R^2 + \sigma_{D_m}^2}, \tag{9}$$

where $j \in \{1, ..., m\}$ and $m \neq M$.

At D_j, SIC will be performed until all signals of D_m are decoded successfully. Thus, the required SINR at D_m to successfully decode the signal by itself is given by

$$\gamma_m^{D} = \frac{G^2 a_m P_S |\hat{h}_{1,b}|^2 |\hat{g}_m|^2}{G^2 \sum_{j=m+1}^{M} a_j P_S |\hat{h}_{1,b}|^2 |\hat{g}_m|^2 + G^2 |\hat{g}_m|^2 \sigma_R^2 + \sigma_{D_m}^2}. \tag{10}$$

We should note that the last user D_M needs to decode all signals of other users before decoding its signals. Consequently, the SINR for D_M to decode its own signals can be expressed as

$$\gamma_M^{D} = \frac{G^2 a_M P_S |\hat{h}_{1,b}|^2 |\hat{g}_M|^2}{G^2 |\hat{g}_M|^2 \sigma_R + \sigma_{D_M}^2}. \tag{11}$$

3. Outage Probability Analysis

In this section, we derive the exact closed-form expression of the outage probability, taking into consideration the imperfect CSI and partial relay selection. It is well-known that the event that D_j can decode the signals of D_m successfully is

$$\Delta_{m,j} = \left\{ \frac{G^2 a_m P_S |\hat{h}_{1,b}|^2 |\hat{g}_j|^2}{G^2 \sum_{j=m+1}^{M} a_j P_S |\hat{h}_{1,b}|^2 |\hat{g}_j|^2 + G^2 |\hat{g}_j|^2 \sigma_R^2 + \sigma_{D_m}^2} > \gamma_{thj} \right\}, \tag{12}$$

where $\gamma_{thj} = 2^{\frac{2r}{1-\alpha}} - 1$ is the predefined outage threshold. This threshold is served as the protected value of the SINR to ensure the quality of service of the system and satisfy the target data rate r of D_j.

Let us denote $X = |\hat{h}_{1,b}|^2$ and $Z = |\hat{g}_j|^2$. Without loss of generality, we assume that the temperature noise $\sigma_R^2 = \sigma_{D_m}^2 = \sigma^2$. Thus from (12), we can rewrite $\Delta_{m,j}$ as

$$\Delta_{m,j} = \left\{ \frac{G^2 a_m P_S X Z}{G^2 \sum_{j=m+1}^{M} a_j P_S X Z + G^2 Z + \sigma^2} > \gamma_{thj} \right\}. \tag{13}$$

From (13) and after some manipulations, we can rewrite (13) as

$$\Delta_{m,j} \stackrel{(\lambda)}{=} \left\{ X > \theta_j, \ Z > \frac{\theta_j}{G^2 (X - \theta_j)} \right\}, \tag{14}$$

where $\theta_j = \dfrac{\gamma_{thj}}{P_S \left(a_m - \sum_{j=m+1}^{M} a_j \gamma_{thj} \right)}$, step λ holds when the condition $a_m > \sum_{j=1+m}^{M} a_j \gamma_{thj}$ is satisfied. It should be noticed that $\theta_j = \dfrac{\gamma_{thj}}{P_S \left(a_m - \sum_{j=m+1}^{M} a_j \gamma_{thj} \right)}$ is a constant and depends on the power allocation coefficient and the target data rate of D_j.

The outage event occurs at D_j when it fails to decode its own signal or unsuccessfully performs SIC for the signals of D_m [32], i.e., $\Lambda_{m,j} = \gamma_{m,j}^D < \gamma_{\text{thj}}, 1 \le j \le m$. Outage probability of the system occurs when the maximum SNRs at D_j falls below the threshold to decode signal. Thus, we have

$$P_{out}^j = \text{Pr}(\gamma_{m,j}^D \le \gamma_{\text{thj}}) = 1 - \text{Pr}(\gamma_{m,j}^D > \gamma_{\text{thj}}), \quad 1 \le j \le m. \tag{15}$$

$$P_{out}^j = 1 - \text{Pr}\{\Delta_{m,1} \cap \Delta_{m,2} \cap ... \cap \Delta_{m,m}\}, \tag{16}$$

where $\Delta_{m,j}$ is the complementary in the set of $\Lambda_{m,j}$.

The condition in (14) always occurs, i.e., the outage probability is equal to one, if $a_m \le \sum_{j=1+m}^M a_j \gamma_{\text{thj}}$. Hence, we need to allocate more power for D_m to satisfy the following condition

$$a_m > \sum_{j=1+m}^M a_j \gamma_{\text{thj}}. \tag{17}$$

Let us denote $\theta^* = \max(\theta_1, \theta_2, \cdots, \theta_m)$ [33], then the outage probability P_{out}^j of D_j can be reformulated as

$$P_{out}^j = 1 - \text{Pr}\left\{Z > \frac{\theta^*}{G^2(X - \theta^*)}, \ X > \theta^*\right\}. \tag{18}$$

Using the conditional probability property [34] with respect to X, and applying the law of joint CDF, we have

$$P_{out}^j = 1 - \int_{\theta^*}^{\infty} \left[1 - F_Z\left(\frac{\theta^*}{G^2(x - \theta^*)}\right)\right] f_X(x)dx. \tag{19}$$

To calculate the expression of the outage probability in (19), we first derive the CDF of Z and the PDF of X as follows.

When the nth relay is selected as the best relay, the PDF of order statistic with respect to $|h_{1,b}|^2$ in a set of N relays is obtained by using the binomial Newton expansion [35], that is,

$$\begin{aligned}
f_{|h_{1,b}|^2}(x) &= N\left[F_{|h_{1,i}|^2}(x)\right]^{N-1} f_{|h_{1,i}|^2}(x) \\
&= \sum_{n=1}^N \binom{N}{n}(-1)^{n-1}\frac{n}{\Omega_{1,n}} \exp\left(-\frac{nx}{\Omega_{1,n}}\right),
\end{aligned} \tag{20}$$

where $\binom{N}{n} = \frac{n!}{n!(N-n)!}$, N and n are non-negative integers, $f_{|h_{1,i}|^2}(x) = \frac{1}{\Omega_{1,i}}\exp\left(-\frac{x}{\Omega_{1,i}}\right)$ and $F_{|h_{1,i}|^2}(x) = 1 - \exp\left(-\frac{x}{\Omega_{1,i}}\right)$ are respectively the CDF and PDF of $|h_{1,i}|^2$, which is the channel gain of each link from source node to relay. According to the probability theory, the PDFs of $|\hat{h}_{1,b}|^2$ and $|h_{1,b}|^2$ which are respectively denoted by $f_{|\hat{h}_{1,b}|^2}(\hat{x})$ and $f_{|h_{1,b}|^2}(x)$ can be calculated by using the joint PDF, i.e., $f_{|\hat{h}_{1,b}|^2}(\hat{x}) = \int_0^{\infty} f_{|\hat{h}_{1,b}|^2,|h_{1,b}|^2}(\hat{x}, x)dx$. Another way to calculate the joint PDF of $|\hat{h}_{1,b}|^2$ is based on the properties of conditional probability, that is,

$$f_{|\hat{h}_{1,b}|^2}(\hat{x}) = \int_0^{\infty} f_{|\hat{h}_{1,b}|^2||h_{1,b}|^2}(\hat{x}|x) f_{|h_{1,b}|^2}(x)dx, \tag{21}$$

where

$$f_{|\hat{h}_{1,b}|^2||h_{1,b}|^2}(\hat{x}|x) = \frac{f_{|h_{1,i}|^2,|h_{1,i}|^2}(\hat{x}, x)}{f_{|h_{1,i}|^2}(x)}. \tag{22}$$

Using the joint PDF which is given in ([36], Equation (9.389)), we can rewrite the numerator of (22) as

$$f_{|\hat{h}_{1,i}|^2,|h_{1,i}|^2}(\hat{x},x) = \frac{\exp\left(-\frac{(\hat{x}+x)}{(1-\rho^2)\Omega_{1,n}}\right)}{(1-\rho^2)\Omega_{1,n}^2} I_0\left(\frac{2\rho\sqrt{\hat{x}x}}{(1-\rho^2)\Omega_{1,n}}\right), \tag{23}$$

where $I_0(x)$ is the modified zero order Bessel function of the first kind [22].

Without loss of generality, all correlation coefficients are assumed to have the same values, that is, $\rho = \rho_1 = \rho_2$. Substituting (20), (22), and (23) into (21), after using the equation $\int_0^\infty e^{-\alpha z} I_0(2\sqrt{\beta z})dz = (1/\alpha)\exp(\beta/\alpha)$ which is given in ([22], Equation (6.614.3)), and then perform some manipulations, we have the PDF of X in the case of imperfect CSI as

$$f_{|\hat{h}_{1,b}|^2}(\hat{x}) = \sum_{n=1}^{N}\binom{N}{n}\frac{n(-1)^{n-1}}{\Omega_{1,n}\Psi(\rho,n)}\exp\left(-\frac{n\hat{x}}{\Omega_{1,n}\Psi(\rho,n)}\right), \tag{24}$$

where $\Psi(\rho,n) = 1 + (n-1)(1-\rho^2)$.

From (24), the CDF of $|\hat{h}_{1,b}|^2$ is given by

$$F_{|\hat{h}_{1,b}|^2}(\hat{x}) = 1 - \sum_{n=1}^{N}\binom{N}{n}(-1)^{n-1}\exp\left(-\frac{n\hat{x}}{\Omega_{1,n}\Psi(\rho,n)}\right). \tag{25}$$

Based on the result of order statistics which is provided in ([34], Equation (7.14), p. 246), and after some similar calculations as above, the PDF of the ordered variable Z is expressed as

$$f_{|\hat{g}_j|^2}(z) = \sum_{j=1}^{M}\binom{M}{j}\frac{(-1)^{j-1}j}{\Omega_z\Psi(\rho,j)}\exp\left(-\frac{jz}{\Omega_z\Psi(\rho,j)}\right), \tag{26}$$

where $\Psi(\rho,j) = 1 + (j-1)(1-\rho^2)$.

From (26), we can derive the CDF of $|\hat{g}_i|^2$ as

$$F_{|\hat{g}_i|^2}(\hat{z}) = 1 - \sum_{j=1}^{M}\binom{M}{j}(-1)^{j-1}\exp\left(-\frac{j\hat{z}}{\Omega_z\Psi(\rho,j)}\right). \tag{27}$$

Plugging (27) and (24) into (19), and after some manipulations, we obtain the expression of the outage probability as

$$P_{out}^j = 1 - \sum_{j=1}^{M}\binom{M}{j}(-1)^{j-1}\sum_{n=1}^{N}\binom{N}{n}\frac{n(-1)^{n-1}}{\Omega_{1,n}\Psi(\rho,n)}\int_{\theta^*}^{\infty}\exp\left(-\frac{j\theta^*}{\Omega_z\Psi(\rho,j)G^2(x-\theta^*)}-\frac{n\hat{x}}{\Omega_{1,n}\Psi(\rho,n)}\right)dx. \tag{28}$$

Let $u = x - \theta^*$, (28) becomes

$$P_{out}^j = 1 - \sum_{j=1}^{M}\binom{M}{j}(-1)^{j-1}\sum_{n=1}^{N}\binom{N}{n}\frac{n(-1)^{n-1}}{\Omega_{1,n}\Psi(\rho,n)}\exp\left(-\frac{n\theta^*}{\Omega_{1,n}\Psi(\rho,n)}\right)$$
$$\times\int_0^{\infty}\exp\left(-\frac{j\theta^*}{\Omega_z\Psi(\rho,j)G^2u}-\frac{nu}{\Omega_{1,n}\Psi(\rho,n)}\right)du. \tag{29}$$

Using ([22], Equation (3.324)), we can rewrite the exact closed-form expression of the outage probability as in (30), where $K_1(.)$ denotes the modified first order Bessel function of the second kind.

$$P_{out}^j = 1 - \sum_{j=1}^{M}\binom{M}{j}(-1)^{j-1}\sum_{n=1}^{N}\binom{N}{n}\frac{(-1)^{n-1}}{\Omega_{1,n}}\exp\left(-\frac{n\theta^*}{\Omega_{1,n}\Psi(\rho,n)}\right)$$
$$\times\sqrt{\frac{4nj\theta^*}{\Omega_z\Psi(\rho,j)\Omega_{1,n}\Psi(\rho,n)G^2}}K_1\left(\sqrt{\frac{4nj\theta^*}{\Omega_z\Psi(\rho,j)\Omega_{1,n}\Psi(\rho,n)G^2}}\right). \tag{30}$$

From the expression of the outage probability which is given in (30), we can see that when the outdated CSI happens, the outage performance is a function of ρ.

4. Ergodic Capacity Analysis

In this section, we analyze the ergodic capacity of the proposed NOMA relaying system in comparison with that of the OMA relaying system. Due to the fact that the hardware complexity and performance degradation of the NOMA system is directly proportional to the number of users, we also set the number of users be equal to three for both NOMA and OMA systems as used in [37]. For the OMA system, we consider orthogonal frequency division multiple access (OFDMA). According to the Shannon theory, the instantaneous rate of D_m is given by

$$\mathcal{R}_{\text{NOMA}}^{mth} = \frac{1-\alpha}{2} \log_2 \left(1 + \gamma_m^D\right).\tag{31}$$

From (10), when the transmission power is high, we can approximate the required SINR at D_m as

$$\gamma_m^D \approx \frac{G^2 a_m P_S |\hat{h}_{1,b}|^2 |\hat{g}_m|^2}{G^2 \sum_{j=m+1}^{M} a_j P_S |\hat{h}_{1,b}|^2 |\hat{g}_m|^2 + \sigma_{D_m}^2}.\tag{32}$$

Substituting (32) into (31), we have

$$\begin{aligned}
\mathcal{R}_{\text{NOMA}}^{mth} &\approx \frac{1-\alpha}{2} \log_2 \left(1 + \frac{G^2 a_m P_S |\hat{h}_{1,b}|^2 |\hat{g}_m|^2}{G^2 \sum_{j=m+1}^{M} a_j P_S |\hat{h}_{1,b}|^2 |\hat{g}_m|^2 + \sigma_{D_m}^2}\right) \\
&= \frac{1-\alpha}{2} \log_2 \left(\frac{G^2 P |\hat{h}_{1,b}|^2 |\hat{g}_m|^2 + 1}{G^2 \sum_{j=m+1}^{M} a_j P |\hat{h}_{1,b}|^2 |\hat{g}_m|^2 + 1}\right),
\end{aligned}\tag{33}$$

where $P = \frac{P_S}{\sigma_{D_m}^2}$.

Based on the properties of the logarithmic function, we can rewrite (33) as

$$\mathcal{R}_{\text{NOMA}}^{mth} = \underbrace{\frac{1-\alpha}{2}\mathbb{E}\left\{\log_2\left(1 + G^2 P |\hat{h}_{1,b}|^2 |\hat{g}_m|^2\right)\right\}}_{\mathcal{I}_1}$$
$$- \underbrace{\frac{1-\alpha}{2}\mathbb{E}\left\{\log_2\left(1 + G^2 \sum_{m=1}^{M-1} a_m P |\hat{h}_{1,b}|^2 |\hat{g}_m|^2\right)\right\}}_{\mathcal{I}_2},\tag{34}$$

then solve its components by using the partial integration, i.e.,

$$\begin{aligned}
\mathcal{I}_u &= \left\{\log_2(1+\Gamma_u)\left[F_{\Gamma_u}(x_u) - 1\right]\right\}_0^\infty - \frac{1}{2\ln 2}\int_0^\infty \frac{1}{1+x_u}\left[F_{\Gamma_u}(x_u) - 1\right]dx_u \\
&= \frac{1}{2\ln 2}\int_0^\infty \frac{1}{1+x_u}\left[1 - F_{\Gamma_u}(x_u)\right]dx_u,
\end{aligned}\tag{35}$$

where $F_{\Gamma_u}(x_u)$ is the CDF of random variable Γ_u with $u \in \{1,2\}$, $\Gamma_1 = G^2 P |\hat{h}_{1,b}|^2 |\hat{g}_m|^2$, and $\Gamma_2 = G^2 \sum_{m=1}^{M-1} a_m P |\hat{h}_{1,b}|^2 |\hat{g}_m|^2$.

Using the condition probability, we have CDF of Γ_1 as

$$\begin{aligned}
F_{\Gamma_1}(x_1) &= \Pr(G^2 P |\hat{h}_{1,b}|^2 |\hat{g}_m|^2 \le x_1) \\
&= \int_0^\infty \Pr\left(|\hat{g}_m|^2 \le \frac{x_1}{G^2 P |\hat{h}_{1,b}|^2}\right) f_{|\hat{h}_{1,b}|^2} d|\hat{h}_{1,b}|^2.
\end{aligned}\tag{36}$$

From (24) and (27) we can calculate $F_{\Gamma_1}(x_1)$ as

$$
\begin{aligned}
F_{\Gamma_1}(x_1) = 1 - &\sum_{j=1}^{M} \binom{M}{j}(-1)^{j-1} \sum_{n=1}^{N} \binom{N}{n} \frac{(-1)^{n-1}}{\Omega_{1,n}} \\
&\times \sqrt{\frac{4njx_1}{\Omega_z \Psi(\rho,j)\Omega_{1,n}\Psi(\rho,n)PG^2}} K_1\left(\sqrt{\frac{4njx_1}{\Omega_z \Psi(\rho,j)\Omega_{1,n}\Psi(\rho,n)PG^2}}\right).
\end{aligned}
\tag{37}
$$

Similarly, for $F_{\Gamma_2}(x_2)$, we have

$$
\begin{aligned}
F_{\Gamma_2}(x_2) = 1 - &\sum_{j=1}^{M} \binom{M}{j}(-1)^{j-1} \sum_{n=1}^{N} \binom{N}{n} \frac{(-1)^{n-1}}{\Omega_{1,n}} \\
&\times \sqrt{\frac{4njx_2}{\Omega_z \Psi(\rho,j)\Omega_{1,n}\Psi(\rho,n)bPG^2}} K_1\left(\sqrt{\frac{4njx_2}{\Omega_z \Psi(\rho,j)\Omega_{1,n}\Psi(\rho,n)bPG^2}}\right).
\end{aligned}
\tag{38}
$$

where $b = \sum_{m=1}^{M-1} a_m$.

Replacing (38) into (35), we obtain $\mathcal{I}_1$ as

$$
\begin{aligned}
\mathcal{I}_1 = &\frac{1}{2\ln 2} \sum_{j=1}^{M} \binom{M}{j}(-1)^{j-1} \sum_{n=1}^{N} \binom{N}{n} \frac{(-1)^{n-1}}{\Omega_{1,n}} \\
&\times \int_0^\infty \frac{1}{1+x_1} \sqrt{\mathcal{A}(n,j)x_1} K_1\left(\sqrt{\mathcal{A}(n,j)x_1}\right) dx_1,
\end{aligned}
\tag{39}
$$

where $\mathcal{A} = \frac{4}{\Omega_z \Psi(\rho,j)\Omega_{1,n}\Psi(\rho,n)PG^2}$.

Based on ([22], Equation (9.343)), we can rewrite (39) as

$$
\mathcal{I}_1 = \frac{1-\alpha}{2\sqrt{2}\ln 2} \int_0^\infty \frac{1}{1+x_1} G_{0\,2}^{2\,0}\left(\frac{x_1}{\Omega_z \Psi(\rho,j)\Omega_{1,n}\Psi(\rho,n)PG^2}\,\middle|\,{\textstyle\frac{3}{4}, -\frac{1}{4}}\right) dx_1.
\tag{40}
$$

Then, using ([22], Equation (7.811.5)) and after some manipulations, we have

$$
\mathcal{I}_1 = \frac{1-\alpha}{2\sqrt{2}\ln 2} G_{1\,3}^{3\,1}\left(\frac{1}{\Omega_z \Psi(\rho,j)\Omega_{1,n}\Psi(\rho,n)PG^2}\,\middle|\,{\textstyle\genfrac{}{}{0pt}{}{0}{0,\frac{3}{4},-\frac{1}{4}}}\right),
\tag{41}
$$

where $G_{pq}^{mn}\left(x\big|_{b_s}^{a_r}\right)$ is the Meijer's G-Function ([22], Equation (9.3)).

Plugging (38) into (35), and doing similar manipulations which were used to derive $\mathcal{I}_1$, we obtain

$$
\mathcal{I}_2 = \frac{1-\alpha}{2\sqrt{2}\ln 2} G_{1\,3}^{3\,1}\left(\frac{1}{\Omega_z \Psi(\rho,j)\Omega_{1,n}\Psi(\rho,n)bPG^2}\,\middle|\,{\textstyle\genfrac{}{}{0pt}{}{0}{0,\frac{3}{4},-\frac{1}{4}}}\right),
\tag{42}
$$

To compare the ergodic capacities of the NOMA and OMA systems, we let β be the bandwidth which is assigned for D_1 and $(1-\beta)/2$ be the remaining bandwidth which is assigned for D_2 and D_3, where $(0 < \beta < 1)$ and the whole bandwidth is 1Hz. From ([38], Equation (7.4)), we can extend the achievable end-to-end ergodic capacity of the OFDMA system with three users as

$$
\begin{aligned}
\mathcal{R}_{\text{OMA}} = &\frac{1-\alpha}{2}\beta \log_2\left(1 + \gamma_{\text{SRD}_1}\right) \\
&+ \frac{(1-\alpha)(1-\beta)}{4} \log_2\left(1 + \gamma_{\text{SRD}_2}\right) \\
&+ \frac{(1-\alpha)(1-\beta)}{4} \log_2\left(1 + \gamma_{\text{SRD}_3}\right),
\end{aligned}
\tag{43}
$$

where $\gamma_{\mathrm{SRD}_m}, m \in \{1,2,3\}$ denotes the instantaneous SINR of each user, which is computed as

$$\gamma_{\mathrm{SRD}_1} = \frac{G^2 P_{\mathrm{S}}^{\mathrm{OMA}} |\hat{h}_{1,b}|^2 |\hat{g}_1|^2}{\beta (G^2 |\hat{g}_1|^2 \sigma_{\mathrm{R}}^2 + \sigma_{\mathrm{D}_1}^2)}, \tag{44}$$

$$\gamma_{\mathrm{SRD}_2} = \frac{2 G^2 P_{\mathrm{S}}^{\mathrm{OMA}} |\hat{h}_{1,b}|^2 |\hat{g}_2|^2}{(1-\beta)(G^2 |\hat{g}_2|^2 \sigma_{\mathrm{R}}^2 + \sigma_{\mathrm{D}_2}^2)}, \tag{45}$$

$$\gamma_{\mathrm{SRD}_3} = \frac{2 G^2 P_{\mathrm{S}}^{\mathrm{OMA}} |\hat{h}_{1,b}|^2 |\hat{g}_3|^2}{(1-\beta)(G^2 |\hat{g}_3|^2 \sigma_{\mathrm{R}}^2 + \sigma_{\mathrm{D}_3}^2)}, \tag{46}$$

where $P_{\mathrm{S}}^{\mathrm{OMA}} = P_S/3$ is the equal power allocated for the signal transmission from S to each user D_m ([38], p. 146) . The factor $\frac{1-\alpha}{2}$ appears in (31) and (43) because source node transmits its signals to all users in two time slots of the transmission period T.

5. Numerical Results

In this section, we provide the numerical results to evaluate the system performance in terms of the outage probability (OP) and ergodic capacity of the proposed EH-NOMA relaying system with three users. We also determine the optimal time switching ratio to minimize the OP and compare the ergodic capacities of the proposed EH-NOMA relaying system with EH-OMA relaying system. Regarding to the evaluating method, we use the common approach in this field, that is, to drive a closed-form mathematical expression to model the system performance and then compare the analysis results with Monte-Carlo simulation results to validate the derived mathematical expressions. Unlike previous works, which only considered EH-NOMA systems with two users and under perfect CSI, our paper focuses on the theoretical analysis of an EH-NOMA system with more than two users, taking into account the effects of AF relaying protocol and the feedback delay of wireless channels on the system performance. Since, there are not many similar parameters, it may be an unfair comparison between our proposed EH-NOMA relaying system with previous NOMA relaying systems. Therefore, we use the same system model of the proposed EH-NOMA relaying sytem but replace the NOMA with OMA to demonstrate the benefits of utilizing the NOMA technique in the proposed EH relaying system. Unless otherwise stated, the parameter settings of EH-NOMA and EH-OMA relaying systems are summarized in Table 2. It is noticed that the average SNR is defined as the ratio of the transmission power of source S to the variance of AWGN, that is, SNR $= P_S/\sigma^2$, ranging from 0 dB to 40 dB.

Table 2. Parameter settings of EH-NOMA and EH-OMA relaying systems.

Description	EH-NOMA	EH-OMA
Allocated transmission power	$P_1 = 0.7 P_S, P_2 = 0.2 P_S, P_3 = 0.1 P_S$	$P_i = P_S/3$
Bandwidth	β for D_1, $(1-\beta)/2$ for D_2 and D_3	$B = 1$ Hz for all users
Target data rate	$r = 0.5$ bpcu	
Time switching ratio	$\alpha = 0.3$	
Average channel gain	$\Omega_{1,n} = 1, \Omega_{\mathrm{R}_n \mathrm{D}_1} = 2, \Omega_{\mathrm{R}_n \mathrm{D}_2} = 3, \Omega_{\mathrm{R}_n \mathrm{D}_3} = 6$	
Energy conversion efficiency	$\eta = 0.85$	

Figure 2 shows the outage probability of each user versus the average SINR in dB. The outage probability of the EH-NOMA relaying system is also compared with that of EH-OMA relaying system. Firstly, we can see that the OP of D_3 is lowest among all users while the OP of D_1 is highest. The reason is that the channel gain from R to D_3 is highest (the decay of the magnitude power signal is proportional to the squared distance in multipath fading) because D_3 is the closest user to R while D_1 is the farthest one. Another important observation is that the OPs of D_2 and D_3 in the EH-NOMA relaying system are better than those of D_2 and D_3 in the EH-OMA relaying system, while the OP of D_1 in the EH-OMA relaying system is better than in the NOMA relaying system. However, the gap is insignificant because the number of time slots for the transmission in the EH-OMA relaying system is higher than in the

EH-NOMA relaying system, thus the probability that outage evens happen in the EH-OMA relaying system is also higher than in the EH-NOMA relaying system. On the other hand, the outage threshold of the OMA user is $\gamma_{th}^{OMA} = 2^{\frac{2r}{v(1-\alpha)}} - 1$, where $v \in \{\beta, (1-\beta)/2\}$. In contrast, the outage threshold of the NOMA user is $\gamma_{th} = 2^{\frac{2r}{(1-\alpha)}} - 1$. Then, obviously the outage threshold of the OMA user is obviously higher than that of the NOMA user. However, the OP not only depends on the outage threshold but also on the received SINR at user. In addition, we also see that in the low SINR regime (less than 15 dB), the OPs of all OMA users always outperform those of NOMA users. However, in the high SINR regime (larger 15 dB) only the OP of D_1 in the EH-OMA relaying system is better than that in the EH-NOMA relaying system. We can also see in Figure 2 that the diversity gain of all users is equal to one.

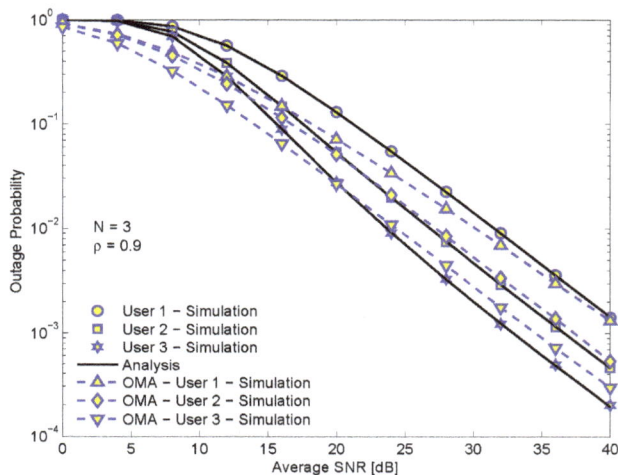

Figure 2. The outage probability of each user in energy harvesting (EH)-NOMA and EH-OMA relaying systems versus the average SINR. $\rho = 0.9$, the number of relays $N = 3$.

Figure 3 plots the OP of D_1 in the EH-NOMA relaying system versus the average SINR in dB for different channel correlation coefficients ρ. Firstly, we see that higher ρ reduces the OP, but the reduction is not remarkable for small ρ. The improvement in OP is only significant when ρ is near to 1. We should remind that ρ indicates the correlation degree between the transmission channel and the feedback channel in time coherent at the transmitter. The analysis results are in excellent agreement with the simulation ones, confirming the correctness of our mathematical analysis.

Figure 4 illustrates the OP of D_1 in the NOMA system versus the average SINR in dB for different numbers of relays N. From Figure 4, we see that when the number of relays increases, the outage performance of the system is improved. It is because increasing the number of relays will provide more opportunity for selecting the connection links from source node to relay, which not only makes the achievable decoding performance better but also increases the amount of harvested energy. In addition, the diversity gains is not significantly improved with N because the diversity order of PRS scheme is always equal to one.

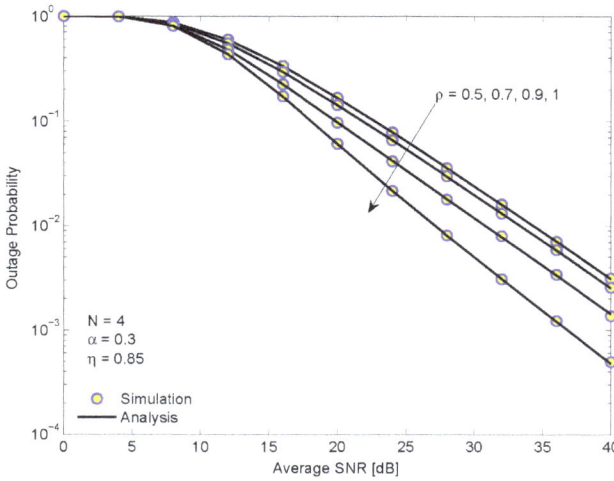

Figure 3. The outage probability of D_1 in the EH-NOMA relaying system versus the average SINR for different correlation coefficients.

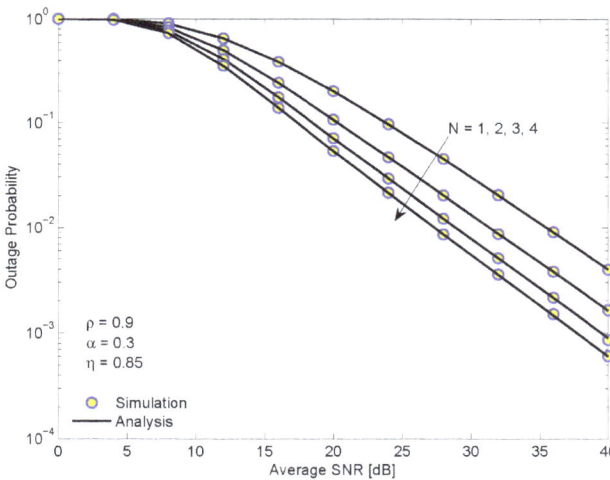

Figure 4. The outage probability of D_1 in the EH-NOMA relaying system versus the average SINR for different numbers of relays.

Figure 5 presents the OP of D_1 in the EH-NOMA relaying system versus time switching ratio α for different numbers of relays. The values of α range from 0 to 0.7 while SINR remains at 15 dB. Firstly, we see that there exists an optimal value of α which minimizes the OP. Moreover, the minimum value of OP depends on the number of relays N, i.e., as N is higher the minimal OP becomes smaller. The reason is that when N increases, the SINR of the first hop will be better because the PRS method is used. Another important observation is that the optimal value of α which minimizes the OP is approximately 0.2 regardless of the number of relays.

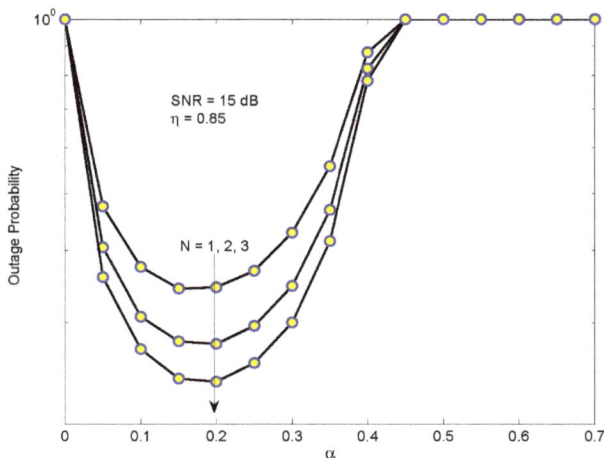

Figure 5. The outage probability of D_1 in EH-NOMA relaying versus the time switching ratio α for different number of relays.

Figure 6 demonstrates the OP of D_1 in the EH-NOMA relaying system versus the correlation coefficient ρ for different average SINR. We can see that the OP reduces as ρ increases. In the worst case $\rho = 0$, the instantaneous CSI at the transmission time does not correlate with the instantaneous CSI at the relay-selection time or at the power-allocation time. In contrast, in the best case $\rho = 1$, the instantaneous CSI at the transmission time closely correlates with the instantaneous CSI at the relay-selection time or at the power-allocation time. The improvement in the CSI leads to better power allocation and signal processing of the system. Figure 6 also shows that when $\rho < 0.8$, the enhancement of OP is not significant and the system performance is only improved when the correlation coefficient ρ is close to 1.

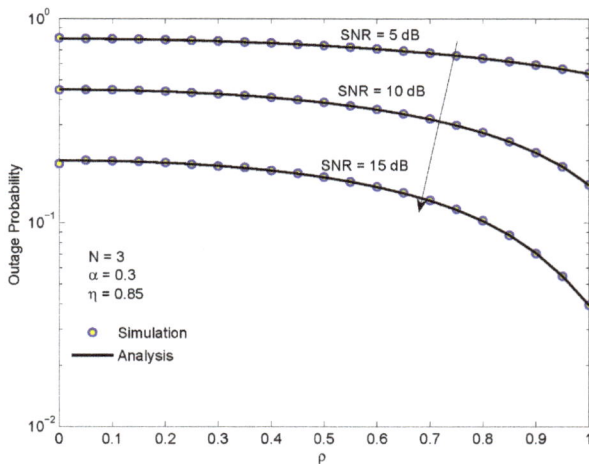

Figure 6. The outage probability of D_1 in the EH-NOMA relaying system versus the correlation coefficient for different average SINRs.

Figure 7 depicts the ergodic capacity of each user in EH-NOMA relaying system versus the average SINR in dB. As observed from Figure 7, the ergodic capacity of D_3 outperforms the ergodic capacities of D_1 and D_2. Moreover, the ergodic capacities of D_1 and D_2 increase slightly in the low SINR region and is saturated in the high SINR region. In contrast, the ergodic capacity of D_3 increases exponentially with respect to the SINR. This reason is that D_1 does not use SIC but only detects the signal of itself. Meanwhile, D_2 must use the first-order SIC first and then D_3 uses the second-order SIC. Thus, the impact of interference on D_1 is higher than D_2 and D_3. However, there exists the trade-off between the complexity and the achievable ergodic capacity of the system. We also see a good match between the analysis results and the simulation results, especially in the high SINR regime. On the other hand, the ergodic capacity in the case of perfect SIC is compared with that in the case of imperfect SIC. We can see that the ergodic capacity in case imperfect SIC is lower. Moreover, the gap between them increases with the SINR. It is because when SINR increases, the interference caused by imperfect SIC also increases. Therefore, the SINR as well as the ergodic capacity become slowly higher. Another feature is that the ergodic capacity of D_1 remains the same in both cases because D_1 does not use SIC when decoding the signals.

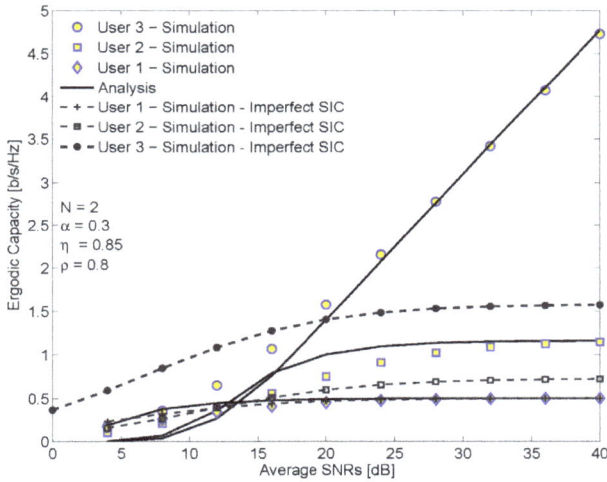

Figure 7. The ergodic capacity of each user and the ergodic capacity in EH-NOMA relaying system versus the average SINR.

Figure 8 provides the simulation results of the ergodic rate of NOMA and OMA systems versus the average SINR in dB. From Figure 8, we see that the ergodic rate of NONA system is always higher than the OMA system as the number of relays increases. It is because the NOMA system uses the whole bandwidth for each user while the OMA system uses individual bandwidth for each user, resulting in higher spectrum usage efficiency. Another important observation is that when N gets higher, the difference gap of the ergodic capacities of these two systems does not increase linearly. Thus, we do not need to use a large number of relays for partial relay selection scheme because it may increase the complexity of the system but not significantly enhance its performance.

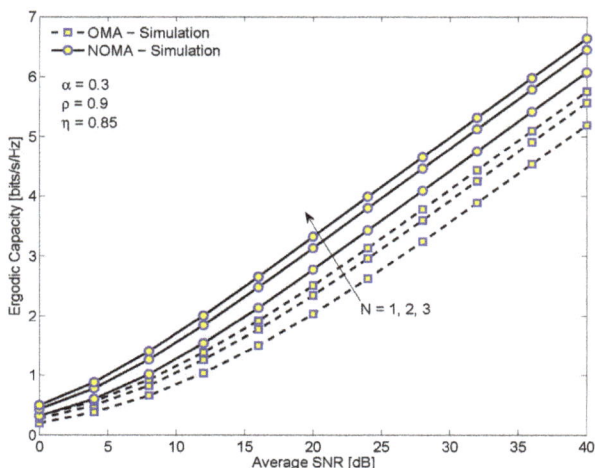

Figure 8. The comparison of the ergodic capacities of the EH-NOMA relaying system and the EH-OMA relaying system versus the average SINR for different numbers of relays.

6. Conclusions

In this paper, we propose a downlink NOMA relaying system with the best RF energy harvesting relay and investigate the impact of CSI imperfection on the performance of the proposed NOMA relaying system over Rayleigh fading channel. Specifically, we provide detailed derivations of the exact closed-form expression of OP and the approximate expression of the ergodic capacity of the proposed NOMA relaying system. Based on the expression of the OP, the optimal energy harvesting duration which minimizes the OP in the condition of imperfect CSI can be determined. The results show that imperfect CSI significantly reduces the system performance. In addition, we show that the spectrum efficiency of our proposed NOMA relaying system outperforms that of the OMA relaying system in the same parameter settings. All analysis results are in excellent agreement with the simulation results, confirming the correctness of the mathematical analysis. The proposed EH-NOMA relaying system can support the communication for multiple users through the best relay without relying on the external power supply. Thus it can be applied in surveillance sensor networks for disaster detection or in Internet of Things (IoT) where installing fixed power lines or frequent battery replacement for a large number of nodes may be not convenient. Using the results in this paper, we can choose an appropriate time switching ratio to balance between the energy harvesting and signal processing so that the outage probability of the proposed EH-NOMA relaying system system can be reduced upto 76.32%. Moreover, compared with the EH-OMA relaying system, the OP of the proposed EH-NOMA relaying system is 9.41% lower and the ergodic capacity is 17.64% higher at the average SNR = 40 dB.

Author Contributions: The main contributions of T.M.H., N.L.V., B.C.N. were to create the main ideas and execute performance evaluation by extensive simulation while L.T.D. worked as the advisors to discuss, create, and advise the main ideas and performance evaluations together.

Funding: This research received no external funding.

Conflicts of Interest: The authors declare no conflict of interest.

References

1. Liu, Y.; Qin, Z.; Elkashlan, M.; Ding, Z.; Nallanathan, A.; Hanzo, L. Nonorthogonal multiple access for 5G and beyond. *IEEE Commun. Surv. Tutor.* **2017**, *105*, 2347–2381. [CrossRef]

2. Liu, Y.; Qin, Z.; Elkashlan, M.; Nallanathan, A.; McCann, J.A. Non-orthogonal multiple access in large-scale heterogeneous networks. *IEEE J. Sel. Areas Commun.* **2017**, *55*, 2667–2680. [CrossRef]

3. Wang, B.; Dai, L.; Xiao, M. Millimeter Wave NOMA. In *Encyclopedia of Wireless Networks*; Springer: Berlin/Heidelberg, Germany, 2018.

4. Lv, T.; Ma, Y.; Zeng, J.; Mathiopoulos, P.T. Millimeter-Wave NOMA Transmission in Cellular M2M Communications for Internet of Things. *IEEE Internet Things J.* **2018**, *5*, 1989–2000. [CrossRef]

5. David, K.; Berndt, H. 6G Vision and Requirements: Is There Any Need for Beyond 5G? *IEEE Veh. Technol. Mag.* **2018**, *99*, 72–80. [CrossRef]

6. Long, D.N. Resource Allocation for Energy Efficiency in 5G Wireless Networks. *EAi Endorsed Trans. Ind. Netw. Intell. Syst.* **2018**, *5*, e1.

7. Le T.D.; Tran, M.H.; Nguyen, T.T.; Choi, S.G. Analysis of partial relay selection in NOMA systems with RF energy harvesting. In Proceedings of the International Conference on Recent Advances in Signal Processing, Telecommunications & Computing (SigTelCom), Ho Chi Minh City, Vietnam, 29–31 January 2018; pp. 13–18.

8. Tran, M.H.; Vu, V.S.; Nguyen, C.D.; Pham, T.H. Optimizing Duration of Energy Harvesting for Downlink NOMA Full-Duplex over Nakagami-*m* fading channel. *AEu-Int. J. Electron. Commun.* **2018**, *5*, 199–206.

9. Tran, H.V.; Kaddoum, G. RF wireless power transfer: Regreening future networks. *IEEE Potential* **2018**, *2*, 35–41. [CrossRef]

10. Tran, H.V.; Kaddoum, G. Robust Design of AC Computing-Enabled Receiver Architecture for SWIPT Networks. *IEEE Wirel. Commun. Lett.* **2019**, *8*, 801–804. [CrossRef]

11. Tran, H.V.; Kaddoum, G.; Truong, T.K. Resource allocation in SWIPT networks under a nonlinear energy harvesting model: Power efficiency, user fairness, and channel nonreciprocity. *IEEE Trans. Veh. Technol.* **2018**, *9*, 8466–8480. [CrossRef]

12. Bariah, L.; Muhaidat, S.; Al-Dweik, A. Error Probability Analysis of NOMA-based Relay Networks with SWIPT. *IEEE Commun. Lett.* **2019**, *23*, 1223–1226. [CrossRef]

13. Moltafet, M.; Azmi, P.; Mokari, N.; Javan, M.R.; Mokdad, A. Optimal and fair energy efficient resource allocation for energy harvesting-enabled-PD-NOMA-based HetNets. *IEEE Trans. Wirel. Commun.* **2018**, *17*, 2054–2067. [CrossRef]

14. Guo, C.; Zhao, L.; Feng, C.; Ding, Z.; Chen, H. Energy Harvesting Enabled NOMA Systems with Full-duplex Relaying. *IEEE Trans. Veh. Technol.* **2019**. [CrossRef]

15. Sun, R.; Wang, Y.; Wang, X.; Zhang, Y. Transceiver Design for Cooperative Non-Orthogonal Multiple Access Systems with Wireless Energy Transfer. *IET Commun.* **2016**, *10*, 1947–1955. [CrossRef]

16. Han, W.; Ge, J.; Men, J. Performance Analysis for NOMA Energy Harvesting Relaying Networks with Transmit Antenna Selection and Maximal-Ratio Combining over Nakagami-*m* Fading. *IET Commun.* **2016**, *10*, 2687–2693. [CrossRef]

17. Liu, Y.; Ding, Z.; Elkashlan, M.; Poor, H.V. Cooperative non-orthogonal multiple access with simultaneous wireless information and power transfer. *IEEE J. Sel. Areas Commun.* **2016**, *34*, 938–953. [CrossRef]

18. Tran, M.H.; Nguyen, T.T.; Nguyen, H.H.; Pham, T.H. Performance analysis of decode-and-forward partial relay selection in NOMA systems with RF energy harvesting. *Wirel. Netw.* **2018**, 1–11. [CrossRef]

19. Yang, Z.; Ding, Z.; Fan, P.; Al-Dhahir, N. The Impact of Power Allocation on Cooperative Non-orthogonal Multiple Access Networks with SWIPT. *IEEE Trans. Wirel. Commun.* **2017**, *16*, 4332–4343. [CrossRef]

20. Dutta, B.; Budhiraja, R.; Koilpillai, D.R. Limited-feedback low-encoding complexity precoder design for downlink of FDD multi-user massive MIMO systems. *IEEE Trans. Commun.* **2017**, *65*, 1956–1971. [CrossRef]

21. Hoydis, J.; Brink, S.T.; Debbah, M. Massive MIMO in the UL/DL of cellular networks: How many antennas do we need? *IEEE J. Sel. Areas Commun.* **2017**, *65*, 1956–1971. [CrossRef]

22. Zwillinger, D. *Table of Integrals, Series, and Products*, 7th ed.; Elsevier: Amsterdam, The Netherlands, 2014.

23. Firag, A.; Smith, P.J.; Suraweera, H.A.; Nallanathan, A. Performance of beamforming in correlated MISO systems with estimation error and feedback delay. *IEEE Trans. Commun.* **2011**, *10*, 2592–2602. [CrossRef]

24. Tran, M.H.; Tran, X.N.; Nguyen, T.; Le, T.D. Performance Analysis of MIMO SWIPT Relay Network with Imperfect CSI. *Mob. Netw. Appl.* **2018**, *24*, 630–642.

25. Suraweera, H.A.; Smith, P.J.; Shafi, M. Capacity limits and performance analysis of cognitive radio with imperfect channel knowledge. *IEEE Trans. Veh. Commun.* **2010**, *59*, 1811–1822. [CrossRef]

26. Bletsas, A.; Khisti, A.; Reed, D.P.; Lippman, A. A simple cooperative diversity method based on network path selection. *IEEE J. Sel. Areas Commun.* **2006**, *24*, 659-672. [CrossRef]

27. Tran, T.D.; Kong, H.Y. Performance analysis of incremental amplify-and-forward relaying protocols with nth best partial relay selection under interference constraint. *Wirel. Pers. Commun.* **2013**, *71*, 2741–2757.

28. Tran, T.D.; Duong, Q.T.; Costa, D.B.; Vo N.Q.B.; Elkashlan, M. Proactive relay selection with joint impact of hardware impairment and co-channel interference. *IEEE Trans. Commun.* **2015**, *63*, 1594–1606.

29. Nasir, A.A.; Zhou, X.; Durrani, S.; Kennedy, R.A. Relaying protocols for wireless energy harvesting and information processing. *IEEE Trans. Wirel. Commun.* **2013**, *12*, 3622–3636. [CrossRef]

30. Ju, M.C.; Kang, K.M.; Hwang, K.S.; Jeong, C. Maximum Transmission Rate of PSR/TSR Protocols in Wireless Energy Harvesting DF-Based Relay Networks. *IEEE J. Sel. Areas Commun.* **2015**, *33*, 2701–2717. [CrossRef]

31. Pedersen, K.I.; Kolding, T.E.; Seskar, I.; Holtzman, J.M. Practical implementation of successive interference cancellation in DS/CDMA systems. In Proceedings of the ICUPC—5th International Conference on Universal Personal Communications, Cambridge, MA, USA, 2 October 1996; pp. 321–325.

32. Men, J.; Ge, J.; Zhang, C. Performance analysis of non-orthogonal multiple access for relaying networks over Nakagami-*m* fading channels. *IEEE Trans. Veh. Technol.* **2016**, *66*, 1200–1208. [CrossRef]

33. Men, J.; Ge, J.; Zhang, C. Performance analysis for downlink relaying aided non-orthogonal multiple access networks with imperfect CSI over Nakagami-*m* fading. *IEEE Access* **2017**, *5*, 998–1004. [CrossRef]

34. Papoulis, A.; Pillai, S.U. *Probability, Random Variables, and Stochastic Processes*, 4th ed.; McGraw-Hill: New York, NY, USA, 2002.

35. Xiong, J.; Tang, Y.; Ma, D.; Xiao, P.; Wong, K.K. Secrecy performance analysis for TAS-MRC system with imperfect feedback. *IEEE Trans. Inf. Forensics Secur.* **2015**, *10*, 1617–1629. [CrossRef]

36. Simon, M.K.; Alouini, M.S. *Digital Communication Over Generalized Fading Channels: A Unified Approach to Performance Analysis*; John Wiley & Sons: Hoboken, NJ, USA, 2000.

37. Bariah, L.; Muhaidat, S.; Al-Dweik, A. Error Probability Analysis of Non-Orthogonal Multiple Access over Nakagami-*m* Fading Channels. *IEEE Trans. Commun.* **2018**, *7*, 1586–1599.

38. Benjebbour, A.; Saito, K.; Li, A.; Kishiyama, Y.; Nakamura, T. Non-Orthogonal Multiple Access (NOMA): Concept and Design. In *Signal Processing for 5G: Algorithms and Implementations*; John Wiley & Sons: Hoboken, NJ, USA, 2016; pp. 143–168.

sensors

MDPI

Article

Energy, Carbon and Renewable Energy: Candidate Metrics for Green-Aware Routing?

Md. Mohaimenul Hossain [1,2,*], Jean-Philippe Georges [1,2], Eric Rondeau [1,2] and Thierry Divoux [1,2]

1   Université de Lorraine, CRAN, UMR 7039, Campus Sciences, BP 70239,
     Vandoeuvre-lès-Nancy CEDEX 54506, France
2   CNRS, CRAN, UMR 7039, France
*   Correspondence: mohaimenul.hossain@univ-lorraine.fr

Received: 28 May 2019; Accepted: 26 June 2019; Published: 30 June 2019

Abstract: There are all sort of indications that Internet usage will go only upwards, resulting in an increase in energy consumption and CO_2 emissions. At the same time, a significant amount of this carbon footprint corresponds to the information and communication technologies (ICT) sector, with around one third being due to networking. In this paper we have approached the problem of green networking from the point of view of sustainability. Here, alongside energy-aware routing, we have also introduced pollution-aware routing with environmental metrics like carbon emission factor and non-renewable energy usage percentage. We have proposed an algorithm based on these three candidate-metrics. Our algorithm provides optimum data and control planes for three different metrics which regulate the usage of different routers and adapt the bandwidth of the links while giving the traffic demand requirements utmost priority. We have made a comparison between these three metrics in order to show their impact on greening routing. The results show that for a particular scenario, our pollution-aware routing algorithm can reduce 36% and 20% of CO_2 emissions compared to shortest path first and energy-based solutions, respectively.

Keywords: green networking; energy aware routing; carbon footprint; adaptive link rate; control and data plane

---

1. Introduction

In this modern era of advancements, the ever-growing information and communication technologies (ICT) sector plays an important role. Improvement in technologies, the availability of inexpensive and extreme capacity optical transmission, increasing popularity of streaming services and Internet of things technology have increased the amount of usage of the Internet exponentially and there is no indication that this course will change. With the increasing dependency on ICT services, the need for generating electricity has also increased substantially in order to power the ICT infrastructure, which is a significant contributor in producing carbon dioxide, a leading greenhouse gas (GHG). Global GHG emissions data shows that the highest contributor to global CO_2 emissions in 2015 was electricity production. Surprisingly, all the ICT devices and infrastructure, excluding smartphones, consumes 8% of the total electricity production and is projected to reach 14% by 2020 [1]. Similar claims are also stated into European Commission's digital agenda report in 2013: "ICT products and services are currently responsible for 8 to 10% of the EU's electricity consumption and up to 4% of its carbon emissions" [2].

Among the various ICT sectors, network architectures and devices are one of the highest energy consumers. According to [3], network devices and architecture are responsible of 29% of the total energy consumption by ICT. The main reason behind this is the increase of Internet users all over the world and secondly, most of the network architectures and resources like bandwidth, processing power and memory are designed bearing in mind the peak hour network usage [4,5]. This results in redundancy and overprovisioning of the resources and consumes extra energy during off peak hours which is unnecessary outside these peaks.

Due to this ever-increasing energy consumption from ICT devices and especially from network infrastructures, it's been a while now that researchers have started working on energy efficiency and hence the term "Green Networking" has emerged. The main idea behind green networking is to improve energy efficiency and reduce undesirable energy consumption which will in fact reduce the carbon footprint produced by ICT devices. Several works have been done considering the problem of energy efficiency in networks and there are mainly two research directions. The first at the network hardware level where the prime focus is on building energy efficient circuits, improving the power draining components like memory or modifying the cooling systems. The second direction is focused more on the network design level where the whole network architecture is considered, and optimization of the routing is done in order to reduce energy consumption also known as energy aware routing (EAR). EAR is implemented by different power optimization techniques like, turning off the unused network devices (nodes and links), Energy Efficient Ethernet (EEE) or Adaptive Link Rate (ALR).

The main motivation behind working on energy efficiency is to reduce carbon emissions. However, it needs to be remembered that energy consumption is not always proportional to carbon emissions. Same amount of energy production can have a very different impact on the environment in terms of carbon emissions. For example, if we assume an average value of 400 g of CO_2 emissions for 1 kWh of electricity production [6], France produces less than one fifth of the average value whereas Poland produces three times the average value. Therefore, for these two countries, consuming the same amount electricity has a quite different result in terms of environmental impact. Which begs the question: *"Is it enough to focus only on energy consumption while trying to achieve green networking?"*

In order to address this question, in this paper, we have introduced the term pollution-aware routing (PAR) which uses energy efficiency techniques and environmental variables together to find a more pollution-aware solution. We have presented two different type of pollution metrics, namely carbon emission factor (CEF) and non-renewable energy usage percentage (NRE). Our idea is to compare and analyze the performance of the introduced pollution indicators with the well-known, widely used indicator, energy efficiency. For energy efficiency, we have considered two green policies. The first one is turning off the network devices (links and nodes) when they are not necessary and the second one is adaptive link rate (ALR). These policies are quite often used by the research community and very prominent in terms of energy savings. Section 2.1 contains a few of the works which have included these policies while designing energy aware routing. As explained in [7], the problem of finding the minimum number of links and nodes to be turned on while fulfilling the traffic constraint is a problem of multi-commodity flow class which is known as NP-hard problem. We have used a genetic algorithm as our heuristic to solve this problem. As a solution, we have provided a data plane and a control plane which contains the information of the routing and the bandwidth distribution of each link (ALR implementation), respectively. One main novelty of our algorithm consists of addressing both planes at the same time, leading to optimal solutions. In this paper, we have formulated the network and the energy consumption model for each of the network devices. By introducing CEF as a pollution indicator we want to address the carbon emission problem more directly. On the other hand, some energy resources have different impacts in addition to carbon emissions such as nuclear power plants, for example, produce nuclear waste. Therefore, our second pollution metric is NRE which will be applied to reduce the non-renewable energy usage percentage. The main difference between the currently used (i.e., energy consumption) and our introduced metric is that ours consider external factors (like power production) whereas energy consumption only reflects network efficiency. In Figure 1 we have depicted this phenomenon. Secondly, CEF and NRE will vary based on the location (i.e., country), in this regard we can say that we have addressed the problem with a more holistic approach considering in fact the political engagement of each country regarding the production of the energy.

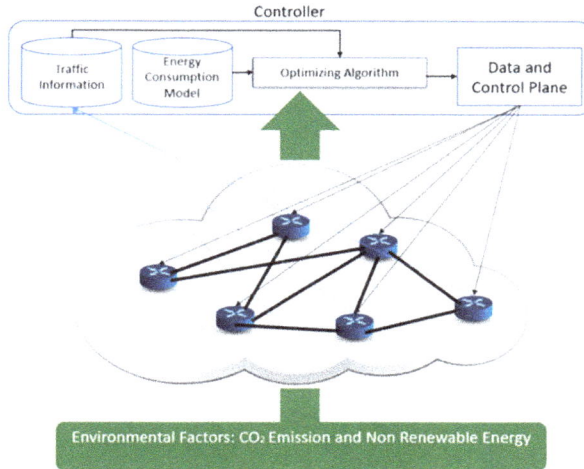

Figure 1. Basic Architecture of the System.

Based on three different indicator/metrics three different objective functions are formulated, which are used as a fitness function for a genetic algorithm which is providing a set of data and control plane as output. These planes are providing operational topology for the network but by focusing either on reducing energy consumption, or total carbon emission or non-renewable energy usage percentage respectively.

We have considered our solution from the point of view of a software defined network (SDN). SDN is fairly a new network architecture where the data plane and control plane are separated in order to have more control over the network. It integrates a controller which orchestrates the underlying forwarding hardware. This centralized approach offers mechanisms allowing network operators to have a global point of view which is suitable for our scenario where a single change in one part of the network topology can have an impact all over the network. The authors of [8] used a centralized approach in order to provide zoning evaluation of voltage distribution. Moreover, a centralized system is always preferable for globally monitoring the system according to network policies and service requirements [9]. The controller receives the traffic information from the network, and based on the geographical position of the nodes, the values of environmental metrics like CO_2 emission and non-renewable energy usage are received. Then the controller, based on the optimizing algorithm, provides a data plane and a control plane for the network. The detail description of how a control plane and a data plane are designed will be described in Section 3. Figure 1 gives an overview of the system.

The paper is organized as follow: Section 2 explains the motivation in the light of related work and sustainable green networking and summarizes the objective of our work. Section 3 contains the concept of EAR and PAR and a description of centralized sustainable routing. Section 4 contains the formalization of the problem and the overview of the heuristic function. In S5 we have a case study and the performance evaluation of the algorithm and lastly Sections 6 and 7 contain the future work and conclusions of the work, respectively.

2. Motivation

2.1. Related Work

For the last decade or so, several approaches are made in order to achieve green networking. The authors of [10] provided a detail discussion on the different research trends in green networking. Research works like [11,12] focus only turning off links and line cards whereas many other papers like [13–15] focus on turning off both links and nodes. Authors in reference [4] proposed a novel heuristic, GreenTE, which maximizes the power savings and satisfies the quality constraints including link utilization and packet delay. The proposed a heuristic solution that is proactive in terms of path calculation. They have defined a set of previously calculated paths and they chose the solution from one of them. Authors in references [12,13] emphasize least link and least flow without link adapting. They have used a greedy algorithm to find out the least used link or devices with the least flow and then turn them off. They did reactive path calculation. This means that based on the heuristic several paths are being calculated instantly and the best one for the particular scenario is chosen. Works like [15], provide a clear idea about how an optical core network behaves in terms of energy consumption. On the other hand [16–18] all talked about implementing ALR in order to reduce energy consumption. Other works like [12,13,19–21] all have considered energy aware routing. Authors in reference [20] provided a greedy algorithm-based heuristic and along with links they have also focused on reducing the energy consumption by turning off nodes. Nevertheless, in their solution they have not considered ALR. In current times alongside these solutions various other solutions have been proposed using SDN in order to benefit from the global knowledge of the central controller. Authors in reference [14] proposed an incremental greedy algorithm-based solution where a controller dynamically adds nodes and links to the initial small topology network in order to satisfy user demand requirements. Authors in reference [22] also proposed an energy efficient solution combining energy aware routing with SDN-based Ethernet networks. There are also a few works which have taken a different approach towards green networking. Authors in references [23–25] all have tried to achieve green networking in data centers by using renewable energy. Authors in references [23,24] changed the destination node based on availability of the renewable energy and on the other hand in [25], alongside with renewable energy, other criteria like geographical load balancing and server speed scaling are considered. The authors of [26,27] proposed a similar eco-friendly routing idea using renewable energy where they have used clustering to choose a cluster head with most renewable energy and then choose the member node accordingly. Authors in reference [26] is for wireless networks and [27] is for IP networks. However, one thing is common in the abovementioned works is that all of them consider green network that same as the energy efficient network. They have all tried actively or passively to reduce the energy consumption in order to get an energy efficient network. However, even if energy has a link to sustainability issues, its direct impact on the environment in terms of air pollution and Earth's resources is not explicitly specified. Therefore, in our work we have introduced two different green metrics–CO_2 emission and non-renewable energy usage percentage–and make a comparison between them and energy efficiency approaches.

2.2. Sustainability and Green Networking

The authors of [28] explained a very important point by mentioning that achieving sustainability is not a straightforward approach of saving resources like energy. Even if from technical point of view, it might seem direct approach, in fact the economical and behavioral perspective of the society needs to be included when trying to achieve a sustainable solution. In [29], they have explained that, in order to build a sustainable system, it is necessary to determine the shared interests of both business and society's point of view. In fact, sustainability is a unified concept which considers environmental, social, and economic aspects as three fundamental pillars. They indicate that, viable development requires the understanding of nature, society and economic capital or colloquially speaking the planet, people, and profits [30–32]. Therefore, in order to design a sustainable green networking architecture, it is very important to somehow consider all the three pillars. In this paper, we have proposed a centralized model of generating sustainable control and data planes considering the three pillars. The details of the model are described in Section 3.2.

2.3. Summarizing the Goal

We have discussed a few works focusing on renewable energy, but almost all of them are focused on topology-aware routing, which is different from energy-aware routing. To the best of our knowledge, carbon emissions are not fully considered as a routing metric. In this regard, the goal of this paper is to compare the performance of these metrics and how algorithms (optimizing both routing and green policies) need to be adapted. Performance is going to be assessed in a centralized sustainable routing model.

3. Green Routing and Metrics

3.1. Concept of EAR and PAR

In order to get a clear view of the proposed system, it is better to start with EAR and PAR. While implementing green networking, one of the most common and effective solutions is to shut down as much network equipment as possible while keeping the operational network, which is also known as energy-aware routing (EAR). This drop in the number of active devices reduces the energy consumption which has in fact a positive impact on the carbon footprint. The fundamental difference between a classical routing approach like shortest path first (SPF) and EAR is that SPF considers the number of hops and tries to reach the destination as quickly as possible without focusing on the energy consumption of the network, whereas, EAR tries to reduce the overall energy consumption of the network, which might result a longer path for some demands and similar length paths for other demands.

In this paper we have talked about a different type of routing strategy which is pollution-aware routing (PAR). The main concept of PAR is to route the demands with different paths based on environmental factors such as carbon emission factors and the non-renewable energy usage percentage of a node and try to reduce the targeted goal of reducing the CO_2 emission or the usage percentage of non-renewable energy of the total network architecture. This routing is applicable only for geographically diverse, distributed network architectures where every node has different means of energy production. The main motivation behind PAR is the variation of values of CO_2 emission in terms of energy production methods as shown in Figure 2. Values are taken from [33].

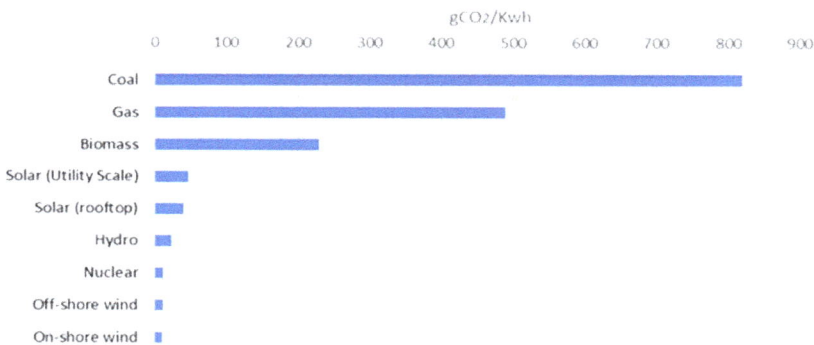

Figure 2. Emissions of selected electricity supply technologies (gCO$_2$/kWh).

It can be clearly seen that for same amount of electricity production the CO_2 emissions can vary a lot. Therefore, in PAR, instead of reducing the energy consumption of the network architecture, the environmental metrics are considered. The first metric is CO_2 emissions. As different countries use different techniques for producing electricity, they all have different carbon emission factors (CEFs). Therefore, in this case, the routing will be based on the CEF. The paths for the demands will be chosen in order to reduce the carbon emissions of the total network architecture. Similarly, here we have considered a second kind of PAR with the second green metric NRE, where nodes will be chosen based on renewable energy (wind, solar, hydro) percentage. The idea is to decrease the overall amount of non-renewable energy usage of the network architecture while fulfilling the demands. Both find the paths for the demands in order to facilitate their targeted goal without considering the path size or amount of energy consumption. The example depicted in Figure 3 will give a better grasp of the concept.

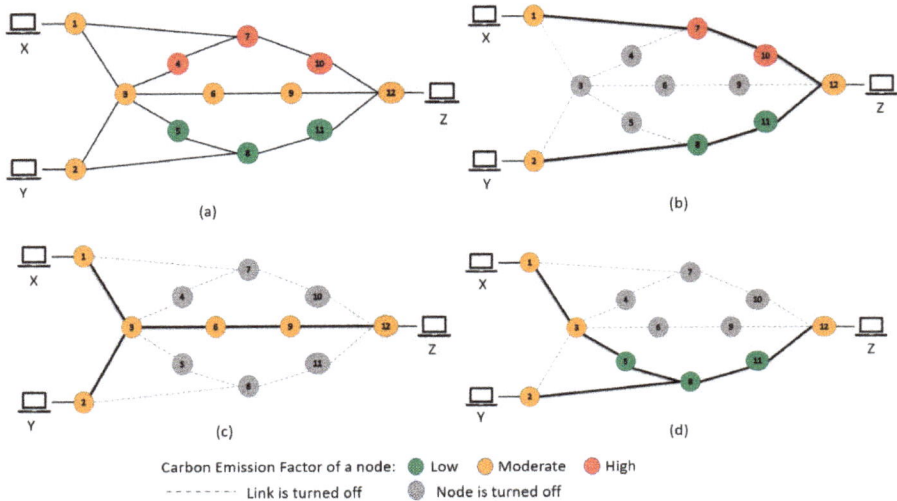

Figure 3. Example of SPF, EAR and PAR. (**a**) shows a topology with 12 nodes, (**b**) shows the topology for SPF, (**c**) shows how EAR will act in this scenario, (**d**) we have a different topology compared to (**b**,**c**).

Figure 3a shows a topology with 12 nodes. For this example, we have considered CO_2 emissions as a factor for PAR. For simplicity there are only two demands. One is from X to Z and the second one is from Y to Z. Again, to avoid complexity, the sum of the two demand requirements is less than any available links' bandwidth. Three different colors are used instead of values for the carbon emission factor. Green means the node has the lowest amount of CEF, orange means moderate and red means it has the highest amount of CEF. A node is grey colored if it is turned off. Figure 3b shows the topology for SPF. The lowest number of hops are considered for each demand. Regardless of the nodes CEF or even without considering the total energy consumption, Figure 3c shows how EAR will act in this scenario. It will choose a common link in order to reduce the number of active nodes and links. It will reduce one node and two links compared to SPF. However, this will also not be concerned with the CO_2 emissions. In Figure 3d we have a different topology compared to Figure 3b,c. It didn't look for the shortest path or lowest number of nodes and links. Rather it chooses a topology where the total amount of CO_2 emissions will be low. It will not consider the total number of turned on nodes and links if the overall CO_2 emissions are on the low side.

3.2. Centralized Sustainable Routing

In this paper, we have also proposed a sustainable solution for the problem of achieving 'green networking'. As explained in Section 2, a solution cannot be sustainable without addressing the three pillars of sustainability. In the section above, EAR and PAR are described. Our solution, depicted in Figure 4, integrates all these elements in order to address the 'green networking' problem. Through our solution we would like to investigate the performance of the proposed approach (PAR) and the available approach (EAR). We have designed the solution keeping in mind the three pillars of sustainability, namely economy, environment and society. Figure 4 shows the architecture of the proposed system. In the following part, how the three pillars are integrated into our solution is described briefly. Firstly, our system takes traffic demand from users and the topology of the network from the Internet service provider (ISP) as input. Both users and ISPs are the stakeholders for our solution. ISP has a direct relation with the profit, whereas users are mainly concerned about the quality of service (QoS) like throughput requirements. We also take geography-based environmental information such as carbon emission factor (CEF) and non-renewable energy usage percentage (NRE) as input. This information is one of the key factors for providing a green outcome. These environmental parameters are acting as the second pillar of sustainability–environment. Lastly, our system provides a user of this system the ability to choose to select a solution focusing on either reducing energy consumption or CO_2 emissions or non-renewable energy usage percentage (the value of Φ in Equation (8)). The social aspect of sustainability focuses on balancing the needs of the individual with the needs of the group. By giving the choosing ability to the Internet service providers, it integrates the social entity of the sustainability into our system.

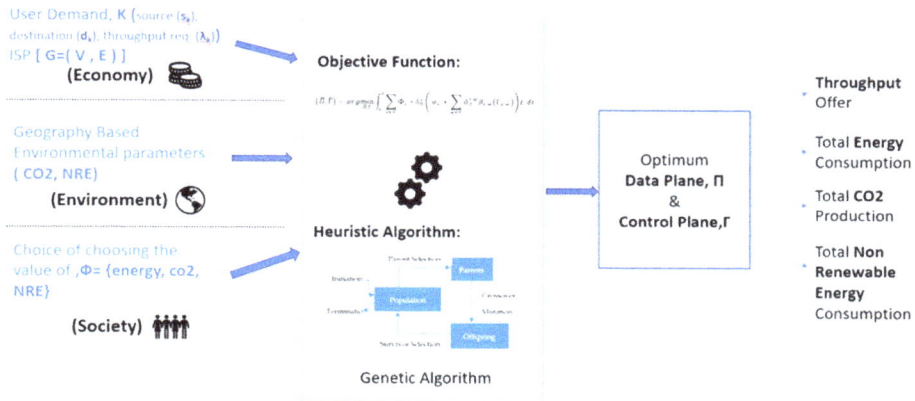

Figure 4. Generation of sustainable data and control plane.

For example, in case of the GÉANT network, any environmental strategy or law defined and imposed by European Union can be implemented in the network by tuning the optimizing equation. After getting these inputs our system provides a solution based on the equation derived in the previous section. It uses a genetic algorithm as the heuristic algorithm in order to solve the optimization problem. The heuristic algorithm is described in the following section. After that an optimum data and control plane are provided to implement the solution into the topology. These data and control planes are calculated based on the inputs given to the system. The main goal of these planes is to minimize the given objective function while fulfilling the QoS constraint. Alongside the solution topology settings, these planes also provide vital information about energy consumption, CO_2 emissions and non-renewable energy usage percentage of the topology. This information is essential, especially for the people pillar of sustainable development because it provides clear information about the impact of network users on the environment, which by default provides them a guideline to act accordingly. This kind of feedback always make the people in society more involved in the system which is a prerequisite for creating a complete sustainable system. Our proposed solution is a sustainable solution towards achieving green networking. The heuristic algorithm used for optimization is explained in the next section.

4. Formalization of the Problem

4.1. Problem Definition

Let us assume a network topology is defined by a directed graph G = (V, E), where V is set of vertices i ∈ {1, 2, . . ., |V| = n} and E is set of edges (i, j), which are 2–elements subsets of V. The adjacency matrix is a square |V| × |V| matrix A, such that the element A_i, ∈ {1,0} has the value of 1 when an edge exists between two vertices i and j, and 0 when there is no edge. The C matrix contains the capacity of each edge that means C_{ij} represents the capacity of the link between node *i* and *j*. Capacity is the only part of G which is controllable. Now let's come to the dynamic part of the network which is traffic demands. Let's suppose K is a set of traffic demands K = {1, 2, ... , k}. A demand consists of three information source (s_k), destination (d_k) and throughput requirement (λ_k) expressed in b/s.

In the framework of Software-Defined Networking (SDN) and more generally speaking network automation, we would consider that the topology is redundant, which means that several paths (i.e., more nodes and links are added—offline—to the physical topology) are available. However, only those paths will be considered which are satisfying both throughput demand and the capacity constraint of the links. The data plane matrix is denoted by Π, which includes all the forwarding decision for all flows over all nodes. Each row of the matrix is dedicated for one demand. $\Pi_{k,v}$ returns the node **v + 1**, which is an adjacent node of **v** into path P_k ($\Pi_{k,v}$ = **v + 1**) or null set if the path does not include node **v**. The path ends when it returns d_k. A path $P_k = \left\{ s_k, \Pi_{k,s_k}, \Pi_{k,\Pi_{k,s_k}}, ..., d_k \right\}$ then can be retrieved. As mentioned earlier, different paths are available for every pair of source and destination therefore for each demand set several data planes are possible. The data plane should respect the following constraint for a demand i:

$$\forall v \in \{s_i\_d_i\}, \quad w = \Pi_{i,v} \quad such\ as \sum_{(k \in K | \Pi_{k,v} = w)} \lambda_k \leq C_{v,w} \tag{1}$$

This ensures that only the links that are in the topology are considered and no link will be assigned more throughput than its capacity. Furthermore, the control plane, denoted by matrix Γ, is used to define the controllability of the link capacity (i.e., ALR mechanism). The capacity can be switched to different discrete values from the maximum capacity defined by C to 0. Zero is considered when both interfaces of a link are turned off. Initially, Γ = C and later from the second flow C is replaced by Γ in Equation (1). The value $\Gamma_{i,j}$ represents the final link capacity required to fulfill every demand requirement.

Figure 5 shows an example of how data and control planes work. Here as a topology a portion of the GÉANT Network is used and let's consider that the topology has three different levels of bandwidth: 10, 100 and 1000 Mbps, respectively. For three demands K = {{s_1 = 12, d_1 = 2, λ_1 = 8 Mb/s}, {s_2 = 4, d_2 = 3, λ_2 = 60 Mb/s}, {s_3 = 3, d_3 = 6, λ_3 = 95 Mb/s}}. The three chosen paths are {12 – 1 – 2}, {4 – 11 – 3} and {3 – 1 – 12 – 6}, respectively. Then the data plane and the control plane will be look like Figure 5b,c. There are several possible solutions. The data plane is created based on the chosen paths and the control plane is created based on the data plane. Even though all the demands have a bandwidth requirement of less than 100 Mbps, we can see in the control plane, edge {1,12} (link 4 on Figure 5) has a final capacity of 1000 Mbps. In fact, this edge is included in the solution path of both demand-1 and demand-3 which have a combined throughput demand of 103 (8 + 95) Mbps. A link capacity of 100 Mbps is hence not anymore enough and as only three levels of bandwidth, namely 10, 100 and 1000 Mbps are considered, it will be replaced by a 1000 Mbps capacity one. Both data plane and control planes are vital for calculating our objective functions that will be discussed in the next part.

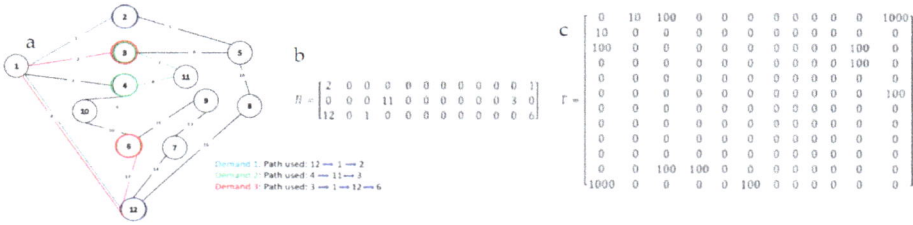

Figure 5. (a) Topology (a portion of the GÉANT network); (b) A possible data plane; (c) control plane.

4.2. Modelling of the Objective Functions

We have three different criteria that need to be minimized, namely energy consumption, CO_2 emissions and the non-renewable energy usage. We have formalized the cost of data and control planes based on the objective function. However, as mentioned earlier, the carbon emissions and non-renewable energy both can be used as a multiplicative factor with the energy model. Therefore, here the energy model is described first. For calculating the energy consumption, a simple model from [34] is used where energy consumption model of a Cisco Ethernet switch is provided. Similar models can also be found in the literature [8,35,36], even for different devices. To simplify, it is assumed here that energy consumption of a node **v**, follows the linear model:

$$\varepsilon_v(t) = \int_0^t \left(\alpha_v + \sum_{w \in v} \delta_{v,w} \beta_{v,w} \right) t \cdot dt \qquad (2)$$

where ε_v in Watt-hour (Wh), α is the static power consumption (when no interfaces are activated), $\delta_{v,w} = 1$ if $C_{v,w} > 0$ and 0 else (to know if the interface to the neighbor w is used) and $\beta_{v,w}$ is the power consumption of the interface port itself. The model consists of two parts: a fixed/static one that must be considered each time the node is turning on and a dynamic one that depends on the control plane. As we want to include two green strategies (turn off nodes when not required and ALR) into the model without hampering QoS constraint, in order to imply the first strategy, two binary variables are included:

$$\delta_\Pi^v = \begin{cases} 1 \ if \ \sum_{k \in K} \Pi_{k,v} > 0 \ or \ v \in \Pi \\ 0 \qquad\qquad\qquad\qquad else \end{cases} \qquad (3)$$

$$\delta_\Pi^{v,w} = \begin{cases} 1 \ if \ \exists_k \in K, \ \Pi_{k,v} = w \ or \ \Pi_{k,v} = v \\ 0 \qquad\qquad\qquad\qquad\qquad else \end{cases} \qquad (4)$$

A node v is used in the data plane if it either forwards ($\sum_{k \in K} \Pi_{k,v} > 0$) and receives ($v \in \Pi$) traffic or it is used as a mediatory node for a selected path. Now, in the case of ALR, the second policy that we have considered, the matrix Γ contains the decisions (Figure 5c), such the power consumption $\beta_{v,w}$ will now vary according to the link capacity. Hence, the objective function in terms of data and control planes while satisfying the demands requirements (elementary paths, throughput, links capacity limitations) is as follows:

$$\left(\hat{\Pi}, \hat{\Gamma}\right) = \underset{\hat{\Pi},\hat{\Gamma}}{argmin} \int_0^t \sum_{v \in V} \delta_\Pi^v \left(\alpha_v + \sum_{w \in V} \delta_\Pi^{v,w} \beta_{v,w} (\Gamma_{v,w}) \right) t.dt \qquad (5)$$

This objective function depicts the problem of minimizing the energy consumption of overall system. Now, let's consider the environmental factors. Suppose, Λ is a set of carbon emission factors (CEFs) for all the nodes. Where, $\Lambda = \{\Lambda_1, \Lambda_2, \ldots, \Lambda_v\}$. And ψ is a set of non-renewable energy usage percentage (NREs) for all the nodes. $\psi = \{\psi_1, \psi_2, \ldots, \psi_v\}$. Here one thing to notice is that for all, the goal is to reduce it.

As mentioned earlier, that both environmental parameters are a multiplicative factor of the original energy consumption model, therefore, objective function can be rewritten for CEF and NRE respectively in the following way:

$$\left(\hat{\Pi}, \hat{f}\right) = \underset{\hat{\Pi}, \hat{f}}{\arg\min} \int_0^t \sum_{v \in V} \Lambda_v * \delta_\Pi^v \left(\alpha_v + \sum_{w \in V} \delta_\Pi^{v,w} \beta_{v,w} (\Gamma_{v,w}) \right) t.dt \tag{6}$$

$$\left(\hat{\Pi}, \hat{f}\right) = \underset{\hat{\Pi}, \hat{f}}{\arg\min} \int_0^t \sum_{v \in V} \psi_v * \delta_\Pi^v \left(\alpha_v + \sum_{w \in V} \delta_\Pi^{v,w} \beta_{v,w} (\Gamma_{v,w}) \right) t.dt \tag{7}$$

The goal is to provide data and control planes which will reduce the energy consumption or carbon emission or non-renewable energy percentage of the overall system. Then from Equations (5)–(7) a generalized version of the objective function for system would be:

$$\left(\hat{\Pi}, \hat{f}\right) = \underset{\hat{\Pi}, \hat{f}}{\arg\min} \int_0^t \sum_{v \in V} \Phi_v * \delta_\Pi^v \left(\alpha_v + \sum_{w \in V} \delta_\Pi^{v,w} \beta_{v,w} (\Gamma_{v,w}) \right) t.dt \tag{8}$$

where, $\Phi_v = 1$ *or* Λ_v *or* ψ_v depending on the minimizing criteria (i.e.,: energy consumption, carbon emission, non-renewable energy percentage) of the system.

4.3. Description of the Heuristic Function

This optimization problem falls into the multi-commodity flow class which is known as NP-hard problems and a heuristic algorithm is required to find the solution. Therefore, in order to solve our optimization problem formulated in the previous section we have used a genetic algorithm (GA) as our heuristic algorithm. It is an evolutionary algorithm which is inspired by the genetic processes of biological organisms. In [37–39] it is shown how this heuristic can be applied for solving NP-complete optimization problem related to networking like finding shortest path or designing an industrial Ethernet infrastructure. A GA has crossover and mutation which help the problem to not to get stuck in local minima which makes it robust than any other enumerative approaches. In the following part different parts of the GA are described in brief.

The crucial part of applying GA to a problem is to design a chromosome according to the problem. Later these will create a population pool. Each chromosome represents a single candidate solution of the search space. In our case, the data plane is converted into a chromosome as a data plane contains a solution set for each demand following the capacity constraint. Therefore, in order to build a chromosome pool, we have created a set of data planes. To do that, we have used another heuristic which is randomized depth first search (RDFS). RDFS randomly selects the next node for going into next level into the tree. Simple DFS is unable to create a diverse solution set. Therefore, RDFS is used to find the path for each demand and then that path is added to create the solution set. RDFS considers the current state of the topology capacity wise. Then, all the solutions that are found, follow the capacity constraint. Once we have a data plane then we can convert the data plane into chromosome. If we have a set of demand k = {1, 2, 3, . . . , k} and if a topology has E number of edges {1, 2, 3, . . . , E} then in a chromosome consists of k*E number of 0's and 1's. Where 0 represents this link is not used for fulfilling this demand and 1 represents this link is turned on for fulfilling this demand. Figure 6 shows the structure of the chromosome. This chromosome is actual representation of the example with three demands that we discussed in Section 4.1. In our initial population pool, we have always included one solution achieved by using SPF. Therefore, a population pool with a size of N has N-1 chromosome gathered by RDFS and the other one is using SPF.

Chromosome		
Demand 1	Demand 2	Demand 3

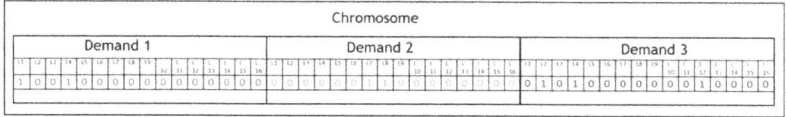

Figure 6. Structure of the chromosome.

For crossover operation we have decided to randomly select half of the bits of strings from one chromosome and rest half from another chromosome instead of using one cut or two cuts where chromosomes are divided into two equal parts or three equal parts and change the parts in between them respectively. By doing single crossover operations the process is generating two new chromosomes. In our case mutation is also done demand wise. Instead of a single bit values for a single demand, the whole strings of bit of a demand is replaced. There are two kinds of mutation percentage. The first one indicates how many of the chromosomes which are previously shortlisted by selection process will go through the mutation process. However, from a chromosome, how many demands will be selected for mutation is denoted by the second one. For example, mutation-1 equals to 10% means 10 of each 100 chromosomes are selected for mutation operation and mutation-2, 1% means 1 of each 100 demands from a chromosome which is selected by mutation-1 will be chosen randomly for mutation operation.

After crossover and mutation, the most important part of genetic algorithm is the fitness function which will after each iteration discard few of the solutions and promote the rest of the solutions for the next round. In our case we have three kind of fitness function based on our objective functions. Based on this fitness function score, different solutions are chosen for different cases.

After running GA, the next part is to convert the chromosome into a data plane and then control a plane so that controller can make appropriate changes in the topology. Even though each chromosome indicates that if a link is used for a demand or not, the direction of the flow of the traffic is not defined. Therefore, the algorithm needs to take care the information regarding direction, when converting from chromosome to data plane and then control plane. The algorithm to retrieve the data plane Π and the control plane Γ is given in Algorithms 1 and 2 respectively. It simply corresponds to the following equation:

$$\Pi_{k,v} = \chi_{k\times(v,w)} \times w \tag{9}$$

Algorithm 2 finds the minimal allocation satisfying the throughput demand gathered by the data plane. To note that the green policy, consists of shutdown **v** switches when $\delta_\Pi^v = 0$ and output ports **w** of switches **v** when $\delta_\Pi^{v,w} = 0$ and decreasing capacities of output ports **w** of switches **v** when $\Gamma_{v,w} < C_{v,w}$. Finally, Algorithm 3 shows the overall algorithm for implementing GA. Here **eval()** function is evaluating the fitness of the chromosome.

Algorithm 1. Retrieval of data and control plane from chromosome.

```
to_planes
  input  : X a chromosome
  output: Π the data plane and Γ the control plane

  initialization of Π such that all elements are null;
  for k_{s,d} ∈ K do
    while s ≠ d do
      foreach v ∈ V do
        if X_{k×(v,w)} = 1 then
          Π_{k,v} ← w;
          s ← w;
          break;

  Γ ← to_data_plane(Π);
  return (Π, Γ);
```

Algorithm 2. Computation of the minimum control plane from given data plane.

---

to_data_plane

   input : Π the data plane and $c_{v,w}$ the vector of the admissible capacities
             for an output port of a node v to a node w
   output: Γ the control plane

   forall $(v,w) \in E$ **do**
      if $\delta_\Pi^{v,w} = 0$ **then**
         | $\Gamma_{v,w} \leftarrow 0$;
      else
         forall $c_{v,w}$ *(in a decreasing order)* **do**
            if $c_{v,w} \geq \max\left(\sum_{k \in \mathcal{K}|\Pi_{k,v}=w} \lambda_k, \sum_{k \in \mathcal{K}|\Pi_{k,w}=v} \lambda_k\right)$ **then**
               /* symetric allocation */
               | $\Gamma_{v,w} \leftarrow c_{v,w}$;
            else
               | break;

   return Γ;

Algorithm 3. The Full Algorithm.

Data: $\mathcal{K}$ the initial ordered set of demands, C the initial adjacency matrix
      weighted by the maximum capacity of each port
Result: $\hat{\Pi}$ the optimal data plane and $\hat{\Gamma}$ the optimal control plane

1   *initialization of the $n-1$ chromosomes through the RDFS algorithm*;
2   **for** $i \leftarrow 1$ to $n-1$ **do**
3      $\mathcal{K}' \leftarrow$ random_order$(\mathcal{K})$;
4      $\mathcal{A} \leftarrow C$;
5   **for** $k \in \mathcal{K}'$ **do**
6          $\Pi_{i_k} \leftarrow$ rdfs$(\mathcal{A}, \lambda_k)$;
7   **forall** $(v,w)|\Pi_{i_{k,v}} = w$ **do**
8              $\mathcal{A}_{v,w} \leftarrow \mathcal{A}_{v,w} - \lambda_k$;

9   *initialization of the n-th chromosome through the SPF algorithm*;
10   $\mathcal{K}' \leftarrow$ random_order$(\mathcal{K})$;
11   $\mathcal{A} \leftarrow C$;
12   **for** $k \in \mathcal{K}'$ **do**
13      $\Pi_{n_k} \leftarrow$ spf_a_k$(\mathcal{A}, \lambda_k)$;
14   **forall** $(v,w)|\Pi_{n_{k,v}} = w$ **do**
15          $\mathcal{A}_{v,w} \leftarrow \mathcal{A}_{v,w} - \lambda_k$;

16   **for** $i \leftarrow 1$ to n **do**
17      $\mathcal{X}_i \leftarrow$ to_chromosome(Π_i);

18   *evolution of the population*;
19   eval$(\mathcal{X})$; sort$(\mathcal{X})$;
20   **repeat**
21      $\mathcal{X}_s \leftarrow \mathcal{X}_s \subset \mathcal{X}$ such that $|\mathcal{X}_s| = p_s|\mathcal{X}|$;   /* selection */
22   **forall** $x_i \in \mathcal{X}_s$ **do**   /* cross-over */
23   **if** random$(1) < p_c$ **then**
24              $\mathcal{K}' \leftarrow \mathcal{K}' \in \mathcal{P}(\mathcal{K})$ such that $|\mathcal{K}'| = |\mathcal{K}|/2$;   /* any $\mathcal{P}(\mathcal{K})$ */
25   **forall** $k \in \mathcal{K}'$ **do** $x'_k = x_i$;
26              $x_j \leftarrow x_j \in \mathcal{X}$ and $x_j \neq x_i$;   /* any x_j */
27   **forall** $k \notin \mathcal{K}'$ **do** $x'_k = x_j$;
28              $\mathcal{X} \leftarrow \mathcal{X} \cup x'$;

29   **forall** $x_i \in \mathcal{X}_s$ **do**   /* mutation */
30   flag$\leftarrow$ false;
31   **if** random$(1) < p_{m_1}$ **then**
32   **forall** $k \in \mathcal{K}$ **do**
33                  $x' \leftarrow x$;
34   **if** random$(1) < p_{m_2}$ **then**
35   flag$\leftarrow$ true;
37                      $z \leftarrow$ a possible subpart (path) of chromosome for demand k;
38   **forall** $(v,w) \in E$ **do**
39                          $x'_{k \times (v,w)} \leftarrow z_{v,w}$;

40   **if** *flag* **then** $\mathcal{X} \leftarrow \mathcal{X} \cup x'$;   3
41   eval$(\mathcal{X})$; sort$(\mathcal{X})$; $\mathcal{X} \leftarrow n$ first elements of $\mathcal{X}$;
42   **until** *time occurred (time over)*;
43   $\left(\hat{\Pi}, \hat{\Gamma}\right) \leftarrow$ to_planes$(\mathcal{X}_1)$;

---

5. Evaluation of the System

5.1. Experiment Setup

For evaluating the performance of the system, the topology of the GÉANT network infrastructure has been considered. The GÉANT project operates the European network for the research and education community. GÉANT has 22 nodes and each node of GÉANT represents a different country. Because of that, the values of CEF and NRE will be different for each node. The network structure and the traffic has been taken from SNDlib [40] published in 2005. However, in order to keep the data compatible with the network equipment as it is now using 100-gbps optical fiber, network's overall traffic has also increased significantly. From [41,42] it can be concluded that, during the last decade or so, traffic has increased almost 28 times. In order to keep up with the times, the real data is been multiplied by the increased traffic coefficient. Experiments are conducted with a set of 25 randomly generated demands. The demand sizes are randomly distributed. Network detail summary can be seen in Table 1.

Table 1. Network topology parameters.

Network	GÉANT
Nodes	22
Links	36
Link type	Full-Duplex
Demand Structure	(Source, Destination, Throughput)
Demand Type	Aggregated (one demand request for one source destination couple)
Bandwidth Capacity	1 Gbps, 10 Gbps, 40 Gbps, 100 Gbps

For power measurement, values are taken from [5,43]. For the links, power consumption of the optical transmission networking (OTN) layer interfaces per port has been considered. Power measurement values are given in Table 2. The values of carbon emission factor and non-renewable energy usage percentage have been taken from [44–46]. The details of all the national characteristics in terms of CEF and NRE are shown in Appendix B Table A2.

Table 2. Power consumption values for node and links.

Type	Power in Watts
Static Node	10000
1-Gbps port	7
10-Gbps port	34
40-Gbps port	160
100-gbps port	360

Lastly, the genetic algorithm has five factors, namely run time, population size, crossover rate, and two types of mutation rate. In order to run the experiment, we have tuned these values. In our work, we have fine-tuned these parameters using design of experiments. For this paper we are using the following values for the parameters of genetic algorithm in all the experiments: run time 30 s, population size 40, crossover rate 40%, mutation-1 rate 80% and mutation-2 rate 4%. The process of selecting the parameter values is briefly described in Appendix A.

5.2. Result Analysis

A set of experiments has been conducted to analyze the performance of the three different objective functions. The results are compared with shortest path first (SPF) with the same green policies that our algorithm has been considered. Twenty five demands are randomly taken for the experiment as we want to see how the algorithm uses the two green policies according to the objective function. This is because, if the number of demands are higher, then all the nodes of the topology will be involved as either source or destination, then all the nodes must be kept turned on and there will be no scope to analyze the performance of the algorithm in terms of first green policy where nodes will be shut down when not in use. Figure 7 shows the graph which gives the comparison of three different approaches. As we can see, every algorithm outperforms other two when their minimizing factor has been chosen.

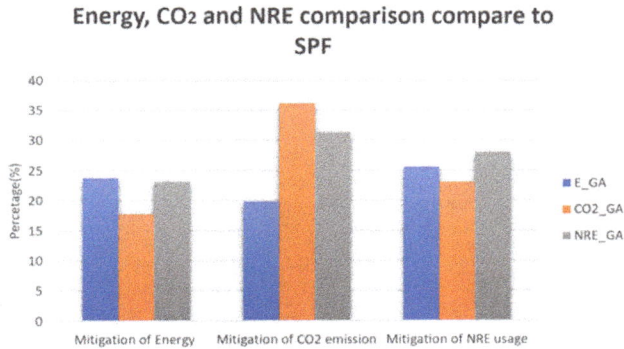

Figure 7. Comparison between three algorithms.

When energy is considered as the minimizing factor, that means for our EAR with GA (E_GA), we have maximum energy savings compared to the other two approaches. EAR saves around 24% of energy consumption compared to SPF for this scenario, whereas both approaches of PAR (CO_GA and NRE_GA) save around 18% and 23% energy consumption, respectively. However, when comes to reducing the CO_2 emissions, both CO_2_GA and NRE_GA reduce an appreciable amount compared to E_GA. CO_2_GA reduces 36% of the CO_2 emissions compared to SPF, whereas, NRE_GA and E_GA are able to reduce them by 31% and 20%, respectively. Even though our EAR (E_GA) has outperformed SPF by a large margin in all three aspects, in terms of reducing CO_2 emissions both pollution-based algorithms performs better. In fact, when the optimizing function focuses on reducing CO_2 emissions CO_2_GA reduces CO_2 emissions by almost double compared to E_GA. Lastly, while for reducing non-renewable energy usage NRE_GA performs better than the other two. NRE_GA reduces the non-renewable energy usage by 28% compared to the shortest path. Table A3 in Appendix B shows the all obtained results for SPF and the other three algorithms. In contains country-wise consumption energy values, CO_2 and non-renewable energy for each algorithm in order to understand the results more clearly. All these results raise the question, as the goal of green networking is to reduce the carbon footprint from the environment, then how much effective will designing a green network by only considering energy efficiency be. Even if a solution focusing on reducing CO_2 emissions consumes more energy but if the total CO_2 emissions of the system are lower that means the system is consuming more green energy than brown energy (energy sources producing a higher amount of CO_2).

Based only on the consumption values used in the paper, Table 3 provides an annual emission of energy CO_2 and non-renewable energy. Here, at first, we have shown that if no green policy is used then what the total emissions will be and then the consumption values for shortest path first and our three algorithms are given. These values only reflect the emissions due to the devices and links. However, the actual amount will be much higher considering the building, cooling and many other criteria.

Table 3. Energy, cost and CO_2 emission analysis for a year.

Method	Energy (MWh)	CO_2 (Tons)	Non-Renewable Energy (MWh)
Without any green policy	2154	1088	1470
SPF	1522	684	1074
E_GA	1160	548	799
CO_2_GA	1252	436	824
NRE_GA	1170	470	777

Figure 8 gives the different topologies for different solutions. The first one in the top left gives the topology for SPF. As is clearly visible, the maximum number of nodes and links are used in this topology. Now if we analyze the differences between different topologies there are some interesting changes in the topology. The energy-based solution has not used node six and node sixteen which are Switzerland and France, and node 14 which is The Netherlands is used. For energy-based solution every node is considered as same as every node consumes same amount of energy, whereas, for the CO_2-based solution the emissions could be completely different for the same number of nodes. As for our example, The Netherlands emits 130 times more CO_2 gm per kw-h compared to Switzerland and six times more compared to France which has huge impact on the overall result. In the same way, node-2 (Poland) is used by E_GA whereas both of the PAR solutions avoid Poland as it has huge CEF and NRE usage percentage. These choices made by E_GA might reduce the overall energy consumption of the network, but at a cost of a high carbon emission.

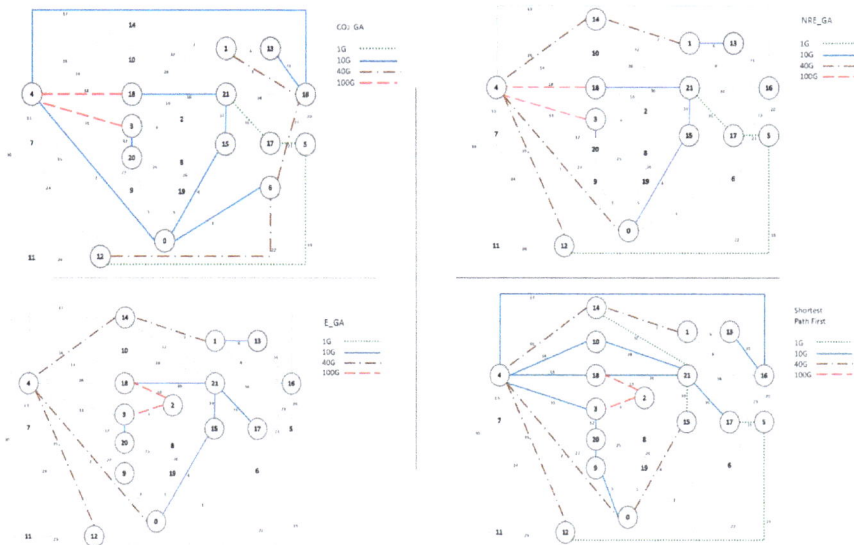

Figure 8. Topology for different algorithm.

Now, if we compare the solutions of the two PAR-based options, as mentioned above they both have avoided node-2 (Poland), however, for NRE_GA a majority of the traffic went through node-4 (Germany) and node-16 (France) is completely avoided, whereas for CO_2_GA a substantial amount of traffic went through France, even though it has very low carbon footprint. It is because NRE_GA only focuses on non-renewable energy percentage and as France uses nuclear power plants for producing electricity, it has very high non-renewable energy usage percentage. For example, for the same topology, let's consider a demand from node 5 to node 14. There are several possible solutions. However, we will consider only the solutions which are shortest that means lowest number of hops. Now, there are four different possible paths. Table 4 gives a summary of all four shortest paths with carbon emission factors and non-renewable energy usage percentages. Now, here only the NRE and CEF values of intermediate nodes are considered. Based on the NRE_GA the optimal solution would be path-3 (with Italy and Germany) even though this path has one of the highest carbon emission factors, whereas, the optimal solution based on CEF would be path-2 but as this path has a maximum amount of NRE percentage therefore this will not be chosen by NRE_GA.

Table 4. Demand paths from node-5 to node-14.

No.	Paths	NRE% Sum of the Intermediate Nodes	CEF Sum of the Intermediate Nodes
1	5-16-4-14	1.535	0.742
2	5-16-1-14	1.659	0.294
3	5-12-4-14	1.337	1.082
4	5-12-11-14	1.602	1.15

1-Belgium, 4-Germany, 5-Spain, 11-Israel, 12-Italy, 14-Netherlands, 16-France.

Based on this analysis of the three algorithms, it can be said that even though the outcome of the algorithm depends on the demand set, it is very much a possibility that in order to achieve a sustainable solution focusing only on energy efficiency might not be a wise decision. Our result shows that for this particular scenario both of our algorithms outperform the energy-based solution in terms of reducing carbon emissions. Specially CO_2_GA reduces carbon emission by more than 20% compared to energy-based solution which is definitely a non-negligible amount.

5.3. Sustainability Discussion

The three pillars of sustainability are a great means of explaining the complete sustainability problem. We have included all the three pillars into our system in order to provide a balanced sustainable solution. Our system provides the information about the throughput of different parts of the network. Whichever objective function is selected it gives an insight about the three parameters for measuring the status of the green networking, namely energy consumption, CO_2 emissions and percentage of non-renewable energy usage, but as the goal of green networking is to reduce the carbon emissions, therefore the objective function minimizing the CO_2 has utmost importance. As we can see, CO_2_GA reduces CO_2 emissions by around 8% compared to our E_GA and by 40% compared to SPF with two green policies. These are definitely non-negligible values, however, a more tangible example might give a better understanding of the situation. For our scenario, if we compare the result of SPF and CO_2_GA from Table 3, in one year it can save up to 248 tons of CO_2. This is in fact equivalent to the carbon emissions if a person were to make a round trip to JFK in New York from CDG in Paris by plane 381 times. Even when comparing with E_GA, the carbon savings is equivalent to 172 round trips by plane for the same source and destination. If we consider the savings in terms of money, even if we consider an average rate for cost of electricity production, our all three can save more than 25% of the cost compared to SPF. Lastly, the proposed system respects both the user traffic demand and the objective function choice and provides a data and control plane and hence integrates the social attribute of the sustainability.

6. Future Works

In our work we have proposed a solution which gives the required control plane and data plane for implementing it on an SDN platform. However, changes in the demand set result in a new topology and the new topology will be disseminated by the controller to the network. This dissemination process also requires energy, and hence has a carbon footprint. Therefore, for the future work we are planning to add a penalty system that will determine if a change in the network topology is necessary or not with a change in the demand set. In fact, this will control the frequency of network changes to an optimum level. Additionally, in this work all the nodes are equally treated whereas, that is not the case in reality. Our equation is flexible enough to add this variance into the system. Therefore, next step will be to adjust the equation accordingly and run the experiment for variable values for the node. Additionally, the carbon emissions and the non-renewable energy usage rate vary all the time. This is due to weather conditions, to the period of the day (night/day), to energy demand peaks requiring using for example more coal for producing electricity. Then, the idea is to include in the fitness function these variations in adding temporal factor in order to get more realistic results. In the same way, energy costs and its evolution could be integrated in the fitness function in order to cover the economy pillar of sustainable development. Finally, these variations (energy costs, carbon emitted for producing energy, etc.) could be estimated by predictive models for efficiently managing the changes of network topology in considering the penalty explained above.

7. Conclusions

In this work, we have proposed a sustainable method for greening the Internet. We have introduced the term "pollution-aware routing" and it has been added to the classical energy-aware routing. While proposing this new term and introducing a new way of looking the problem of green networking we have at the same time tried to answer whether the question concerning only energy efficiency while trying to achieve green networking is enough or not. We have two different approaches, one works based on the carbon emission factor of the nodes whereas the second one works based on the non-renewable energy usage percentage. We show that our pollution-aware routing can have significant impact on CO_2 emissions compared to energy-aware solutions. Our system provides a holistic approach towards attaining sustainability. At the same time, it provides a control plane and data plane so that the system can be implemented in a centralized system using SDN.

Author Contributions: Conceptualization, M.M.H., J.-P.G., E.R., and T.D.; methodology, M.M.H., J.-P.G., E.R., and T.D.; software, M.M.H. and J.-P.G.; validation, M.M.H., J.-P.G., E.R., and T.D.; formal analysis, M.M.H. and J.-P.G.; investigation, M.M.H.; resources, M.M.H. and J.-P.G.; data curation, M.M.H.; writing—original draft preparation, M.M.H.; writing—review and editing, M.M.H., J.-P.G., E.R. and T.D.; visualization, M.M.H.; supervision, J.-P.G., E.R. and T.D.; project administration, E.R., J.-P.G. and T.D.; funding acquisition, E.R., J.-P.G. and T.D.

Funding: This research received no external funding.

Conflicts of Interest: The authors declare no conflict of interest.

Appendix A

As our heuristic algorithm is based on genetic algorithm, we have to do an extensive set of experiments for tuning the parameters. Five parameters are needed to be adjusted which are time, population size, crossover percentage, mutation-1 percentage and lastly mutation-2 percentage. We have selected few fixed values for each parameter for design of experiment. Each value of each parameter is preliminarily tested before being used in DoE. In Table A1, different level of values for each parameter is given. All the parameters are tuned in order to maximize objective function.

Table A1. DoE parameter summary.

Factor	Values
Time	10 s, 20 s, 30 s
Population Size	20, 40, 80
Crossover Percentage	20%, 40%, 80%
Mutation-1 Percentage	20%, 40%, 80%
Mutation-2 Percentage	4%, 8%, 12%

Running the full factorial experiment on our parameters using the above mentioned settings, provides us significant information about the parameter tuning. For both 25 demands and 400 demands DoE shows similar kinds of patterns. From these experiments we have achieved an optimum setting for all the genetic algorithm parameters which are, run time 30s, population size 40, crossover rate 40%, mutation-1 rate 80% and mutation-2 rate 4%.

Appendix B

Table A2 shows the national characteristics in terms of carbon emission factor and non-renewable energy usage factor.

Table A2. List of nodes of the GEANT network with carbon and non-renewable energy factors.

Node No.	Country	CEF	NRE Factor
0	Austria	0.176	0.257
1	Belgium	0.224	0.834
2	Poland	1.196	0.863
3	Czech Republic	0.938	0.85
4	German	0.672	0.71
5	Spain	0.342	0.619
6	Switzerland	0.003	0.602
7	Greece	1.921	0.726
8	Croatia	0.386	0.348
9	Hungary	0.589	0.899
10	Ireland	0.521	0.753
11	Israel	0.74	0.975
12	Italy	0.41	0.327
13	Luxembourg	0.276	0.792
14	Netherlands	0.413	0.879
15	United States	0.547	0.72
16	France	0.07	0.825
17	Portugal	0.4	0.465
18	Sweden	0.023	0.429
19	Slovenia	0.578	0.694
20	Slovakia	0.282	0.755
21	United Kingdom	0.508	0.721

Table A3 shows the obtained results for each countries' consumption of energy CO_2 and non-renewable energy for our scenario.

Table A3. Countrywise energy, CO_2 and non-renewable energy consumption values.

Algorithm Country	SPF Energy (Watt)	SPF CO₂ (gCO₂/kWh)	SPF NRE (Watt)	GA_E Energy (Watt)	GA_E CO₂ (gCO₂/kWh)	GA_E NRE (Watt)	GA_CO₂ Energy (Watt)	GA_CO₂ CO₂ (gCO₂/kWh)	GA_CO₂ NRE (Watt)	GA_NRE Energy (Watt)	GA_NRE CO₂ (gCO₂/kWh)	GA_NRE NRE (Watt)
Austria	10354	1822.30	2660.97	10194	1794.14	2619.85	10102	1777.95	2596.21	10194	1794.14	2619.85
Belgium	10160	2275.84	8473.44	10194	2283.45	8501.79	10160	2275.84	8473.44	10194	2283.45	8501.79
Poland	10720	12821.1	9251.36	10720	12821.1	9251.36	0	0	0	0	0	0
Czech Republic	10428	9781.46	8863.8	10294	9655.77	8749.9	10394	9749.57	8834.9	10394	9749.57	8834.9
German	10616	7133.95	7537.36	10480	7042.56	7440.8	10788	7249.53	7659.48	11200	7526.4	7952
Spain	10014	3424.78	6198.66	0	0	0	10014	3424.78	6198.66	10014	3424.78	6198.66
Switzerland	0	0	0	0	0	0	10354	31.062	4162.30	0	0	0
Greece	0	0	0	0	0	0	0	0	0	0	0	0
Croatia	0	0	0	0	0	0	0	0	0	0	0	0
Hungary	10068	5930.05	9051.13	0	0	0	0	0	0	0	0	0
Ireland	10068	5245.42	7581.20	0	0	0	0	0	0	0	0	0
Israel	0	0	0	0	0	0	0	0	0	0	0	0
Italy	10167	4168.47	6374.70	10160	4165.6	6370.32	10167	4168.47	6374.70	10167	4168.47	6374.70
Luxembourg	10034	2769.38	7946.92	10034	2769.38	7946.92	10034	2769.38	7946.92	10034	2769.38	7946.92
Netherlands	10327	4265.05	9077.43	10320	4262.16	9071.28	0	0	0	10320	4262.16	9071.28
United States	10167	5561.34	8672.45	10068	5507.19	8588.00	10068	5507.19	8588.00	10068	5507.19	8588.00
France	10068	704.76	8306.1	0	0	0	10388	727.16	8570.1	0	0	0
Portugal	10041	4016.4	4669.06	10034	4013.6	4665.81	10014	4005.6	4656.51	10014	4005.6	4656.51
Sweden	10428	239.844	4473.61	10394	239.062	4459.02	10394	239.062	4459.02	10394	239.062	4459.02
Slovenia	0	0	0	0	0	0	0	0	0	0	0	0
Slovakia	10068	2839.17	7601.34	10034	2829.58	7575.67	10034	2829.5	7575.67	10034	2829.5	7575.67
United Kingdom	10116	5138.9	7293.63	10102	5131.816	7283.542	10075	5118.1	7264.075	10075	5118.1	7264.075

References

1. Pickavet, M.; Vereecken, W.; Demeyer, S.; Audenaert, P.; Vermeulen, B.; Develder, C.; Demeester, P. Worldwide energy needs for ICT: The rise of power-aware networking. In Proceedings of the 2nd International Symposium on Advanced Networks and Telecommunication Systems, Mumbai, India, 15–17 December 2008.
2. Digital Agenda: Global Tech Sector Measures Its Carbon Footprint. 2013. Available online: http://europa.eu/rapid/press-release_IP-13-231_en.htm (accessed on 28 June 2019).
3. Peter, C.; Andrae, A.S.G. Emerging Trends in Electricity Consumption for Consumer ICT. Available online: https://wireless.kth.se/wp-content/uploads/sites/19/2014/08/Emerging-Trends-in-Electricity-Consumption-for-Consumer-ICT.pdf (accessed on 28 June 2019).
4. Zhang, M.; Yi, C.; Liu, B.; Zhang, B. GreenTE: Power-aware traffic engineering. In Proceedings of the 18th IEEE International Conference on Network Protocols (ICNP'10), Kyoto, Japan, 5–8 October 2010; pp. 21–30.
5. Addis, B.; Capone, A.; Carello, G.; Gianoli, L.G.; Sanso, B. Energy management through optimized routing and device powering for greener communication networks. *IEEE/ACM Trans. Netw.* **2014**, *22*, 313–325. [CrossRef]
6. Gunaratne, C.; Christensen, K.; Suen, S.W. Ethernet Adaptive Link Rate (ALR): Analysis of a buffer threshold policy. In Proceedings of the IEEE Global Communications Conference (GLOBECOM 2006), San Francisco, CA, USA, 27 November–1 December 2006.
7. Chiaraviglio, L.; Mellia, M.; Neri, F. Reducing Power Consumption in Backbone Networks. In Proceedings of the IEEE International Conference on Communications (ICC 2009), Dresden, Germany, 14–18 June 2009.
8. Kim, H.; Feamster, N. Improving network management with software defined networking. *IEEE Commun. Mag.* **2013**, *51*, 114–119. [CrossRef]
9. Di Fazio, A.R.; Russo, M.; De Santis, M. Zoning Evaluation for Voltage Optimization in Distribution Networks with Distributed Energy Resources. *Energies* **2019**, *12*, 390. [CrossRef]
10. Fang, C.; Yu, F.R.; Huang, T.; Liu, J.; Liu, Y. A survey of green information-centric networking: Research issues and challenges. *IEEE Commun. Surv. Tutor.* **2015**, *17*, 1455–1472. [CrossRef]
11. Chabarek, J.; Sommers, J.; Barford, P.; Estan, C.; Tsiang, D.; Wright, S. Power awareness in network design and routing. In Proceedings of the 27th Conference on Computer Communications (IEEE INFOCOM 2008), Phoenix, AZ, USA, 13–18 April 2008; pp. 457–465.
12. Mumey, B.; Tang, J.; Hashimoto, S. Enabling green networking with a power down approach. In Proceedings of the 2012 IEEE International Conference on Communications (ICC), Ottawa, ON, Canada, 10–15 June 2012; pp. 2867–2871.
13. Markiewicz, A.; Tran, P.N.; Timm-Giel, A. Energy consumption optimization for software defined networks considering dynamic traffic. In Proceedings of the 2014 IEEE 3rd International Conference on Cloud Networking (CloudNet), Luxembourg, 8–10 October 2014; pp. 155–160.
14. Wang, R.; Jiang, Z.; Gao, S.; Yang, W.; Xia, Y.; Zhu, M. Energy-aware routing algorithms in software-defined networks. In Proceedings of the IEEE 15th International Symposium on A World of Wireless, Mobile and Multimedia Networks (WoWMoM), Sydney, Australia, 19 June 2014; pp. 1–6.
15. Lorincz, J.; Mujaric, E.; Begusic, D. Energy consumption analysis of real metro-optical network. In Proceedings of the 38th International Convention on Information and Communication Technology, Electronics and Microelectronics (MIPRO), Opatija, Croatia, 25–29 May 2015; pp. 550–555.
16. Lemay, M.; Nguyen, K.K.; Arnaud, B.S.; Cheriet, M. Toward a zero-carbon network: Converging cloud computing and network virtualization. *IEEE Internet Comput.* **2012**, *16*, 51–59. [CrossRef]
17. Gupta, M.; Singh, S. Using Low-Power Modes for Energy Conservation in Ethernet LANs. In Proceedings of the 26th IEEE International Conference on Computer Communications (INFOCOM 2007), Barcelona, Spain, 6–12 May 2007; pp. 2451–2455.
18. Nedevschi, S.; Popa, L.; Iannaccone, G.; Ratnasamy, S.; Wetherall, D. Reducing Network Energy Consumption via Sleeping and Rate-Adaptation. In Proceedings of the 5th USENIX Symposium on Networked Systems Design and Implementation (NDSI2008), San Francisco, CA, USA, 16–18 April 2008.
19. Avallone, S.; Ventre, G. Energy efficient online routing of flows with additive constraints. *Comput. Netw.* **2012**, *56*, 2368–2382. [CrossRef]

20. Awad, M.K.; Rafique, Y.; Alhadlaq, S.; Hassoun, D.; Alabdulhadi, A.; Thani, S. A greedy power-aware routing algorithm for software-defined networks. In Proceedings of the 2016 IEEE International Symposium on Signal Processing and Information Technology (ISSPIT), Limassol, Cyprus, 12–14 December 2016; pp. 268–273.

21. Manjate, J.A.; Hidell, M.; Sjödin, P. Can Energy-Aware Routing Improve the Energy Savings of Energy-Efficient Ethernet? *IEEE Trans. Green Commun. Netw.* **2018**, *2*, 787–794. [CrossRef]

22. Maaloul, R.; Taktak, R.; Chaari, L.; Cousin, B. Energy-Aware Routing in Carrier-Grade Ethernet using SDN Approach. *IEEE Trans. Green Commun. Netw.* **2018**, *2*, 844–858. [CrossRef]

23. Gattulli, M.; Tornatore, M.; Fiandra, R.; Pattavina, A. Low-emissions routing for cloud computing in IP-over-WDM networks with data centers. *IEEE J. Sel. Areas Commun.* **2014**, *32*, 28–38. [CrossRef]

24. Singh, D.; Chandwani, G. Carbon-aware routing in software defined inter data center network. In Proceedings of the 2015 IEEE International Conference on Advanced Networks and Telecommuncations Systems (ANTS), Kolkata, India, 15–18 December 2015; pp. 1–6.

25. Zhou, Z.; Liu, F.; Xu, Y.; Zou, R.; Xu, H.; Lui, J.C.; Jin, H. Carbon-aware load balancing for geo-distributed cloud services. In Proceedings of the 2013 IEEE 21st International Symposium on Modelling, Analysis and Simulation of Computer and Telecommunication Systems, San Francisco, CA, USA, 14–16 August 2013; pp. 232–241.

26. Jo, S.K.; Wang, L.; Kangasharju, J.; Muehlhaeuser, M. Cost-Effective and Eco-Friendly Green Routing Using Renewable Energy. In Proceedings of the 27th International Conference on Computer Communication and Networks (ICCCN), Hangzhou, China, 30 July–2 August 2018; pp. 1–8.

27. Mineraud, J.; Wang, L.; Balasubramaniam, S.; Kangasharju, J. Hybrid renewable energy routing for ISP networks. In Proceedings of the 35th Annual IEEE International Conference on Computer Communications (IEEE INFOCOM 2016), San Francisco, CA, USA, 10–14 April 2016; pp. 1–9.

28. Hilty, L.; Lohmann, W.; Huang, E. Sustainability and ICT—An overview of the field. *Politeia* **2011**, *27*, 13–28.

29. Calabrese, A.; Costa, R.; Levialdi, N.; Menichini, T. Integrating sustainability into strategic decision-making: A fuzzy AHP method for the selection of relevant sustainability issues. *Technol. Forecast. Soc. Chang.* **2019**, *139*, 155–168. [CrossRef]

30. Elkington, J. *Cannibals with Forks: The Triple Bottom Line of 21st Century Business*; Capstone Publishing: Oxford, UK, 1997.

31. Kajikawa, Y. Research core and framework of sustainability science. *Sustain. Sci.* **2008**, *3*, 215–239. [CrossRef]

32. Hansmann, R.; Mieg, H.A.; Frischknecht, P. Principal sustainability components: Empirical analysis of synergies between the three pillars of sustainability. *Int. J. Sustain. Dev. World Ecol.* **2012**, *19*, 451–459. [CrossRef]

33. Schlömer, S.; Bruckner, T.; Fulton, L.; Hertwich, E.; McKinnon, A.; Perczyk, D.; Roy, J.; Schaeffer, R.; Sims, R.; Smith, P.; et al. Annex III: Technology-specific cost and performance parameters. *Clim. Change* **2014**, 1329–1356.

34. Hossain, M.M.; Rondeau, E.; Georges, J.P.; Bastogne, T. Modeling the power consumption of ethernet switch. In Proceedings of the International Sustainable ecological engineering design for society conference (SEEDS 2015), Ritz, UK, 17–18 September 2015.

35. Fithritama, A.; Rondeau, E.; Bombardier, V.; Georges, J.P. Modeling fuzzy rules for managing power consumption of ethernet switch. In Proceedings of the 23rd International Conference on Software, telecommunications and computer networks (SoftCOM), Split, Croatia, 16–18 September 2015; pp. 42–47.

36. Gupta, M.; Grover, S.; Singh, S. A Feasibility Study for Power Management in LAN Switches. In Proceedings of the 12th IEEE International Conference on Network Protocols (ICNP 2004), Berlin, Germany, 8 October 2004; pp. 361–371.

37. Krommenacker, N.; Rondeau, E.; Divoux, T. Genetic algorithms for industrial Ethernet network design. In Proceedings of the 4th IEEE International Workshop on Factory Communication Systems, Vasteras, Sweden, 28–30 August 2002; pp. 149–156.

38. Bui, T.N.; Moon, B.R. Genetic algorithm and graph partitioning. *IEEE Trans. Comput.* **1996**, *45*, 841–855.

39. Ahn, C.W.; Ramakrishna, R.S. A genetic algorithm for shortest path routing problem and the sizing of populations. *IEEE Trans. Evol. Comput.* **2002**, *6*, 566–579.

40. Sndlib-Library of Test Instances for Survivable Fixed Telecommunication Network Design. Available online: http://sndlib.zib.de (accessed on 27 June 2019).

41. VNI Global Fixed and Mobile Internet Traffic Forecast. Available online: https://www.cisco.com/c/en/us/solutions/collateral/service-provider/visual-networking-index-vni/vni-hyperconnectivity-wp.html (accessed on 27 June 2019).
42. White paper on Global IP Traffic Forecast and Methodology, 2006–2011. Available online: http://www.hbtf.org/files/cisco_IPforecast.pdf (accessed on 27 June 2019).
43. Van Heddeghem, W.; Idzikowski, F. Equipment power consumption in optical multilayer networks-source data. *J. Photonic Netw. Commun.* **2012**, *24*, 1–28.
44. Brander, M.; Sood, A.; Wylie, C.; Haughton, A.; Lovell, J. Technical Paper - Electricity-specific emission factors for grid electricity. Available online: https://ecometrica.com/assets/Electricity-specific-emission-factors-for-grid-electricity.pdf (accessed on 28 June 2019).
45. List of Countries by Electricity Production from Renewable Sources. Available online: https://en.wikipedia.org/wiki/List_of_countries_by_electricity_production_from_renewable_sources (accessed on 27 June 2019).
46. Koffi, B.; Cerutti, A.; Duerr, M.; Iancu, A.; Kona, A.; Janssens-Maenhout, G. *Covenant of Mayors for Climate and Energy: Default Emission Factors for Local Emission Inventories*; European Commission, JRC Technical Report; Publications Office of the European Union: Luxembourg, 2017.

sensors

MDPI

Article

Optimizing Charging Efficiency and Maintaining Sensor Network Perpetually in Mobile Directional Charging

Xianghua Xu [1,*], Lu Chen [1] and Zongmao Cheng [2]

[1] School of Computer Science and Technology, Hangzhou Dianzi University, Hangzhou 310018, China; chenlu0154@hdu.edu.cn
[2] School of Science, Hangzhou Dianzi University, Hangzhou 310018, China; zmcheng@hdu.edu.cn
* Correspondence: xhxu@hdu.edu.cn; Tel.: +86-571-8691-9113

Received: 25 April 2019; Accepted: 8 June 2019; Published: 12 June 2019

Abstract: Wireless Power Transfer (WPT) is a promising technology to replenish energy of sensors in Rechargeable Wireless Sensor Networks (RWSN). In this paper, we investigate the mobile directional charging optimization problem in RWSN. Our problem is how to plan the moving path and charging direction of the Directional Charging Vehicle (DCV) in the 2D plane to replenish energy for RWSN. The objective is to optimize energy charging efficiency of the DCV while maintaining the sensor network working continuously. To the best of our knowledge, this is the first work to study the mobile directional charging problem in RWSN. We prove that the problem is NP-hard. Firstly, the coverage utility of the DCV's directional charging is proposed. Then we design an approximation algorithm to determine the docking spots and their charging orientations while minimizing the number of the DCV's docking spots and maximizing the charging coverage utility. Finally, we propose a moving path planning algorithm for the DCV's mobile charging to optimize the DCV's energy charging efficiency while ensuring the networks working continuously. We theoretically analyze the DCV's charging service capability, and perform the comprehensive simulation experiments. The experiment results show the energy efficiency of the DCV is higher than the omnidirectional charging model in the sparse networks.

Keywords: wireless power transfer; directional charging vehicle; charging efficiency; RWSN

1. Introduction

Wireless power transfer is a promising technology to replenish energy to sensors in Rechargeable Wireless Sensor Networks (RWSN), to keep the network working continuously [1]. Wireless Power Transfer (WPT) is mainly using magnetic resonance coupling [1–3] or RF radiation technology [4,5]. To achieve efficient energy transfer in RF radiation technology, it generally requires directional transmission by using high-gain and directional antennas for power transmitters and receivers to focus the energy in narrow energy beams [6]. It has a more stable and higher efficiency of power transfer compared with omnidirectional charging [7]. Consequently, in the mobile directional charging scenario in RWSN, a rechargeable sensor can only receive power from a mobile charging vehicle equipped with a directional power transmitting antenna, or called directional charging vehicle (DCV), when they are located in the covered sector of the DCV's directional antennas.

Products from Powercast [8] carry out wireless charging by leveraging the electromagnetic radiation technique, with which energy transmitters broadcast the RF energy and receivers capture the energy and convert it to DC. Applications of the electromagnetic radiation technique for wireless charging have been reported in References [9–11]. As more and more applications of wireless charging technology have been envisioned, the Wireless Power Consortium [12] has been established to start the efforts of setting an international standard for interoperable wireless charging.

Recently, most research works of mobile charging in RWSN adopted the omnidirectional power transfer model [13–17]. Although some works have studied the directional charger's deployment problem in RWSN [9,18–20], to the best of our knowledge, there is no literature that has studied the mobile directional charging problem. However, a directional antenna provides significant enhancement over the omnidirectional antenna in terms of direction beam [21]. Moreover, when charging distributed sensor nodes, a directional antenna, rather than an omnidirectional antenna, is more energy-efficient because of the smaller proportion of off-target radiation [22]. Inspired by the research issues in the literature on mobile omnidirectional wireless charging scheduling [13–17], and directional charger's deployment [9,18–20], we propose the directional wireless charging optimization problem in this paper. The complex factors of RF power transmission in practical environment are simplified in our research problem.

In this paper, we investigate the mobile directional charging optimization problem in wireless sensor networks. As shown in Figure 1, the data collection sensor network is deployed in a 2D plane area. The sensors transmit data to the sink node through multiple hops route. The charger's base station serves for the DCV. The DCV starts from the base station and moves along the planned docking spots and path to replenishing energy for all sensors in a charging cycle. The mobile directional charging optimization problem is how to determine the DCV's docking spots and charging directions in the 2D plane, and plan the moving path through all docking spots to replenish energy for the sensor network. The objective is to optimize Energy Charging Efficiency (ECE) of the DCV while maintaining the sensor network working continuously. The ECE is the ratio of the energy received by all sensors to the energy consumed by the DCV in a charging cycle. This problem is named as Charging Efficiency Optimization Problem (CEOP) of mobile directional charging in RWSN.

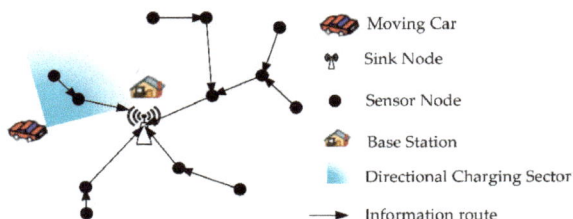

Figure 1. Directional mobile charging scenario for the data collection network in RWSN (Rechargeable Wireless Sensor Networks).

The CEOP problem has two main technical challenges. The first challenge is that since both the DCV's docking spots and its charging orientations are continuous values, it is hard to determine the DCV's docking spots and charging orientations to meet the charging coverage for all sensors. The second challenge is how to plan a DCV's moving path that ensures no sensors will run out of energy during the charging cycle.

The CEOP is an NP-hard problem and it is difficult to design a global optimal solution. We consider dividing the CEOP problem into two sub-problems: (1) How to determine the appropriate docking spots of the DCV in the 2D plane and the DCV's charging direction at each docking spot; (2) How to plan the DCV's moving path and charging time at each docking spot to meet the network's energy requirements and optimize the DCV's energy charging efficiency.

We model the charging docking point planning on the 2D plane as a location optimization of mobile charging with the objective of minimizing the number of docking points under the constraints and of maximizing the charging coverage utility locally. Then, we use the TSP optimization to minimize the charging path loop and maximizing energy charging efficiency for the whole network.

The main contributions are as follow:

- As far as we know, this is the first work investigating the mobile directional charging problem in WRSN aiming to maximize the energy charging efficiency and maintain the networks working continuously.
- We prove that the problem is NP-hard.
- We propose the coverage utility of the DCV's directional charging, and design an approximation algorithm to determine the docking spots and their charging orientations while minimizing the number of the DCV's docking spots and maximizing the charging coverage utility. It ensures the mobile charging coverage for all the sensors in the network and improves the energy charging efficiency locally.
- We propose a moving path planning algorithm for the DCV's mobile charging to optimize the DCV's energy charging efficiency while ensuring the networks working continuously.
- We theoretically analyze the DCV's charging service capability, and perform the comprehensive simulation experiments. The experiment results show that energy charging efficiency is higher than omnidirectional charging model in the data collection network.

The remainder of the paper is organized as follows: In Section 2, we review the related work of RWSN; In Section 3, we present the description of directional charging model, network energy consumption and problem definition; In Section 4, we propose the optimization algorithms; In Section 5, we give analysis of network size and area size that one DCV can serve; In Section 6, we present simulation result; Section 7 concludes this paper.

2. Related Works

The existing wireless energy transfer can be divided into Single-Input Single-Output energy transfer model [23–34] and Single-Input Multiple-Output energy transfer model [13–17,31–35]. Energy transfer optimization problems can be divided into static charging stations' deployment [11,18–20,35–38] and mobile charging vehicles' dispatching problems [13–17,23–34].

Mobile omnidirectional wireless charging problem. All existing works considering the mobile wireless charging adopt the omnidirectional power transfer model. Unlike the omnidirectional charging problem, we should not only determine the charging stop point and plan the charging path, but also determine the charging direction at each charging stop point. Yi et al. [13] investigate how to schedule the omnidirectional charging vehicle to maximize its vacation time and achieve higher charging efficiency of sensor networks. Xie et al. [17] investigate the mobile charging problem of co-locating the mobile base station on the wireless charging vehicle. Wu et al. [15] studied the omnidirectional charger vehicle dispatch problem to maximize the network lifetime and improve the energy efficiency for large-scale WSNs. Khelladi et al. [14] modeled the omnidirectional charger dispatching problem as a charging path optimization problem, and aimed to minimize the number of stop locations in the charging path and reducing the total energy consumption of the mobile charger. Jiang et al. [16] consider the on-demand mobile charging problem which schedules the omnidirectional charger to maximize the covering utility.

Directional wireless chargers deployment problem. All existing charging works which adopt the directional power transfer model only concern the directional chargers' deployment problem in RWSN, rendering them not applicable to our problem. Dai et al. [9] investigated directional chargers' deployment problem to optimize charging utility for the sensor network. Dai et al. [18] proposed the notion of omnidirectional charging and studied the omnidirectional chargeability under the deterministic deployment of chargers and random deployment of chargers. The goal is to achieve that at any position in the area with any orientation can be charged by directional chargers with power being no smaller than a given threshold. Jiang et al. [19] studied the wireless charger deployment optimization problem, which is to deploy as few as possible chargers to make the WRSN sustainable. Ji et al. [20] further investigated the deployment optimization problem of wireless chargers equipped with 3D beamforming directional antennas, and achieve the deployment of as few as possible chargers to make the WRSN sustainable.

To best of our knowledge, this is the first work to study the mobile directional charging problem in RWSN. The closest to our work is mobile omnidirectional charging and deployment of directional charger. Compared with omnidirectional power transfer model, there are two strengths to introduce directional power transfer model in mobile charging application in RWSN. The first is that in the sparse sensor networks, using high gained RF radio directional power transfer antenna can reduce energy transmission waste and improve energy charging efficiency. The second is that the directional charger can cover longer distance and transfer more stable energy.

3. Problem Formation

Table 1 describes the symbols used in this paper.

Table 1. Symbol and Notations.

Symbol	Meaning
s_k	Coordinate of docking spot k
o_i	Coordinate of sensor node i
$\overrightarrow{\theta_{s_k}}$	DCV's charging orientation at docking spot k
$d(s_k, o_i)$	Euclidean distance between sensor node o_i and the docking spot s_k
$P_{k,i}(s_k, o_i)$	DCV's energy transfer function at docking spot s_k for sensor node o_i
A	Charging angle of DCV (°)
v	The moving speed of DCV (m/s)
D	Effective charging distance of DCV (m)
P_{out}	Energy transmit power of DCV (J/m)
ω_c	Moving energy consumption of DCV (J/m)
C_{max}	Energy capacity of DCV
ω_{o_i}	Energy consumption of sensor node i
e^s	Energy consumption for sensing one unit data
e^t	Energy consumption for transmitting one unit data
e^r	Energy consumption for receiving one unit data
R_{o_i}	Sensing data generation rate of sensor node i
$L \times L$	Size of the area

3.1. Directional Charging Model

As shown in Figure 2, we introduce the DCV's directional power transfer model as follows. When the effective charging distance of directional charger is D and charging coverage angle is A, the effective charging coverage area is a sector determined by its docking spot s_k and charging orientation vector $\overrightarrow{\theta_{kj}}$.

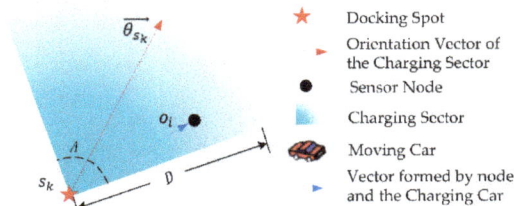

Figure 2. Directional Charging Model.

For a sensor node o_i is located at z_{o_i}, in order to determine whether the node o_i can be charged by the DCV stopped at docking spot s_k with charging orientation vector $\overrightarrow{\theta_{s_k}}$, we have two judgment conditions: (1) The node o_i is within the coverage angle A of the charger, denotes as inequality (1); and, (2) The distance between node o_i and charger is less than D, denotes as inequality (2).

$$\left(s_k \overset{\rightarrow}{-} z_{o_i} \right) \times \overset{\rightarrow}{\theta_{s_k}} \geq \parallel s_k z_{o_i} \parallel \times \cos\left(\frac{A}{2}\right) \tag{1}$$

where $\parallel s_k z_{o_i} \parallel$ denotes the distance between the location of the charger s_k and the location of sensor node z_{o_i}.

$$\parallel s_k z_{o_i} \parallel \leq D \tag{2}$$

We refer the RF wireless charging model in Reference [11] to calculate a node's energy received from a wireless charger:

$$P_r = \frac{G_s G_r \eta}{L}\left(\frac{\lambda}{4\pi(d+\beta)}\right)^2 P_{out} \tag{3}$$

where d is the distance between a sensor node and a wireless charger, P_{out} is the charger's transmission power, G_s is the transmitting antenna gain, G_r is the node's receiving antenna gain, L is polarization loss, λ is the wavelength, η is rectifier efficiency, and β is a parameter to adjust the Friis' free space equation for short distance transmission. Except for distance d, all other parameters in Equation (3) are constant values based on the environment and device settings. Therefore, we simplify the charging model in Equation (3) as Equation (4).

$$P_r = \frac{\alpha}{(d+\beta)^2} \tag{4}$$

where d is the distance from a sensor node to the DCV, and α represents other constant environmental parameters including P_{out}, G_s, G_r, L, λ and η in Equation (3).

From Equation (4), we can deduce $P_{k,i}(s_k, o_i)$, the effective charging power of the sensor node o_i received from the DCV which stopped at docking spot s_k with charging orientation vector $\overset{\rightarrow}{\theta_{s_k}}$:

$$P_{k,i}(s_k, o_i) = \begin{cases} \frac{\alpha}{(d(s_k,o_i)+\beta)^2}, & \parallel s_k z_{o_i} \parallel \leq D \text{ and} \\ & \left(s_k \overset{\rightarrow}{-} z_{o_i} \right) \times \overset{\rightarrow}{\theta_{s_k}} \geq \\ & \parallel s_k z_{o_i} \parallel \times \cos\left(\frac{A}{2}\right) \\ 0, & others \end{cases} \tag{5}$$

3.2. Network Energy Consumption Model

We consider that each sensor node consumes energy for data sensing, transmission, and reception. We assume sensor node o_i generates sensing data with a rate R_{o_i}(b/s). Assuming $PSN(o_i)$ is the set of previous sensor nodes that use sensor node o_i on the routing path to the sink node. Equation (6) shows the total energy consumption of sensor node o_i.

$$\omega_{o_i} = \sum_{o_l \in PSN(o_i)} \left(e^t + e^r \right) \times R_{o_l} + \left(e^t + e^s \right) \times R_{o_i} \tag{6}$$

Here e^s, e^t, and e^r represent the energy consumption of one unit data for sensing, transmitting, and receiving respectively [15].

Then we determine the data routing of the network through the minimum energy routing [39]. As shown in Figure 3, nodes o_1, o_2, o_3, o_4, and o_5 sending data to the sink node through node o_6. Then, we have $PSN(o_6) = \{o_1, o_2, o_3, o_4, o_5\}$.

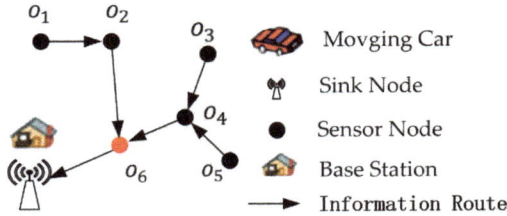

Figure 3. The data routing path to the Sink in the data collection network.

3.3. Problem Formulation

We consider a set of wireless rechargeable sensor nodes $O = \{o_1, o_2, \ldots, o_N\}$ randomly distributed on a $L \times L$ 2D area, each sensor node o_i generates sensing data with a rate $R_{o_i}(b/s)$, $i \in N$. There is a sink node located at Base Station which gathers the data from all sensors in the sensor network. A Multi-hop data routing tree is constructed for forwarding all sensing data to the sink node, as shown in Figure 4.

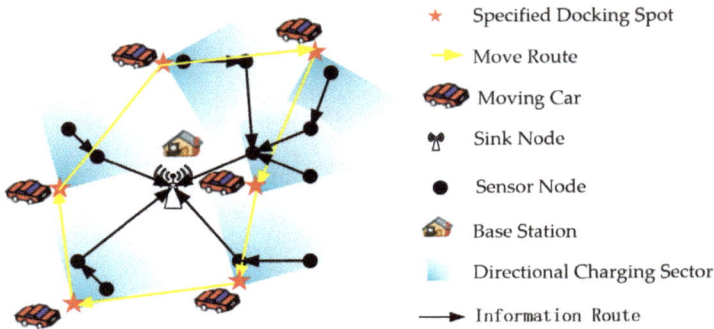

Figure 4. The DVC's docking spots, charging orientations and moving path.

Aiming to keeping the network working continuously, a DCV with an energy capacity of C_{max} is periodic dispatched to travel through a set of *Docking Spots* ($DS = \{s_1, s_2, \cdots, s_M\}$), M denotes the number of docking spots. The DCV stops at each docking spot and rotates its RF charging antenna to a specific orientation to charging the nearby sensors.

In a charging cycle, the DCV starts from the base station, moves through each docking spot and finally returns back to the base station to wait for the next charging cycle. The charging cycle T consists of the moving time T_{mov}, the charging time T_{cha}, and the time rest at the base station T_{res}. The moving time T_{mov} is determined by the length L_c of the DCV's moving path and moving speed v. The charging time T_{cha} is the sum of the dwell times at all docking spots, denote as $T_{cha} = \{t_1 + \ldots + t_k + \cdots + t_M\}$. The remaining time of each cycle is the DCV's rest time T_{res}.

$$T = T_{res} + T_{mov} + T_{cha} = T_{res} + \frac{L_c}{v} + \sum_{k=1}^{M} t_k \qquad (7)$$

Here $\frac{L_c}{v}$ denotes moving time of the DCV, t_k denotes the DCV's charging time at docking spot s_k, the sum of t_k denotes the total charging time of the DCV.

Assume the DCV travels through DS to charging the sensor network. The DCV stops at a docking spot s_k and rotates to a specific charging orientation vector $\vec{\theta}^l_{s_k}$. The sensor nodes which are effectively covered by the DCV denote as $SNC^l_k\left(s_k, \vec{\theta}^l_{s_k}\right)$. The DCV's dwell time is t_k at docking spot s_k.

For a DCV's charging Path, $CP = BSs_{p1} \cdots s_{pk} \cdots s_{pM}\}, s_{pk} \in DS$, we define E_{ECE}, the *Effective Charging Energy* received by all sensor nodes from DCV in a charging cycle T as follows:

$$E_{ECE} = \sum_{s_k \in DS} \sum_{o_i \in SNC^l_k(s_k, \vec{\theta}^l_{s_k})} P_{k,i}(s_k, o_i) \times t_k \tag{8}$$

Here $P_{k,i}(s_k, o_i)$ denotes the receiving power of the sensor node o_i when the DCV is at docking spot s_k; and $SNC^l_k\left(s_k, \vec{\theta}^l_{s_k}\right)$ denotes the sensor set covered by the DCV at s_k and charging direction $\vec{\theta}^l_{s_k}$.

In a charging cycle T, the DCV's energy consumption includes moving and charging energy, denote as E_{mov} and E_{cha} respectively. Charging consumption is determined by charging time and the DCV's output power P_{out}. Moving energy consumption is determined by the length of path L_c and its energy consumption per unit of moving length ω_c. Then the DCV's Energy Consumption, E_{DCV} is denoted as Equation (9).

$$E_{DCV} = E_{mov} + E_{cha} = P_{out} \sum_{k=1}^{M} t_k + \omega_c \times L_c \tag{9}$$

Here t_k denotes the DCV's charging time at docking spot s_k, L_c denotes the DCV's length of moving path, ω_c denotes DCV's consumption power of moving.

We define the DCV's Energy Charging Efficiency as follows.

Energy Charging Efficiency η: the ratio of effective charging energy received by the network to the DCV's total energy consumption in a charging cycle T, denoted as Equation (10):

$$\eta = \frac{E_{ECE}}{E_{DCV}} \tag{10}$$

Here E_{ECE} denotes the *Effective Charging Energy* received by all sensor nodes from DCV in a charging cycle which can be calculated by Equation (8), and E_{DCV} denotes the DCV's energy consumption in a charging cycle which can be calculated by Equation (9).

We define the residual energy value of node o_i at the time τ as $e_{o_i}(\tau)$ in a charging cycle. The node's residual energy value at any time should be not lower than minimum value E_{min}, and not greater than maximum value E_{max}.

The variation of node's residual energy value in a cycle is divided into three stages: 1) before charging; 2) charging stage; 3) after charging. For a sensor node o_i, $e_{o_i}(\tau)$ varied in a charging cycle T as shown in Figure 5. c_k denotes the arrival time of the DCV at docking spot s_k in the first cycle T, t_k denotes the DCV's dwell time at docking spot s_k.

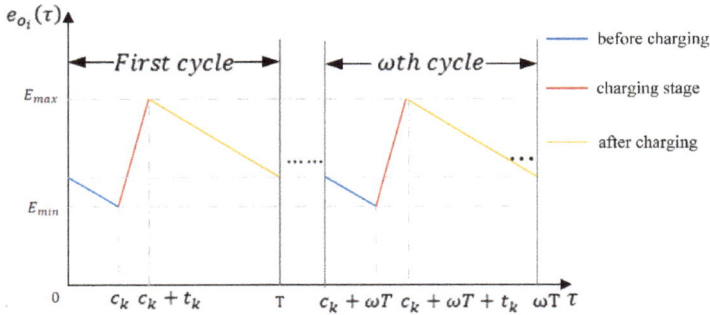

Figure 5. The variation of energy value function of node o_i.

The DCV carries limited energy C_{max}, so we have to make sure that the energy consumed by the charging car is no more than C_{max} in a cycle.

We define the CEOP problem of mobile directional charging as follow:

For the set of wireless rechargeable sensor nodes $O = \{o_1, o_2, \ldots, o_N\}$ randomly distributed on a $L \times L$ 2D area, how to plan the charging docking spots and charging path where the DCV moves along the path to replenishing energy for all sensors and maintains the sensor network working continuously. The objective is to maximize the DCV's energy charging efficiency while maintaining the network working continuously.

CEOP problem is formulated as follow:

$$max \quad \eta = {^{E_{ECE}}}\!/_{E_{DCV}}$$

$$s.t. \quad E_{min} \le e_{o_i}(\tau) \le E_{max}, o_i \in O, 0 \le \tau \le \omega T$$

$$e_{o_i}(\tau) = \begin{cases} e_{o_i}(\omega T) - \omega_{o_i} \times \tau, \tau \in [\omega T, \omega T + c_k] \\ E_{min} + \left(P_{k,i}(s_k, o_i) - \omega_{o_i}\right) \times \tau, \tau \in (\omega T + c_k, \omega T + c_k + t_k] \\ E_{max} - \omega_{o_i} \times \tau, \tau \in (\omega T + c_k + t_k, (\omega + 1)T] \end{cases}$$

$$E_{ECE} = \sum_{s_k \in DS} \sum_{o_i \in SNC\left(s_k, \vec{\theta_{s_k}^i}\right)} P_{k,i}(s_k, o_i) \times t_k$$

(11)

$$E_{DCV} = P_{out} \sum_{k=1}^{M} t_k + \omega_c \times L_c$$

$$E_{DCV} \le C_{max}$$

Here η denotes Energy Charging Efficiency, E_{max} denotes the maximum capacity of node, E_{min} denotes minimum energy value of node. $e_{o_i}(\tau)$ denotes residual energy value of node o_i at the time τ, when $\tau \in [\omega T, \omega T + c_k]$, $e_{o_i}(\tau) = e_{o_i}(\omega T) - \omega_{o_i} \times \tau$, $e_{o_i}(\tau)$ denotes the remaining energy of the node before charging, when $\tau \in (\omega T + c_k, \omega T + c_k + t_k]$, $e_{o_i}(\tau) = E_{min} + \left(P_{k,i}(s_k, o_i) - \omega_{o_i}\right) \times \tau$, $e_{o_i}(\tau)$ denotes the remaining energy of the node during charging, when $\tau \in (\omega T + c_k + t_k, (\omega + 1)T]$, $e_{o_i}(\tau) = E_{max} - \omega_{o_i} \times \tau$, $e_{o_i}(\tau)$ denotes the remaining energy of the node after charging. E_{ECE} denotes the Effective Charging Energy received by all sensor nodes from the DCV in a charging cycle, E_{DCV} denotes the DCV's energy consumption in a charging cycle, C_{max} denotes maximum energy capacity of the DCV.

4. Design and Analysis of Algorithms

It is difficult for the CEOP problem to be solved directly. We solve the problem in two steps and divide it into two sub-problems:

(1)   First, we find the set of Docking Spots ($DS = \{s_1, s_2, \cdots, s_M\}$) and their corresponding Charging Orientation ($CO = \{\vec{\theta}_{s_1}, \vec{\theta}_{s_2}, \cdots, \vec{\theta}_{s_M}\}$) to maximize the charging coverage utility and ensure the mobile charging coverage of the network (Section 3.1).

(2)   Second, we plan the DCV's charging path to travel through all docking spot in DS and the charging residence time at each docking spot to optimize the overall energy charging efficiency while maintaining the sensor network working continuously (Section 3.2).

4.1. Find Charging Docking Spots and Charging Directions

For the 2D plane on which the sensors are randomly deployed, we divide it into grids, and take grid vertices as the DCV's possible docking spots. Then we find the minimum number of the DCV's candidate docking spots and their charging directions to optimize the charging coverage utility locally while achieving mobile charging coverage for the whole network.

We define the DCV's Charging Coverage Utility at docking spot s_k on the charging orientation $\vec{\theta}_{s_k}^l$ as the sum of received power of the charging covered nodes:

$$U\left(s_k, \vec{\theta}_{s_k}^l\right) = \sum_{o_i \in SNC\left(s_k, \vec{\theta}_{s_k}^l\right)} P_{k,i}(s_k, o_i) \tag{12}$$

where $SNC\left(s_k, \vec{\theta}_{s_k}^l\right)$ denotes the sensor nodes covered at docking spot s_k in charging orientation $\vec{\theta}_{s_k}^l$.

Suppose at the docking spot s_k, the DCV has Q_k optional charging directions, i.e., $\{\vec{\theta}_{s_k}^1, \vec{\theta}_{s_k}^2, \cdots, \vec{\theta}_{s_k}^{Q_k}\}$. The maximum charging coverage utility at docking spot s_k is $U_{max}(s_k)$:

$$U_{max}(s_k) = max\left\{ U\left(s_k, \vec{\theta}_{s_k}^1\right), U\left(s_k, \vec{\theta}_{s_k}^2\right) \cdots U\left(s_k, \vec{\theta}_{s_k}^{Q_k}\right) \right\} \tag{13}$$

Here $U\left(s_k, \vec{\theta}_{s_k}^{Q_k}\right)$ denotes charging coverage utility at docking spot s_k in charging orientation $\vec{\theta}_{s_k}^l$.

For *num* grid points on the discrete 2D plane, we get the vertex set of grids: $CS = \{cds_1, \cdots, cds_k, \cdots, cds_{num}\}$, cds_k is coordinates of vertexes. We have to choose a set of candidate docking spots $S = \{s_1, \cdots, s_k, \cdots, s_M\}$, $s_k \in CS$, and their corresponding charging direction $\theta = \{\theta_{s_1}, \cdots, \theta_{s_k}, \cdots, \theta_{s_M}\}$, $\theta_{s_k} \in \{\theta_{s_k}^1, \cdots, \theta_{s_k}^{Q_k}\}$, where s_k has Q_k possible charging directions. We use $U_{sum}^{max}(s)$ denotes the maximum coverage utility of the set S of candidate docking spots as Equation (14).

$$U_{sum}^{max}(S) = \sum_{k=1}^{M} U_{max}(s_k)$$
$$S = \{s_1, \cdots, s_k, \cdots, s_M\}, \ s_k \in CS \tag{14}$$
$$\theta = \{\theta_{s_1}, \cdots, \theta_{s_k}, \cdots, \theta_{s_M}\}, \ \theta_{s_k} \in \{\theta_{s_k}^1, \cdots, \theta_{s_k}^{Q_k}\}$$

As shown in Figure 6, there are three candidate docking spots s_1, s_2, and s_3. The docking spot s_1 can choose two possible orientation $\{\vec{\theta}_{s_1}^1, \vec{\theta}_{s_1}^2\}$, the docking spot s_2 can choose two possible orientation $\{\vec{\theta}_{s_2}^1, \vec{\theta}_{s_2}^2\}$, and the docking spot s_3 can choose orientation $\{\vec{\theta}_{s_3}^1\}$. Therefore, there are five different coverage utility of different combinations of docking spots and orientation vectors. We can calculate the possible coverage utilities at s_1, s_2, and s_3 according to Equation (12). For s_1, two possible coverage utilities are presented as Equation (15). For s_2, two possible coverage utilities are presented as Equation (16). For s_3, one possible coverage utility is presented as Equation (17).

$$\begin{cases} U\left(s_1, \overrightarrow{\theta_{s_1}^2}\right) = P_{1,4}(s_1, o_4) \\ U\left(s_1, \overrightarrow{\theta_{s_1}^1}\right) = P_{1,1}(s_1, o_1) + P_{1,1}(s_1, o_2) \end{cases} \tag{15}$$

$$\begin{cases} U\left(s_2, \overrightarrow{\theta_{s_2}^1}\right) = P_{2,1}(s_2, o_4) + P_{2,5}(s_2, o_5) \\ U\left(s_2, \overrightarrow{\theta_{s_2}^2}\right) = P_{1,6}(s_2, o_6) \end{cases} \tag{16}$$

$$U\left(s_3, \overrightarrow{\theta_{s_3}^1}\right) = P_{3,2}(s_3, o_2) + P_{3,3}(s_3, o_3) + P_{3,6}(s_3, o_6) \tag{17}$$

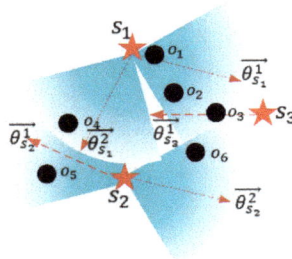

Figure 6. Combination of candidate docking spots and orientations.

As shown in Figure 6, we can get $U_{max}(s_1)$ and $U_{max}(s_2)$ respectively, as Equations (18) and (19).

$$U_{max}(s_1) = U\left(s_1, \overrightarrow{\theta_{s_1}^1}\right) \tag{18}$$

$$U_{max}(s_2) = U\left(s_2, \overrightarrow{\theta_{s_2}^1}\right) \tag{19}$$

Then we get the maximum coverage utility of candidate set $S = \{s_1, s_2, s_3\}$ and their related charging directions $\theta = \{\theta_{s_1}^1, \theta_{s_2}^1, \theta_{s_3}^1\}$, $U_{sum}^{max}(S) = U\left(s_1, \overrightarrow{\theta_{s_1}^1}\right) + U\left(s_2, \overrightarrow{\theta_{s_2}^1}\right) + U\left(s_3, \overrightarrow{\theta_{s_3}^1}\right)$.

The DCV's energy loss includes charging energy loss and moving energy costs. At each docking spot, we aim to reduce the DCV's charging loss and get higher charging effectiveness. By minimizing the number of docking spots, we can reduce the DCV's moving energy cost in the process of mobile charging. Additionally, maximizing charging coverage utility can reduce the charging energy loss at each docking spot. Hence it finally improves the energy charging efficiency in mobile directional charging.

To find the candidate docking spots and their charging directions for improvement of the mobile charging energy efficiency, we propose the two-objective optimization problem as Equation (20), that is Minimizing the number of Stop points and Maximizing charging Coverage Utility under the constraint of charging coverage of all sensors, called the MSMCU (Minimizing the number of Stop points and Maximizing charging Coverage Utility) problem.

$$\min \qquad \sum_{k=1}^{num} a_k$$

$$\max \qquad \sum_{k=1}^{num} U\left(s_k, \overrightarrow{\theta_{s_k}}\right)$$

$$s.t. \qquad \sum_{k=1}^{num} a_k \times x_{i,k} \geq 1, 1 \leq i \leq N \qquad\qquad (20)$$

$$x_{i,k} = \begin{cases} 1, if\ o_i \in SNC\left(s_k, \overrightarrow{\theta_{s_k}^l}\right) \\ 0, if\ o_i \notin SNC\left(s_k, \overrightarrow{\theta_{s_k}^l}\right) \end{cases}$$

where a_k is a binary decision variable that is equal to 1 if region s_k belongs to the minimum stops, and to 0 otherwise. Additionally, the n inequality constraints ensure that every node must belong to at least one stop region in the minimum stops. We analysis Equation (20), give Theorem 1 and the proof of Theorem 1.

Theorem 1. *The MSMCU problem of finding specified docking spots and orientations with minimum the number of stops and maximum coverage utility is NP-hard.*

Proof of Theorem 1. We prove Theorem 1 by giving a special instance of the problem and explaining that the instance is NP-hard.

Instance. We assume that the coverage utility is the maximum as long as a sensor is covered, then the problem can be reduced to solve the Minimum Set Covering Problem. Because the Minimum Set Covering Problem is NP-hard, the MSMCU problem is also NP-hard. □

Then we propose a Greedy approximation algorithm of Maximum Coverage Utility (GMCU). GMCU algorithm firstly divides a 2D plane into grids. Secondly, it takes each grid vertex as a possible stop point and computes its optimal charging Direction and Maximum Coverage Utility (DMCU). Finally, it selects a set of candidate stop points to achieve overall maximum utility and network charging coverage. Let us first introduce the GMCU algorithm, and then introduce the DMCU algorithm.

(1) GMCU algorithm

In the GMCU algorithm, we divide the plane into grids, and take each vertex as a possible docking spot. The coverage of the charger is a 90° sector with radius D. The DCV only chooses one orientation to charge each time it stops, so if the grid's size is too large, some nodes will be missed. The grid's size d must satisfy Equation (21)

$$d \leq \sqrt{2}/2 \times D \qquad\qquad (21)$$

The GMCU algorithm firstly divides a 2D plane into grids, take each grid vertex as a possible docking spot, denoted as CS, and cds_i represents coordinates of vertexes. Put each cds_i into the DMCU algorithm to calculate the maximum coverage utility and the covered nodes set. Choose the docking stops with the maximum value of coverage utility until all nodes are covered. The outputs are the docking spot set (DSS) and the set of covered nodes set ($SANC$) at corresponding directions.

The procedure of GMCU algorithm is presented in Table 2.

(2) The DMCU algorithm

The DMCU algorithm is used to find the charging direction with maximum coverage utility at each docking spot.

Take Figure 7 as an example to illustrate the process of DMCU algorithm: (1) The DCV rotates counter-clockwise with each different node as initial boundary; (2) Calculate the coverage utility of each orientation.

Table 2. The Procedure of the GMCU (Greedy approximation algorithm of Maximum Coverage Utility) Algorithm.

GMCU algorithm: find candidate docking spots and their charging directions
1. **Input:** The length of area: L; Farthest distance DCV can reach: D; Charging angle of DCV: A
2. Discrete the $L \times L$ plane into grids, get the vertex set of grids: $CS = \{cds_1, cds_2, \cdots, cds_k, \cdots, cds_{num}\}$, cds_k is coordinates of vertexes
3. $DDS = \varnothing$, $SANC = \varnothing$, $k = 0$
4. //DDS candidate docking spots
5. //SANC set of cover set which associated with DDS
6. **While** $O \neq \varnothing$ // O set of sensor nodes
7. $\quad SNC_{temp} = \varnothing$, $U_{temp} = 0$, $CDS_{temp} = 0$
8. $\quad$ //find a stop point with max cover utility
9. $\quad$ **While** $k < len(CS)$
10. $\qquad$ Call DMCU(cds_k) to get max coverage utility
11. $\qquad$ $U_{max}(cds_k)$, cover set SNC_k at docking
12. $\qquad$ point cds_k with charging direction $\overrightarrow{\theta_{s_k}}$
13. $\qquad$ **If** $U_{max}(cds_k) > U_{temp}$
14. $\qquad$ $U_{temp} = U_{max}(cds_k)$
15. $\qquad$ $SNC_{temp} = SNC_k$
16. $\qquad$ $CDS_{temp} = cds_k$
17. $\qquad$ **End If**
18. $\qquad$ $k = k + 1$
19. $\quad$ **End While**
20. $\quad$ $SANC = SANC \cup \{SNC_{temp}\}$
21. $\quad$ $DDS = DDS \cup \{CDS_{temp}\}$
22. $\quad$ $O = O - SNC_{temp}$
23. $\quad$ $CS = CS - \{CDS_{temp}\}$
24. $\quad$ $k = 0$
25. **End While**
26. **Output:** set of docking points DDS and set of charging cover sets $SANC$ at related charging directions

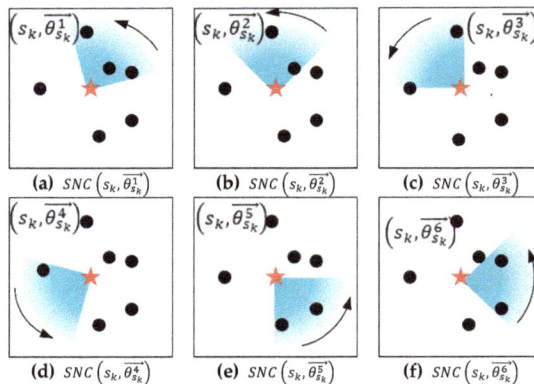

(a) $SNC\left(s_k, \overrightarrow{\theta_{s_k}^1}\right)$ (b) $SNC\left(s_k, \overrightarrow{\theta_{s_k}^2}\right)$ (c) $SNC\left(s_k, \overrightarrow{\theta_{s_k}^3}\right)$

(d) $SNC\left(s_k, \overrightarrow{\theta_{s_k}^4}\right)$ (e) $SNC\left(s_k, \overrightarrow{\theta_{s_k}^5}\right)$ (f) $SNC\left(s_k, \overrightarrow{\theta_{s_k}^6}\right)$

Figure 7. An example for showing the procedure of DMCU (Direction and Maximum Coverage Utility) algorithm.

Six different coverage utility values can be obtained; the output is orientation with maximum coverage utility of a docking spot and the sensor nodes set that the combination of docking spot and this charging orientation can cover. Figure 7a–f show the set of nodes covered by the DCV at dock spot s_k in each orientations ($\overrightarrow{\theta^1_{s_k}} \sim \overrightarrow{\theta^6_{s_k}}$), $SNC\left(s_k, \overrightarrow{\theta^1_{s_k}}\right) \sim SNC\left(s_k, \overrightarrow{\theta^6_{s_k}}\right)$ represent the corresponding nodes sets. The procedure of DMCU algorithm is presented in Table 3.

Table 3. The Procedure of the DMCU Algorithm.

DMCU algorithm: Find the max utility, cover set and charging orientation at s_k
1. **Input:** Sensor node set: $O = \{o_1, o_2, \cdots, o_i, \cdots, o_N\}$; Coordinates of certain docking spot s_k: (c_x, c_y); Farthest charging distance DCV can reach: D; Charging angle of DCV: A
2. $OCS = \varnothing$ // OCS sensors' set possible covered by s_k
3. $i = 0$
4. **While** $i < N$: //find sensors' set OCS at docking spot s_k
5.   Calculate Euclidean distance between sensor o_i
6.   and docking spot d_i
7.   **If** $d_i < D$:
8. $OCS = OCS \cup \{o_i\}$
9.   **End If**
10. $i = i + 1$
11. **End While**
12. **If** $OCS \neq \varnothing$:
13. $L = len(OCS)$
14.   Calculate the all possible charging angles:
15. $\varphi = \{\gamma_1, \gamma_2, \cdots, \gamma, \cdots, \gamma_L\}$
16.   Sort sensors in set OCS in ascending order
17.   according to the value of angles.
18. $DCS = \varnothing, k = 0$
19.   // calculate L directions' cover sets
20.   **While** $k < L$
21. $m = 0, SNC_{tmp} = \varnothing$
22.   **While** m < L
23.   **If** $\gamma_k \leq \gamma_m \leq (\gamma_k + A)\%360$:
24. $SNC_{tmp} = SNC_{tmp} \cup \{o_m\}$
25.   **End If**
26. $m = m + 1$
27.   **End while**
28. $DCS = DCS \cup \{SNC_{tmp}\}$
29. $k = k + 1$
30.   **End while**
31. $CUS_{tmp} = \varnothing, SNC_{tmp} = \varnothing, j = 0, \gamma_{tmp} = 0$
32.   // find the cover set with max utility
33.   **While** $j < len(DCS)$
34.   Calculate cover utility $CUS[j]$ of $DCS[j]$
35.   **If** $CUS[j] > CUS_{tmp}$
36. $CUS_{tmp} = CUS[j]$
37. $SNC_{temp} = DCS[j]$
38. $\gamma_{tmp} = \varphi[j]$
39.   **End If**
40. $j = j + 1$
41.   **End While**
42. $U_{max} = CUS_{tmp}, SNC = SNC_{temp}, \gamma = \gamma_{tmp}$
43. **End If**
44. **Output:** max utility U_{max}, covered set SNC, direction γ

OCS represents a coverage set of a candidate docking spot, the initial value is null. If the distance between the node and the candidate stop is not greater than D, then add the node into *OCS*. γ_j in φ represents the angle formed by each node in *OCS* at each candidate dock spot. DCS_k represents a coverage set of candidate stop with kth charging direction, and the CUS_j indicates the corresponding coverage utility value. The DMCU algorithm finally outputs the maximum value $U_{s_k}^{max}$ in CUS_j and the set of covered nodes $SNC\left(s_k, \overrightarrow{\theta_{s_k}^l}\right)$ covered at this docking spot s_k with corresponding direction $\overrightarrow{\theta_{s_k}^l}$.

As shown in Figure 8, we randomly deploy 100 nodes in the 20×20 m^2 area and run the GMCU algorithm to determine specified docking spots and orientations with maximum coverage utility and minimum the number of docking spots.

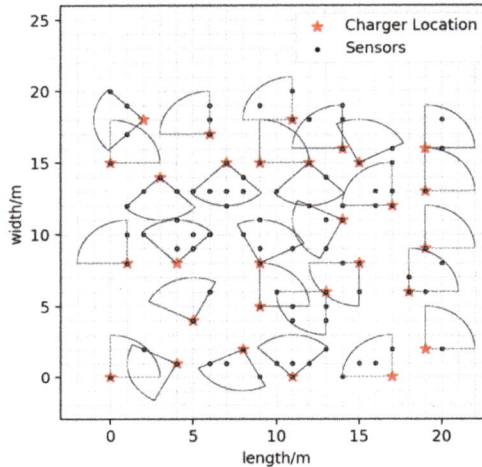

Figure 8. Illustration of the GMCU algorithm's example result: DCV's candidate charging locations and charging directions.

4.2. Plan Moving Path and Charging Residence Time

In this section, we plan the DCV's charging moving path to travel through all candidate docking spots chosen by the GMCU algorithm and the charging residence times at each docking spot to maintain the network's continuous working and optimize the overall charging energy efficiency.

Firstly, we introduce the charging cycle T. As shown in Figure 9, the charging cycle T consists of the DCV's moving time, the charging residence time at each docking spot, and the rest time at the base station. The moving time is determined by the length of charging path. The charging residence time at each docking spot is determined by charging energy requirement of sensors covered by the DCV.

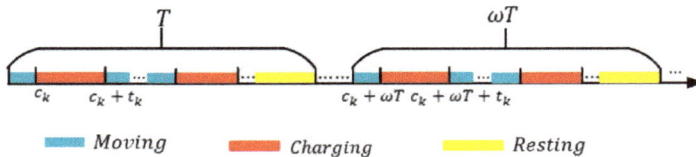

Figure 9. Periodic behavior of the charging car.

To achieve the goal of maintaining network perpetually, the charging process can be repeated periodically. Then this periodical charging cycle must meet two requirements:

(1) The energy received by a sensor is greater or equal to the energy consumed in a charging cycle;

(2)   The residual energy value of a node will not be lower than E_{min} during a charging cycle.

The cover sets charged by the DCV at dock spot s_k on charging orientations $\overrightarrow{\theta^l_{s_k}}$: $SNC\left(s_k, \overrightarrow{\theta^l_{s_k}}\right) = \{o^{k_l}_1, o^{k_l}_2, \cdots, o^{k_l}_m\}$. We can derive the minimal charging residence time t_k according to the charging cover sets $SNC\left(s_k, \overrightarrow{\theta^l_{s_k}}\right)$ at docking spot s_k:

$$t_k = \max_{o_i \in SNC(s_k, \overrightarrow{\theta^l_{s_k}})} \left\{ \frac{\omega_{o_i}}{P_{k,i}(s_k, o_i)} \right\} \times T \tag{22}$$

Here ω_{o_i} denotes the energy consumption of sensor node o_i, $P_{k,i}(s_k, o_i)$ denotes the receiving power of the sensor node o_i when the DCV is at docking spot s_k.

We denote c_k as the arrival time of the DCV at docking spot k in the first cycle. Denote $d_{0,1}$ as the distance between the base station and the first docking spot, $d_{l,l+1}$ as the distance between *l*th and $(l + 1)$ *th* docking spot.

$$c_k = \sum_{l=0}^{k-1} \frac{d_{l,l+1}}{v} + \sum_{l=1}^{k-1} t_l \tag{23}$$

Here t_l denotes the DCV's charging time at docking spot s_l.

According to Figure 5, we can derive from Equation (22):

$$E_{max} - \min_{o_i \in SNC(s_k, \overrightarrow{\theta^l_{s_k}})} \{e_{o_i}(c_k)\} \geq E_{max} - \min_{o_i \in SNC(s_k, \overrightarrow{\theta^l_{s_k}})} \{e_{o_i}(T)\} \tag{24}$$

That is to say $e_{o_i}(m \times T + c_k)$ is the minimum value of $e_{o_i}(\tau)$. To have $e_{o_i}(\tau) \geq E_{min}$, it is sufficient to have:

$$e_{o_i}(m \times T + c_k) = e_{o_i}(m \times T) - c_k \times \omega_{o_i} \geq E_{min}, o_i \in SNC\left(s_k, \overrightarrow{\theta^l_{s_k}}\right) \tag{25}$$

while $m \geq 1$:

$$
\begin{aligned}
e_{o_i}(m \times T + c_k) &= e_{o_i}(m \times T) - c_k \times \omega_{o_i} \\
&= e_{o_i}((m-1) \times T + c_k + t_k) - \{m \times T - [(m-1) \times T + c_k + t_k]\} - c_k \times \omega_{o_i} \\
&= e_{o_i}((m-1) \times T + c_k + t_k) - (T - t_k) \times \omega_{o_i} = E_{max} - (T - t_k) \times \omega_{o_i}
\end{aligned}
\tag{26}
$$

Therefore, if Equation (27) holds, we have $e_{o_i} \geq E_{min}$, the sensor s_k can working continuously.

$$E_{max} - (T - t_k) \times \omega_{o_i} \geq E_{min} \tag{27}$$

We can get the Charging Cycle T when the two periodical charging requirements are met. Then we plan the DCV's charging moving path.

When the DCV moves along the shortest Hamiltonian circle, we can achieve the maximum energy efficiency η.

We can proof this based on contradiction. Suppose the shortest travel route for the Hamilton Circle is $L = \{s_1 s_2 \cdots s_M\}$, and there exists a move route $\hat{L} = \{s_3 s_2 \cdots s_M s_1\}$. Assume that $\hat{\eta} \geq \eta$ is established.

$$\hat{\eta} = \frac{E}{\hat{E}_{DCV}} = \frac{\sum_{k=1}^{M} \sum_{i=1}^{N} P_{k,i}(s_k, o_i) \times \hat{t}_k}{P_{out} \times \sum_{k=1}^{M} t_j + \omega_c \times \hat{L}} \tag{28}$$

$$\eta = \frac{E}{E_{DCV}} = \frac{\sum_{k=1}^{M} \sum_{i=1}^{N} P_{k,i}(s_k, o_i) \times t_k}{P_{out} \times \sum_{k=1}^{M} t_k + \omega_c \times L} \tag{29}$$

The energy received by the node in this cycle is equal to the energy consumed. The numerator of Equations (28) and (29) are equal. Because $L \leq \hat{L}$, $\hat{\eta} \leq \eta$, thus leading to a contradiction. Therefore, we can dispatch the DCV moving along the shortest Hamiltonian circle to achieve the maximum energy efficiency.

We redefine Equation (11) as Equation (30)

$$max \quad \eta$$

$$s.t. \quad t_k = \max_{o_i \in SNC(s_k, \theta^l_{s_k})} \left\{ \frac{\omega_{o_i}}{P_{k,i}(s_k, o_i)} \right\} \times T$$

$$E_{max} - (T - t_k) \times \omega_{o_i} \geq E_{min} \tag{30}$$

$$T = T_{res} + T_{TSP} + \sum_{k=1}^{M} t_k$$

$$P_{out} \sum_{k=1}^{M} t_k + \omega_c \times L_c \leq C_{max}$$

Here η denotes Energy Charging Efficiency, ω_{o_i} denotes the energy consumption of sensor node o_i, $P_{k,i}(s_k, o_i)$ denotes the receiving power of the sensor node o_i when the DCV is at docking spot s_k, t_k denotes the DCV's charging time at docking spot s_k, T_{res} denotes rest time of the DCV, T_{TSP}. denotes the moving time of the DCV, the sum of t_k denotes the total charging time of the DCV.

Finally, we get the charging residence time at each docking spot and energy efficiency by solving the planning problem.

5. Analysis of the DCV's Service Capability

We use only one DCV with energy capacity of E_{max} to maintain WRSN perpetually. Therefore, the network size and area size are limited. This section will specifically analyze the service capability of the DCV.

Assume that the number of stops is M, the charging time of each stop is t_k, the distance between adjacent stops is $d_{k-1,k}$, the length of the return route is d_{back}. Two constraints must be satisfied for each round of charging: (1) the energy received by each node is not less than the energy consumed, formulated as Equation (31); and (2) the DCV should not run out of energy in a round, formulated as Equation (32).

$$\min_{o_i \in SNC(s_k, \theta^l_{s_k})} \left\{ P_{k,i}(s_k, o_i) \right\} \times t_k \geq \max_{o_i \in SNC(s_k, \theta^l_{s_k})} \left\{ \omega_{o_i} \right\} \times \left((t_1 + \cdots + t_M) + \frac{d_{1,2} + \cdots + d_{M-1,M} + d_{back}}{v} \right), 0 \leq k \leq M \tag{31}$$

$$P_{out} \times (t_1 + \cdots + t_k + \cdots + t_M) + (d_{1,2} + d_{2,3} + \cdots + d_{M-1,M} + d_{back}) \times \omega_c \leq C_{max} \tag{32}$$

Here, P_{out} is the charger's transmission power, v denotes the moving speed of the DCV, ω_c denotes DCV's consumption power of moving, C_{max} denotes maximum energy capacity of the DCV.

We first analyze the maximum size of area. Assuming that there are only two nodes in the network and they are on the diagonal line of the network, the consuming power is the minimum ω_{min}, the DCV stops at the nodes respectively, and the receiving power of the nodes is both P_{out}. Then the number of stops is two ($M = 2$), the shortest distance of moving route is $2\sqrt{2} * l$, l denotes length of the network ,then we can get Equation (33)

$$l_{max} = \frac{C_{max} \times (P_{out} - 2 \times \omega_{min})}{2\sqrt{2} \times \left(\frac{P_{out} \times \omega_{min}}{v} + \omega_c \times (P_{out} - 2 \times \omega_{min}) \right)} \tag{33}$$

Secondly, we analyze the minimum size of area. Assume that the nodes are evenly distributed in the network, the consuming power is the maximum ω_{max}, the DCV stops at the nodes respectively, and the receiving power is all the minimum P_{min}. Then the number of stops is formulated as Equation (34).

$$\left[\frac{l}{\frac{\sqrt{2}}{2} \times D}\right]^2 \tag{34}$$

The longest distance of move route is formulated as Equation (35).

$$2 \times (M-1) \times \sqrt{2} \times D \tag{35}$$

We bring Equations (34) and (35) into Equations (31) and (32) to get Equations (36).

$$M = \frac{-b \pm \sqrt{b^2 - 4 \times a \times c}}{2 \times a}$$

$$\begin{cases} a = 2\sqrt{2} \times \omega_{max} \times D \times P_{min} \\ b = 2\sqrt{2} \times D \times \omega_c \times v \times (P_{min} - \omega_{max}) \\ \quad -2\sqrt{2} \times \omega_{max} \times D \times P_{min} \\ c = v \times (P_{min} - \omega_{max}) \times (C_{max} + 2\sqrt{2} \times D \times \omega_c) \end{cases} \tag{36}$$

Therefore, the minimum length of area is L_{min}, formulated as Equation (37).

$$L_{min} = \left| \sqrt{d^2 \times M} \right| \tag{37}$$

When the network area is the smallest, assuming that the charger can charge CN nodes simultaneously at most, the number of nodes can reach the maximum. Then the maximum number of nodes is CNS, formulated as Equation (38).

$$CNS = M \times CN \tag{38}$$

In summary, when the size of the area is between L_{min} and L_{max} and the size of network is less than CNS, the proposed charging model and approximate algorithm can satisfy the two constraints: 1) the energy received by each node is not less than the energy consumed; and 2) the DCV should not run out of energy in a round.

6. Simulation Experiments

In this section, we describe comprehensive simulation experiments to investigate the algorithms' performance under different influence factors, such as grid size, area size, and network size. In the existing literature, there are no related works that study mobile directional charging problem in WRSN. Therefore, we conducted simulations experiments and compared charging efficiency with mobile omnidirectional charging models [14]. The simulation experiments were performed on a 64-bit Windows 10 system; the programming languages were C++ and Python. The algorithms were realized in the C++ language. Additionally, the visualization of deployment results was realized in Python. In the simulation experiments, we set up the parameters of the DCV and rechargeable sensor network, as in Table 4.

Table 4. Parameter Setting.

Parameter	Value
E_{max}	10,000 J
P_{ou}	3 J/s
ω_c	0.3 J/m
D	3 m
v	0.5 m/s
R_{o_i}	randomly generated in References [1,10] b/s
e^s	0.01 mJ/b
e^t	0.06 mJ/b
e^r	0.05 mJ/b
α, β	10
The number of sensor nodes	20, 40, 60, 80, 100, 120, 140, 160, 180,200
The size of area	15×15 m^2, 20×20 m^2, 25×25 m^2, 30×30 m^2, 35×35 m^2,

6.1. Comparison Experiments on Different Grid Size

In our approach, we discretized the continuous 2D plane with gridding. We investigated how grid size affects the algorithm's performance. We randomly deployed 20, 40 and 60 nodes in the 15×15 m^2 area, changed the grid size, and explored the variation of energy efficiency and docking spots number. Figure 10 shows that with the decrease of grid size, the energy efficiency of the DCV increase. Additionally, a stable grid size tends to be 0.2 m. Figure 11 shows that with the decrease of grid size, the number of specified docking spots decreases. Additionally, it tends to be stable when grid size is .2 m.

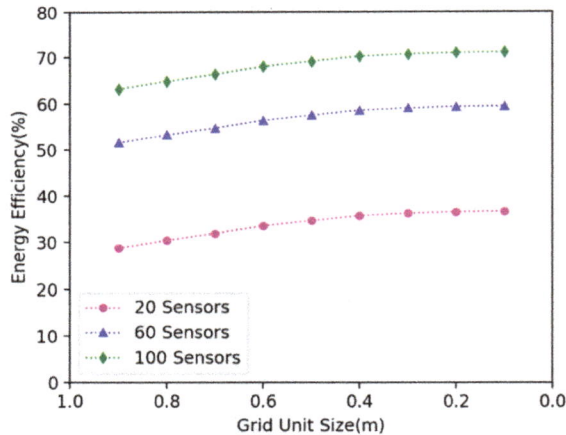

Figure 10. The effect of different grid sizes on the energy efficiency.

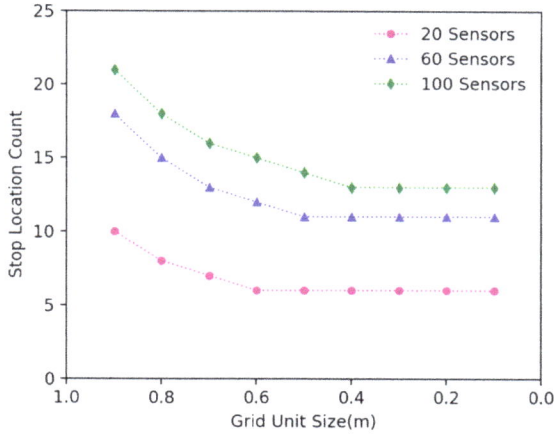

Figure 11. The effect of different grid sizes on the number of specified docking spots.

6.2. Comparison Experiments on Different Network Size And Area Size

We investigated how network size affects the algorithm's performance. We randomly deployed 20, 40, 60, 80, 100, 120, 140, 160, 180, 200 nodes in 15×15 m^2, 20×20 m^2, 25×25 m^2, 30×30 m^2, 35×35 m^2 plane areas respectively, and explored the variation in the energy efficiency of DCV. It can be seen in Figure 12 that as the number of nodes increases, the energy efficiency increases; because the number of nodes increases in the same area, the number of nodes can be covered by the DCV increases, so more energy is received by the nodes, and the energy efficiency is improved. As shown in Figure 13, when the number of nodes remains unchanged and the area becomes larger, the energy efficiency decreases. This is because as the area becomes larger, the distance between nodes becomes larger, the moving path of the DCV becomes longer, and the energy consumed on moving increases, which leads to the decrease of the energy efficiency of the DCV.

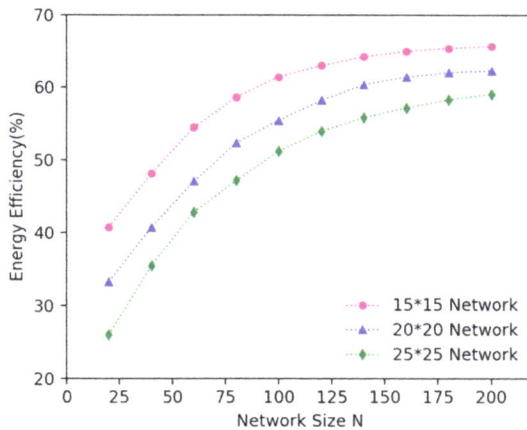

Figure 12. The influence of network size N on energy efficiency of the DCV.

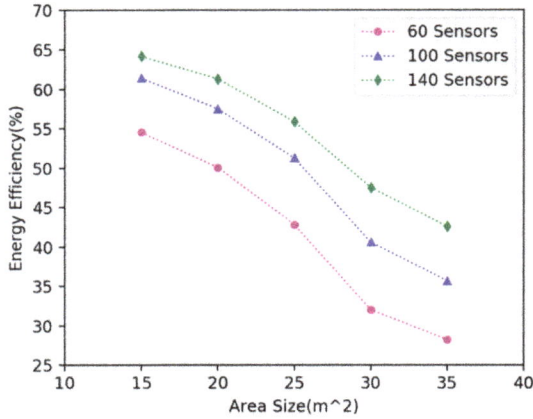

Figure 13. The influence of area size on energy efficiency of the DCV.

6.3. Comparison Experiments on Mobile Omnidirectional and Directional Charging

In the existing literature, there are no related works that use directional charging model for mobile charging in WRSN. Therefore, we conducted simulation experiments and compared charging efficiency with mobile omnidirectional charging [14]. We randomly deployed 20, 40, 60, 80, 100, 120, 140, 160, 180 and 200 nodes in 15×15 m^2, 20×20 m^2 and 25×25 m^2 areas. In experiments, we used DCV and omnidirectional charging vehicle respectively to charge the network according to the algorithms proposed in this paper, and compare their energy efficiency. Figures 14–16 show the variation of energy efficiency in different area size and network size. The experiments show that the energy efficiency of DCV is higher than that of omnidirectional charging vehicle in the network with sparse nodes. As the node density increases, the energy efficiency of DCV and omnidirectional charging vehicle will gradually converge. Hence our mobile directional charging algorithm is more suitable in a network with sparse nodes compared with mobile omnidirectional charging.

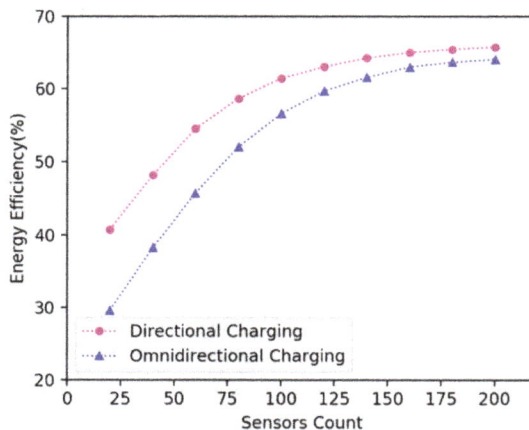

Figure 14. Energy efficiency of the DCV and omnidirectional charging vehicle in 15×15 m^2 area.

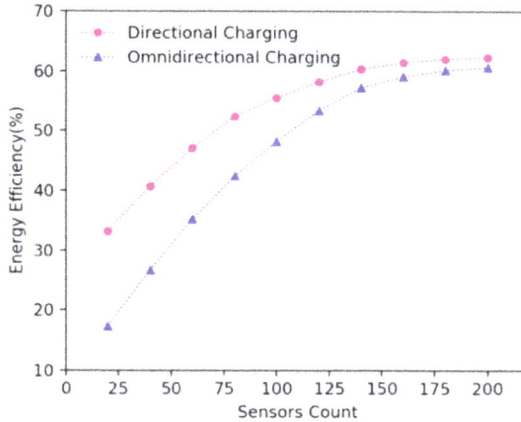

Figure 15. Energy efficiency of the DCV and omnidirectional charging vehicle in 20×20 m^2 area.

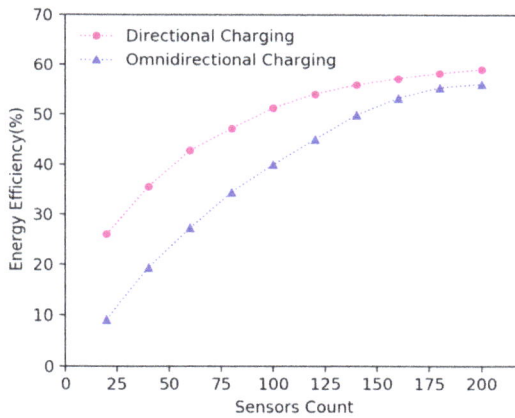

Figure 16. Energy efficiency of the DCV and omnidirectional charging vehicle in 25×25 m^2 area.

7. Conclusions

In this paper, we investigated the DCV's charging efficiency optimization problem in RWSN while maintaining sensor network working continuously. We proved that the problem is NP-hard. Firstly, we proposed the coverage utility of directional charging. Then we transformed the finding of candidate docking spots and their charging directions on the 2D plane into a two-objective optimization problem of minimizing number of stop points and maximizing charging coverage utility. Additionally, we proposed a greedy approximation algorithm to solve the two-objective optimization problem and find the set of candidate stop points of the DCV. Finally, we planned the DCV's charging moving path to travel through all candidate docking spots to maintain the network's continuous working and optimize the overall energy charging efficiency. We theoretically analyzeed the DCV's charging service capability, and performed the comprehensive simulation experiments. The simulation experiment results show that energy charging efficiency is higher than omnidirectional charging model in the sparse networks.

As stated in the literature [40], WPT has several limitations when applied to a WSN. First, it has very low energy transfer efficiency as distance increases. Second, it is sensitive to obstruction between an energy source and a receiver. Therefore, this technology is only suitable in the ultra-low-power

WSN scenario. In future work, we will further investigate more practical energy replenishment optimization problem in WSN, in which we can use a hybrid energy replenishing scheme, such as wireless charging for ultra-low-power sensor nodes and solar energy harvesting for high-power sensor nodes in the network.

Author Contributions: This paper was prepared through a collective effort of all of the authors. In particular: X.X. contributed the concept of the research, the state-of-the-art and references, as well as the optimization algorithm concept; L.C. contributed the objective function definition, the constraints definition, performed the simulations, the analysis of the results, the preparation of the diagrams and figures and the preparation of the text; Z.C. contributed the interpretation of the simulation results, the description of the optimization algorithm, and the formal definition of the optimization problem.

Funding: This work is supported by the Key Science-Technology Program of Zhejiang Province, China (2017C01065) and the National Natural Science Foundation of China (61370087).

Conflicts of Interest: The authors declare no conflict of interest.

References

1. Kurs, A.; Moffatt, R.; Soljacic, M. Simultance mid-range power transfer to multiple devices. *Appl. Phys. Lett.* **2010**, *96*, 34. [CrossRef]
2. André, K.; Aristeidis, K.; Robert, M.; Joannopoulos, J.D.; Peter, F.; Marin, S. Wireless power transfer via strongly coupled magnetic resonances. *Science* **2007**, *317*, 83–86.
3. Valenta, C.R.; Durgin, G.D. Harvesting Wireless Power: Survey of Energy-Harvester Conversion Efficiency in Far-Field, Wireless Power Transfer Systems. *IEEE Microw. Mag.* **2014**, *15*, 108–120.
4. Sample, A.P.; Yeager, D.J.; Powledge, P.S.; Mamishev, A.V.; Smith, J.R. Design of an RFID-based battery-free programmable sensing platform. *IEEE Trans. Instrum. Meas.* **2008**, *57*, 2608–2615. [CrossRef]
5. Xie, L.; Yi, S.; Hou, Y.T.; Lou, A. Wireless power transfer and applications to sensor networks. *IEEE Wirel. Commun.* **2013**, *20*, 140–145.
6. Zhang, R.; Ho, C.K. MIMO Broadcasting for Simultaneous Wireless Information and Power Transfer. *IEEE Trans. Wirel. Commun.* **2013**, *12*, 1989–2001. [CrossRef]
7. Ding, Z.; Zhong, C.; Ng, D.W.K.; Peng, M.; Suraweera, H.A.; Schober, R.; Poor, H.V. Application of Smart Antenna Technologies in Simultaneous Wireless Information and Power Transfer. *IEEE Commun. Mag.* **2015**, *53*, 86–93. [CrossRef]
8. Available online: http://www.powercastco.com/ (accessed on 10 June 2019).
9. Dai, H.; Wang, X.; Liu, A.X.; Ma, H.; Chen, G. Optimizing wireless charger placement for directional charging. In Proceedings of the IEEE INFOCOM 2017-IEEE Conference on Computer Communications, Atlanta, GA, USA, 1–4 May 2017.
10. Peng, Y.; Li, Z.; Zhang, W.; Qiao, D. Prolonging sensor network lifetime through wireless charging. In Proceedings of the 2010 31st IEEE Real-Time Systems Symposium, San Diego, CA, USA, 30 November–3 December 2010.
11. He, S.; Chen, J.; Jiang, F.; Yau, D.K.; Xing, G.; Sun, Y. Energy Provisioning in Wireless Rechargeable Sensor Networks. *IEEE Trans. Mob. Comput.* **2011**, *12*, 1931–1942. [CrossRef]
12. Wireless Power Consortium. Available online: http://www.wirelesspowerconsortium.com/ (accessed on 10 June 2019).
13. Yi, S.; Xie, L.; Hou, Y.T.; Sherali, H.D. Multi-Node Wireless Energy Charging in Sensor Networks. *IEEE/ACM Trans. Netw.* **2015**, *23*, 437–450.
14. Khelladi, L.; Djenouri, D.; Lasla, N.; Badache, N.; Bouabdallah, A. MSR: Minimum-Stop Recharging Scheme for Wireless Rechargeable Sensor Networks. In Proceedings of the 2014 IEEE 11th Intl Conf on Ubiquitous Intelligence and Computing and 2014 IEEE 11th Intl Conf on Autonomic and Trusted Computing and 2014 IEEE 14th Intl Conf on Scalable Computing and Communications and Its Associated Workshops, Bali, Indonesia, 9–12 December 2014.
15. Wu, G.; Chi, L.; Ying, L.; Lin, Y.; Chen, A. A Multi-node Renewable Algorithm Based on Charging Range in Large-Scale Wireless Sensor Network. In Proceedings of the International Conference on Innovative Mobile & Internet Services in Ubiquitous Computing, Blumenau, Brazil, 8–10 July 2015.

16. Jiang, L.; Wu, X.; Chen, G.; Li, Y. Effective on-Demand Mobile Charger Scheduling for Maximizing Coverage in Wireless Rechargeable Sensor Networks. *Mob. Netw. Appl.* **2014**, *19*, 543–551. [CrossRef]
17. Xie, L.; Shi, Y.; Hou, Y.T.; Lou, W.; Sherali, H.D. On traveling path and related problems for a mobile station in a rechargeable sensor network. In Proceedings of the Fourteenth ACM International Symposium on Mobile Ad Hoc Networking & Computing, Bangalore, India, 29 July–1 August 2013.
18. Dai, H.; Wang, X.; Liu, A.X.; Zhang, F.; Yang, Z.; Chen, G. Omnidirectional chargability with directional antennas. In Proceedings of the IEEE International Conference on Network Protocols, Singapore, Singapore, 8–11 November 2016.
19. Jiang, J.; Liao, J. Efficient Wireless Charger Deployment for Wireless Rechargeable Sensor Networks. *Energies* **2016**, *9*, 696. [CrossRef]
20. Ji, H.L.; Jiang, J.R. Wireless Charger Deployment Optimization for Wireless Rechargeable Sensor Networks. In Proceedings of the International Conference on Ubi-media Computing & Workshops, Ulaanbaatar, Mongolia, 12–14 July 2014.
21. Cho, J.; Lee, J.; Kwon, T.; Choi, Y. Directional antenna at sink (DAaS) to prolong network lifetime in wireless sensor networks. In Proceedings of the Wireless Conference -enabling Technologies for Wireless Multimedia Communications, Athens, Greece, 2–5 April 2006.
22. Moraes, C.; Myung, S.; Lee, S.; Har, D. Distributed Sensor Nodes Charged by Mobile Charger with Directional Antenna and by Energy Trading for Balancing. *Sensors* **2017**, *17*, 122. [CrossRef] [PubMed]
23. Ouadou, M.; Zytoune, O.; Aboutajdine, D. Wireless charging using mobile robot for lifetime prolongation in sensor networks. In Proceedings of the 2014 Second World Conference on Complex Systems (WCCS), Agadir, Morocco, 10–12 November 2014.
24. Chen, S.H.; Chang, Y.C.; Chen, T.Y.; Cheng, Y.C.; Wei, H.W.; Hsu, T.S.; Shih, W.K. Prolong Lifetime of Dynamic Sensor Network by an Intelligent Wireless Charging Vehicle. In Proceedings of the Vehicular Technology Conference, Boston, MA, USA, 6–9 September 2015.
25. Xu, W.; Liang, W.; Jia, X.; Xu, Z.; Li, Z.; Liu, Y. Maximizing Sensor Lifetime with the Minimal Service Cost of a Mobile Charger in Wireless Sensor Networks. *IEEE Trans. Mob. Comput.* **2018**, *17*, 2564–2577. [CrossRef]
26. Tu, W.; Xu, X.; Ye, T.; Cheng, Z. A Study on Wireless Charging for Prolonging the Lifetime of Wireless Sensor Networks. *Sensors* **2017**, *17*, 1560. [CrossRef] [PubMed]
27. Xu, W.; Liang, W.; Lin, X.; Mao, G.; Ren, X. Towards Perpetual Sensor Networks via Deploying Multiple Mobile Wireless Chargers. In Proceedings of the International Conference on Parallel Processing, Minneapolis, MN, USA, 9–12 September 2014.
28. Xie, L.; Shi, Y.; Hou, Y.T.; Sherali, H.D. Making Sensor Networks Immortal: An Energy-Renewal Approach with Wireless Power Transfer. *IEEE/ACM Trans. Netw.* **2012**, *20*, 1748–1761. [CrossRef]
29. Ye, X.; Liang, W. Charging utility maximization in wireless rechargeable sensor networks. *Wirel. Netw.* **2017**, *23*, 1–13. [CrossRef]
30. Liang, W.; Xu, W.; Ren, X.; Jia, X.; Lin, X. Maintaining Large-Scale Rechargeable Sensor Networks Perpetually via Multiple Mobile Charging Vehicles. *ACM Trans. Sens. Netw.* **2016**, *12*, 1–26. [CrossRef]
31. Dai, H.; Wu, X.; Chen, G.; Xu, L.; Lin, S. Minimizing the number of mobile chargers for large-scale wireless rechargeable sensor networks. *Comput. Commun.* **2014**, *46*, 54–65. [CrossRef]
32. Shi, Y.; Xie, L.; Hou, Y.T.; Sherali, H.D. On renewable sensor networks with wireless energy transfer. In Proceedings of the IEEE INFOCOM, Shanghai, China, 10–15 April 2011; Volume 2, pp. 1350–1358.
33. Liang, H.; Kong, L.; Yu, G.; Pan, J.; Zhu, T. Evaluating the On-Demand Mobile Charging in Wireless Sensor Networks. *IEEE Trans. Mob. Comput.* **2015**, *14*, 1861–1875.
34. Tsoumanis, G.; Aissa, S.; Stavrakakis, I.; Oikonomou, K. Performance Evaluation of a Proposed On-Demand Recharging Policy in Wireless Sensor Networks. In Proceedings of the 2018 IEEE 19th International Symposium on "A World of Wireless, Mobile and Multimedia Networks" (WoWMoM), Chania, Greece, 12–15 June 2018.
35. Sheng, Z.; Qian, Z.; Kong, F.; Jie, W.; Lu, S. P3: Joint optimization of charger placement and power allocation for wireless power transfer. In Proceedings of the 2015 IEEE Conference on Computer Communications (INFOCOM), Kowloon, Hong Kong, 26 April–1 May 2015.
36. Sheng, Z.; Qian, Z.; Jie, W.; Kong, F.; Lu, S. Wireless Charger Placement and Power Allocation for Maximizing Charging Quality. *IEEE Trans. Mob. Comput.* **2018**, *17*, 1483–1496.

37. Zorbas, D.; Raveneau, P.; Ghamri-Doudane, Y. On Optimal Charger Positioning in Clustered RF-power Harvesting Wireless Sensor Networks. In Proceedings of the ACM International Conference on Modeling, Valletta, Malta, 13–17 November 2016.

38. Tong, B.; Zi, L.; Wang, G.; Zhang, W. How Wireless Power Charging Technology Affects Sensor Network Deployment and Routing. In Proceedings of the IEEE International Conference on Distributed Computing Systems, Genova, Italy, 21–25 June 2010.

39. Zi, L.; Yang, P.; Zhang, W.; Qiao, D. J-RoC: A Joint Routing and Charging scheme to prolong sensor network lifetime. In Proceedings of the IEEE International Conference on Network Protocols, Vancouver, BC, Canada, 17–20 October 2011.

40. Lu, X.; Wang, P.; Niyato, D.; Dong, I.K.; Han, Z. Wireless Networks with RF Energy Harvesting: A Contemporary Survey. *IEEE Commun. Surv. Tutor.* **2017**, *17*, 757–789. [CrossRef]

sensors

MDPI

Article

Deep-Learning-Based Physical Layer Authentication for Industrial Wireless Sensor Networks

Run-Fa Liao [1], Hong Wen [2,*], Jinsong Wu [3*], Fei Pan [1], Aidong Xu [4], Yixin Jiang [4], Feiyi Xie [1] and Minggui Cao [1]

[1] National Key Laboratory of Science and Technology on Communications, University of Electronic Science and Technology of China, Chengdu 611731, China; runfa.liao@std.uestc.edu.cn (R.-F.L.); panfeivivi@hotmail.com (F.P.); helloyuiki@foxmail.com (F.X.); cmjcmjcmj@163.com (M.C.)

[2] School of Aeronautics and Astronautics, University of Electronic Science and Technology of China, Chengdu 611731, China

[3] Department of Electrical Engineering, Universidad de Chile, Santiago 8370451, Chile; wujs@ieee.org

[4] EPRI, China Southern Power Grid Co., Ltd., Guangzhou 510080, China; xuad@csg.cn (A.X.); jiangyx@csg.cn (Y.J.)

[*] Correspondence: sunlike@uestc.edu.cn or wujs@ieee.org

Received: 18 April 2019; Accepted: 27 May 2019; Published: 28 May 2019

Abstract: In this paper, a deep learning (DL)-based physical (PHY) layer authentication framework is proposed to enhance the security of industrial wireless sensor networks (IWSNs). Three algorithms, the deep neural network (DNN)-based sensor nodes' authentication method, the convolutional neural network (CNN)-based sensor nodes' authentication method, and the convolution preprocessing neural network (CPNN)-based sensor nodes' authentication method, have been adopted to implement the PHY-layer authentication in IWSNs. Among them, the improved CPNN-based algorithm requires few computing resources and has extremely low latency, which enable a lightweight multi-node PHY-layer authentication. The adaptive moment estimation (Adam) accelerated gradient algorithm and minibatch skill are used to accelerate the training of the neural networks. Simulations are performed to evaluate the performance of each algorithm and a brief analysis of the application scenarios for each algorithm is discussed. Moreover, the experiments have been performed with universal software radio peripherals (USRPs) to evaluate the authentication performance of the proposed algorithms. Due to the trainings being performed on the edge sides, the proposed method can implement a lightweight authentication for the sensor nodes under the edge computing (EC) system in IWSNs.

Keywords: PHY-layer; light-weight authentication; neural network; WSN; industrial

1. Introduction

With the development of Industry 4.0, wireless sensor networks (WSNs) have great application prospects for industrial scenarios due to their advantages over traditional wired networks [1–4]. However, fully-automated mechanized operations and the wireless communication environments make the industrial wireless sensor networks (IWSNs) have stronger requirements for high security and low latency [5]. M.Luvisotto et al. [6] mentioned that the response delay in IWSNs should be in milliseconds. Moreover, under the edge computing (EC) system in IWSNs, some sensor nodes are in some completely security-free environments because there are no redundant computing resources and transmission resources. Therefore, lightweight authentication is urgently needed to enhance the security of IWSNs while ensuring low latency. The encrypted methods [7,8] are too heavy to support the nodes due to complex computing. I. Bhardwaj et al. [9] did some lightweight processing on the password, but their method still cannot meet some specific requirements. Some other researchers proposed a fast cross-authentication scheme that combines non-cryptographic and cryptographic algorithms to solve the security and latency issues [10]. In addition, the heterogeneous nature of the IWSNs makes traditional encryption-based authentication methods more complex to implement or manage. However, physical (PHY) layer methods provide some new approaches to protect the lightweightIWSNs. The high authentication rate and low cost are especially valued for such applications. By introducing deep learning (DL) into the PHY-layer authentication method, under the EC system, the training is performed under the edge devices, and the sensor nodes almost do not bear any extra costs.

D. Christin et al. [1] surveyed related WSN technologies dedicated to industrial automation from the aspects of security and quality of service (QoS). The work in [4] presented a QoS framework for IWSNs guaranteeing the delay bound and the target reliability. N. Neshenko et al. [11] surveyed the challenges and research problems in the Internet of Things (IoT) including intrusion detection systems, threat modeling, and emerging technologies. However, the papers mentioned above only address the security and reliability issues from the perspective of the system architecture or simply give a direction for future research. L. Xiao et al. [12] proposed a method to enhance the security of underwater sensor networks exploiting the power delay profile of the underwater acoustic channel to discriminate the sensors. The article [13] presented a two-factor user authentication protocol using the hash function that protects against other attacks in wireless sensor networks, with the exception of denial of service (DoS) and node attacks. However, the traditional security methods have relatively large requirements on computing resources and communication resources, which cannot meet the requirement of low latency.

PHY-layer authentication can achieve lightweight authentication and effectively address the tradeoff between the security and low latency requirement of the wireless sensor networks in industrial scenarios. The PHY-layer authentication methods can distinguish the legitimate sensor nodes and illegal ones by physical layer channel information, such as channel state information (CSI) [14–17], received signal strength indicator (RSSI) [18–20], received signal strength (RSS) [21], and the radio frequency (RF) fingerprint [22,23]. However, the PHY-layer authentication methods mentioned above based on the hypothesis test are mostly compared with a threshold to distinguish users, which makes it difficult to discriminate multi-nodes at the same time. Authenticating multi-nodes simultaneously is a multi-classification problem, which needs to be solved urgently.

Deep learning has a large number of applications, such as computer vision, image classification, pattern recognition [24–26], and so on. There are considerable research works using deep learning in wireless communications, such as in channel estimation and channel prediction. P. Illy et al. used machine learning to enhance the security of edge computing by implementing intrusion detection [27]. The paper [28] used the deep neural network to estimate the CSIs in orthogonal frequency division multiplexing (OFDM) systems. The work in [29] proposed a Raleigh fading channel prediction scheme with a deep learning method. N. Wang et al. [30] proposed a physical-layer authentication scheme based on extreme learning machine to detect spoofing attack. The DL-based PHY-layer authentication methods proposed in this paper can achieve multi-user authentication in a short time.

Unlike the traditional test-threshold-based PHY-layer authentication, the DL-based PHY-layer authentication methods can distinguish multiple sensor nodes simultaneously and maintain excellent performance. In the EC system, multi-sensor nodes need to be authenticated simultaneously, which is suitable for using the DL-based methods. The DL-based authentication methods are usually divided into the offline training phase and online authentication phase. The PHY-layer authentication framework we proposed in this paper also includes an online retraining process. In summary, the DL-based sensor nodes' authentication algorithms proposed in this paper, utilizing the spatial diversity of wireless channels, can discriminate the sensor nodes without the test thresholds and have more practical application values. The main contributions of our work can be summarized as follows:

- We propose a DL-based PHY-layer authentication framework to enhance the security of industrial sensor networks. We also briefly explore the applications of the framework for practical industrial scenarios.
- Three different algorithms are adopted to implement the PHY-layer authentication in IWSNs, including the deep neural network (DNN)-based sensor nodes' authentication method, the convolutional neural network (CNN)-based sensor nodes' authentication method, and the convolution preprocessing neural network (CPNN)-based sensor nodes' authentication method.
- Simulation results show that the proposed algorithms can achieve better performance. In addition, the experiments in the engineering center with USRPs validate their utility in practical industrial environments.

The rest of this paper is organized as follows. We present the preliminaries and system model in Sections 2 and 3, respectively. The DL-based PHY-layer authentication method in industrial wireless sensor networks is proposed in Section 4. We provide numerical experiments in Section 5. The experiment in a practical environment and conclusions are presented in Sections 6 and 7, respectively.

The symbols used in this article are briefly described as follows. Uppercase bold letters are used for the matrix (e.g., H, W) and lowercase bold letters for vectors (e.g., x, y). The elements are represented by the letters with subscripts and not bold (e.g., x_i, ω_{1i}).

2. Preliminaries

2.1. Channel State Information

Due to the inherent characteristics of the wireless channels, the transmitted signals may experience a series of attenuations, such as, multipath effects, fading, shadowing, and delay distortion. The channel state information (CSI) provides us the channel variations experienced during propagations. In wireless communications, CSI represents the channel properties of a communication link. The CSI needs to be estimated by the receiver to detect the transmitted signals.

In the wireless fading channel, the system is modeled as:

$$y = Hx + n, \tag{1}$$

where y and x represents the receive and transmit signal, respectively. H denotes the channel matrix, which is the CSI we mentioned above. n denotes the additive white Gaussian noise vector, which follows a complex standard normal distribution. $n \sim \mathcal{CN}(0, \sigma_0)$, where the mean value is zero and the noise covariance matrix σ_0 is known. H represents the channel's frequency response, which can be estimated by y and x in the receiving end.

2.2. Deep Neural Network

Generally speaking, DNN is a deeper version of the artificial neural network (ANN) through increasing the number of hidden layers in order to enhance the ability in representation or classification. As shown in Figure 1, it is a typical deep neural network with an input layer, multiple hidden layers, and an output layer. Each layer has a large number of neurons. The input of each neuron is the output of the upper neuron multiplied by the corresponding coefficient, and the output of each neuron is the input activated by activation functions. For example, the output of the first neuron in the first hidden layer is:

$$z_1^1 = f_a \left(\sum_i \omega_{1i} x_i + \xi_1 \right), \tag{2}$$

where ω_{1i} denotes the weight coefficient of links z_1^1 and x_i. ξ_1 denotes the threshold coefficient of z_1^1. $f_a(\cdot)$ represents the activation function. Common activation functions are the sigmoid function, the rectified linear unit (ReLU) function, and the soft-max function, defined as $f_{sigmoid}(x) = \frac{1}{1-e^{-x}}$, $f_{ReLU}(x) = \max(0, x)$, $f_{softmax}(x) = \frac{e^x}{\|e^x\|_1}$, respectively, where x is a vector and $\|\cdot\|_1$ denotes the ℓ_1-norm. Usually, the hidden layer and output layer use the ReLU function and the soft-max function, respectively. The output of lth layer is given by:

$$z^l = f_a^l \left(W^l \cdot z^l + \xi^l \right). \tag{3}$$

We use the $\Psi^l(\cdot)$ to represent the operation of each layer of neurons. Then, we have the output of the deep neural network,

$$\hat{y} = \Psi(W, \Xi) = \Psi^L \left(\Psi^{L-1} \left(\cdots \Psi^1(x) \right) \right). \tag{4}$$

The application of the neural network is executed in two steps, a training phase and an identification phase. When in the training phase, the input data (i.e., CSI) of the input layer and the corresponding label y are known. Then, we train the parameters W and ξ by minimizing the cost function $\mathcal{L}$ by the gradient descent method, which is formulated as:

$$\hat{W}, \hat{\xi} = \arg\min_{W, \xi} (\mathcal{L}), \tag{5}$$

where $\mathcal{L}$ represents the value of the loss function. The loss function usually uses a mean squared error function or a cross entropy function, which is given by:

$$\mathcal{L}_{mean-square} = \|y - \hat{y}\|_2^2, \tag{6}$$

or:

$$\mathcal{L}_{cross-entropy} = y^T \cdot \log(\hat{y}), \tag{7}$$

where $(\cdot)^T$ denotes the transpose of the matrix or vector.

In the identification phase, the label of the input data (i.e., CSI) is unknown. By inputting CSI to the neural network, its corresponding output $\hat{y}$ will be used to identify and classify the input CSI.

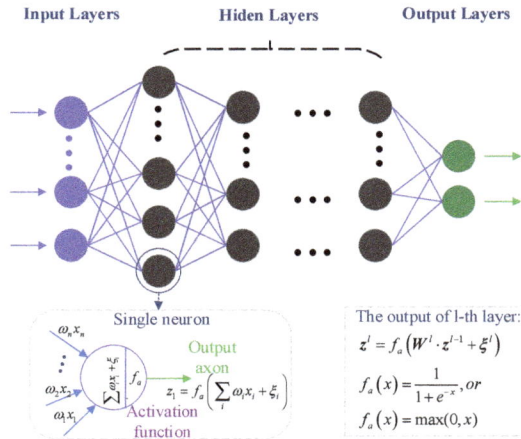

Figure 1. The deep neural network.

2.3. Convolutional Neural Network

The convolutional neural network (CNN) is part of the feedforward neural network with convolutional computation and a deep structure [11]. CNN includes convolutional layers, pooling layers, and fully-connected layers compared with ordinary neural network. The convolutional layer computes multiple convolutions in parallel to produce a set of linear activation responses. Further, the convolution operation can effectively extract features form the original signal (e.g., CSI). The output of the convolutional layer is given by:

$$Z^l = f_a^l \left(Z^{l-1} \otimes W^l + \xi^l \right), \tag{8}$$

where Z^l denotes the output of the lth layer and W^l and ξ^l denote the convolution kernel and threshold in the lth layer, respectively. $\otimes$ represents the convolution operation. $f_a^l (\cdot)$ denotes the activation function in CNN, often using the ReLU function.

Following the convolutional layer is the pooling layer, which effectively reduces the data dimension without losing valid information. The pooling function replaces the output of the network at that location using the overall statistical characteristics of the adjacent outputs at a location; for example, the maximum value in the adjacent rectangular region. Other commonly-used pooling functions include the average value in an adjacent rectangular region, the ℓ_2 norm, and the weighted average in adjacent regions. The main goal is to reduce the dimension or the resolution of feature maps. The pooling operation, which is a subsampling, can facilitate the extraction of high-level features.

The fully-connected layer of CNN is more like a hidden layer in DNN. There can be one fully-connected layer or multiple in CNN. We convert CSI into a matrix and use different colors to represent different values. As shown in Figure 2, it is a typical convolutional neural network with two convolutional layers, two pooled layers, and one fully-connected layer. We can see that the CSI converts to a matrix of 32 by 32 in size. The size of the convolution kernel in the first convolutional layer is four by four. After the convolution and activation, the average pooling operation is performed with a kernel of four by four in size. Then, there is another convolution, activation, and pooling operation. The final two layers are the fully-connected layer and the output layer activated with soft-max. The output of CNN can be formulated as:

$$\hat{y} = Y(w, \xi) = Y^L \left(Y^{L-1} \left(\cdots Y^1 (X) \right) \right). \tag{9}$$

Like DNN, CNN is also executed in two steps, a training phase and an identification phase. During the training phase, the input data (i.e., CSI) and corresponding labels y will be used to train the parameters w and ξ in CNN, which is formulated as:

$$\hat{w}, \hat{\xi} = \arg \min_{w, \xi} (\mathcal{L}), \tag{10}$$

where $\mathcal{L}$ denotes the value of the loss function in CNN. In the identification phase, the well-trained CNN will be used to perform the PHY-layer authentication.

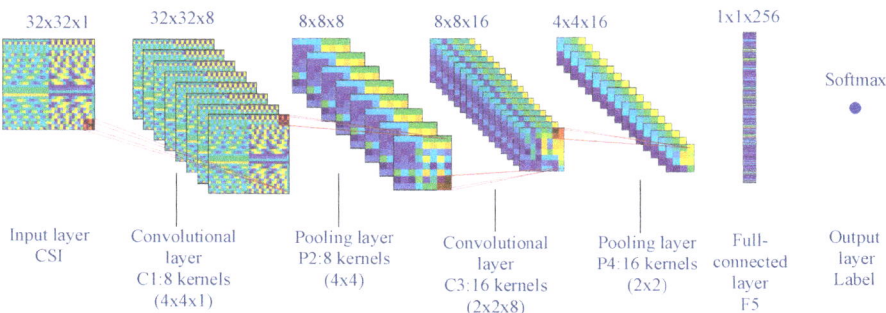

Figure 2. The convolutional neural network. CSI, channel state information.

3. System Model

We propose a DL-based PHY-layer authentication for an industrial wireless sensor network that can resist the spoofing attack. The methods we propose can enhance the security of the industrial wireless network without sacrificing communication resources. As shown in Figure 3, we placed many sensor nodes in the different locations of the industrial scene. The wireless sensor nodes send the pilot to the base station (BS) with time division duplexing (TDD) mode. First of all, each node needs to be identified by the upper layer authentication to facilitate labeling the corresponding CSI. In the initialization phase, we trained our neural networks through the training data (i.e., CSIs) and corresponding labels. Then, we authenticated the legitimate and illegal sensor nodes with newly-estimated CSI in the authentication phase. In the retraining phase, we updated the CSIs' training set with the new channel information of certified sensor nodes and retrained the neural network for the next authentication. The authentication processing of the industrial wireless sensor network is shown in Figure 4.

The DL-based PHY-layer authentication we propose can dynamically adjust system parameters over time. It can further improve the accuracy of authentication and has higher practicality.

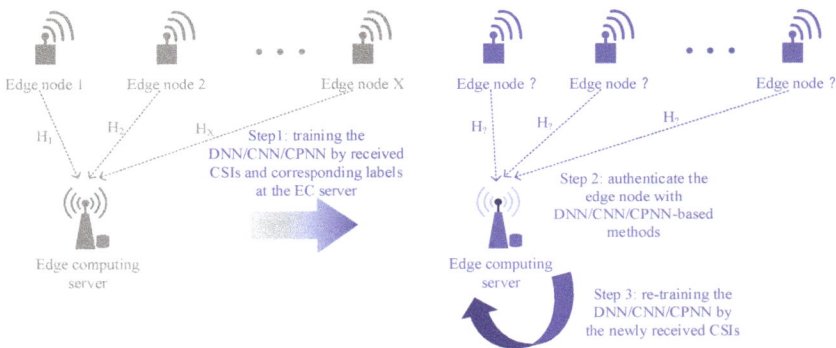

Figure 3. The system model of DL-based PHY-layer authentication in IWSNs.

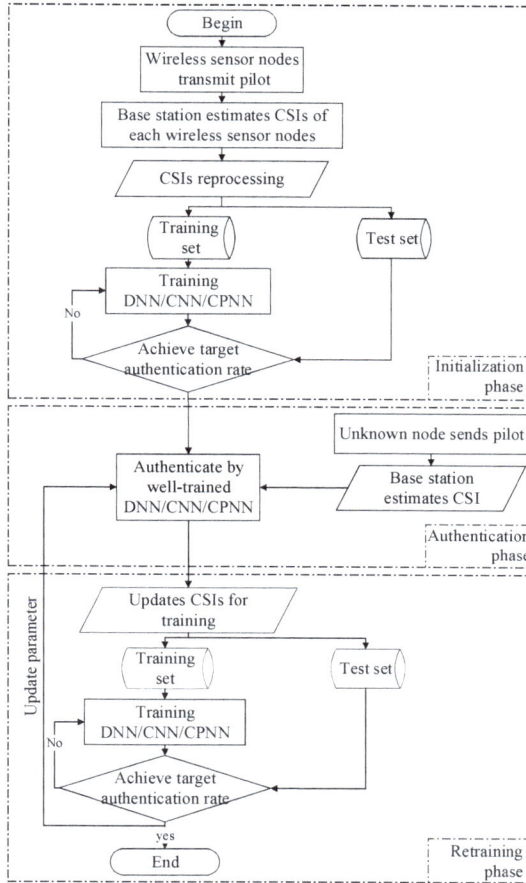

Figure 4. DL-based PHY-layer authentication flow chart.

4. Deep Learning-Based Sensor Nodes' Authentication Algorithms

In our previous work, we briefly introduced the physical layer channel authentication based on CNN [31]. This paper will further improve the CNN algorithm and propose a rapid-DNN-based PHY-layer authentication algorithm to meet the low latency requirements of industrial wireless sensor networks.

4.1. DNN-Based Sensor Nodes' Authentication

The DNN-based PHY-layer authentication in industrial wireless sensor networks uses the DNN to implement sensor nodes' authentication. In the initialization phase, the base station collects channel state information of each sensor node and performs the corresponding labeling according to the upper layer protocol authentication (e.g., EAP, AKA). The DNN was trained by the collected information to obtain the initial neural network parameters. In the authentication phase, the CSI of the unknown sensor node will be authenticated by the well-trained DNN in the initialization phase. After the new CSI has been authenticated, the dataset will be trained again for the next authentication.

Algorithm 1 DNN-based sensor nodes' authentication.

Input: The ith CSI to authenticate $x^{(i)}$

Output: The label of unknown CSI $\hat{y}^{(i)}$, the new weight matrix $W^\dagger$, and threshold vector $\xi^\dagger$ of DNN

1: Initialize all connection weights $W_0^\dagger$, and thresholds $\xi_0^\dagger$ in the network will be obtained through the training of DNN, using the pre-acquired dataset $D^\dagger = \{(x_k, y_k)\}_{k=1}^m$;
2: Compute $\hat{y}^{(i)}$ by well-trained DNN;
3: Update the training set $D^\dagger = \{(x_k, y_k)\}_{k=1}^m$ by $\left(x^{(i)}, \hat{y}^{(i)}\right)$;
4: Retrain the DNN by the new dataset and get new weight matrix $W^\dagger$ and threshold vector $\xi^\dagger$;
5: Return $\hat{y}^{(i)}$, $W^\dagger$, $\xi^\dagger$.

4.2. CNN-Based Sensor Nodes' Authentication

The CNN-based sensor nodes' authentication method is more like the DNN-based sensor nodes' authentication. They have the same steps except that the authenticated neural network changes from DNN to CNN. In the initialization phase, the CNN will be trained by the pre-acquired dataset of different sensor nodes. Then, the ith CSI will be authenticated by the well-trained CNN. At last, the CNN will be retrained after the dataset is updated.

Algorithm 2 CNN-based sensor nodes' authentication.

Input: The ith CSI to authenticate $x^{(i)}$

Output: The label of unknown CSI $\hat{y}^{(i)}$, the new weight matrix $W^\diamond$, and threshold vector $\xi^\diamond$ of CNN

1: Initialize all connection weights $W_0^\diamond$, and thresholds $\xi_0^\diamond$ in the network will be obtained through the training of CNN, using the pre-acquired dataset $D^\diamond = \{(x_k, y_k)\}_{k=1}^m$;
2: Compute $\hat{y}^{(i)}$ by the well-trained CNN;
3: Update the training set $D^\diamond = \{(x_k, y_k)\}_{k=1}^m$ by $\left(x^{(i)}, \hat{y}^{(i)}\right)$;
4: Retrain the CNN by the new dataset and get new weight matrix $W^\diamond$ and threshold vector $\xi^\diamond$;
5: Return $\hat{y}^{(i)}$, $W^\diamond$, $\xi^\diamond$.

4.3. Convolution Pre-Processing Neural Network-based Sensor Nodes' Authentication

The convolution pre-processing neural network-based sensor nodes' authentication method we propose in this paper has shorter training time and higher authentication accuracy. The core idea is to perform offline convolution preprocessing on the CSIs before training the neural network. The convolution preprocessing can effectively reduce the data dimension and extract the feature information of the CSIs, while the convolution kernels are trained by pre-obtained CSIs and corresponding labels. After convolution, activation, and pooling by the convolution kernels, the CSIs x_k become $\tilde{x}_k$. The latter's dimensions are much smaller than the former's. For the CPNN-based sensor node authentication method, the convolution kernels $V^\perp$ need to be calculated by the pre-obtained CSIs. Then, the neural network is trained by the new dataset $D^\perp = \{(\tilde{x}_k, y_k)\}_{k=1}^m$ in the initialization phase to get the weight matrix $W_0^\perp$ and threshold vector $\xi_0^\perp$.

Algorithm 3 CPNN-based sensor nodes' authentication.

Input: The ith CSI to authenticate $x^{(i)}$

Output: The label of unknown CSI $\hat{y}^{(i)}$, the new weight matrix $W^{\perp}$, and threshold vector $\xi^{\perp}$ of CPNN

1: Initialize: training the CNN by the pre-acquired CSIs to obtain the convolution kernels $V^{\perp}$; the dataset $D^{\perp} = \{(\bar{x}_k, y_k)\}_{k=1}^{m}$ obtained by convolution; the weights $W_0^{\perp}$ and thresholds $\xi_0^{\perp}$ in the neural network will be obtained through the training of CPNN, using dataset $D^{\perp} = \{(\bar{x}_k, y_k)\}_{k=1}^{m}$;

2: Convolution pre-processing of the CSI $x^{(i)}$ into $\bar{x}^{(i)}$;

3: Compute $\hat{y}^{(i)}$ by the well-trained CPNN;

4: Update the training set $D^{\perp} = \{(\bar{x}_k, y_k)\}_{k=1}^{m}$ by $\left(x^{(i)}, \hat{y}^{(i)}\right)$;

5: Retrain the CPNN by the new dataset, and get new weight matrix $W^{\perp}$ and threshold vector $\xi^{\perp}$;

6: Return $\hat{y}^{(i)}, W^{\perp}, \xi^{\perp}$.

4.4. Complexity Analysis

We compare the computational complexity of each sensor nodes' authentication methods in this section. The initialization phase was performed offline, and we will not discuss its computational resources and latency issues. In the authentication phase, the DNN-based sensor nodes' authentication method needs to perform:

$$b^l = W^l \cdot z^{l-1} + \xi^l. \tag{11}$$

As shown in Table 1, the computational complexity of the mathematical operation of DNN-based method is almost $O\left(\max\left(n^1 \times n^2, n^2 \times n^3, \cdots, n^{L-1} \times n^L\right)\right)$, where n^l denotes the number of neurons in the lth layer in DNN. In our numerical experiments, we used a five-layer DNN in which the number of neurons in each layer was 1024, 120, 60, 25, 4. Therefore, the computational complexity is almost 1×10^5. The CNN-based sensor nodes' authentication method needs to perform:

$$B^l = Z^{l-1} \otimes W^l + \xi^l. \tag{12}$$

The computational complexity of the mathematical operation of the CNN-based method is almost $O\left(\max\left(n^1 \times n_{ker}^1 \times n_{num}^1, n^2 \times n_{ker}^2 \times n_{num}^2, \cdots, n_{full}^{L-1} \times n^L\right)\right)$, where n^l indicates the number of convolution operations in the lth layer. n_{ker}^l and n_{num}^l denote the dimensions and the number of convolution kernels in the lth layer. n_{full}^{L-1} and n^L represent the number of neurons in the fully-connected and output layers. In our numerical experiments, we used eight convolution kernels with dimensions of $4 \times 4 \times 1$ and 16 convolution kernels with dimensions of $2 \times 2 \times 8$. The dimensions of the input layer and fully-connected layer were $32 \times 32 \times 1$ and $1 \times 1 \times 256$, respectively. Therefore, the computational complexity of the CNN-based method was almost 5×10^5. The CPNN-based sensor nodes' authentication method needs to perform convolution pre-processing on CSI, and the computation complexity of pre-processing was relatively small. The computational complexity of the CPNN-based method is $O\left(\max\left(n^0 \times n_{ker}^0 \times n_{num}^0, n^1 \times n^2, \cdots, n^{L-1} \times n^L\right)\right)$, which is almost the same as that of the DNN-based method, where n^0 denotes the number of convolution operations in pre-processing and n^l denotes the dimensions of the CSI after being processed in the lth layer. n_{ker}^0 and n_{num}^0 denote the dimension and number of convolution kernels in pre-processing, respectively. There were 16 convolution kernels of size $4 \times 4 \times 1$ in the pre-processing of the CPNN-based method. There were four convolution steps. The computational complexity of the CPNN-based method was almost 2×10^4.

During the retraining phase, the number of parameters that needed to be trained is shown in Table 2. The DNN-based sensor nodes' authentication method needs to train weight matrix $W^{\dagger}$ and threshold vector $\xi^{\dagger}$, in which it needs to train $(n^1 \times n^2 + n^2 \times n^3 + \cdots + n^{L-1} \times n^L)$

parameters. There were almost 1×10^5 parameters for the DNN-based sensor nodes' authentication method in our numerical experiments. The CNN-based sensor nodes' authentication method needs to train convolution kernels $W^\diamond$ and threshold vector $\xi^\diamond$, which needed to train $\left(n^1_{kernel} \times n^1_{num} + n^2_{kernel} \times n^2_{num} + n^{L-1}_{full} \times n^L \right)$ parameters. In our numerical experiments, only 1×10^3 parameters needed to be trained. The CPNN-based authentication method needed to train weight matrix $W^\perp$ and threshold vector $\xi^\perp$. Like the DNN-based method, the parameters of CPNN-based method depended on the number of neurons in each layer. However, the dimension of the input in the CPNN-based method was much smaller than the DNN-based method. The number of neurons in each layer of CPNN was 256, 50, 25, 12, and 4. There were almost 1×10^4 parameters that needed to be trained in the retraining phase.

Table 1. The computational complexity in the authentication phase.

Algorithms	Computational Complexity	Simulation
DNN-based	$O\left(\max\left(n^1 \times n^2, n^2 \times n^3, \cdots, n^{L-1} \times n^L\right)\right)$	1×10^5
CNN-based	$O\left(\max\left(n^1 \times n^1_{ker} \times n^1_{num}, n^2 \times n^2_{ker} \times n^2_{num}, \cdots, n^{L-1}_{full} \times n^L\right)\right)$	5×10^5
CPNN-based	$O\left(\max\left(n^0 \times n^0_{ker} \times n^0_{num}, n^1 \times n^2, \cdots, n^{L-1} \times n^L\right)\right)$	2×10^4

Table 2. The number of parameters in the retraining phase.

Algorithms	Number of Parameters	Simulation
DNN-based	$\left(n^1 \times n^2 + n^2 \times n^3 + \cdots + n^{L-1} \times n^L\right)$	1×10^5
CNN-based	$\left(n^1_{kernel} \times n^1_{num} + n^2_{kernel} \times n^2_{num} + n^{L-1}_{full} \times n^L\right)$	1×10^3
CPNN-based	$\left(n^1 \times n^2 + n^2 \times n^3 + \cdots + n^{L-1} \times n^L\right)$	1×10^4

5. Numerical Experiments

Simulations have been performed to evaluate the performance of DL-based PHY-layer authentication for industrial wireless sensor networks. We performed the simulations under different nodes and analyzed the impact of the number of sensor nodes on the authentication success rate. We also compared the performance of different algorithms under different numbers of sensor nodes. Cost J denotes the value of the loss function, which is calculated by (6) or (7). The authentication rate P_a is defined as the probability of discriminating the wireless sensor nodes.

We considered the tapped delay line (TDL) model to simulate Raleigh fading channels with multipath delays [32]. The TDL model uses a set of non-frequency selective fading generators (such as the FWGN model or the Jakes model), where each generator is independent of each other and has an average power of one. The channel state information of different transmitters can be generated by:

$$y(n) = \sum_{d=0}^{N_D-1} h_d(n)x(n-d). \tag{13}$$

where N_D denotes the number of taps of the filters. We set the normalized Doppler shift $f_d = 0.125$, and six paths with different power delays were selected to synthesize the channels of different wireless sensor nodes. For more realistic consideration, the time delay of the first five paths of the sensor nodes was the same, which was 0 second (s), 5×10^{-6} s, 1×10^{-5} s, 1.5×10^{-5} s, and 2×10^{-5} s, respectively. When there were twelve sensor nodes, the time delay of the sixth path of each sensor node was as shown in Table 3.

Table 3. The time delay of the sixth path of 12 sensor nodes.

Sensor Node 1	Sensor Node 2	Sensor Node 3	Sensor Node 4	Sensor Node 5	Sensor Node 6
6.6×10^{-5} s	6.2×10^{-5} s	5.8×10^{-5} s	5.4×10^{-5} s	5.0×10^{-5} s	4.6×10^{-5} s
Sensor Node 7	**Sensor Node 8**	**Sensor Node 9**	**Sensor Node 10**	**Sensor Node 11**	**Sensor Node 12**
4.2×10^{-5} s	3.8×10^{-5} s	3.4×10^{-5} s	3.0×10^{-5} s	2.6×10^{-5} s	2.2×10^{-5} s

When there were four sensor nodes, the sixth paths of each sensor node were 6.6×10^{-5} s, 4.6×10^{-5} s, 3.4×10^{-5} s, 2.2×10^{-5} s, respectively. Sampling interval $t_{sampling} = 5 \times 10^{-6}$ s; the signal to noise ratio (SNR) of the simulation channel was 4 dB; the number of subcarriers was $n_{sub\_carrier} = 256$; the number of pilot intervals and of the cyclic prefix length were $n_{pilot\_inteval} = 256$ and $l_{cp\_length} = 30$, respectively.

We used a five-layer neural network for the DNN-based sensor nodes' authentication method, where the numbers of neurons in the hidden layer were 120, 60, and 25. The size of the input layer was determined by the CSI dimension, and the size of the output layer was determined by the number of sensor nodes. The convolutional neural network used in the CNN-based algorithm had seven layers, which were an input layer, two convolution layers, two pooling layers, one fully-connected layer, and an output layer. The two convolutional layers respectively used eight convolution kernels of size $4 \times 4 \times 1$ and 16 convolution kernels of size $2 \times 2 \times 8$, respectively. For the CPNN-based algorithm, it had 16 convolution kernels of size $4 \times 4 \times 1$ for the convolution pre-processing. In the authentication phase and retraining phase, we used a five-layer neural network for the CPNN-based algorithm, where the numbers of neurons in the hidden layer were 50, 25, and 12. Moreover, the adaptive moment estimation (Adam) accelerated gradient algorithm and minibatch skill was used for the accelerating of the neural networks' training.

As shown in Figure 5a, the x-axis is the number of neural network iterations and the y-axis is the cost function value. As the number of iterations increased, the cost function value decreased. In addition, the fewer the sensor nodes, the faster the cost function value dropped. We can visually see the authentication rate under different sensor nodes from Figure 5b. After 30 iterations, the authenticate rates tended to be stable. However, the authentication rate of four sensor nodes was higher than that of six sensor nodes, and the authentication rate of 12 sensor nodes was the lowest. Specifically, after 30 iterations, the authentication rates of 4 sensor nodes, 6 sensor nodes, 8 sensor nodes, and 12 sensor nodes was 95.5%, 80.83%, 77.25%, and 66.5%, respectively.

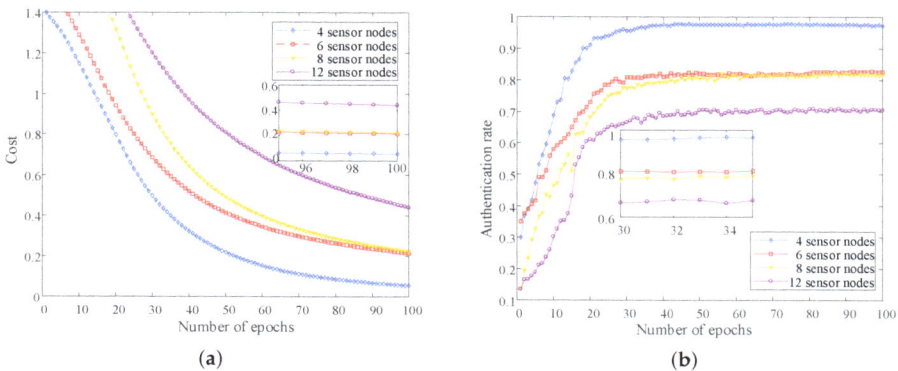

Figure 5. The authentication performance with different sensor nodes. (**a**) The cost value under different numbers of sensor nodes with the DNN-based method; (**b**) The authentication rate under different numbers of sensor nodes with the DNN-based method.

By discussing the authentication success rate under different numbers of hidden layers, we researched the robustness of the DL-based authentication method. The DNN-based algorithm had the most excellent performance. Therefore, we considered the influence of different hidden layer numbers on the authentication rate under the DNN-based method. As shown in Figure 6a, the authentication rate of the DNN-based method with different numbers of hidden layers increased as the iterations increased. The greater the number of hidden layers, the faster the convergence of the neural network's performance. The authentication success rate of the DNN-based method with different hidden layers after the training was stabilized are shown in Figure 6b. As the number of hidden layers increased, the authentication success rate increased. However, due to the inherent characteristics of the specific wireless channels, the performance of the DNN-based method did not continue to grow and tended to be stable, after the number of hidden layers was increased to a certain number.

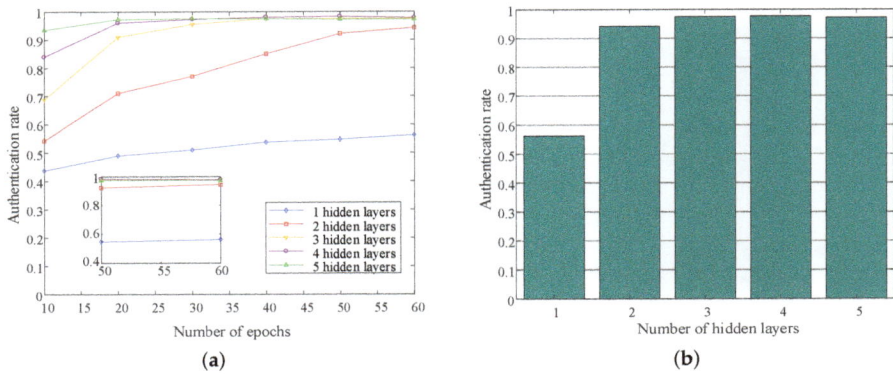

Figure 6. The authentication performance with different numbers of hidden layers. (**a**) The authentication rate of different numbers of hidden layers; (**b**) The authentication rate of different numbers of hidden layers after training was stabilized.

In addition, we performed simulation analysis on the authentication performance of different algorithms under different numbers of sensor nodes. As shown in Figure 7a, the DNN-based method had the best performance, because it had many parameters. For example, the authentication rates of the DNN-based method were 95.5% and 77.25% under four sensor nodes and eight sensor nodes, respectively. The CNN-based algorithm had the worst performance, because of the convolution and pooling and more or less lost some information of CSIs. For example, the authentication rates of the CNN-based method were 86.25% and 67.87% under four sensor nodes and eight sensor nodes. Another CPNN-based method we proposed in this paper was similar in performance to the CNN-based method. The authentication rates of CPNN-based algorithm were 85.25% and 66.75% under four sensor nodes and eight sensor nodes. However, the CPNN-based method had the shortest training time compared to the DNN-based algorithm and CNN-based algorithm, as shown in Figure 7b. Therefore, it has a better application prospect in the actual industrial wireless sensor network. We can see that the CNN-based method had the longest training time, followed by the DNN-based method.

In summary, the DNN-based sensor nodes' authentication had the best authenticate performance and a relatively limited training time. However, its training parameters will grow exponentially as the dimensions of CSI become larger. Therefore, the DNN-based algorithm is suitable for a shorter CSI authentication scheme. The CNN-based sensor nodes' authentication method effectively reduced the parameters that the neural network needed to train. However, due to the convolution operation and the pooling operation, it did not meet the requirements of saving training time, especially when the dimension of CSI was relatively small. At last, the CPNN-based sensor nodes' authentication method can effectively solve the problem of training time and authentication performance. It has an unparalleled advantage in practical industrial wireless sensor network applications.

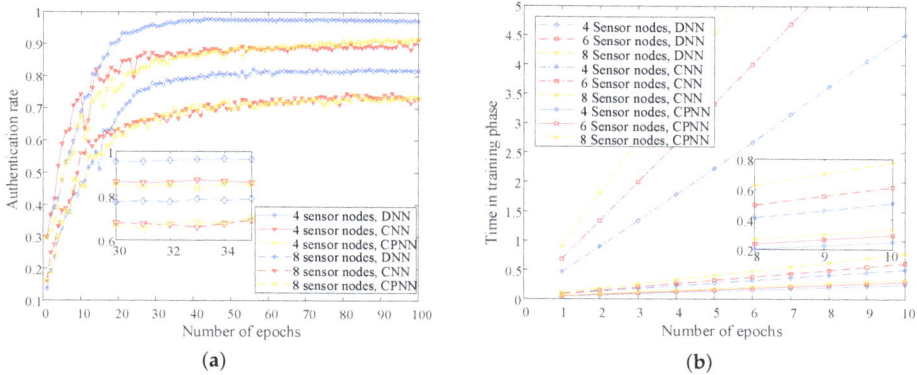

(a)   **(b)**

Figure 7. The authentication performance with different algorithms. (**a**) The authentication rate of different algorithms under different numbers of sensor nodes; (**b**) The time in the training phase of different algorithms under different numbers of sensor nodes.

6. Experiments In Practical Environment

Experiments have been performed with universal software radio peripherals (USRPs) to evaluate the authentication performance of the proposed DL-based PHY-layer authentication algorithms in industrial wireless sensor networks. The experimental simulations were performed at the school's engineering center, which has a large number of industrial facilities, such as computer numerical control (CNC) engraving and milling, CNC lathe, and so on. As shown in Figure 8, five radio sensor nodes equipped with industrial computer and USRPs were placed in a $43.56 \times 38.84 \times 6.5 \text{m}^3$ factory. The base station was equipped with 8 antennas in Position 2, and the other sensor nodes were equipped with 2, 4, or 8 antennas in Positions 1, 3, 4, and 5. The distances between sensor nodes and the base station varied from 5–25 meters (m). In this experiment, we set the carrier frequency $f_c = 3$ gigahertz (GHz), the interval of subcarriers $f_{interval\_subcarrier} = 15$ kilohertz (kHz), and the number of subcarriers $n_{subcarrier} = 128$. The transmitting power of USRPs was 15 dBm, and the transmission gain was 20 dB. The practical view of the engineering center is shown in Figure 9.

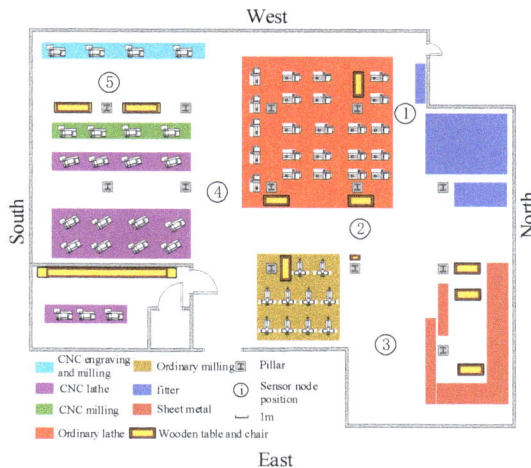

Figure 8. The network topology.

Figure 9. The location of the wireless sensor nodes in the practical industrial scenario.

We tested the authentication rates of sensor nodes with different antennas in different locations. As shown in Figure 10, as the number of antennas increased, the authentication success rate increased correspondingly. For example, the authentication rate of the DNN-based algorithm with 2 antennas was 92%, while the authentication rate of the DNN-based algorithm with 4 antennas and 8 antennas was 99.5% and 99.5%, respectively. From the histograms of different colors, we can see that the DNN-based sensor nodes' authentication method had the best performance. For example, the authentication rate of DNN-based algorithm with 8 antennas was 99.5%, while the authentication rate of the CNN-based algorithm with 8 antennas was 85%. In addition, the CPNN-based algorithm had the same performance as the CNN-based algorithm. However, the retraining time of the CPNN-based method was much shorter than that of the CNN-based algorithm.

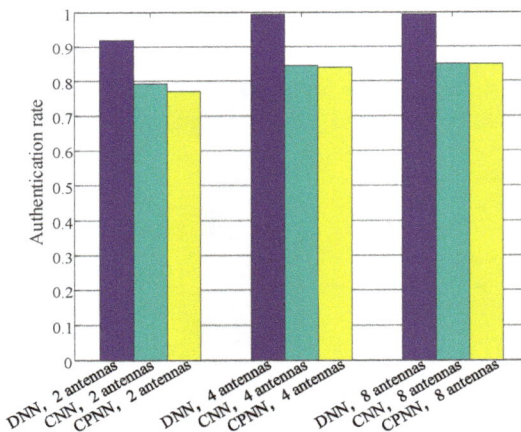

Figure 10. The authentication rate with USRPs.

7. Conclusions

The DL-based PHY-layer authentication method in industrial wireless sensor networks we proposed in this paper has a strong practical significance. It can both achieve lightweight authentication and authenticate multiple nodes simultaneously. Especially for the CPNN-based sensor nodes' authentication algorithm, it had good authentication performance and an ultra-short retraining time. The DNN-based sensor nodes' authentication had the best authenticate performance and a relatively limited training time. However, its training parameters will grow exponentially as the dimensions of CSI become larger. Therefore, the DNN-based algorithm is suitable for a shorter CSI authentication scheme. As shown in Table 2, the CNN-based algorithm and CPNN-based algorithm effectively reduced the parameters that the neural network needed to train. However, due to the convolution operation and the pooling operation, the CNN-based algorithm did not meet the requirements of saving training time, especially when the dimension of CSI was relatively small. At last, the CPNN-based sensor nodes' authentication method can effectively solve the problem of training time and authentication performance. It has an unparalleled advantage in practical industrial wireless sensor network applications.

Author Contributions: The work was realized with the collaboration of all the authors. R.-F.L. and M.C. contributed to the main results and code implementation. H.W. and J.W. organized the work, provided the funding, and revised the draft of the paper. R.-F.L. and H.W. designed the experiments. R.-F.L., F.P., and F.X. performed the experiments. R.-F.L. and H.W. analyzed the experimental results. R.-F.L., H.W., F.P., A.X., Y.J., F.X., and M.C. discussed the results. R.-F.L. wrote the original manuscript.

Funding: This work was supported by NSFC (No. 61572114), the National Major R & D Program (2018YFB0904900, 2018YFB0904905), the Sichuan Sci & Tech Basic Research Condition Platform Project (No.2018TJPT0041); and the Sichuan Sci & Tech Service Development Project (No.18KJFWSF0368). This work was also supported in part by the Hunan Provincial Nature Science Foundation Project 2018JJ2535 and the Chile CONICYT FONDECYT Regular Project 1181809.

Acknowledgments: Special thanks to the Engineering Center of UESTCfor the experimental scene they provided.

Conflicts of Interest: The authors declare no conflict of interest.

Abbreviations

The following abbreviations are used in this manuscript:

AKA	Authentication and key agreement
ANN	Artificial neural network
CNN	Convolutional neural network
CPNN	Convolution preprocessing neural network
CSI	Channel state information
DL	Deep learning
DNN	Deep neural network
EAP	Extensible authentication protocol
IWSNs	Industrial wireless sensor networks
MHz	Megahertz
OFDM	Orthogonal frequency division multiplexing
PHY	Physical
QoS	Quality of service
ReLU	Rectified linear unit
RSS	Received signal strength
RSSI	Received signal strength indicator
TDD	Time division duplexing
USRPs	Universal software radio peripherals
WSNs	Wireless sensor networks

References

1. Christin, D.; Mogre, P.S.; Hollick, M. Survey on wireless sensor network technologies for industrial automation: The security and quality of service perspectives. *Future Internet* **2010**, *2*, 96–125. [CrossRef]
2. Low, K.S.; Win, W.N.N.; Er, M.J. Wireless sensor networks for industrial environments. In Proceedings of the International Conference on Computational Intelligence for Modelling, Control and Automation and International Conference on Intelligent Agents, Web Technologies and Internet Commerce (CIMCA-IAWTIC'06), Vienna, Austria, 28–30 November 2005; pp. 271–276.

3.   Bal, M. Industrial applications of collaborative wireless sensor networks: A survey. In Proceedings of the 2014 IEEE 23rd International Symposium on Industrial Electronics (ISIE), Istanbul, Turkey, 1–4 June 2014, pp. 1463–1468.

4.   Zoppi, S.; Van Bemten, A.; Gürsu, H.M.; Vilgelm, M.; Guck, J.; Kellerer, W. Achieving Hybrid Wired/Wireless Industrial Networks with WDetServ: Reliability-Based Scheduling for Delay Guarantees. *IEEE Trans. Ind. Informat.* **2018**, *14*, 2307–2319. [CrossRef]

5.   Pan, F.; Pang, Z.; Luvisottò, M.; Xiao, M.; Wen, H. Physical-Layer Security for Industrial Wireless Control Systems: Basics and Future Directions. *IEEE Ind. Electron. Mag.* **2018**, *12*, 18–27. [CrossRef]

6.   Luvisotto, M.; Pang, Z.; Dzung, D. Ultra high performance wireless control for critical applications: Challenges and directions. *IEEE Trans. Ind. Informat.* **2017**, *13*, 1448–1459. [CrossRef]

7.   Zhao, G.; Yang, X.; Zhou, B.; Wei, W. RSA-based digital image encryption algorithm in wireless sensor networks. In Proceedings of the 2nd International Conference on Signal Processing Systems, Dalian, China, 5–7 July 2010.

8.   Aysal, T.C.; Barner, K.E. Sensor data cryptography in wireless sensor networks. *IEEE Trans. Inf. Forensics Secur.* **2008**, *3*, 273–289. [CrossRef]

9.   Bhardwaj, I.; Kumar, A.; Bansal, M. A review on lightweight cryptography algorithm for data security and authentication in IoTs. In Proceedings of the 4th International Conference on Signal Processing, Computing and Control (ISPCC), Solan, India, 21–23 September 2017 ; pp. 504–509.

10.   Moreira, C.M.; Kaddoum, G.; Bou-Harb, E. Cross-layer authentication protocol design for ultra-dense 5g hetnets. In Proceedings of the 2018 IEEE International Conference on Communications (ICC), Kansas City, MO, USA, 20–24 May 2018.

11.   Neshenko, N.; Bou-Harb, E.; Crichigno, J.; Kaddoum, G.; Ghani, N. Demystifying IoT Security: An Exhaustive Survey on IoT Vulnerabilities and a First Empirical Look on Internet-scale IoT Exploitations. *IEEE Commun. Surveys Tut.* **2019**. [CrossRef]

12.   Xiao, L.; Sheng, G.; Wan, X.; Su, W.; Cheng, P. Learning-Based PHY-Layer Authentication for Underwater Sensor Networks. *IEEE Commun. Lett.* **2019**, *23*, 60–63. [CrossRef]

13.   Das, M.L. Two-factor user authentication in wireless sensor networks. *IEEE Trans. Wireless Commun.* **2009**, *8*, 1086–1090. [CrossRef]

14.   Liu, H.; Wang, Y.; Liu, J.; Yang, J.; Chen, Y.; Poor, H.V. Authenticating users through fine-grained channel information. *IEEE Trans. Mobile Comput.* **2018**, *17*, 251–264. [CrossRef]

15.   Liu, T.Y.; Lin, P.H.; Lin, S.C.; Hong, Y.W.P.; Jorswieck, E.A. To avoid or not to avoid CSI leakage in physical layer secret communication systems. *IEEE Commun. Mag.* **2015**, *53*, 19–25. [CrossRef]

16.   Zou, Y.; Wang, X.; Shen, W. Optimal relay selection for physical-layer security in cooperative wireless networks. *J. Sel. Areas Commun.* **2013**, *31*, 2099–2111. [CrossRef]

17.   Tugnait, J.K. Wireless user authentication via comparison of power spectral densities. *IEEE J. Sel. Areas Commun.* **2013**, *31*, 1791–1802. [CrossRef]

18.   Pei, C.; Zhang, N.; Shen, X.S.; Mark, J.W. Channel-based physical layer authentication. In Proceedings of the 2014 IEEE Global Communications Conference, Austin, TX, USA, 8–12 December 2014; pp. 4114–4119.

19.   Bhargava, V.; Sichitiu, M.L. Physical Security Perimeters for Wireless Local Area Networks. *Int J. Netw. Secur.* **2006**, *3*, 73–84.

20.   Chen, Y.; Yang, J.; Trappe, W.; Martin, R.P. Detecting and localizing identity-based attacks in wireless and sensor networks. *IEEE Trans. Veh. Technol.* **2010**, *59*, 2418–2434. [CrossRef]

21.   Yang, J.; Chen, Y.; Trappe, W.; Cheng, J. Detection and localization of multiple spoofing attackers in wireless networks. *IEEE Trans. Parallel Distrib. Syst.* **2013**, *24*, 44–58. [CrossRef]

22.   Xie, F.; Wen, H.; Li, Y.; Chen, S.; Hu, L.; Chen, Y.; Song, H. Optimized Coherent Integration-Based Radio Frequency Fingerprinting in Internet of Things. *IEEE Internet Things J.* **2018**, *5*, 3967–3977. [CrossRef]

23.   Chen, Y.; Wen, H.; Song, H.; Chen, S.; Xie, F.; Yang, Q.; Hu, L. Lightweight one-time password authentication scheme based on radio-frequency fingerprinting. *IET Commun.* **2018**, *12*, 1477–1484. [CrossRef]

24.   LeCun, Y.; Bengio, Y.; Hinton, G. Deep learning. *Nature* **2015**, *521*, 436–444. [CrossRef] [PubMed]

25.   Arulampalam, G.; Bouzerdoum, A. A generalized feedforward neural network architecture for classification and regression. *Neural Netw.* **2003**, *16*, 561–568. [CrossRef]

Sensors **2019**, *19*, 2440

26.   Carpenter, G.A. Neural network models for pattern recognition and associative memory. *Neural Netw.* **1989**, *2*, 243–257. [CrossRef]

27.   Illy, P.; Kaddoum, G.; Moreira, C.M.; Kaur, K.; Garg, S. Securing Fog-to-Things Environment Using Intrusion Detection System Based On Ensemble Learning. *arXiv* **2019**, arXiv:1901.10933.

28.   Ye, H.; Li, G.Y.; Juang, B.H. Power of deep learning for channel estimation and signal detection in OFDM systems. *IEEE Wireless Commun. Lett.* **2018**, *7*, 114–117. [CrossRef]

29.   Liao, R.F.; Wen, H.; Wu, J.; Song, H.; Pan, F.; Dong, L. The Rayleigh Fading Channel Prediction via Deep Learning. *Wireless Commun. Mobile Comput.* **2018**, *2018*, 1–11. [CrossRef]

30.   Wang, N.; Jiang, T.; Lv, S.; Xiao, L. Physical-layer authentication based on extreme learning machine. *IEEE Commun. Lett.* **2017**, *21*, 1557–1560. [CrossRef]

31.   Liao, R.F.; Wen, H.; Pan, F.; Song, H.; Xu, A.; Jiang, Y. A Novel Physical Layer Authentication Method with Convolutional Neural Network. *IEEE ICAICA* **2019**, accepted.

32.   Clarke, R. A statistical theory of mobile-radio reception. *Bell Syst. Tech. J.* **1968**, *47*, 957–1000. [CrossRef]

Article

Social-Aware Peer Discovery for Energy Harvesting-Based Device-To-Device Communications

Zelalem Legese Hailemariam [1], Yuan-Cheng Lai [1], Yen-Hung Chen [2,*], Yu-Hsueh Wu [1] and Arthur Chang [3]

[1] Department of Information Management, National Taiwan University of Science and Technology, Taipei 106, Taiwan; zolacool2@gmail.com (Z.L.H.); laiyc@cs.ntust.edu.tw (Y.-C.L.); a0926622943@gmail.com (Y.-H.W.)

[2] Department of Information Management, National Taipei University of Nursing and Health Sciences, Taipei 112, Taiwan

[3] Bachelor Program in Interdisciplinary Studies, National Yunlin University of Science and Technology, Yunlin 640, Taiwan; changart@yuntech.edu.tw

* Correspondence: pplong@gmail.com

Received: 7 March 2019; Accepted: 16 May 2019; Published: 18 May 2019

Abstract: In Device-to-Device (D2D) communications, the first step is to find all of the neighboring peers in the network by performing a peer discovery process. Most previous studies use the social behaviors of the users to adjust the sending rates of the peer discovery messages (i.e., beacons) under the constraint of consumed power for increasing the Peer Discovery Ratio (PDR). However, these studies do not consider the potential for energy harvesting, which allows for the User Equipments (UEs) to procure additional power within charging areas. Accordingly, this paper proposes an Energy-Ratio Rate Decision (ERRD) algorithm that comprises three steps, namely Social Ratio Allocation (SRA), Energy Ratio Allocation (ERA), and Beacon Rate Decision (BRD). The SRA step determines the allocated power quantum for each UE from the total budget power based on the social behavior of the UE. The ERA step then adjusts this allocated power quantum in accordance with the power that is harvested by the UE. Finally, the BRD step computes the beacon rate for the UE based on the adjusted power quantum. The simulation results show that ERRD outperforms the previously-reported Social-Based Grouping (SBG) algorithm by 190% on the PDR for a budget power of one watt and 8% for a budget power of 20 watts.

Keywords: Device-to-Device (D2D); peer discovery; energy harvesting; social awareness

---

1. Introduction

The demands that are placed on wireless communications have exponentially increased in recent years due to the proliferation of User Equipments (UEs) and the unceasing development of new mobile services. Consequently, the lack of spectrum resources has emerged as a significant concern for communication operators. Currently, any UE that wishes to transmit data to another UE must transmit this data via the Base Station (BS) in the cellular network. However, if the distance between the two UEs is sufficiently small, then the potential exists for the UEs to communicate directly, thereby saving cellular spectrum resources. Accordingly, a new communication paradigm designated as Device-to-Device (D2D) communication has been proposed as a means of achieving short-distance transmissions in 5G networks with improved resource utilization efficiency.

In implementing D2D communications, the first, and most critical, step is that of peer discovery, in which the UEs attempt to identify all of their neighbors in the network [1]. The peer discovery process is generally performed using a beacon mechanism and it aims to maximize the Peer Discovery Ratio (PDR), which is defined as the number of peers that are found in the discovery process divided by the total number of peers in the network. Most previous studies on peer discovery focus on the problems of improving the efficiency of the search process, minimizing the number of collisions, and determining a suitable beacon rate for the UEs [2–14].

In practice, a large proportion of D2D communications stems from the interaction between users over social media, such as Facebook, Twitter, Plurk, and so on [15,16]. Consequently, many studies have investigated the problem of social-aware peer discovery [17–21]. In general, the results have shown that the social behavior of the users provides a useful tool for adjusting the beacon rates of the UEs in such a way to improve the overall efficiency of the peer discovery process [17–21].

Energy (Energy and power are used interchangeably in this paper) is a critical concern in D2D communications, since the UEs generally only have limited energy resources and they often consume energy extremely rapidly when running user applications. Furthermore, the devices (e.g., mobile phones/sensors) in mine pine or oil well also encounter the energy issue, because they cannot recharge the power with any metal contact, hence avoiding the sparkles and reducing the explosion probabilities. Energy harvesting has thus attracted growing interest in recent years as a means of enabling UEs to scavenge energy from surrounding energy sources, especially for limited capacity energy storage electrical devices and systems [22–29]. Broadly speaking, the scavenged energy can be classified as either renewable energy or nonrenewable energy. In the former case, the energy is obtained from natural sources (e.g., thermal, solar, and wind), while in the latter case, the energy is obtained from artificial sources (e.g., electromagnetic resonance, electromagnetic induction, and radio frequency). The energy harvesting mainly adopts the technology of wireless power transfer (WPT) via radio frequency (RF) to transmit power, and then uses the rectifying antenna to convert the received radio signal to direct current (DC) and charge the battery [23]. In light of the promising flexibility of energy harvesting, many organizations [24] (i.e., International Telecommunications Union, European Cooperation in Science and Technology, Wireless Power Transfer Consortium for Practical Applications) and scholars [25,26] have been involved in variable application developments, which include drone powered wireless sensor network [27], emergency ubiquitous power source system [28], and wireless power transfer in electric vehicles environments [29]. Some studies have focused on energy harvesting-based D2D networks, which use harvested energy to promote the D2D data communications [30–32].

As stated above, the literature contains many studies on the use of social behavior mechanisms to enhance the performance of D2D communications. However, while some of these studies actively address the problem of limiting the energy that is consumed in the peer discovery process, none of them consider an environment in which the UEs are able to procure additional energy from the environment while using energy-harvesting techniques. Nonetheless, such a strategy is of considerable benefit in improving the performance of the peer discovery process. In particular, UEs that acquire additional energy can send a greater number of beacon messages (thereby increasing the PDR), while those that acquire no additional energy can reduce their beacon rate in order to minimize the out-of-energy risk and prolong their participation in the discovery process.

Consequently, the present study proposes a novel peer discovery algorithm for energy-harvesting environments, designated the Energy-Ratio Rate Decision (ERRD) algorithm. ERRD comprises three steps, namely Social Ratio Allocation (SRA), Energy Ratio Allocation (ERA), and Beacon Rate Decision (BRD). SRA determines an initial allocation power quantum for each UE that is based on its social behavior. ERA then adjusts this power quantum based on the harvested power of the UE. Finally, BRD computes the beacon rate for the UE based on the adjusted power quantum. For UEs with a high harvested power, ERRD increases the beacon rate, thereby increasing the PDR. By contrast, for UEs with low (or no) harvested power, ERRD reduces the beacon rate, and hence prolongs the lifetime of the UE, thereby increasing the time for which the UE can perform beacon discovery.

This study attempts to determine the suitable beacon rates of all UEs to maximize PDR for D2D communications in an energy-harvesting environment. The contributions of this paper are across three orientations: (1) regarding the problem orientation, to the best of our knowledge, this study is the first paper to consider the social-aware peer discovery problem for D2D communications in an energy-harvesting environment, (2) regarding the solution orientation, we propose ERRD, which determines the beacon rate according to UE's social behavior and the harvested power quantum to increase its PDR, and (3) regarding the evaluation orientation, some of the simulations are conducted to investigate many important parameters along with significant observations.

The remainder of this paper is organized, as follows. Section 2 introduces the related work in the field. Section 3 describes the system model that is considered in the present study and formulates the related problem. Section 4 introduces the ERRD algorithm and describes its detailed operation. Section 5 presents and discusses the simulation results. Finally, Section 6 provides some brief concluding remarks and indicates the intended direction of future research.

2. Related Works

The problem of peer discovery in D2D communications has attracted considerable attention in the literature. Broadly speaking, existing approaches can be classified as either social-aware or social-unaware (see Table 1). Within each classification, the proposed mechanisms can be further divided as autonomous, network-assisted, or network-controlled. In autonomous mechanisms, the UEs find nearby peers by themselves, and hence the major issue lies in determining efficient methods for broadcasting the beacons with a minimum number of collisions. By contrast, in network-assisted solutions, the BS collects surrounding the information, determines a suitable beacon rate for each UE, and allocates a proper amount of resources to each UE to perform its transmissions. Finally, in network-controlled mechanisms, the BS directly helps the UEs to find peers by locating UEs.

Table 1. Comparison of related works.

Category	Ref.	Type	Main Goal	Approach
Social-unaware	[2]	Autonomous	Min resource	UEs periodically and synchronously send beacons using FDM.
	[3]	Autonomous	Min resource	UEs use trellis tone modulation multiple-access scheme.
	[4]	Autonomous	Quick recovery	UEs use a common channel and a group of channels to send beacons.
	[5]	Autonomous	LTE compatible	UEs listen to SRS channel to identify nearby UEs.
	[6]	Autonomous	LTE compatible	UEs listen to SRS channel to detect active UEs.
	[7]	Autonomous	Min energy	UEs determine the beacon rate based on their state.
	[8]	Network-Assisted	Min collision	BS determines the beacon rate for UEs depending on the number of requests sent by UE.
	[9]	Network-Assisted	Min collision	UEs authorized to perform discovery contend to transmit beacons.
	[10]	Network-Assisted	Min interference	Similar to [9], but interference from cellular UEs imposed on D2D pairs is also considered.
	[11]	Network-Assisted	Min resource	UEs send preamble to nearby UEs and BS allocates uplink RBs for UEs.
	[12]	Network-Assisted	Beacon schedule	BS roughly estimates the location of UEs by measuring channel components.
	[13]	Network-Controlled	Min resource	BSs locate UEs by AOA.
	[14]	Network-Controlled	Min energy	Wifi scans are first used to determine the UE locations and BS then sends D2D broadcast.
Social-aware	[17]	Autonomous	Trust	UEs find trusted UEs.
	[18]	Autonomous	Two-hop pairing	UEs send requests to trusted UEs, which forward request to all one-hop UEs.
	[19]	Autonomous	Hybrid attributes	UEs use three key social attributes to construct neighbor lists.
	[20]	Autonomous	Max content delivery	Social relationship is used as a weight for D2D pair formation and content sharing.
	[21]	Network-Assisted	Max PDR	BS determines the beacon rate based on contact rate.
	ERRD	Network-Assisted	Max PDR	BS adjusts beacon rate based on harvested energy amount.

FDM: Frequency Division Multiplexing, SRS: Sounding Reference Signal, AOA: Angle of Arrival.

The literature contains several autonomous mechanisms for peer discovery using a social-unaware approach [2–7]. For example, the FlashLinQ protocol that was proposed in [2] uses frequency division multiplexing (FDM) to propagate the beacons through the network [2]. Notably, the beacons are transmitted both periodically and synchronously, and hence FlashLinQ provides an effective means of estimating the amount of consumed resources and timing the resource consumption to minimize resource snatching. The scheme that is presented in [3] replaces the FDM-based peer discovery process in [2], with a non-orthogonal multiple-access scheme that is referred to as Trellis Tone Modulation Multiple-Access (TTMMA). TTMMA uses single-tone transmissions and achieves long-distance discovery due to its low Peak-to-Average Power Ratio (PAPR). Furthermore, it makes a higher discovery capacity through its use of a non-orthogonal resource assignment mechanism possible. Based on the assumption of a synchronized superframe structure among the UEs, the peer discovery mechanism in [4] uses both a common channel and a group of channels to send beacons subject to certain rules and procedures that are designed to minimize the discovery time. To ensure compatibility with the standard LTE protocol, the UEs contained in [5] perform peer discovery by listening to the Sounding Reference Signal (SRS) channel, being originally designed for data uploading purposes in LTE networks and peer UEs can access it. However, the method is only capable of detecting the presence of peers, i.e., not active peers that are interested in D2D discovery and communications. Accordingly, the mechanism in [6] exploits the standardized uplink signal structure in the SRS channel to accomplish both the detection of the active peers and the identification of their beacons. Finally, in the peer discovery method that was proposed in [7], the UEs reside in one of five different states, namely, *Keep Alive, Advertise, Discover, Follow,* and *Passive*, depending on their behavior. The UEs in each state then employ a particular beacon rate that is chosen in advance in such a way as to maximize the power saving in the network.

The literature contains various proposals for network-assisted peer discovery methods [8–12]. In [8], the UEs advertise their presence using a random access mechanism and the BS accepts the D2D requests, allocates resource blocks (RB), and chooses an appropriate beacon rate for each UE, depending on the number of requests that they produce. In the peer discovery method that is proposed in [9], the UEs authorized to perform discovery in a given discovery interval contend to transmit their beacons in a time-frequency multiplexed pool of network-allocated resources. In [9], it is assumed that the transmitted beacons are always successful, i.e., other peers can successfully receive them. However, the method in [10] considers a more realistic network model that is based on the Poisson Point Process (PPP), in which the effect of the Channel State Information (CSI) on the performance of the D2D discovery process is taken into account by considering the interference that is imposed on the D2D pairs by nearby cellular users. In [11], each UE sends a preamble to the nearby UEs via a newly-introduced physical channel, and the UEs that receive this preamble send a corresponding report message to the BS by means of a normal random access procedure. The BS then allocates an uplink RB for each reported preamble, such that the UEs that initially sent the preamble can send a further report message to the BS. Finally, the BS, by comparing their reported preambles, identifies pairs of UEs that are in close proximity to one another. The method in [12] exploits the fact that some of the channel components of the UEs are spatially correlated to enable the BS to make a rough estimate of the UE locations by measuring these components. The BS then schedules the transmissions of the beacons that are sent by nearby UEs, such that nearby UEs transmit their beacons at similar times.

Network-controlled methods have several important advantages over Autonomous and Network-Assisted methods, including low power consumption, reduced interference, and a low beacon transmission cost. Consequently, the authors in [13] proposed a peer discovery method based on the Angle of Arrival (AOA) measurements that were obtained by multiple BSs and further analyzed the performance of network-controlled D2D discovery in random spatial networks. The authors in [14] proposed a centralized novel approach, called ROOMMATEs, which utilizes the ubiquitous WiFi network, which combines with BS for indoor peer discovery. ROOMMATEs is an unsupervised approach that can provide different granularity location information. However, none of the studies in [2–14] consider the potential for improving the peer discovery performance by exploiting the social behaviors of the UEs in the network.

In fact, the literature contains very few proposals for social-aware peer discovery mechanisms [17–21]. Among those methods that have been proposed, three schemes [17–19] adopt an autonomous approach. The method in [17] focuses on the security of the data transmissions and it chooses UEs with high social ties (i.e., high trust) to perform D2D communications. However, by adopting such an approach, it is possible that no UEs may be available for pairing. Consequently, the authors in [18] proposed a two-hop pairing process, in which any UE failing to find a trusted UE with its wanted contents for D2D communications sends a request to all the one-hop neighbors of its trusted UEs. The method in [19] uses three key social attributes, namely the trust degree, the similarity degree between UEs, and the center degree of each UE, to construct a unified metric with which to construct neighbor lists for peer discovery. Reference [20] addresses the content delivery problem that is related to optimization of peer discovery and resource allocation by combining both the social and physical layer information in D2D networks. The social relationship, which is modeled as the probability of selecting similar contents, is used as a weight to characterize the impact of social features on D2D pair formation and content sharing. The peer discovery in the physical layer depends on the received signal power by UEs, while the social relationship is mainly used to optimize the context delivery. The method in [21], which is called Social-Based Grouping (SBG), adopts a network-assisted approach to perform peer discovery. UEs are grouped based on their social feature and centrality, and the UEs in the same group are assigned the same beacon rate. The BS, in accordance with the rate at which each UE contacts the other UEs in the network, determines the beacon rate. The UE with more contacts will send beacons at a higher rate.

Similar to SBG, our study also focuses on the network-assisted approach to perform peer discovery. However, between [21] and our study, there are some big differences: (1) SBG determines the beacon rate with only considering the social feature, while ERRD determines it with not only considering the social feature, but also the harvested power. (2) The UEs in SBG are grouped and the UEs in the same group are assigned the same beacon rate due to its high complexity, while each UE in ERRD has its individual beacon rate due to its low complexity. (3) Since ERRD extra considers the harvested power to determine more suitable beacon rate, its PDR can be significantly better than that of SBG. This can be easily observed in Section 5.

3. Problem Description

This section commences by introducing the system model and associated notations. The problem statement is then formally defined.

3.1. System Model

The interval between the time t at which UE i comes within range of UE j and the time at which it was last within range of UE j, t_0, is referred to as the D2D contact interval of the two UEs, $CI_{i,j}$, and it is defined as

$$CI_{i,j} = \min_t \left\{ (t - t_0) : \|L_i - L_j\| \le R_{i,j}, \ t > t_0 \right\}, \tag{1}$$

where $\|.\|$ denotes the distance between the two UEs; L_i and L_j are the locations of UEs i and j, respectively; and, $R_{i,j}$ is the coverage range between the two UEs.

Based on the contact interval between the two UEs, the D2D contact rate between them, denoted as $\lambda_{i,j}$, is defined as

$$\lambda_{i,j} = \frac{1}{E\left[CI_{i,j}\right]}, \tag{2}$$

where $E[.]$ denotes the expectation. The average contact rate of UE i, denoted as λ_i, is then computed as

$$\lambda_i = \frac{\sum_{j=1, j \ne i}^{N} \lambda_{i,j}}{N - 1}, \tag{3}$$

where N is the total number of UEs in the network. Let $CI_{i,j}$ follow a cumulative distribution function (CDF) of $F_{CI_{i,j}}(x)$, with rate $\lambda_{i,j}$. Assume further that $F_{CI_{i,j}}(x)$ is a uniform distribution, and can thus be written as

$$F_{CI_{i,j}}(x) = \begin{cases} 0, & x < 0, \\ \frac{x}{2E\left[CI_{i,j}\right]}, & 0 < x < 2E\left[CI_{i,j}\right], \\ 1, & x > 2E\left[CI_{i,j}\right]. \end{cases} \tag{4}$$

Let the social ratio of UE i be defined as the square root of its average contact rate over the sum of the square root of each average contact rate, i.e.,

$$r_i = \frac{\lambda_i^{\frac{1}{2}}}{\sum_{j=1}^{N} \lambda_j^{\frac{1}{2}}}. \tag{5}$$

In modeling the energy-harvesting environment, it is assumed that the charging devices (CDs) and BS convert their power into RF signals. Furthermore, any UEs within the coverage of these CDs or the BS acquire this RF signal and then convert it into power through special equipment. The power that is obtained by UE i from energy harvesting, denoted as OP_i, is thus calculated as

$$OP_i = \sigma(CDP_m || L_i - CDL_{m, \, i \in CDR_m} ||^{-v}), \tag{6}$$

where σ is an energy harvesting efficiency factor that reflects the ability of UE i to change the RF signal into power; v is the path loss exponent that is caused by interference, which increases with an increasing distance; and, L_i is the location of UE i. In addition, CDP_m, CDR_m, and CDL_m are the transmitted power, coverage, and location, respectively, of CD m, when UE i lies within its coverage [30–32].

3.2. Problem Statement

Before formally defining the problem statement, the used notations are listed in Table 2. As shown, the notations fall into six categories that relate to the system, range, power, location, contact, and beacon, respectively. The system parameters define the number of UEs in the network and the total peer discovery time, respectively, while the range parameters describe the coverage of the devices. The power parameters define the transmission powers of the devices and the various power variables that are used in the ERRD model. The location parameters define the positions of the devices. The contact parameters describe the contact behaviors of the UEs. Finally, the beacon parameters define the beacon rates of the UEs.

Table 2. Notation table.

Category	Notation	Description	Property
System	N	Number of UEs	Input
	T	Total time of peer discovery	Input
Range	$R_{i,j}$	Converge between UE i and j	Input
	CDR_m	Coverage of CD m or BS ($m = 0$ represents BS)	Input
Power	BP	Budget power	Input
	TP	Consumed power in sending a beacon	Input
	CDP_m	Transmission power of CD m or BS ($m = 0$ represents BS)	Input
	$\sigma \in (0,1]$	Energy harvesting efficiency factor	Input
	v	Path loss exponent	Input
	P_i	Owned power of UE i	Input
	OP_i	Obtained power of UE i	Variable
	GP_i	Allocated power of UE i	Variable
	AP	Total power allocated to UEs	Variable
Location	L_i	Location of UE i	Input
	CDL_m	Location of CD m or BS ($m = 0$ represents BS)	Input
Contact	$CI_{i,j}$	Contact interval of UEs i and j	Input
	$\lambda_{i,j}$	Contact rate of UEs i and j	Variable
	λ_i	Average contact rate of UE i	Variable
	r_i	Social ratio of UE i	Variable
	OPR_i	Ratio of OP_i over r_i	Variable
Beacon	μ_i	Beacon rate of UE i	Output

The energy harvesting technology that is considered in this study is wireless power transfer (WPT) via radio frequency (RF), as shown in Figure 1 [23]. The basic idea is that the RF transmitter transmits radio signal towards the receiving antenna at the desired frequency and power level. The RF receiver then applies electromagnetic radiation to charge the battery, that is, the receiving antenna receives the traveling signal and the rectifier converts the alternating current (AC) to direct current (DC) to charge the battery. The energy carrier can be located at 2.4 GHz or 5 GHz frequency band under the considerations that these bands are internationally reserved for Industrial, Scientific, and Medical (ISM) purposes. Energy harvesting can explore sufficient power sources in significant radio coverage by increasing the transmitting power under the regulation of the government.

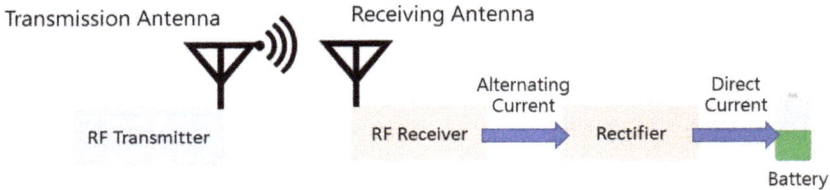

Figure 1. The energy harvesting technology considered in this study.

This study attempts to determine the suitable beacon rates of all UEs to explore most peers in a D2D network under the constraints of budget power, the limitation of overall power consumption. This study further introduces a new scenario, *i.e.*, energy harvesting, which previous studies did not consider. The energy harvesting enables UEs within charging areas to procure additional power from surrounding energy sources. Thus, under considering the UE may be out-of-energy and the study determines the suitable beacon rates of all UEs to maximize PDR according to the budget power and harvested power.

The problem that we investigated is formally described, as follows. First, for a beacon rate of UE i equal to μ_i, the peer discovery ratio (PDR) of the UE can be calculated as [33]

$$\text{PDR} = 1 - 2\mu_i^2 \int_0^{\frac{1}{\mu_i}} \int_0^y F_{CI_{i,j}}(x)dxdy. \tag{7}$$

The second term denotes the missing probability and the probability UE i cannot detect the contact. In other words, it is the probability that the UE i cannot find other peers.

According to the arrival contact rate of each UE, the aim of the peer discovery process is to maximize the total PDR, as

$$\sum_{i=1}^{N} \lambda_i \left(1 - 2\mu_i^2 \int_0^{\frac{1}{\mu_i}} \int_0^y F_{CI_{i,j}}(x)dxdy \right).s.t. \sum_{1}^{N} \mu_i \times TP \le BP, \tag{8}$$

where TP is the transmission power used to send a beacon message and BP is the total budget power quantum of the network. However, if a UE is out-of-energy, it cannot find any peers in the network and any other peers cannot find it. Thus, in performing the peer discovery process, the objective given in Equation (8) should be modified, as follows:

$$\text{max PDR} = \sum_{i=1}^{N} \begin{cases} \lambda_i \left(1 - 2\mu_i^2 \int_0^{\frac{1}{\mu_i}} \int_0^y F_{CI_{i,j}}(x)dxdy \right), & P_i > 0 \\ 0, & P_i \le 0 \end{cases} \tag{9}$$

$$s.t. \ \textstyle\sum_1^N \mu_i \times TP \le BP,$$

where P_i is the power reserved for peer discovery of UE i. Differing from Equation (8), the UE cannot find other peers and it cannot be found by any other peers when $P_i \le 0$ in Equation (9), because it is a more reasonable condition. Also note that the constraint denotes that the total power consumption must be less than the budget power quantum.

Thus, the problem statement is formally given, as follows.

Sensors **2019**, *19*, 2304

Input: the parameters marked input in Table 2.

Output: the beacon rate of UE i, μ_i, $\forall i$.

Objective: max PDR $= \sum\limits_{i=1}^{N} \begin{cases} \lambda_i\left(1 - 2\mu_i^2 \int_0^{\frac{1}{\mu_i}} \int_0^y F_{CI_{i,j}}(x)dxdy\right), & P_i > 0 \\ 0, & P_i \leq 0. \end{cases}$,

Constraint: $\sum\limits_{1}^{N} \mu_i \times TP \leq BP$.

4. Solutions

Before introducing the ERRD algorithm in detail, the overall operations of UEs and BS are first described. The peer discovery is conducted per time period T. At the beginning of time period, each UE will send its ID and its harvested power quantum to BS. Thus, BS will know which UEs in its coverage and immediately extracts their social features from the database. According to UEs' social features and the harvested power quantum, and the pre-determined budget power quantum, BS can run ERRD to determine the beacon rate for each UE, and then sends the determined beacon rate to the corresponding UE. After receiving this rate, the UE will send the beacons accordingly. Since each UE only calculates the harvested power quantum, according to Equation (8), its operation is very simple. Therefore, below, we focus on describing the algorithm in BS, i.e., ERRD.

In D2D communications, the beacon rate is proportional to the amount of consumed power. Hence, the ERRD algorithm that is proposed in this study first virtually allocates a proper power quantum to each UE and then determines the corresponding beacon rate according to this allocated power quantum. As described in the following sub-sections, ERRD comprises three steps, namely *Social Ratio Allocation* (SRA), *Energy Ratio Allocation* (ERA), and *Beacon Rate Decision* (BRD). SRA first allocates the budget power among the UEs, depending on their social ratios. For each UE, ERA then adjusts this allocated power quantum according to the amount of energy that is harvested by the UE. Finally, BRD computes the beacon rate of each UE, depending on the adjusted allocated power quantum.

Below, the concept of ERRD is first explained. As ERRD is composed of three steps, we describe SRA, ERA, and BRD in sequence in Section 4.1. After describing the concept of ERRD, we formally exhibit the pseudo code of ERRD in Section 4.2. Finally, an example to illustrate the overall ERRD operation is given in Section 4.3.

4.1. ERRD Algorithm

4.1.1. Social Ratio Allocation

Let AP be the total allocated power quantum and GP_i be the amount of power that is allocated to UE i. Since the peer discovery process is subject to the constraint that the total consumed power must be less than or equal to budget power BP, AP is initialized as BP. According to [21], when $F_{CI_{i,j}}(x)$ has a uniform distribution (as shown in Equation (4)), the maximum PDR is achieved when the beacon rate of each UE is set proportional to its social ratio, i.e., $\mu_i \propto r_i$. Thus, the initially allocated power quantum for every UE is given by $GP_i = AP \times r_i$.

4.1.2. Energy Ratio Allocation

The ERA step is the most critical step in ERRD. In order to properly explain the step, it is appropriate to introduce the following intuitive thought regarding the approach for allocating the power quantum, depending on the amount of power that is obtained from energy harvesting. The smart approach employed by ERRD is then introduced.

- Intuitive thought

The SRA step in the ERRD algorithm allocates the budget power quantum among the UEs based on their social ratios. However, some UEs can obtain power from energy harvesting, and hence their obtained power quantum OP_i may exceed the allocated power quantum GP_i. In this case, the UEs can utilize the obtained power quantum, rather than the originally allocated power quantum to perform their beacon transmissions. The unused portion of the allocated power quantum can then be returned to AP for the re-allocation to other UEs. Conversely, if OP_i is less than GP_i, i.e., the UE only acquires little (or no) energy via harvesting, the UE uses the original allocated power quantum GP_i and returns its obtained power quantum OP_i to AP. Combining these two cases, the unused power quantum of UE i is given by the minimum of OP_i and GP_i, i.e., $\min(OP_i, GP_i)$.

Since AP and GP_i are dependent, it is necessary to obtain them alternately while using an iterative approach. Let x^k denote the value of x in the k-th iteration and $k = 0$ denote the initial value. Therefore, the initial value of AP, AP^0, is set as the budget power quantum, BP, and it is allocated to each UE in accordance with r_i, i.e., $P_i^0 = AP^0 \times r_i$. Any unused power quantum must be returned to AP. Consequently, AP is equal to the sum of BP and the returned power quantum of all the UEs, i.e.,

$$AP^{k+1} = BP + \sum_{i=1}^{N} \min\left(OP_i, GP_i^k\right), \ k \geq 0. \tag{10}$$

Once AP^{k+1} is calculated, the new allocated power quantum of each UE can be obtained as $GP_i^{k+1} = AP^{k+1} \times r_i$, and used to calculate AP^{k+2} accordingly. As shown in Lemma 1, AP is non-decreasing as the number of iterations increases.

Lemma 1: $AP^{k+1} \geq AP^k, \ \forall k \geq 0$.

Proof.

1. As $AP^0 = BP$ and $AP^1 = BP + \sum_{i=1}^{N} \min(OP_i + GP_i^0)$, it follows that $AP^1 \geq AP^0$ since all OP_i and GP_i^0 are non-negative. Thus, $k = 1$ holds.

2. Suppose that $k = n - 1$ holds, i.e., $AP^n \geq AP^{n-1}$. $AP^n \geq AP^{n-1}$ implies that $GP_i^n \geq GP_i^{n-1}$, $1 \leq i \leq N$ since $GP_i^n = AP^n \times r_i$. Hence,

$$BP + \sum_{i=1}^{N} \min(OP_i, GP_i^n) \geq BP + \sum_{i=1}^{N} \min(OP_i, GP_i^{n-1})$$
$$\Rightarrow AP^{n+1} \geq AP^n$$

   Thus, $k = n$ holds.
3. From mathematical induction, $AP^{k+1} \geq AP^k, \ \forall k \geq 0$. $\quad \square$

Intuitively, AP^k can be iteratively calculated until no further change in its value is obtained, i.e., $AP^{k+1} = AP^k$. However, while such an approach is technically feasible, it requires many iterations to converge, causing high complexity, and it is hence impractical for real-world peer discovery applications. Consequently, the following smart approach is proposed instead.

- Smart approach

According to the relationship between OP_i and GP_i, ERRD classifies the UEs into two groups, namely G_1 when $OP_i < GP_i$ and G_2 when $OP_i \geq GP_i$. Therefore, Equation (10) can be re-formulated as

$$AP^{k+1} = BP + \sum_{i \in G_1^k} OP_i + \sum_{i \in G_2^k} GP_i^k. \tag{11}$$

Lemma 2: *In each iteration, the UEs in G_2 may be shifted to G_1, but the UEs in G_1 cannot be shifted to G_2.*

Proof.

1. From Lemma 1, AP is non-decreasing and GP_i is also non-decreasing, i.e., $GP_i^{k+1} \geq GP_i^k$, $\forall i$.
2. Since OP_i is fixed, the condition $OP_i < GP_i^k$ implies that $OP_i < GP_i^{k+1}$. Therefore, any member in G_1 will not be shifted to G_2.
3. Consequently, in each iteration, only the UEs in G_2 may be shifted to G_1, but no UEs in G_1 will be shifted to G_2. □

From Lemma 2, the ERA process seeks to shift any UEs belonging to G_2 to G_1 if possible in each iteration in order to reduce the convergence time. However, to achieve this, two issues must first be addressed, namely (1) which UE should be considered first to be shifted and (2) whether this UE can actually be shifted.

ERA determines the answer to the first issue by inspecting the ratio of the obtained power quantum over the corresponding social ratio, i.e., $OPR_i = OP_i/r_i$. A smaller value of OPR_i implies that OP_i is more likely to be less than GP_i. In other words, UE i is more likely to belong to G_1. Thus, in the k-th iteration, according to the current grouping, UE m^k, whose OPR_{m^k} is the smallest among those of all the UEs in G_2^k, is chosen as the pivot UE, and it is most likely to be shifted from G_2^k to G_1^k, as described in the following lemma.

Lemma 3: *The pivot UE m^k among G_2^k is most likely be shifted to G_1^k.*

Proof.

1. $OPR_i = \frac{OP_i}{r_i} \Rightarrow \frac{OPR_i}{AP^k} = \frac{OP_i}{AP^k \times r_i} \Rightarrow \frac{OPR_i}{AP^k} = \frac{OP_i}{GP_i^k} \Rightarrow OPR_i = \frac{OP_i}{GP_i^k} AP^k$.

2. In the k-th iteration, as AP^k is fixed, UE m^k has the smallest OPR in G_2, i.e., $\frac{OP_{m^k}}{GP_{m^k}^k}$ is the smallest, and hence OP_{m^k} is most likely to be less than $GP_{m^k}^k$. Consequently, UE m^k is most likely to be shifted from G_2^k to G_1^k. □

ERA considers whether or not this UE can actually be shifted after determining the pivot UE. To achieve this, a virtual critical point, defined as $OP_{m^k} = GP_{m^k}$, is considered. At this critical point, the virtual total allocated power quantum, VAP^k, obtained using m^k as the baseline, is given as

$$
\begin{aligned}
VAP^k &= BP + \sum_{i \in G_1^k} OP_i + \sum_{i \in G_2^k} GP_i^k \\
&= BP + \sum_{i \in G_1^k} OP_i + \sum_{i \in G_2^k} GP_{m^k} \times \frac{r_i}{r_{m^k}} \\
&= BP + \sum_{i \in G_1^k} OP_i + \sum_{i \in G_2^k} OP_{m^k} \times \frac{r_i}{r_{m^k}}.
\end{aligned}
\tag{12}
$$

The virtual allocated power quantum for m^k is then obtained as $VGP_{m^k}^k = VAP^k \times r_{m^k}$. The virtual critical point represents the threshold between G_2^k and G_1^k. Thus, checking whether $VGP_{m^k}^k$ is more than OP_{m^k} provides an efficient means of determining whether UE m^k belongs to G_2^k or G_1^k. If $VGP_{m^k}^k$ is larger than OP_{m^k}, the pivot UE m^k should be shifted from G_2^k to G_1^k and the iteration process should continue to the next round. Otherwise, the pivot UE m^k should remain in G_2^k. As the pivot UE m^k among G_2^k is the most likely be shifted to G_1^k, the other UEs belonging to G_2^k also remain in G_2^k. In other words, the grouping process is complete and no further changes in the memberships of G_1 and G_2 are required.

Once the grouping process is finished (in iteration *), GP_{m^*} can be computed as

$$\frac{GP^*_{m^*}}{r_{m^*}} = BP + \sum_{i \in G^*_1} OP_i + \sum_{i \in G^*_2} GP^*_{m^*} \times \frac{r_i}{r_{m^*}}. \tag{13}$$

After solving Equation (13), $GP^*_{m^*}$ can be obtained as

$$GP^*_{m^*} = \frac{r_{m^*} \times \left(BP + \sum_{i \in G^*_1} OP_i \right)}{1 - \sum_{i \in G^*_2} r_i}. \tag{14}$$

AP^* can then be computed as $AP^* = GP^*_{m^*}/r_{m^*}$. In addition, each GP^*_i can be computed with AP^*. Finally, the allocated power quantum for each UE is set as the maximum between GP^*_i and OP_i.

4.1.3. Beacon Rate Decision

In the BRD step, since OP_i is considered, the sum of all the allocated power quanta is equal to the budget power quantum plus all the obtained power quanta, as shown in Lemma 4.

Lemma 4: $\sum_{i=1}^{N} \max\left(OP_i, GP^*_i\right) = BP + \sum_{i=1}^{N} OP_i$.

Proof.

1. For UE i in G_2, the allocated power quantum is GP^*_i. As the UE has an obtained power quantum of OP_i, the allocated power quantum occupying the budget power quantum is equal to $GP^*_i - OP_i$. Thus, it follows that $\sum_{i \in G^*_2} \left(GP^*_i - OP_i \right) = BP$.

2. $$\sum_{i=1}^{N} \max\left(OP_i, GP^*_i\right) = \sum_{i \in G^*_1} OP_i + \sum_{i \in G^*_2} GP^*_i$$
    $$= \sum_{i \in G^*_1} OP_i + \sum_{i \in G^*_2} \left(GP^*_i - OP_i \right) + \sum_{i \in G^*_2} OP_i$$
    $$= \sum_{i=1}^{N} OP_i + \sum_{i \in G^*_2} \left(GP^*_i - OP_i \right) = BP + \sum_{i=1}^{N} OP_i \qquad \square$$

Due to this value exceeding the BP quantum, the final allocated power quantum should be normalized through multiplication by R, i.e., the ratio of BP over the sum of the allocated power quanta. That is,

$$R = \frac{BP}{\sum_{i=1}^{N} \max\left(OP_i, GP^*_i\right)} = \frac{BP}{BP + \sum_{i=1}^{N} OP_i}. \tag{15}$$

Finally, the μ_i of each UE i is computed as the allocated power quantum multiplied by R and divided by TP.

4.2. Pseudo Code

Algorithm 1 shows the pseudo code of the ERRD algorithm, where lines 1–7 correspond to the SRA step, lines 8–24 describe the ERA step, and lines 25–29 are the BRD step. In the SRA step, lines 1–7 initialize the variables and compute r_i, GP_i, and OPR_i for all the UEs. In the ERA step, lines 8–12 classify the UEs into two groups. In particular, UEs with an OP_i greater than or equal to GP_i are shifted into group G_2. Lines 15–17 then select UE m whose OPR_m is the smallest among those of all the UEs in G_2 as the pivot UE and compute VAP and VGP_m accordingly. Line 18 checks whether the pivot UE meets the condition that OP_m is larger than VGP_m. If the condition holds, the final AP is computed in lines 19 and 20. Otherwise, lines 14–23 are repeated iteratively until this condition is reached. Finally, in the BRD step, R is calculated in line 25. Line 27 then computes GP_i based on the final AP obtained in the ERA process and line 28 normalizes the allocated power quantum and computes the normalized μ_i.

Algorithm 1 Energy Ratio Rate Allocation Algorithm

---

1 : $AP \leftarrow BP$, $G_1 \leftarrow \{1 \sim N\}$, $G_2 \leftarrow \varnothing$

2 : Calculate all λ_i from $CI_{i,j}$; $\lambda_{ALL} \leftarrow \sum\limits_{i=1}^{N} \lambda_i^{\frac{1}{2}}$

3 : **for** $i = 1$ to N **do**

4 : $r_i \leftarrow \frac{\lambda_i^{\frac{1}{2}}}{\lambda_{ALL}}$

5 : $GP_i \leftarrow AP \times r_i$

6 : $OPR_i \leftarrow \frac{OP_i}{r_i}$

7 : **end for**

8 :   **for** $i = 1$ to N **do**

9 :   **if** $OP_i \geq GP_i$ **then**

10 : $G_2 \leftarrow G_2 \cup i$, $G_1 \leftarrow G_1 - i$

11 : **end if**

12 : **end for**

13 : **if** $G_2 \neq \varnothing$ **then**

14 :   **repeat**

15 : $m \leftarrow \text{argmin } OPR_i$, $\forall i \in G_2$

16 : $VAP \leftarrow BP + \left(\sum\limits_{i \in G_1} OP_i \right)$

         $+ \left(\sum\limits_{i \in G_2} OP_m \times \frac{r_i}{r_m} \right)$

17 : $VGP_m \leftarrow VAP \times r_m$

18 :   **if** $OP_m \geq VGP_m$ **then**

19 : $GP_m \leftarrow \frac{r_m \times \left(BP + \sum_{i \in G_1} OP_i \right)}{1 - \sum_{i \in G_2} r_i}$

20 : $AP \leftarrow \frac{GP_m}{r_m}$

21 :   **end if**

22 : $G_1 \leftarrow G_1 \cup m$, $G_2 \leftarrow G_2 - m$

23 : **until** $OP_m \geq VGP_m$ **or** $G_2 = \varnothing$

24 : **end if**

25 : $R \leftarrow \frac{BP}{BP + \sum_{i=1}^{N} OP_i}$

26 : **for** $i = 1$ to N **do**

27 : $GP_i \leftarrow AP \times r_i$

28 : $\mu_i \leftarrow \max(OP_i, GP_i) \times \frac{R}{TP}$

29 : **end for**

Note that the pseudo code of ERRD is run in BS at the beginning of each time period T. All allocating, returning, and re-allocating power quanta from the BS to the UEs are virtually calculated in BS, rather than the real transfer between BS and UEs. Therefore, the communication between BS and UEs happens when each UE sends its ID and its harvested power quantum to BS at the beginning of each time period, and BS sends the determined beacon rate to the corresponding UE after it has executed ERRD.

The time complexity of ERRD is calculated, as follows. The lines 1–2, lines 3–7, lines 8–12, line 25, and lines 26–28 individually requires O(N). The repeat loop of lines 14–23 are executed as most N times, because at least a UE will be shifted from G_2 into G_1 in each iteration. The time complexities of calculating m in line 15, VAP in line 16, and GP_m in line 19 are O(N). Thus, the time complexity of the repeat loop is O(N^2). Therefore, ERRD has low time complexity O(N^2) and it can be implemented in a real-time environment.

4.3. Illustrative Example

The following discussions present an illustrative example to demonstrate the detailed operational steps of ERRD. It is assumed that the network contains five UEs with average social contacts, λ_i, of 1, 4, 9, 16, and 25, respectively. It is further assumed that the OP_i values of the five UEs are 5, 4, 5, 3, and 1, respectively. Finally, the total BP is assumed to be 15 and TP is set as 1.

The SRA step first computes r_i in accordance with the average social contacts, λ_i, i.e., $r_1 = \frac{\sqrt{1}}{\sqrt{1} + \sqrt{4} + \sqrt{9} + \sqrt{16} + \sqrt{25}} = \frac{1}{15}$. Thus, r_1, r_2, r_3, r_4, and r_5 are obtained as $\frac{1}{15}$, $\frac{2}{15}$, $\frac{3}{15}$, $\frac{4}{15}$, and $\frac{5}{15}$, respectively (see Table 3). For each UE, GP_i is then computed as the product of AP, which is initialized as BP, and r_i, i.e., $GP_i = AP \times r_i$. In other words, GP_1, GP_2, GP_3, GP_4 and GP_5 are obtained as 1, 2, 3, 4, and 5, respectively. In addition, OPR_i is calculated as OP_i/r_i. For example, $OPR_1 = \frac{5}{\frac{1}{15}} = 75$. The UEs are then classified into two groups, namely G_1 or G_2, by comparing OP_i with GP_i. In the present example, OP_1, OP_2, and OP_3 are greater than GP_1, GP_2, and GP_3, respectively, while OP_4 and OP_5 are less than GP_4 and GP_5, respectively. Consequently, group G_1 is determined to be {4, 5}, while group G_2 is obtained as {1, 2, 3}.

Table 3. Initialized values.

UE ID	r_i	OP_i	GP_i	OPR_i	Set
1	$\frac{1}{15}$	5	1	75	G_2
2	$\frac{2}{15}$	4	2	30	G_2
3	$\frac{3}{15}$	5	3	25	G_2
4	$\frac{4}{15}$	3	4	$\frac{45}{4}$	G_1
5	$\frac{5}{15}$	1	5	3	G_1

Based on the results that are presented in Table 3, the ERA step selects UE 3 as the pivot UE, since the value of OPR_3 is the smallest among all of the OPR values in G_2. To check whether the grouping process is finished, ERA uses the ratio of r_i over r_3 to compute VAP as $BP + OP_5 + OP_4 + OP_3 \times \frac{r_3}{r_3} + OP_3 \times \frac{r_2}{r_3} + OP_3 \times \frac{r_1}{r_3} = 15 + 3 + 1 + \frac{5}{3} + \frac{10}{3} + 5 = 29$. The VAP result is then used to compute VGP_3 as $29 \times \frac{3}{15} = \frac{29}{5}$. OP_3 is less than VGP_3, that is, the virtually allocated power quantum exceeds the obtained power quantum. Hence, UE 3 is shifted from group G_2 to group G_1.

The procedure that is described above is iteratively repeated until the grouping process is complete. For the present example, G_1^* is obtained as {3, 4, 5} and G_2^*, as {1, 2}. Once the grouping process is finished, GP_2^* is GP_m^* and in accordance with Equation (14) (line 19 in the algorithm), GP_2^* is computed as $\frac{\frac{2}{15} \times (15+5+3+1)}{1-\left(\frac{1}{15}+\frac{2}{15}\right)} = 4$. Thus, AP^* is obtained as $\frac{4}{\frac{2}{15}} = 30$ (line 20). Finally, the values of GP_i^* are computed using the determined value of AP^*. Choosing the larger value between OP_i and GP_i^* for each UE, the total consumed power quantum is obtained as $5 + 4 + 6 + 8 + 10 = 33$, which is equal to the sum of BP and all OP_i (15+5+4+5+3+1). As the total consumed power quantum is larger than BP, normalization by $R = \frac{15}{33} = \frac{5}{11}$ is performed and used to compute μ_i. For example, $\mu_1 = 5 \times \frac{5}{11} \times \frac{1}{TP} = \frac{25}{11}$. Table 4 shows the final results for all the UEs.

Table 4. Final results of the illustrative example.

UE ID	r_i	OP_i	GP_i	Maximum	μ_i
1	$\frac{1}{15}$	5	2	OP_i: 5	$\frac{25}{11}$
2	$\frac{2}{15}$	4	4	OP_i: 4	$\frac{20}{11}$
3	$\frac{3}{15}$	5	6	GP_i: 6	$\frac{30}{11}$
4	$\frac{4}{15}$	3	8	GP_i: 8	$\frac{40}{11}$
5	$\frac{5}{15}$	1	10	GP_i: 10	$\frac{50}{11}$

5. Performance Evaluation

The PDR performance of the proposed ERRD algorithm was compared with that of the previously reported Social-Based Grouping (SBG) algorithm [21]. In SBG, the UEs are grouped based on their social feature: centrality and the UEs in the same group are assigned the same beacon rate. In the present simulations, SBG classifies the UEs into three groups.

5.1. Dataset and Environment

The simulations were performed using the *Infocom06* user mobility trace [34], which consisted of the D2D communication contacts of 98 individuals that were recorded over the *IEEE Infocom Conference* in 2006. The first half of the dataset was used to calculate the beacon rates from the contact rates of the UEs, while the second half was used to evaluate and compare the performance of the two schemes (ERRD and SBG). The dataset contains no information regarding the actual physical locations of the users and the BS. Consequently, in performing the simulations, an artificial environment for energy harvesting was created, with dimensions of 600 × 600 m^2. The simulation field was partitioned into a 3 × 3 grid containing a BS with a 10-watt power in the center and eight CDs with a five-watt power distributed around the outside (see Figure 2). The UEs were uniformly deployed in the 3 × 3 grid initially and their locations were then randomly moved as the simulations proceeded to simulate the mobility of UEs. Thus, the UEs obtained harvested power from different CDs (or the BS) at different points in the simulation process.

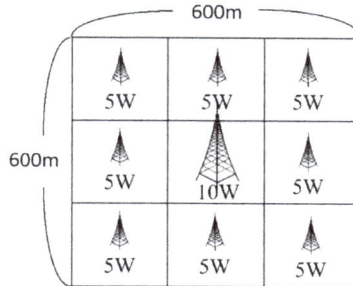

Figure 2. Simulation environment.

The adopted dataset is the same as the dataset used in [21], because it is the most popular realistic dataset used in D2D communications, although it does not include the actual locations of UEs. To the best of our knowledge, there is no typical dataset that includes UE contacts and UE locations. Thus, in this study, we adopted the dataset in [21] and simulated the UE locations and their mobility by ourselves.

Each simulation was run for 1000 seconds. The plotted simulation results were then computed as the average value obtained over 100 simulations that were performed under identical conditions. The default environmental parameters were set as shown in Table 5. In accordance with the *Infocom06* dataset, the number of UEs was set as $N = 98$. Moreover, the path loss exponent, v, of the BS and CD RF transmissions was set as 2, the energy harvesting efficiency factor, σ, was set as 0.7, and the transmission power of the beacon messages was set as 20 mW. The *BP* of the peer discovery process (with a duration of 1000 s) was set as 10 W. Finally, the mobility (moving speed) of the UEs was set as 1 under the assumption that the mobility in the *Infocom06* dataset is 1. In general, as the moving speed of the UEs increases, the contact interval between them reduces, and *vice versa*. Thus, for a mobility value that is equal to 2, the contact interval between the UEs is equal to half that in the *Infocom06* dataset.

Some assumptions are made for simplifying the simulations: (1) The harvested power quantum is according to Equation (6), although this quantum will be affected by interferences or other factors in a real environment. (2) The locations of UEs are limited in this area 600×600 m^2. That is, if a UE moves outside this area, then its location will be randomly located within this area. (3) The beacon is perfectly transmitted, i.e., it will not encounter any collision or be interfered by other noises.

The simulations compared the performance of the ERRD algorithm with that of SBG under different settings of the *BP*, *CDP*, and *mobility* parameters. We compared ERRD and SBG, rather than other solutions because of two points. (1) As described in Section 2, ERRD and SBG belong to the same type: social-aware and network-assisted, but other solutions belong to different types. (2) As SBG used social features to adjust the beacon rate, it always outperforms other solutions without social-awareness. The evidence was exhibited in [21]. The UEs were classified into two types to facilitate the comparison between the two schemes, namely those with an increased beacon rate (IBR) and those with a decreased beacon rate (DBR), respectively. In the former case, the UEs using ERRD had a higher beacon rate than those using SBG, while, in the latter case, the beacon rate of the UEs using ERRD was lower than that of those using SBG.

Table 5. Default parameter settings in performance evaluation simulations.

Parameter	Default Value
N	98
T	1000 seconds
v	2
σ	0.7
CDP_m	10 watts (m = 0, i.e., BS) 5 watts (m = 1–8)
CDR_m	100 meters
BP	10 watts
TP	20 milliwatts
P_i	300 milliwatts
Mobility	1

5.2. Effect of BP

Figure 3 shows the PDR that was obtained under the two schemes for various values of the *BP* in the range of 0–20 W. For both schemes, the PDR exhibits a logarithmic-type increase as *BP* increases. This result is reasonable, since, intuitively, as the *BP* initially increases, the beacon rates of the UEs also increase, and hence a greater number of peers can be found. However, as the BP continues to increase, the number of originally unfound peers reduces, and hence the improvement in the PDR also reduces. The contact intervals in the *Infocom06* dataset are not uniformly distributed and some of the intervals are extremely short. Thus, if the beacon rate is assigned a very high value in an attempt to increase the number of discovered peers, the consumed *BP* significantly increases. Therefore, in practical implementations, the *BP* should be set in such a way as to achieve a satisfactory tradeoff between the PDR and the consumed *BP*.

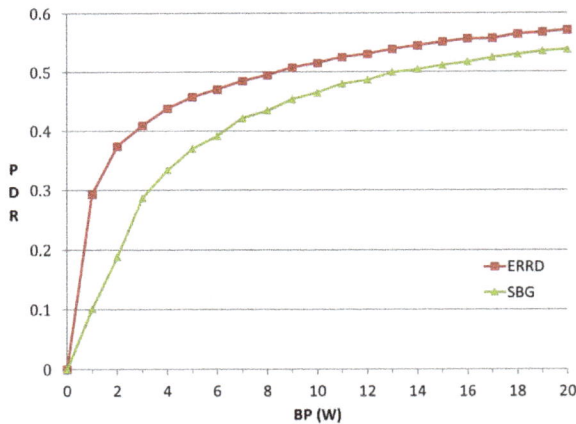

Figure 3. Peer Discovery Ratio (PDR) vs. total budget power quantum of the network (BP).

For a given value of the *BP*, ERRD consistently achieves a higher PDR than SBG. For example, the PDR achieved under ERRD for a *BP* of 1 W is around 190% higher than that obtained under SBG, while for a *BP* of 20 W, the PDR is approximately 8% higher. This performance improvement can be attributed to two main factors. First, the IBR UEs using ERRD can send more beacons than those using SBG, and hence achieve a higher PDR. Second, although the DBR UEs using ERRD send fewer beacons than those using SBG, they have a longer lifetime. Consequently, the UEs can still achieve a higher PDR than those using SBG. However, the performance improvement that is offered by ERRD reduces with an increasing BP. This finding is reasonable, since, under a large *BP*, the amount of harvested energy is relatively smaller than that directly allocated by the BS. In other words, for most UEs, $OP_i < GP_i$, and hence the value of GP_i under ERRD is similar to that under SBG. Consequently, the PDR performance gap between the two schemes reduces.

Some intermediate results can be observed to better understand the differences between ERRD and SBG. First, note that the overall power consumptions for ERRD and SBG are the same as the overall power consumption is always limited by the budget power quantum; so, showing this value is unnecessary. However, the power consumption of each UE is quite different for these two approaches. Thus, we observe two intermediate results: Number of out-of-energy UEs (NoE) and the coefficient of variation (CV) of power in UEs. The NoE represents the number of UEs that power quantum reserved for peer discovery has been exhausted, that is, the power consumption exceeds the value of the power quantum reserved for peer discovery plus the harvested power quantum. The out-of-energy UEs cannot find other peers and cannot be found by other peers. The CV represents the power distribution among the UEs. The lower CV means a better balance of UE battery power.

Figure 4 shows the NoE and CV that were obtained under the two schemes for various values of the *BP*. Observing this figure, NoE increases as *BP* exceeds a threshold. When BP increases, the UE can send more beacons. Therefore, the probability that a UE exhausts its battery power becomes larger, resulting in the increase of NoE. The out-of-energy UE appears when *BP* is 6 for SBG, while the out-of-energy UE appears when *BP* is 16 for ERRD. This is because SBG only considers the social ratio to send the beacons, while ERRD not only considers the social ratio, but also harvested power. When a UE has low harvested power and a high social ratio, ERRD reduces its beacon rate to reduce the probability of out-of-energy. However, in this case, SBG still lets this UE send beacons at a high rate, so it is very likely to be out-of-energy. On the other hand, CVs for ERRD and SBG almost linearly increase as *BP* increases, but SBG has a sharper slope. SBG generates many out-of-energy UEs and many UEs having much power, as they obtain much harvested power and send few beacons, resulting in more unbalanced battery power distribution among UEs, i.e., a higher CV.

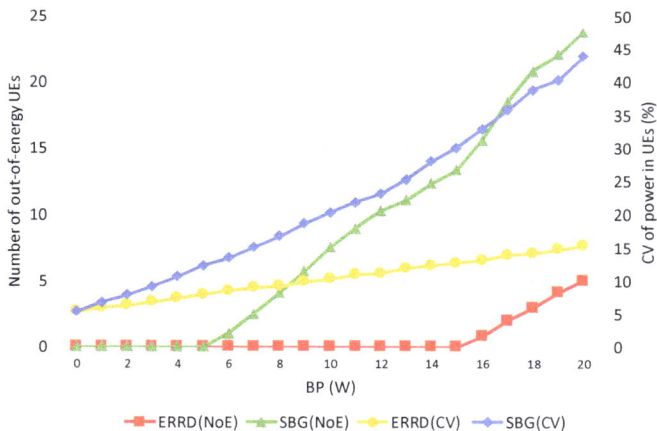

Figure 4. Number of out-of-energy User Equipments (NoE) and coefficient of variation (CV) vs. BP.

5.3. *Effect of CDP*

Figure 5 shows the variation of the PDR with the *CDP*. In general, a larger *CDP* indicates that more UEs can harvest energy, or individual UEs can acquire a greater amount of energy. However, the total consumed power in the network is limited to *BP*, irrespective of the value assigned to *CDP*. In other words, for a larger *CDP*, ERRD cannot consume more power than *BP*, but can only allocate the power quantum to each UE more precisely, such that all of the UEs can more efficiently send their beacons. As shown in Figure 5, the PDR obtained under ERRD rapidly increases as the *CDP* first rises since the beacon rate decision made by ERRD reduces the probability that the UEs run out of energy. For a *CDP* value greater than 2, almost none of the UEs are out of energy. In this case, the PDR slightly increases, since the IBR UEs can find a greater number of peers. However, the PDR is not guaranteed to continuously increase when the number of UEs with an energy-harvesting capability exceeds a certain threshold. Therefore, in implementing the ERRD algorithm, a threshold should be set, whereby when the number of UEs with an energy-harvesting capability is greater than this threshold, ERRD should let some of the UEs store the harvested energy in their batteries, rather than expending it on beacon transmissions in order to provide power for other applications.

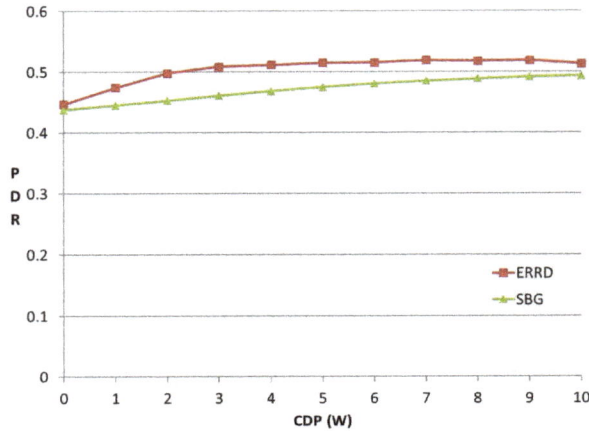

Figure 5. PDR vs. CDP.

For the SBG scheme, the PDR also increases with an increasing *CDP*. However, it increases at a slower rate than under ERRD, since, even though SBG does not explicitly consider the power obtained from energy harvesting, the number of out-of-energy UEs still decreases as the number of UEs having an energy-harvesting capability increases.

Figure 6 shows the NoE and CV that were obtained under the two schemes for various values of the *CDP*. Observing this figure, NoE decreases as *CDP* increases, because UEs can harvest more power. Therefore, the probability that a UE exhausts its battery power becomes lesser, resulting in the decrease of NoE. However, the NoE of SBG is significantly larger than that of ERRD, because of the reasons that are described in Figure 4. On the other hand, CVs for ERRD and SBG almost linearly decrease as *CDP* increases. When *CDP* is small, the UE in the BS coverage still can harvest much power, but the UEs in the CD coverage only harvest less power, resulting in a larger CV. Similar to Figure 4, we can see SBG has a sharper slope of CV than ERRD.

5.4. Effect of Mobility

Figure 7 shows the effect of the UE mobility on the PDR under the two schemes. As expected, the PDR reduces with an increasing mobility for both schemes, since, as the UEs move more rapidly, the contact rate between them increases, and hence the peers are less easily found for a given beacon rate. Nonetheless, the ERRD algorithm consistently outperforms SBG by around 10% for all the values of the UE mobility. Although the duration available for energy harvesting from one particular CD reduces as the UE mobility increases, the chance of harvesting energy from the other CDs increases. As a result, the UE mobility has no significant effect on the amount of harvested energy under the ERRD scheme. A similar tendency also occurs under the SBG scheme. Consequently, the performance advantage of ERRD over SBG is maintained, irrespective of the value of the UE mobility.

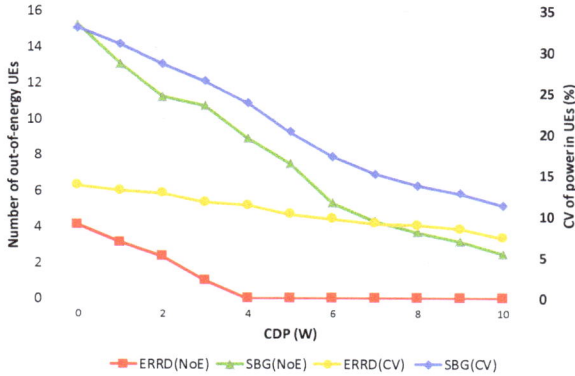

Figure 6. NoE and CV vs. CDP.

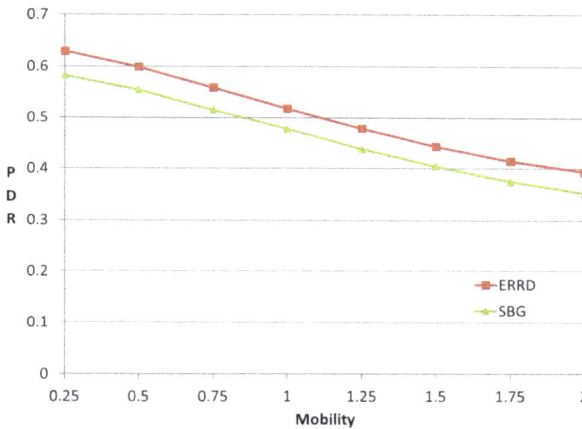

Figure 7. PDR vs. mobility.

6. Conclusions and Future Works

This study has proposed a social-aware peer discovery mechanism, designated as ERRD, for energy harvesting-based D2D communications. ERRD first allocates the budget power of the network among the UEs, depending on their social ratios and then adjusts the allocated power quantum of each UE, depending on its harvested energy. Finally, ERRD sets the beacon rates of the UEs based on their adjusted power quanta subject to the constraint that the total power allocated to the UEs may not exceed the *BP* quantum of the network. ERRD improves the PDR performance by increasing the beacon rates of the high-energy-harvesting UEs, thereby increasing the number of peers that they can discover; and, reducing the beacon rates of the low-energy-harvesting UEs, thereby extending their lifetimes and prolonging the period for which they can participate in the peer discovery process. The simulation results have shown that ERRD outperforms the SBG scheme that is reported in the literature by around 8–190%, depending on the *BP* quantum of the network. The PDR performance of ERRD does not significantly increase as the *BP* increases beyond 10 W or the *CDP* exceeds 2. However, in mobile environments, ERRD retains a 10% performance advantage over SBG, irrespective of the moving speed of the UEs.

Currently, 5G D2D communications and energy-harvesting devices are not so popular to give a realistic example of use case. Therefore, the paper is more research-oriented, rather than system-oriented. However, we believe that D2D communications and energy-harvesting devices will become more popular in the future. The scenario that is considered in the study will actually happen and our proposed solution, ERRD, can be applied.

A complete D2D communication not only includes peer discovery, but also mode selection and resource allocation. Future studies will aim to establish a more comprehensive energy harvesting-based D2D communication model that includes mode selection and resource allocation. In its current form, ERRD considers the past behavior (sociality) and present condition (energy harvesting) of the UEs, but it does not consider the impact of the surrounding information, such as the number of peers in the network or the interference. Therefore, in future studies, ERRD will be extended to take such information into account in order to obtain a more robust estimation of the most appropriate beacon rate for D2D communications. Finally, the performance of ERRD in a realistic environment should be further investigated, as our study proves the outperformance of ERRD by simulations. For example, a UE who has a social-network account and, at the same time, can harvest energy can enhance how much of PDR. To observe this, ERRD should be realistically implemented in BS. We will pay the efforts on this implementation in the future.

Author Contributions: Conceptualization, Z.L.H. and Y.-C.L.; methodology, Z.L.H. and Y.-C.L.; software, Y.-H.W.; validation, Y.-H.W.; formal analysis, Y.-C.L.; writing—original draft preparation, Y.-H.C.; writing—review and editing, Y.-C.L. and A.C.

Funding: This research was funded by Ministry of Science and Technology of Taiwan, MOST 107-2221-E-011-083.

Conflicts of Interest: The authors declare no conflict of interest.

References

1. Wei, L.; Hu, R.Q.; Qian, Y.; Wu, G. Enabling device-to-device communications underlying cellular networks: Challenges and research aspects. *IEEE Commun. Mag.* **2014**, *52*, 90–96. [CrossRef]
2. Baccelli, F.; Khude, N.; Laroia, R.; Li, J.; Richardson, T.; Shakkottai, S.; Tavildar, S.; Wu, X. On the design of device-to-device autonomous discovery. In Proceedings of the International Conference on Communication Systems and Networks, Bangalore, India, 3–7 January 2012.
3. Lim, C.; Jang, M.; Kim, S.H. Trellis tone modulation multiple-access for peer discovery in D2D networks. *Sensors* **2018**, *18*, 1228. [CrossRef] [PubMed]
4. Li, H.B.; Miura, R.; Kojima, F. Channel access proposal for enabling quick discovery for D2D wireless networks. In Proceedings of the International Conference on Computing, Networking and Communications (ICNC), Santa Clara, CA, USA, 26–29 January 2017.
5. Tang, H.; Ding, Z.; Levy, B.C. Enabling D2D communications through neighbor discovery in LTE cellular networks. *IEEE Trans. Signal Process.* **2014**, *62*, 5157–5170. [CrossRef]
6. Nasraoui, L.; Atallah, L.N. SRS-based D2D neighbor discovery scheme for LTE cellular networks. In Proceedings of the IEEE International Symposium on Personal, Indoor, and Mobile Radio Communications (PIMRC), Montreal, QC, Canada, 8–13 October 2017.
7. Doppler, K.; Ribeiro, C.B.; Kneckt, J. Advances in D2D communications: Energy efficient service and device discovery radio. In Proceedings of the IEEE Wireless VITAE, Chennai, India, 28 February–3 March 2011.
8. Yang, Z.J.; Huang, J.C.; Chou, C.T.; Hsieh, H.Y.; Hsu, C.W.; Yeh, P.C.; Hsu, C.C. Peer discovery for device-to-device (D2D) communication in LTE-A networks. In Proceedings of the IEEE Globecom Workshops, Atlanta, GA, USA, 9–13 December 2013.
9. Nguyen, P.; Wijesinghe, P.; Palipana, R.; Lin, K.; Vasic, D. Network-assisted device discovery for LTE-based D2D communication systems. In Proceedings of the IEEE ICC, Sydney, Australia, 10–14 June 2014.
10. Naslcheraghi, M.; Marandi, L.; Ghorashi, S.A. A novel device-to-device discovery scheme for underlay cellular networks. In Proceedings of the Iranian Conference on Electrical Engineering (ICEE), Tehran, Iran, 2–4 May 2017.
11. Choi, K.W.; Han, Z. Device-to-device discovery for proximity-based service in LTE-Advanced system. *IEEE J. Sel. Areas Commun.* **2015**, *33*, 55–66. [CrossRef]
12. Lee, W.; Kim, J.; Choi, S.W. New D2D peer discovery scheme based on spatial correlation of wireless channel. *IEEE Trans. Veh. Technol.* **2016**, *65*, 10120–10125. [CrossRef]

13. Xenakis, D.; Kountouris, M.; Merakos, L.; Passas, N.; Verikoukis, C. Performance analysis of network-assisted D2D discovery in random spatial networks. *IEEE Trans. Wirel. Commun.* **2016**, *15*, 5695–5707. [CrossRef]

14. Nguyen, N.T.; Choi, K.W.; Song, L.; Han, Z. ROOMMATEs: An unsupervised indoor peer discovery approach for LTE D2D communications. *IEEE Trans. Veh. Technol.* **2018**, *67*, 5069–5083. [CrossRef]

15. Nitti, M.; Stelea, G.A.; Popescu, V.; Fadda, M. When social networks meet D2D communications: A survey. *Sensors* **2019**, *19*, 396. [CrossRef] [PubMed]

16. Li, Y.; Wu, T.; Hui, P.; Jin, D.; Chen, S. Social-aware D2D communications: Qualitative insights and quantitative analysis. *IEEE Commun. Mag.* **2014**, *52*, 150–158. [CrossRef]

17. Zhang, Z.; Wang, L.; Liu, D.; Zhang, Y. Peer discovery for D2D communications based on social attribute and service attribute. *J. Netw. Comput. Appl.* **2017**, *86*, 82–91. [CrossRef]

18. Tan, Z.; Li, X.; Ji, H.; Wang, K.; Zhang, H. Social-aware peer discovery and resource allocation for device-to-device communication. In Proceedings of the Digital Media Industry & Academic Forum, Santorini, Greece, 4–6 July 2016.

19. Wu, H.; Wang, Y. Socially-inspired peer discovery for D2D communications. In Proceedings of the IEEE SmartWorld, Ubiquitous Intelligence & Computing, Advanced & Trusted Computing, Scalable Computing & Communications, Cloud & Big Data Computing, Internet of People and Smart City Innovation, Guangzhou, China, 8–12 October 2018.

20. Xu, C.; Gao, C.; Zhou, Z.; Chang, Z.; Jia, Y. Social network-based content delivery in device-to-device underlay cellular networks using matching theory. *IEEE Access* **2017**, *5*, 924–937. [CrossRef]

21. Zhang, B.; Li, Y.; Jin, D.; Hui, P.; Han, Z. Social-aware peer discovery for D2D communications underlaying cellular networks. *IEEE Trans. Wirel. Commun.* **2015**, *14*, 2426–2439. [CrossRef]

22. Tang, X.; Wang, X.; Cattley, R.; Gu, F.; Ball, A.D. Energy harvesting technologies for achieving self-powered wireless sensor networks in machine condition monitoring: A review. *Sensors* **2018**, *18*, 4113. [CrossRef] [PubMed]

23. Tran, L.-G.; Cha, H.-K.; Park, W.-T. RF power harvesting: A review on designing methodologies and applications. *Micro Nano Syst. Lett.* **2017**, *5*, 14. [CrossRef]

24. ITU. Applications of Wireless Power Transmission via Radio Frequency Beam. Available online: https://www.itu.int/pub/R-REP-SM.2392 (accessed on 18 May 2019).

25. Bouchouicha, D.; Dupont, F.; Latrach, M.; Ventura, L. Ambient RF energy harvesting. In Proceedings of the International Conference on Renewable Energies and Power Quality, Granada, Spain, 23–25 March 2010.

26. Pinuela, M.; Mitcheson, P.D.; Lucyszyn, S. Ambient RF energy harvesting in urban and semi-urban environments. *IEEE Trans. Microw. Theory Tech.* **2013**, *61*, 2715–2726. [CrossRef]

27. Shinohara, N.; Kamiyoshikawa, N. Study of flat beam in near-field for beam-type wireless power transfer via microwaves. In Proceedings of the 2017 11th European Conference on Antennas and Propagation (EUCAP), Paris, France, 19–24 March 2017.

28. Mitani, T.; Yamakawa, H.; Shinohara, N.; Hashimoto, K.; Kawasaki, S.; Takahashi, F.; Yonekura, H.; Hirano, T.; Fujiwara, T.; Nagano, K.; et al. Demonstration experiment of microwave power and information transmission from an airship. In Proceedings of the 2nd International Symposium on Radio System and Space Plasma, Sofia, Bulgaria, 25–27 August 2010.

29. Huang, C.-C.; Lin, C.-L. Wireless power and bidirectional data transfer scheme for battery charger. *IEEE Trans. Power Electron.* **2018**, *33*, 4679–4689. [CrossRef]

30. Lakhlan, P.; Trivedi, A. Energy harvesting-based two-hop D2D communication in cellular networks. In Proceedings of the International Conference on Advances in Computing, Communications and Informatics, Jaipur, India, 21–24 September 2016.

31. Yao, Y.; Huang, S.; Beaulieu, N.C.; Yin, C. Cooperative transmission in cognitive and energy harvesting-based D2D networks. In Proceedings of the IEEE Wireless Communications and Networking Conference, San Francisco, CA, USA, 19–22 March 2017.

32. Sakr, A.H.; Hossain, E. Cognitive and energy harvesting-based D2D communication in cellular networks: Stochastic geometry modeling and analysis. *IEEE Trans. Commun.* **2015**, *63*, 1867–1880. [CrossRef]

33. Wang, W.; Srinivasan, V.; Motani, M. Adaptive contact probing mechanisms for delay tolerant applications. In Proceedings of the ACM International Conference on Mobile Computing Networking, Montréal, QC, Canada, 9–14 September 2007.
34. Chaintreau, A.; Hui, P.; Scott, J.; Gass, R.; Crowcroft, J.; Diot, C. Impact of human mobility on opportunistic forwarding algorithms. *IEEE Trans. Mob. Comput.* **2007**, *6*, 606–620. [CrossRef]

![sensors logo] *sensors*

MDPI

Article

Joint Optimization of Interference Coordination Parameters and Base-Station Density for Energy-Efficient Heterogeneous Networks

Yanzan Sun, Han Xu, Shunqing Zhang, Yating Wu, Tao Wang *, Yong Fang and Shugong Xu

Shanghai Institute for Advanced Communication and Data Science, Key laboratory of Specialty Fiber Optics and Optical Access Networks, Joint International Research Laboratory of Specialty Fiber Optics and Advanced Communication, Shanghai University, Shanghai 200072, China; yanzansun@shu.edu.cn (Y.S.); xuhan_shu@shu.edu.cn (H.X.); shunqing@shu.edu.cn (S.Z.); ytwu@shu.edu.cn (Y.W.); yfang@shu.edu.cn (Y.F.); shugong@shu.edu.cn (S.X.)
* Correspondence: twang@shu.edu.cn; Tel.: +86-139-1698-9806

Received: 16 March 2019; Accepted: 6 May 2019; Published: 9 May 2019

Abstract: Heterogeneous networks (HetNets), consisting of macro-cells and overlaying pico-cells, have been recognized as a promising paradigm to support the exponential growth of data traffic demands and high network energy efficiency (EE). However, for two-tier heterogeneous architecture deployment of HetNets, the inter-tier interference will be challenging. Time domain further-enhanced inter-cell interference coordination (FeICIC) proposed in 3GPP Release-11 becomes necessary to mitigate the inter-tier interference by applying low power almost blank subframe (ABS) scheme. Therefore, for HetNets deployment in reality, the pico-cell range expansion (CRE) bias, the power of ABS and the density of pico base stations (PBSs) are three important factors for the network EE improvement. Aiming to improve the network EE, the above three factors are jointly considered in this paper. In particular, we first derive the closed-form expression of the network EE as a function of pico CRE bias, power reduction factor of low power ABS and PBS density based on stochastic geometry model. Then, the approximate relationship between pico CRE bias and power reduction factor is deduced, followed by a linear search algorithm to get the near-optimal pico CRE bias and power reduction factor together at a given PBS density. Next, a linear search algorithm is further proposed to optimize PBS density based on fixed pico CRE bias and power reduction factor. Due to the fact that the above pico CRE bias and power reduction factor optimization and PBS density optimization are optimized separately, a heuristic algorithm is further proposed to optimize pico CRE bias, power reduction factor and PBS density jointly to achieve global network EE maximization. Numerical simulation results show that our proposed heuristic algorithm can significantly enhance the network EE while incurring low computational complexity.

Keywords: HetNets; interference coordination; energy efficiency; stochastic geometry

---

1. Introduction

The exponential growth of data traffic demand, huge energy consumption and large amounts of global carbon dioxide emissions severely restrict the sustainable development of wireless cellular networks. According to the statistics, the data traffic volume demand in the fifth-generation (5G) wireless communication network will increase of $1000\times$ by 2020. Moreover, the limited spectrum resources also constrain the network capacity improvement [1]. Therefore, the network energy efficiency (EE) which considers both spectral efficiency (SE) and energy consumption has been valued not only as an important network performance indicator for modern wireless networks, but also for the operational expenditure reduction and sustainable development [2].

Heterogeneous networks (HetNets) consisting of macro base stations (MBSs) and low-power pico base stations (PBSs) can improve SE by reusing the spectrum geographically [3] and enhance the wireless link quality by shortening the distance between the transmitter and the receiver [4]. Therefore, HetNets are deemed as a promising technique to support the deluge of data traffic with high network EE. Nonetheless, due to the complex two-tier heterogeneous architecture deployment of HetNets, the challenging inter-tier interference and PBS deployment density will deteriorate the network EE if they are not treated carefully, which are the concerns of this paper for aiming to improve the network EE.

1.1. Motivation

In HetNets consisting of MBSs and PBSs, as shown in Figure 1, the great difference of transmitter power leads to load imbalance between macro tier and pico tier. To address this issue, PBSs adopt cell range expansion (CRE) technology to enlarge the PBS coverage area by adding a positive bias on the reference signal received power (RSRP) of PBSs without increasing transmission power [5,6]. Unfortunately, CRE user equipments (UEs) located in the pico CRE region will suffer serious downlink interference from dominating MBSs, even causing the outage of control signals. As a result, it is important to mitigate the downlink interference to improve the wireless link quality between transmitter and receiver. Then the network EE can be improved.

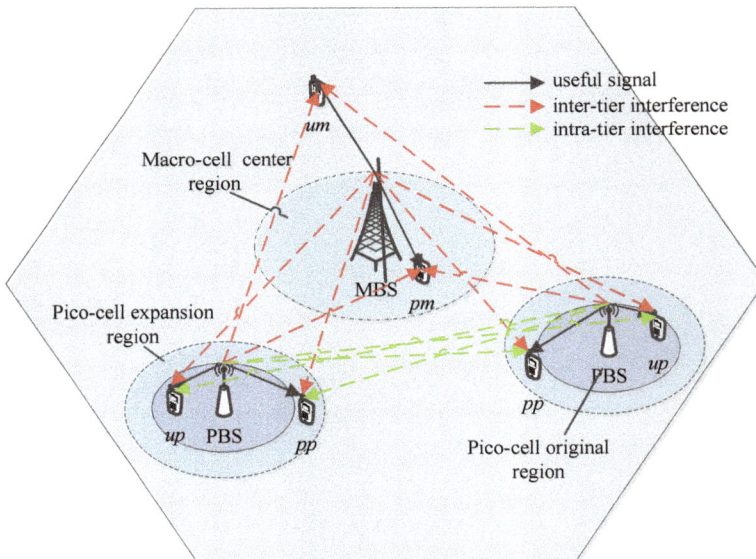

Figure 1. The network scenario.

To mitigate the downlink inter-tier interference for HetNets, further enhanced inter-cell interference coordination (FeICIC) scheme has been specified in 3GPP Release 11 [7]. In FeICIC scheme, MBSs can transmit data to macro center UEs with low transmission power over certain subframes, termed as low power almost blank subframes (ABS), over which PBSs can schedule CRE UEs with reduced interference [8]. On the basis of FeICIC technique, for user association, the pico CRE bias will directly affect the value of user received RSRP from PBSs, which will eventually affect the number of CRE UEs associated to PBSs. As CRE UEs are scheduled in the subframes corresponding to the ABS of MBSs, the transmission power of MBSs in ABS, i.e., ABS power, will decide the interference degree suffered by CRE UEs. Therefore, pico CRE bias and ABS power are two important factors for the wireless link quality of UEs, especially for CRE UEs, which eventually have significant effect on the network EE performance.

To meet the super-large capacity demand in 5G wireless communication networks, more and more base stations (BSs), especially small base stations (SBSs), are deployed in HetNets [9,10]. On the one hand, irregular deployment of massive BSs causes additional and intractable inter-tier interference. On the other hand, although high-density PBSs are deployed to satisfy the peak traffic volume, highly dynamic wireless traffic may deteriorate EE if the capacity gains of numerous PBSs are utilized insufficiently. In short, the BSs deployment strategy based on network load is also one of the key issues to realize 5G green cellular network [11].

In addition, the network EE of HetNets is also affected by some other aspects of factors. For instance, it is proved that the reasonable adjustments of BS transmit power, inter-site distances and the number of MBSs or SBSs contribute to the improvement of EE of HetNets [12]. In [13], assuming that SBSs have performed traffic offloading from MBS, the authors investigated the MBS and SBS power allocation scheme to improve network EE. In [14], the authors investigated the user scheduling and resource allocation method to optimize the network EE for HetNets.

To sum up, for complex HetNets, network EE is affected by many different aspects of factors, e.g., MBS transmit power in low power ABS, MBS transmit power in non-ABS, MBS density, ABS ratio, pico CRE bias, PBS density, PBS transmit power in low power ABS, PBS transmit power in non-ABS, etc. Thus, it will be a very challenging work to analyze the effects of all of these factors together on network EE. In this paper, for HetNets deployment in reality with inter-tier interference coordination by adopting FeICIC scheme, the pico CRE bias, the power reduction factor of low power ABS, and the density of PBSs are three more related factors for the network EE improvement compared with others. Therefore, the above three factors are focused and jointly optimized for the network EE maximization in this paper.

1.2. Contributions

In this paper, we investigate the joint optimization of FeICIC parameters and PBS density in HetNets for network EE improvement. Initially, the system model for two-tier HetNets by using stochastic geometry is described. Then, an analytical expression of the network EE as a function of pico CRE bias, power reduction factor and PBS density is derived. At last, heuristic algorithms are proposed to obtain the optimal values of pico CRE bias, power reduction factor and PBS density to maximize the network EE. The main contributions of this paper are summarized as follows: (1) the closed-form expression of network EE as a function of pico CRE bias, power reduction factor and PBS density is deduced by stochastic geometry theory. (2) The equivalent relationship between pico CRE bias and the power reduction factor is obtained by an approximation calculation. Based on the equivalent relationship, an efficient optimization algorithm is designed to get the near-optimal values of pico CRE bias and power reduction factor at a given PBS density. (3) To achieve the global optimization of network EE, a low computational complexity heuristic algorithm is proposed to jointly optimize pico CRE bias, power reduction factor, and PBS density.

1.3. Organization

The rest of this paper is organized as follows. The system model and user association strategy are described in Section 3. In Section 4, we deduce the closed-form expression of network EE. A low complexity heuristic algorithm is proposed to optimize pico CRE bias, power reduction factor, and PBS density jointly for the network EE maximization in Section 5. Numerical results and discussions are presented in Section 6. Concluding remarks are given in Section 7.

2. Related Works

Early literatures mainly focused on pico CRE bias, ABS ratio, and ABS power optimization for the network capacity maximization [15–18] and recent works began to investigate the network EE improvement from different perspectives including resource management [19–23], FeICIC parameters optimization [24–31] and BS deployment strategy [32–39]. As for resource management, centralized resource allocation algorithms based on convex-optimization [19,20], graph-theory [21] or even game-theory [22,23] were proposed to achieve the maximum EE gain.

As for FeICIC parameter optimization, the ABS ratio dynamical optimization algorithm based on network load was proposed to enhance the network EE in [24]. For improving the network capacity and EE, pico CRE bias optimization problem was further developed combined with power control in [29]. Using the stochastic geometric approach, the expressions for SE and cell-edge throughputs have been derived as a function of the power reduction factor of low power ABS in [30]. To move one step further, pico CRE bias, ABS ratio and ABS power are jointly optimized by a robust EE optimization framework in [25]. In [31], it was proved that FeICIC can achieve a better gain in view of network EE, cell-edge throughput and user fairness compared with eICIC. In addition, the distributed algorithms based on the exact potential game framework for both eICIC and FeICIC optimizations were proposed to offer better network performance. The authors of [26] deducted and analyzed the EE coverage performance of FeICIC with extra pico CRE bias. In [28], FeICIC technique was applied to mitigate the inter-tier interference in HetNets deployed with unmanned aerial vehicles (UAVs). In such a scenario, the locations of UAVs, pico CRE bias and inter-cell interference coordination parameters were jointly optimized by using a genetic algorithm. In [27], the EE of HetNets with joint FeICIC and adaptive spectrum allocation was analyzed by the stochastic geometric approach. The research on FeICIC parameters optimization on the basis of stochastic geometry is relatively few in the existing literatures.

As for BS deployment strategy, general linear power consumption models were developed by means of linear regression in [38]. Meanwhile, the effects of MBS transmission power and on/off switching on instantaneous MBS power consumption were analyzed. A threshold of SBS density in ultra-dense HetNets was investigated in consideration of the backhaul network capacity and network EE in [32]. In [33], the authors came up with an approximation algorithm to solve the intractable user association problem by controlling the PBS density dynamically. The relationship between PBS density and network EE was analyzed under different UE density in [34]. It was proved that both PBS density and MBS density have a notable impact on the network EE in [35]. In [36,37], PBS density and MBS density were jointly optimized based on traffic-aware sleeping strategies and stochastic geometry to enhance the network EE, respectively. In [39], taking the traffic pattern variations into account, the BSs can not only adaptively switch on/off states but also can dynamically scale its transmit power according to network capacity demands. In this way, network energy consumption is reduced.

Despite the aforementioned research works, few works in the literatures focused on FeICIC parameters and PBS density joint optimization for network EE improvement in HetNets, which will be investigated and developed in this paper.

3. System Model

The traditional network models with a hexagonal grid cannot accurately match the actual network deployment. Under such deterministic grid models, Monte Carlo simulations consume huge amounts of time and resources to obtain the statistical results. Recently, a stochastic geometry model was proven to be a tractable analytical model for homogeneous networks and HetNets, where the location distribution of BSs is modeled as a spatial Poisson point process (PPP) [40]. Using PPP, the network performance, like signal to interference plus noise ratio (SINR) coverage [41], rate coverage [8], average rate [42], can be analyzed conveniently by theoretical derivation. Thus, we adopt a stochastic geometry model to model a two-tier HetNets consisting of MBSs and PBSs in this paper.

Let $k \in \{m, p\}$ denote the subscripts of a tier, where $k = m$ represents macro tier and $k = p$ denotes pico tier. MBSs and PBSs are modeled as two identically independent distributions (iid.) PPPs Φ_m and Φ_p with density λ_m and λ_p in the Euclidean plane, respectively. The UEs are also distributed according to a different iid PPP Φ_u with density λ_u. The total spectrum bandwidth is defined as W. To mitigate the downlink interference over the control channel from MBSs to CRE UEs, MBSs adopt the

FeICIC scheme, where all the subframes are divided into protected subframes (PSF), i.e., low power ABS, and unprotected subframes (USF), i.e., non-ABS. The MBSs transmit data at reduced power ρP_m on PSF, where $0 \leq \rho < 1$ is the power reduction factor. In fact MBS transmit power in USF and PBS transmit power will also have effects on network EE. However, to focus on the effects of PBS density, pico CRE bias and power reduction factor on network EE and also for analysis simplification, we assume that MBSs transmit power in USF and PBSs transmit power are set to be the maximum fixed power P_m and P_p, respectively. Let θ to be PSF ratio, which is defined as the proportion between the number of PSF subframes and that of all subframes. Each user is associated with the strongest BS according to the biased received reference signal received power (RSRP) at the user. In this paper, the association bias for MBS is assumed to be unity ($B_m = 1 = 0$ dB) and that of PBS is pico CRE bias depicted as B_p, where $B_p \geq 0$ dB.

Based on the user association strategy, all UEs can be divided into four different types: the type of PSF macro-cell UEs (MUEs) contains the users within the macro-cell center region, the type of USF MUEs includes the users outside the macro-cell center region, the type of PSF pico-cell UEs (PUEs) correspond the users located in the pico CRE region and the type of USF PUEs comprises the users scattered in the original coverage of pico-cell. As shown in Figure 1, we adopt the index $l \in \mathbf{L} = \{pm, um, pp, up\}$ to denote the indication of the above four types of users, respectively, where *pm* represents PSF MUEs, *um* denotes USF MUEs, *pp* signifies PSF PUEs, and *up* indicates USF PUEs. In HetNets with FeICIC, the UEs scheduling strategy for MBS and PBS is shown in Figure 2. The USF MUEs and USF PUEs are scheduled by MBSs and PBSs on USF, respectively. Each MBS schedules PSF MUEs on PSF with reduced power. Then PBS can schedule PSF PUEs in the corresponding subframes with reduced interference.

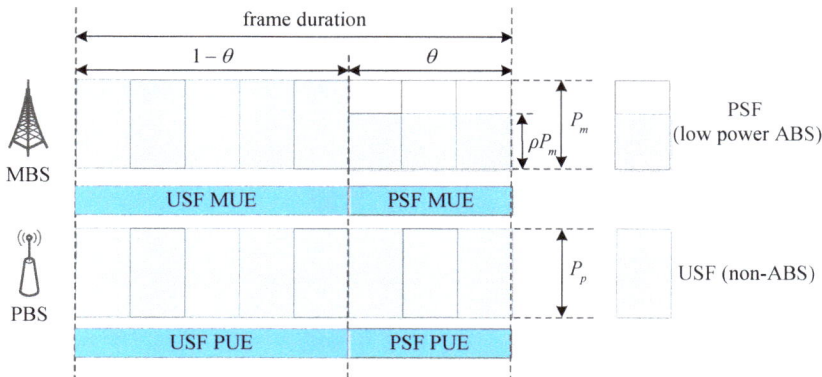

Figure 2. The user equipments (UEs) scheduling strategy for macro base station (MBS) and pico base stations (PBS) with the further-enhanced inter-cell interference coordination (FeICIC) scheme.

According to the Slivyak theorem, there is no difference in properties observed either at a point of the PPP or at an arbitrary point [8]. Therefore, we can simply analyze a typical UE located at the origin. The received signal power of a typical user l from a BS of k tier at a distance of r_k can be represented as $P_k h r_k^{-\alpha}$, where P_k is the full transmission power of BS in k tier, h represents the channel fast fading gain, which is defined as Rayleigh distributed with average unit power, i.e., $h \sim \exp(1)$. The term α denotes the large-scale path loss exponent, which is assumed to be the same in both macro tier and pico tier for convenient analysis. Hence, the SINR of a typical user l based on its user type can be depicted as:

$$\gamma_l = \begin{cases} \dfrac{\rho P_m h r_m^{-\alpha}}{\rho I_m + I_p + \sigma^2}, \text{if } l = pm \\[2mm] \dfrac{P_p h r_p^{-\alpha}}{\rho I_m + I_p + \sigma^2}, \text{if } l = pp \\[2mm] \dfrac{P_m h r_m^{-\alpha}}{I_m + I_p + \sigma^2}, \text{if } l = um \\[2mm] \dfrac{P_p h r_p^{-\alpha}}{I_m + I_p + \sigma^2}, \text{if } l = up \end{cases}, \tag{1}$$

where I_m and I_p denote the full power aggregate interference from the macro tier and pico tier to UE l, respectively, σ^2 represents the thermal noise, ρ is the power reduction factor of MBS transmit power in PSF. P_m and P_p are the full transmission power of MBSs and PBSs, respectively. r_m and r_p are the nearest distance from MBSs and PBSs to a typical UE l, respectively.

The notations used in this paper are shown in Table 1.

Table 1. Notations summary.

Notation	Description
$\lambda_m, \lambda_p, \lambda_u$	Density of MBS, PBS and UE
L, l	Set of user types, indication of the user type
W	Total spectrum bandwidth
h	Channel fast fading gain
α	Large-scale path loss exponent
σ^2	Thermal noise
P_m, P_p	Maximum transmission power of MBS and PBS
ρ	Power reduction factor
θ	PSF ratio
B_p	Pico CRE bias
γ_l	SINR of a typical UE l
I_m, I_p	Full power aggregate interference from macro tier and pico tier
r_m, r_p	Distance from a UE to its nearest MBS and its nearest PBS
A_l	Probability of a typical UE belongs to the user type l
k_c, k_e, k_p	Factor of macro-cell center region, pico CRE region and pico-cell original coverage region
r_l	Distance from a typical UE l and its serving BS
$f_l(r)$	PDF of the distance between a UE and its serving BS
λ_l	Density of BSs associated with user type l
R_l	Mean achievable downlink data rate of a typical UE l
W_l	Spectrum bandwidth allocated to a typical UE l
N_l, N_l^{total}	Mean number of UEs with user type l in a Voronoi cell, total number of UEs with user type l
$P_{m,s}, P_{p,s}$	Static power of MBS and PBS
$\hat{P}_m$	Proportion between maximum transmission power of MBS and that of PBS
$\hat{P}_p$	Proportion between maximum transmission power of PBS and that of MBS
$\lambda_{p,m}$	Proportion between PBS density and MBS density
R^{total}	Total network throughput
p^{total}	Total network power consumption
ρ'	Approximate value of power reduction factor
$\hat{B}_p^*, \hat{\rho}^*, \hat{\lambda}_p^*$	Near-optimal value of pico CRE bias, power reduction factor and PBS density
$B_p^*, \rho^*, \lambda_p^*, EE^*$	Optimal value of pico CRE bias, power reduction factor, PBS density and EE

3.1. User Type Probability

Normally, the user type of a typical UE l can be decided by the relationship between the biased RSRP from its nearest MBS and its nearest PBS as follow:

$$l = \begin{cases} pm, \text{if } \rho P_m h r_m^{-\alpha} > B_p P_p h r_p^{-\alpha} \\ pp, \text{if } P_p h r_p^{-\alpha} < P_m h r_m^{-\alpha} < B_p P_p h r_p^{-\alpha} \\ um, \text{if } \rho P_m h r_m^{-\alpha} < B_p P_p h r_p^{-\alpha} < P_m h r_m^{-\alpha} \\ up, \text{if } P_m h r_m^{-\alpha} < P_p h r_p^{-\alpha} \end{cases}, \tag{2}$$

where B_p is the pico CRE bias. In Equation (2), the conditions for determining user type can be further translated from biased RSRP based inequation into distance based inequation, which is shown as Equation (3).

$$l = \begin{cases} pm, \text{if } k_c r_m < r_p \\ pp, \text{if } k_p r_m < r_p < k_e r_m \\ um, \text{if } k_e r_m < r_p < k_c r_m \\ up, \text{if } k_p r_m > r_p \end{cases} \tag{3}$$

where $k_c = \left[B_p P_p / (P_m \rho) \right]^{1/\alpha}$, $k_e = \left(B_p P_p / P_m \right)^{1/\alpha}$ and $k_p = \left(P_p / P_m \right)^{1/\alpha}$ are defined as macro-cell center region factor, pico CRE region factor and pico-cell original coverage region factor, respectively [8], determining the coverage bound of the macro-cell center region, the pico CRE coverage region and the pico-cell original coverage region, respectively.

In order to obtain the probabilities of four user types, the following lemma is proposed.

Lemma 1. *Due to the locations of MBSs and PBSs follow two iid. PPPs, given two arbitrary coefficient values of n_a and n_b, the probability of $n_a r_m < r_p < n_b r_m$ can be expressed as:*

$$\text{Prob}\left(n_a r_m < r_p < n_b r_m \right) = \text{Prob}\left(r_p > n_a r_m \right) - \text{Prob}\left(r_p > n_b r_m \right)$$
$$= \frac{\lambda_m}{\lambda_m + n_a^2 \lambda_p} - \frac{\lambda_m}{\lambda_m + n_b^2 \lambda_p}, \tag{4}$$

where λ_m and λ_p are the MBS density and PBS density.

Proof. The proof of Lemma 1 is presented in Appendix A. □

Then, the probabilities of the PSF MUEs and the USF PUEs can be calculated similarly based on Lemma 1 and can be expressed by $\text{Prob}\left(r_p > n_a r_m \right) = \frac{\lambda_m}{\lambda_m + n_a^2 \lambda_p}$ and $\text{Prob}\left(r_p < n_b r_m \right) = 1 - \text{Prob}\left(r_p > n_b r_m \right) = 1 - \frac{\lambda_m}{\lambda_m + n_b^2 \lambda_p}$, respectively. Combining Equation (4) with the user association strategy in Equation (3), the probability of this typical UE belonging to the user type l can be defined as $A_l = \text{Prob}(l \in \mathbf{L})$, which is expressed as:

$$A_l = \begin{cases} \dfrac{\lambda_m}{\lambda_m + k_c^2 \lambda_p}, \text{if } l = pm \\[2ex] \dfrac{\lambda_m \lambda_p \left(k_e^2 - k_p^2 \right)}{\left(\lambda_m + k_p^2 \lambda_p \right)\left(\lambda_m + k_e^2 \lambda_p \right)}, \text{if } l = pp \\[2ex] \dfrac{\lambda_m \lambda_p \left(k_c^2 - k_e^2 \right)}{\left(\lambda_m + k_e^2 \lambda_p \right)\left(\lambda_m + k_c^2 \lambda_p \right)}, \text{if } l = um \\[2ex] \dfrac{k_p^2 \lambda_p}{\lambda_m + k_p^2 \lambda_p}, \text{if } l = up \end{cases} \tag{5}$$

In particular, if $\rho = 0$, then $k_c = \infty$ and $A_{pm} = 0$, which means that the PSFs are zero power ABS. The association probabilities of USF MUEs and PSF MUEs versus ρ with different B_p are simulated according to Equation (5) in Figure 3. As shown in Figure 3, with the increase of power reduction factor ρ, the transmission power of MBSs over PSF will increase, resulting in the association probability of USF MUEs decreasing and that of PSF MUEs increasing. It can also be found that the sum of A_{pm} and A_{um} will decrease with the growth of B_p, because more UEs will be offloaded into pico-cells with larger B_p.

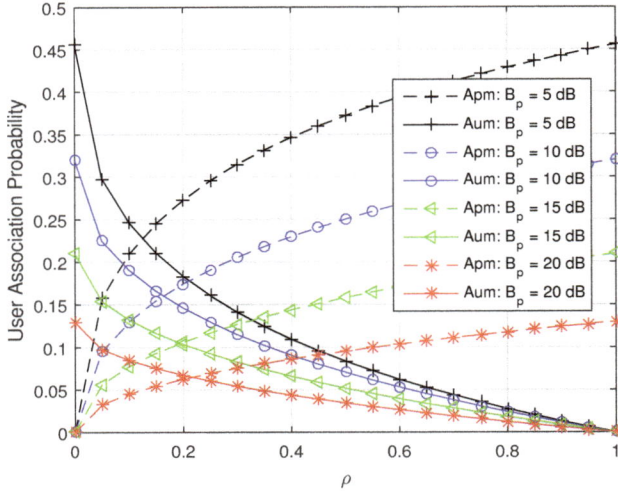

Figure 3. The user association probability of unprotected subframes (USF) micro-cell user equipments (MUEs) and protected subframes (PSF) MUEs versus ρ with fixed $\lambda_p = 3\lambda_m$ and fixed $\lambda_u = 0.0018$.

3.2. Distribution of Serving BS Distance

After a typical UE type is classified according to the user association strategy, the probability density function (PDF) of distance r between this typical UE and its serving BS can be obtained according to Lemma 2 as below.

Lemma 2. *On the basis of user association probability deduced in Equation (5), the probability density function (PDF) $f_l(r)$ of the distance r between the typical UE l and its serving BS can be derived as:*

$$f_l(r) = \begin{cases} \dfrac{2\pi r \lambda_m}{A_{pm}} \exp\left[-\pi r^2 \left(\lambda_m + k_e^2 \lambda_p\right)\right], & \text{if } l = pm \\[3mm] \dfrac{2\pi r \lambda_p}{A_{pp}} \left\{ \exp\left[-\pi r^2 \left(\lambda_m/k_e^2 + \lambda_p\right)\right] - \exp\left[-\pi r^2 \left(\lambda_m/k_p^2 + \lambda_p\right)\right] \right\}, & \text{if } l = pp \\[3mm] \dfrac{2\pi r \lambda_m}{A_{um}} \left\{ \exp\left[-\pi r^2 \left(\lambda_m + k_e^2 \lambda_p\right)\right] - \exp\left[-\pi r^2 \left(\lambda_m + k_c^2 \lambda_p\right)\right] \right\}, & \text{if } l = um \\[3mm] \dfrac{2\pi r \lambda_p}{A_{up}} \exp\left[-\pi r^2 \left(\lambda_m/k_p^2 + \lambda_p\right)\right], & \text{if } l = up \end{cases} \tag{6}$$

Proof. The proof is shown in Appendix B. □

4. Derivation of Energy Efficiency Expression

This section introduces our main analysis model and derives the closed-form expressions of the average achievable downlink rate, the network power consumption and the network EE, respectively.

4.1. The Ratio of PSF

PSF ratio can be denoted to be the proportion between the association probability of PSF PUE and the sum of the association probability of PSF PUE and that of USF PUE, as shown in Equation (7).

$$\theta = \frac{A_{pp}}{A_{pp} + A_{up}} = \frac{\lambda_m \left(k_e^2 - k_p^2 \right)}{k_e^2 \left(\lambda_m + k_p^2 \lambda_p \right)}, \tag{7}$$

where the expressions of A_{pp} and A_{up} are obtained according to Equation (5).

Referring to Equation (7), the PSF ratio versus B_p with different PBS densities λ_p is depicted in Figure 4. As shown in Figure 4, with B_p increasing, the pico CRE area will be enlarged. As a result, the PSF ratio will rise. Moreover, with the PBS density λ_p increasing, the distance between PBSs will be smaller, which will limit the further expansion of pico CRE area, so that the effect of B_p on θ will be weakened with λ_p increasing.

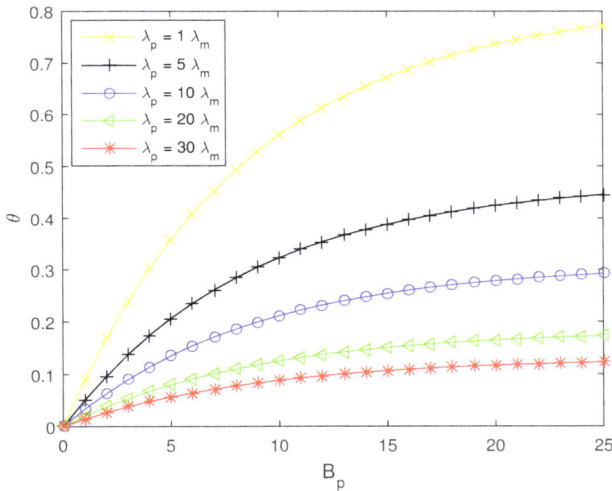

Figure 4. The PSF ratio versus B_p with fixed $\lambda_u = 0.0018$.

4.2. Average Achievable Downlink Rate

Assume that the network system adopts full buffer model and the frequency resource is allocated to all UEs in the coverage of a BS equally. Thus, the mean achievable downlink data rate of a typical UE l can be denoted as:

$$R_l = \frac{W_l}{E[N_l]} E \left[\log_2 \left(1 + \gamma_l \right) \right] \tag{8}$$

where W_l is the spectrum bandwidth allocated to UE l. Specifically, when $l \in \{pm, pp\}$, $W_l = \theta W$ and when $l \in \{um, up\}$, $W_l = (1 - \theta) W$. N_l is the mean number of serving UEs with user type l in a Voronoi cell and its expectation is $E[N_l] = (A_l \lambda_u / \lambda_l) + 1$. If $l \in \{pm, um\}$, $\lambda_l = \lambda_m$, otherwise $\lambda_l = \lambda_p$.

According to the analysis above, we get Lemma 3 as follow:

Lemma 3. *The average achievable downlink rate of a typical UE l can be further represented by*

$$R_l = \frac{2\pi\lambda_l W_l}{A_l N_l} \int_0^\infty \int_0^\infty \exp\left(-\varphi_l - \pi r^2 C_l\right) f_l\left(r\right) dr dt \tag{9}$$

where $\tau = 2^t - 1$, $\varphi_l = -\tau\sigma^2 r_l^\alpha \rho_l^{-1} P_l^{-1}$,

$$
C_l = \begin{cases}
\lambda_m Z\left(\tau, \alpha, 1\right) + \lambda_p \left(\hat{P}_p/\rho\right)^{2/\alpha} Z\left(\tau, \alpha, B_p\right), \text{when } l = pm \\
\lambda_m \left(\rho \hat{P}_m\right)^{2/\alpha} Z\left(\tau, \alpha, B_p^{-1}\rho^{-1}\right) + \lambda_p Z\left(\tau, \alpha, 1\right), \text{when } l = pp \\
\lambda_m Z\left(\tau, \alpha, 1\right) + \lambda_p (\hat{P}_p)^{2/\alpha} Z\left(\tau, \alpha, B_p\right), \text{when } l = um \\
\lambda_m \left(\hat{P}_m\right)^{2/\alpha} Z\left(\tau, \alpha, 1\right) + \lambda_p Z\left(\tau, \alpha, 1\right), \text{when } l = up
\end{cases},
$$

where $Z\left(\tau, \alpha, \beta\right) = \tau^{2/\alpha} \int_{(\beta/\tau)^{2/\alpha}}^\infty \frac{1}{1+x^{\alpha/2}} dx$, $\hat{P}_m = P_m/P_p$ *and* $\hat{P}_p = P_p/P_m$.

Proof. The proof is presented in Appendix C. □

Corollary 1. *To further simplify the analysis, we ignore the noise, i.e.,* $\sigma^2 = 0$ *and set the large-scale path loss exponent* $\alpha = 4$. *In that case, corresponding to the user type, the average achievable downlink rate of four user types can be expressed as:*

$$
\begin{aligned}
R_{pm} &= \int_0^\infty \frac{\theta W / A_{pm} N_{pm}}{\lambda_{p,m} Q\left(\hat{P}_p \tau/\rho, B_p \hat{P}_p/\rho\right) + Q\left(\tau, 1\right)} dt \\
R_{pp} &= \int_0^\infty \frac{\theta W / A_{pp} N_{pp}}{\lambda_{p,m}^{-1} Q\left(\rho \hat{P}_m \tau, B_p^{-1} \hat{P}_m\right) + Q\left(\tau, 1\right)} - \frac{\theta W / A_{pp} N_{pp}}{\lambda_{p,m}^{-1}\left[Q\left(\rho \hat{P}_m \tau, B_p^{-1} \hat{P}_m\right) - k_e^{-2} + k_p^{-2}\right] + Q\left(\tau, 1\right)} dt \\
R_{um} &= \int_0^\infty \frac{(1-\theta) W / A_{um} N_{um}}{\lambda_{p,m} Q\left(\hat{P}_p \tau, B_p \hat{P}_p\right) + Q\left(\tau, 1\right)} - \frac{(1-\theta) W / A_{um} N_{um}}{\lambda_{p,m}\left[Q\left(\hat{P}_p \tau, B_p \hat{P}_p\right) - k_e^2 + k_c^2\right] + Q\left(\tau, 1\right)} dt \\
R_{up} &= \int_0^\infty \frac{(1-\theta) W / A_{up} N_{up}}{\lambda_{p,m}^{-1} Q\left(\hat{P}_m \tau, \hat{P}_m\right) + Q\left(\tau, 1\right)} dt,
\end{aligned} \tag{10}
$$

where $Q\left(\tau, x\right) = \sqrt{x} + \sqrt{\tau} \arctan\left(\sqrt{\tau/x}\right)$, $\lambda_{p,m} = \lambda_p/\lambda_m$, $\hat{P}_p = P_p/P_m$, $\hat{P}_m = P_m/P_p$.

Proof. Set $\alpha = 4$ and $\sigma^2 = 0$, then we can get $\varphi_l = 0$ and $Z\left(\tau, \alpha, \beta\right) = \sqrt{\tau} \int_{\sqrt{\beta/\tau}}^\infty \frac{1}{1+x^2} dx = \sqrt{\tau} \arctan\left(\sqrt{\tau/\beta}\right)$. Combining with Equation (6), the desired results in Equation (9) can be obtained. □

4.3. Network Power Consumption

Generally, the BS power consumption comprises static power consumption and transmit power consumption [35]. The static power consumption is caused by signal processing, battery backup, as well as site cooling, and independent with the BS transmit power consumption. The transmit power consumption is determined by the transmission power of this BS and the load-dependent power consumption coefficient of this BS which is denoted as the number of its serving UEs. Define $P_{m,s}$ and $P_{p,s}$ are the static power consumption of each MBS and each PBS, respectively. With FeICIC scheme, the power consumptions of each MBS in PSF and USF can be expressed as $P_m^{PFS} = P_{m,s} + N_{pm}\rho P_m$ and $P_m^{UFS} = P_{m,s} + N_{um}P_m$, respectively. Similarly, the power consumptions of each PBS in PSF and USF can be given as $P_p^{PFS} = P_{p,s} + N_{pp}P_p$ and $P_p^{UFS} = P_{p,s} + N_{up}P_p$, respectively.

In PFS, the unit area mean power consumption P^{PFS} can be expressed as:

$$P^{PFS} = \lambda_m P_m^{PFS} + \lambda_p P_p^{PFS} = \lambda_m P_{m,s} + \lambda_p P_{p,s} + A_{pm}\lambda_u \rho P_m + A_{pp}\lambda_u P_p. \tag{11}$$

Similarly, the unit area mean power consumption in UFS P^{UFS} can be obtained as follow:

$$P^{UFS} = \lambda_m P_m^{UFS} + \lambda_p P_p^{UFS} = \lambda_m P_{m,s} + \lambda_p P_{p,s} + A_{um}\lambda_u P_m + A_{up}\lambda_u P_p. \tag{12}$$

Hence, the network power consumption can be expressed as:

$$
\begin{aligned}
P^{total} &= \theta P_{PFS} + (1-\theta) \, P_{UFS} \\
&= \lambda_m P_{m,s} + \theta \left(A_{pm} \lambda_u \rho P_m + A_{pp} \lambda_u P_p \right) + \lambda_p P_{p,s} + (1-\theta) \left(A_{um} \lambda_u P_m + A_{up} \lambda_u P_p \right).
\end{aligned}
\tag{13}
$$

4.4. Network Energy Efficiency

The network EE can be defined as the ratio of the achievable network throughput to the network power consumption [37]. For the convenience of derivation, we set $\sigma^2 = 0$ and $\alpha = 4$. Based on Equations (10) and (13), we can get the closed-form expression of network EE in the following:

$$
\begin{aligned}
EE = \frac{R^{total}}{P^{total}} &= \frac{\lambda_u \left(R_{pm} A_{pm} + R_{pp} A_{pp} + R_{um} A_{um} + R_{up} A_{up} \right)}{P^{total}} \\
&= \frac{\lambda_u}{P^{total}} \int_0^\infty \frac{\theta W / N_{pm}}{\lambda_{p,m} Q \left(\hat{P}_p \tau / \rho, B_p \hat{P}_p / \rho \right) + Q(\tau,1)} \\
&\quad + \frac{\theta W / N_{pp}}{\lambda_{p,m}^{-1} Q \left(\rho \hat{P}_m \tau, B_p^{-1} \hat{P}_m \right) + Q(\tau,1)} - \frac{\theta W / N_{pp}}{\lambda_{p,m}^{-1} \left[Q \left(\rho \hat{P}_m \tau, B_p^{-1} \hat{P}_m \right) - k_e^2 + k_p^2 \right] + Q(\tau,1)} \\
&\quad + \frac{(1-\theta) W / N_{um}}{\lambda_{p,m} Q \left(\hat{P}_p \tau, B_p \hat{P}_p \right) + Q(\tau,1)} - \frac{(1-\theta) W / N_{um}}{\lambda_{p,m}^{-1} \left[Q \left(\hat{P}_p \tau, B_p \hat{P}_p \right) - k_e^2 + k_c^2 \right] + Q(\tau,1)} \\
&\quad + \frac{(1-\theta) W / N_{up}}{\lambda_{p,m}^{-1} Q \left(\hat{P}_m \tau, \hat{P}_m \right) + Q(\tau,1)} \, dt
\end{aligned}
\tag{14}
$$

5. Joint Optimization of FeICIC Parameters and Base-Station Density

Due to the fact that MBSs are usually deployed by network operators, MBS density will change slowly and can be assumed to be constant for analysis simplification. Further simplification of the problem analysis, MBS transmit power in USF and PBS transmit power can also be assumed to be constant without considering power control. Moreover, the PSF ratio can be calculated according to Equation (7). Hence, the network EE is mainly impacted by pico CRE bias, power reduction factor, and PBS density under different UE density, i.e., network load. As MBS density is a constant value, we can first obtain the optimal value of the ratio between PBS density and MBS density $\lambda_{p,m}$, denoted as $\lambda_{p,m}^*$, to maximize the network EE. Then the optimal PBS density λ_p^* can be calculated by $\lambda_p^* = \lambda_m \lambda_{p,m}^*$. Thus, the joint optimization problem with the object of network EE maximization can be formulated as follows:

$$
\begin{aligned}
&\underset{\rho, B_p, \lambda_{p,m}}{\arg \max} \; EE \\
&\text{s.t. } 0 < B_p \leq 25 \text{ dB} \\
&\quad\quad 0 \leq \rho < 1 \\
&\quad\quad 0 < \lambda_{p,m} \leq 30
\end{aligned}
\tag{15}
$$

However, the network EE is nonlinear with $\lambda_{p,m}$, B_p and ρ, which is difficult to obtain the optimal $\lambda_{p,m}$, B_p and ρ at the same time with reasonable complexity. Note that the value ranges of $\lambda_{p,m}$, B_p and ρ are limited, which make it possible to seek out the optimal values of $\lambda_{p,m}$, B_p and ρ through a linear search algorithm by fixing two of these three variables, respectively. Therefore, we propose a heuristic algorithm to obtain the sub-optimal solution of the joint optimization problem in Equation (15). The proposed heuristic algorithm decomposes the original optimization problem into two sub-problems including FeICIC parameters optimization and PBS density optimization. For FeICIC parameters optimization, we first derive the approximate relation between B_p and ρ. Then, we get the sub-optimal values of B_p and ρ with given $\lambda_{p,m}$ by an alternating algorithm. For PBS density optimization, the optimal value of $\lambda_{p,m}$ can be obtained by a linear search method based on fixed B_p and ρ. Finally, we alternately solve two sub-problems to achieve globally optimal values of these variables.

5.1. Joint Optimization of Pico CRE Bias and Power Reduction Factor

In order to get the optimal values of B_p and ρ for network EE maximization, suppose that $\lambda_{p,m}$ and λ_u are given. Thus, the pico CRE bias and power reduction factor joint optimization problem can be formulated as follow:

$$\rho^*, B_p^* = \underset{\rho, B_p}{\arg\ \max}\ EE | \lambda_{p,m}$$
$$\text{s.t. } 0 < B_p \leq 25 \text{ dB} \tag{16}$$
$$0 \leq \rho < 1$$
$$\lambda_{p,m} \text{ is an arbitrary constant between 0 and 30,}$$

where B_p^* and ρ^* are the optimal values of pico CRE bias and power reduction factor, respectively. To simplify the solving process, we assume that the overall SINR of PSF MUEs is identical to that of the USF MUEs. Hence, the result of resource allocation will have a direct influence on the network EE. In view of user fairness, the optimal network EE can be achieved when the relationship of association probabilities of PSF MUEs and USF MUEs obey the Equation (17).

$$\theta = \frac{A_{pm}}{A_{pm} + A_{um}}. \tag{17}$$

Combining with Equation (7), we can get the approximate relation of ρ and B_p as:

$$\rho' = \frac{B_p P_p}{P_m \left(k_e^2 / A_{pp} - \lambda_m / \lambda_p \right)^2}, \tag{18}$$

where ρ' denotes the approximate near-optimal value of ρ. The relationships between ρ and B_p under different λ_p are shown in Figure 5. We can find that ρ' is a strictly increasing function with respect to B_p at a given PBS density.

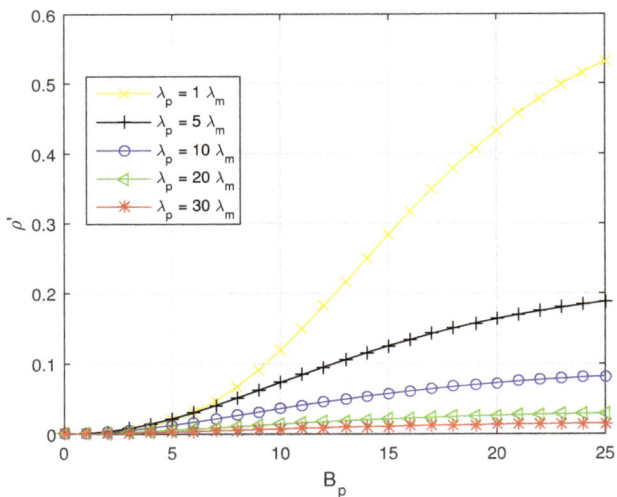

Figure 5. The near-optimal power reduction factor ρ' versus B_p with fixed $\lambda_u = 0.0018$.

Substituting ρ' into Equation (14), we can get the near-optimal value of B_p, denoted as $\hat{B}_p^*$ by solving the following univariate problem.

$$
\begin{aligned}
\hat{B}_p^* = \arg\ \max_{B_p} EE | \lambda_{p,m}, \rho = \rho' \\
\text{s.t. } 0 < B_p \leq 25 \text{ dB} \\
\lambda_{p,m} \text{ is an arbitrary constant between 0 and 30}
\end{aligned} \tag{19}
$$

Then, combining the value of $\hat{B}_p^*$ with Equation (14), the near-optimal value of ρ, denoted as $\hat{\rho}^*$ can be obtained by following univariate problem.

$$
\begin{aligned}
\hat{\rho}^* = \arg\ \max_{\rho} EE | \lambda_{p,m}, \hat{B}_p^* \\
\text{s.t. } 0 \leq \rho < 1 \\
\lambda_{p,m} \text{ is an arbitrary constant between 0 and 30}
\end{aligned} \tag{20}
$$

Thus, $\hat{B}_p^*$ and $\hat{\rho}^*$ are obtained by an alternating algorithm, which is shown in Algorithm 1. Δ_1 and Δ_2 represent the step lengths of B_p and ρ, which are set to be 0.1 and 0.05, respectively. EE^* denotes the optimal value of the network EE. The main idea of this altering algorithm is to obtain the near-optimal values of pico CRE bias and power reduction factor by a two-step linear search approach on the basis of the approximate relationship between them. In line 1 of Algorithm 1, the network scenario and relative parameters are initialized. From line 2 to line 10, the optimal pico CRE bias is obtained by a linear search way on the basis of approximate relationship between pico CRE bias and power reduction factor. From line 11 to line 18, the optimal power reduction factor is calculated based on Equation (14).

Algorithm 1: Joint pico CRE bias and power reduction factor optimization (JBPO) algorithm.

1: **Initialization**: Initialize the network scenario and the values λ_u, λ_m and $\lambda_{p,m}$, where
   $\lambda_{p,m} \in (0, 30]$. Set $\Delta_1 = 0.5$ dB, $\Delta_2 = 0.05$, $\hat{B}_p^* = B_p = 0.5$ dB and $EE^* = 0$.
2: **while** $B_p \leq 25$ dB **do**
3:     Substituting B_p into Equation (18), the approximate near-optimal value of the power
       reduction factor ρ can be obtained as ρ'.
4:     $\rho = \rho'$.
5:     Substituting B_p and ρ into Equation (14) with given λ_u, λ_m and $\lambda_{p,m}$, the current network
       energy efficiency EE' can be obtained as: $EE' = EE\left(B_p, \rho\right) | \lambda_u, \lambda_m, \lambda_{p,m}$.
6:     **if** $EE' > EE^*$ **then**
7:         $\hat{B}_p^* = B_p$, $EE^* = EE'$.
8:     **end if**
9:     $B_p = B_p + \Delta_1$.
10: **end while**
11: $\hat{\rho}^* = \rho = 0$, $B_p = \hat{B}_p^*$.
12: **while** $\rho < 1$ **do**
13:     Substituting ρ into Equation (14) with given λ_u, λ_m, $\lambda_{p,m}$ and B_p, the current network
       energy efficiency EE' can be obtained as: $EE' = EE\left(\rho\right) | \lambda_u, \lambda_m, \lambda_{p,m}, B_p$.
14:     **if** $EE' > EE^*$ **then**
15:         $\hat{\rho}^* = \rho$, $EE^* = EE'$.
16:     **end if**
17:     $\rho = \rho + \Delta_2$.
18: **end while**

5.2. Optimization of PBS Density

Similarly, suppose that B_p, ρ and λ_u are known. Thus, the PBS density optimization problem can be formulated as follow:

$$\lambda_{p,m}^* = \arg\max_{\lambda_{p,m}} EE|\rho, B_p$$

$$\text{s.t. } 0 < \lambda_{p,m} \leq 30 \tag{21}$$

$$B_p \text{ is an arbitrary constant between 0 and 25}$$

$$\rho \text{ is an arbitrary constant between 0 and 1}$$

Assume that the step length of $\lambda_{p,m}$ is Δ_3, which is set to be 0.03. Thus, the optimal PBS density can be obtained by a linear search algorithm to maximize the network EE, which is described in Algorithm 2. Line 1 of Algorithm 2 indicates the network scenario and parameters initialization. From line 2 to line 8, the optimal PBS density is acquired by a linear search method to maximize the network *EE*.

Algorithm 2: Pico base stations (PBS) density optimization (PDO) algorithm.

1: **Initialization:** Initialize the network scenario and the values λ_u, λ_m, B_p and ρ, where
   $B_p \in (0, 25]$ and $\rho \in [0, 1)$. Set $\Delta_3 = 0.03$, $\lambda_{p,m} = 0.3$, $\lambda_p^* = \lambda_{p,m}\lambda_m$ and $EE^* = 0$.
2: **while** $\lambda_{p,m} \leq 30$ **do**
3: Substituting $\lambda_{p,m}$ into Equation (14) with given λ_u, λ_m, B_p and ρ, the current network
    energy efficiency EE' can be obtained as: $EE' = EE\left(\lambda_{p,m}\right)|\lambda_u, \lambda_m, B_p, \rho$.
4:   **if** $EE' > EE^*$ **then**
5:    $\lambda_p^* = \lambda_{p,m}\lambda_m$, $EE^* = EE'$.
6:   **end if**
7:   $\lambda_{p,m} = \lambda_{p,m} + \Delta_3$.
8: **end while**

5.3. Joint Optimization of Pico CRE Bias, Power Reduction Factor and PBS Density

The FeICIC parameter optimization sub-problem and the PBS density optimization sub-problem are solved independently by the aforementioned optimization algorithms. Due to the fact that these variables are affected by each other, we further propose a heuristic pico CRE bias, power reduction factor and PBS density joint optimization algorithm to globally optimize network EE based on the joint pico CRE bias and power reduction factor optimization (JBPO) algorithm and PDO algorithm. The detailed procedure of our proposed heuristic algorithm is summarized in Algorithm 3. ε represents the positive tolerance value. N_{loop} is the iteration times of the algorithm. Line 1 of Algorithm 3 signifies the network scenario and parameters initialization. From line 3 to line 4, the JBPO algorithm is executed to obtain the current optimal pico CRE bias and power reduction factor at a given PBS density. From line 5 to line 6, the PDO algorithm is performed to acquire the current optimal PBS density based on the above obtained pico CRE bias and power reduction factor. From line 2 to line 9, the network EE is iteratively optimized until it cannot improve further within an arbitrary value ε. As a result, the optimal pico CRE bias, power reduction factor and PBS density are obtained and the network EE is maximized.

Algorithm 3: Joint pico pico-cell range expansion (CRE) bias, power reduction factor and PBS density optimization (JBPDO) algorithm.

1: **Initialization:** Initialize the network scenario and the values λ_u, λ_m, B_p, ρ and $\lambda_{p,m}$, where $B_p \in (0, 25]$, $\rho \in [0, 1)$ and $\lambda_{p,m} \in (0, 30]$. Set $EE^* = 0$, $N_{loop} = 0$ and $\varepsilon > 0$.

2: **repeat**

3:     Solving the optimization problem in Equation (16), the near-optimal pico CRE bias $\hat{B}_p^*$ and the near-optimal power reduction factor $\hat{\rho}^*$ at given $\lambda_{p,m}$ can be obtained based on the JBPO algorithm.

4:     $B_p = \hat{B}_p^*$, $\rho = \hat{\rho}^*$.

5:     Solving the optimization problem in Equation (21), the near-optimal PBS density $\hat{\lambda}_p^*$ at given B_p and ρ can be obtained based on the PDO algorithm.

6:     $\lambda_p = \hat{\lambda}_p^*$, $\lambda_{p,m} = \hat{\lambda}_p^* / \lambda_m$.

7:     Substituting B_p, ρ and $\lambda_{p,m}$ into Equation (14) with given λ_u and λ_m, the current network energy efficiency EE' can be obtained as: $EE' = EE\left(B_p, \rho, \lambda_{p,m}\right) | \lambda_u, \lambda_m$.

8:     $B_p^* = B_p$, $\rho^* = \rho$, $\lambda_p^* = \lambda_p$, $EE^* = EE'$, $N_{loop} = N_{loop} + 1$.

9: **until** $|EE' - EE^*| < \varepsilon$

5.4. Computational Complexity

The computational complexity of the JBPO algorithm can be calculated as $O\left(n_{B_p} + n_\rho\right)$, where n_{B_p} and n_ρ are the space sizes of the linear search for B_p and ρ, respectively. The computational complexity of the PDO algorithm is $O\left(n_{\lambda_p}\right)$, where n_{λ_p} is the space size of the linear search for λ_p. Thus, the computational complexity of the JBPDO algorithm is $O\left[\left(n_{B_p} + n_\rho + n_{\lambda_p}\right) \times N_{loop}\right]$.

Due to the fact that there does not an exist effective algorithm for solving Equation (15), we compare the computational complexity of our proposed JBPDO algorithm with that of a traversal algorithm. As for solving Equation (15), the optimal values of pico CRE bias, power reduction factor and PBS density can be obtained by a traversal way, i.e., traversal pico CRE bias, power reduction factor and PBS density optimization (TBPDO) algorithm, which refers to traversing all possible values of these three parameters to maximize the objective function in Equation (15). Thus, the computational complexity of TBPDO algorithm will be $O\left(n_{B_p} \times n_\rho \times n_{\lambda_p}\right)$, which signifies that the computational complexity of our proposed JBPDO algorithm is reduced effectively.

As for solving the objective function in Equation (16), the optimal values of pico CRE bias and power reduction factor can also be obtained by a traversal way, i.e., traversal pico CRE bias and power reduction factor optimization (TBPO) algorithm, which refers to traversing all possible values of these two parameters to maximize the objective function in Equation (16). Indeed, TBPO algorithm consists of two nested ergodic sub-processes: (1) traversal pico CRE bias optimization (TBO) algorithm, which is executed by traversing all pico CRE bias to maximize the objective function in Equation (19) under fixed power reduction factor and PBS density; (2) traversal power reduction factor optimization (TPO) algorithm, which is executed by traversing all power reduction factor to maximize the objective function in Equation (20) under fixed pico CRE bias and PBS density. Therefore, the computational complexities of the TBO and TPO algorithms are $O\left(n_{B_p}\right)$ and $O\left(n_\rho\right)$, respectively. As a result, the computational complexity of the TBPO algorithm will be $O\left(n_{B_p} \times n_\rho\right)$, which is obviously higher than that of our proposed JBPO algorithm.

6. Numerical Results and Analysis

In this section, we not only provide theoretical simulation, but also verify the effectiveness of proposed heuristic algorithms by Monte Carlo simulation. In the theoretical simulation, we considerws a network coverage area within a square region of 1000 m × 1000 m. The deployments of PBS and

MBS follow the PPP model and the typical UE is deployed in the origin. The simulation parameters used in this paper are summarized in Table 2. We took the average results from 30 times of network implementations as Monte Carlo simulations results. In each network implementation, the locations of MBSs, PBSs and UEs were modeled as spatial PPP, respectively. Then, the network EEs was calculated for all the different combination of pico CRE bias, power reduction factor, and PBS density values within their value ranges based on wireless channel quality. Finally, the maximum network EE and the optimal pico CRE bias, power reduction factor and PBS density can be obtained by comparing the calculated network EEs under all combinations. Indeed, the results of Monte Carlo simulation, including the maximal network EE, the optimal pico CRE bias, the optimal power reduction factor and the optimal PBS density, were obtained by a traversal way in each network implementation. Meanwhile, the performances of algorithms shown in the following simulation figures were just simulated based on theoretical derived Equation (14). Therefore, Monte Carlo simulation results had the best performance and can be referred to as a baseline for valuing the performances of those algorithms.

Table 2. Network scenario parameters.

Parameters	Value
Carrier frequency f	2 GHz
Total spectrum bandwidth W	10 MHz
Path loss exponent α	4
Path loss L	$L = 10 \log (L_0) + \alpha 10 \log (r)$, where $L_0 = (4\pi f/c)^2$, $c = 3 \times 10^8$ m/s
MBS transmission power P_m	43 dBm
PBS transmission power P_p	30 dBm
MBS static power $P_{m,s}$	800 W
PBS static power $P_{p,s}$	130 W
MBS density λ_m	0.00003

At first, the network EE performances of the JBPO algorithm were compared with those of the TPO algorithm, TBO algorithm, theoretical simulation and Monte Carlo simulation, as shown in Figures 6 and 7, respectively. As shown in Figure 6, with the increase of PBS density, network EE increased accordingly. With PBS density increase, more UEs were offloaded into the coverage of low power PBSs. Then the distance between the transmitter and the receiver was shorted, resulting in network EE improvement. As shown in Figure 7, with the increase of UE density, the curves of network EE also rose accordingly. As the UE density increased, more UEs can be offloaded into the coverage of low power PBSs by adjusting pico CRE bias and power reduction factor via network EE optimization algorithms. Then network EE can be improved. For the theoretical simulation curve, because the network EE was not optimized and just calculated according to Equation (14) with pico CRE bias and power reduction factor being set to be fixed values 5 dB and 0.25, respectively, so it had the worst performance. The TPO algorithm and the TBO algorithm optimized power reduction factor and CRE bias, respectively. Therefore, the performances of these two algorithms with one parameter optimized were better than that of the theoretical simulation. Our proposed JBPO algorithm can jointly optimize pico CRE bias and power reduction factor together. Therefore, it can further improve the network EE and match the Monte Carlo simulation results very well. Furthermore, the performance of the TPO algorithm was far less than that of the TBO algorithm, which signifies that the influence of pico CRE bias was more than that of the power reduction factor on the network EE, especially in low PBS density. In addition, in Figure 7 the performance gap between the JBPO algorithm and theoretical simulation increases accordingly with the growth of UE density, which further indicates the importance of joint pico CRE bias and power reduction factor optimization for the heavy network load scenario.

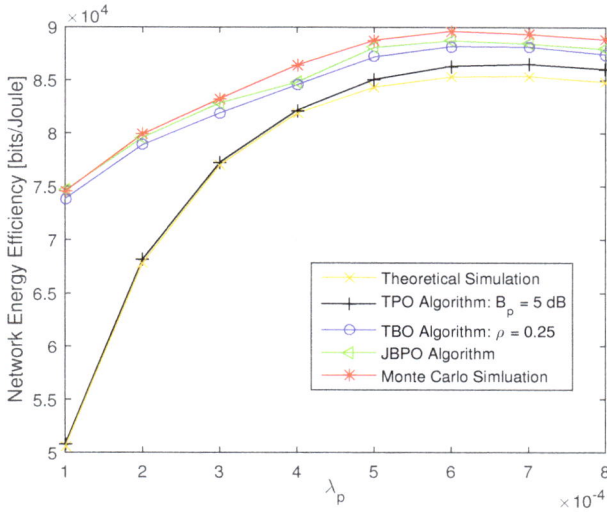

Figure 6. The network energy efficiency (EE) versus λ_p with fixed $\lambda_u = 0.0018$.

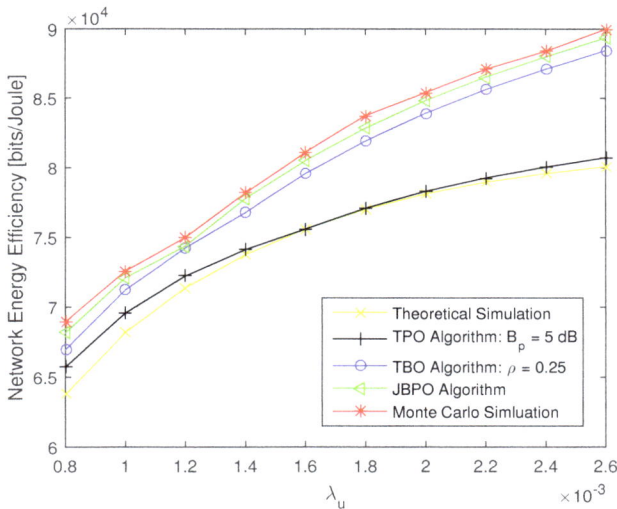

Figure 7. The network EE versus λ_u with fixed $\lambda_p = 10\lambda_m$.

Then, the performances of our proposed JBPO algorithm were compared with that of the TBPO algorithm in Figures 8 and 9 from different aspects, respectively. The relationship between network EE and λ_p with different λ_u are shown in Figure 8. We can see that the performance of our proposed JBPO algorithm was just slightly worse than that of the TBPO algorithm, but the computational complexity of JBPO was much lower than that of the TBPO algorithm. In addition, all curves of network EE increased first and then fell down slightly with PBS density increase, which illustrates that increasing PBS density can improve the network EE significantly within a certain network load. Nonetheless, when the PBS density exceeded a certain limit, further increasing will cause more complex interference and more power consumption, which will result in the network EE deterioration.

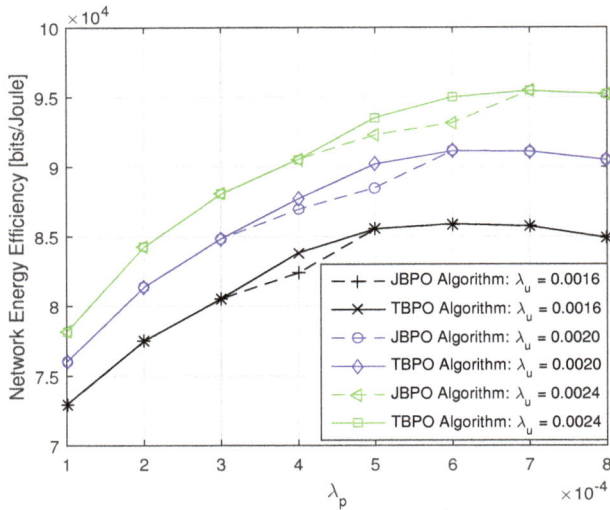

Figure 8. The network EE versus λ_p with different λ_u.

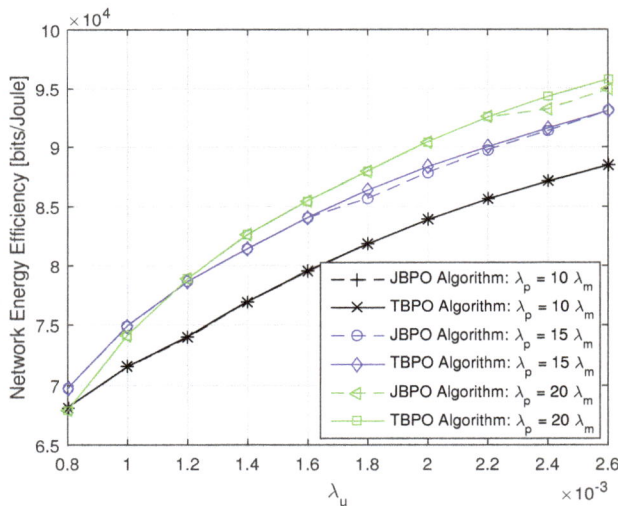

Figure 9. The network EE versus λ_u with different λ_p.

The relationship between network EE and λ_u with different λ_p are depicted in Figure 9. Due to the curves cross with each other under different UE densities, the PBS density should be carefully adjusted according to the network load fluctuation. In addition, for a given PBS density, we can see that the network EE become greater with a higher network load. Referring to Figures 8 and 9 together, although the network EE performance of the TBPO algorithm is just slightly better than that of the JBPO algorithm, the computational complexity of it is $O(n_{B_p} \times n_\rho)$, which is much higher than that of the JBPO algorithm.

Finally, the network EE performances of the JBPDO algorithm were compared with those of the PDO algorithm, JBPO algorithm with fixed $\lambda_p = 10\lambda_m$, TBPDO algorithm, and Monte Carlo simulation in Figure 10. In the TBPDO algorithm, pico CRE bias, power reduction factor and PBS density are jointly optimized to maximize network EE by an exhaustive traversal algorithm based on Equation (14). The Monte Carlo simulation results show the maximum network EE within the value range of pico CRE bias, power reduction factor and PBS density at different network load.

As shown in Figure 10, the proposed JBPDO algorithm can obtain better network EE than that of the JBPO algorithm since the JBPDO algorithm further optimizes the PBS density on the basis of the JBPO algorithm. Meanwhile, the accuracy and effectiveness of our proposed JBPDO algorithm are once again verified by Monte Carlo simulation results. In addition, although the network EE of the TBPDO algorithm outperforms our proposed JBPDO algorithm, the computational complexity of TBPDO algorithm is $O\left(n_{B_p} \times n_\rho \times n_{\lambda_p}\right)$, which is far higher than that of the JBPDO algorithm. The convergence of JBPDO algorithm is provided in Figure 11. From Figure 11, we can find that JBPDO algorithm can converge after three iterations. It is proved that the computational complexity of the JBPDO algorithm is much lower than that of the TBPDO algorithm and more suitable for the real-time network.

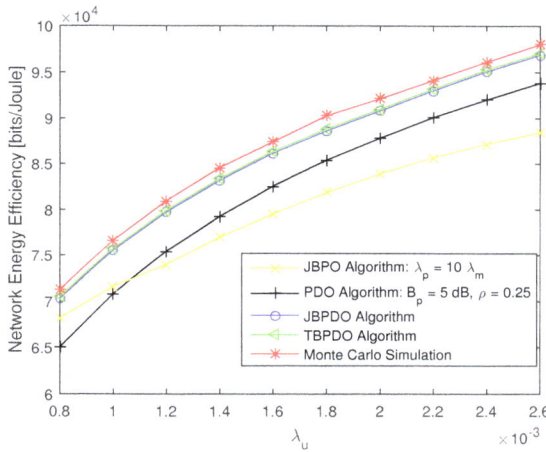

Figure 10. The network EE versus λ_u with different optimization algorithms.

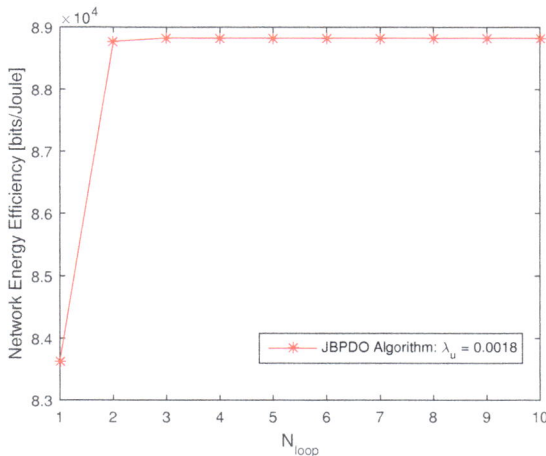

Figure 11. The network EE versus N_{loop} for JBPDO algorithm.

7. Conclusions

In this paper, pico CRE bias, PSF power reduction factor and PBS density are jointly optimized to maximize network EE for a two-tier HetNets with FeICIC. First, we derive the closed-form expression of network EE based on stochastic geometry theory. Then, the near-optimal values of pico CRE bias and power reduction factor are obtained by an alternating algorithm based on the equivalence relation between them at a given PBS density deployment. With fixed pico CRE bias and power reduction factor, the PBS density is optimized by a linear search method. Finally, a heuristic algorithm is proposed to optimize the pico CRE bias, power reduction factor, and PBS density jointly for network EE maximization. Extensive simulation results show the accuracy of network EE theoretical deduction and the effectiveness of our proposed low-complexity heuristic algorithm for network EE improvement.

Author Contributions: Y.S. conceived of the study and proposed the idea; H.X. designed the simulations and wrote the paper; S.Z. and T.W. proposed analysis model and corrected the manuscript; Y.W., Y.F. and S.X. participated in the coordination of the research, corrected the manuscript and searched for funding to support the work.

Acknowledgments: This work was supported by research grants from the National Natural Science Foundation of P.R. China (No. 61701293, 61671011, 61673253, 61420106011) and The National Key R&D Program of China (No. 2017YFE0121400).

Conflicts of Interest: The authors declare no conflict of interest.

Appendix A. Proof of Lemma 1

Considering a typical UE, it is at a distance r_k away from its nearest BS in the k tier, that is to say, there is no BS closer than r_k in the k tier. Due to the locations of MBSs and PBSs follow two iid. PPPs, the cumulative distribution function (CDF) of two-point distance r_k is

$$
\begin{aligned}
F_{r_k}(r) &= 1 - \text{Prob}\,(r_k > r) \\
&= 1 - \text{Prob}\,(no\ BS\ closer\ than\ r_k) \\
&= 1 - \exp\left(-\pi r^2 \lambda_k\right),
\end{aligned}
\tag{A1}
$$

where λ_k is the BS density of k tier. We can obtain the PDF of the distance by the differential of Equation (A1) over r as follows:

$$
f_{r_k}(r) = 2\pi r \lambda_k \exp\left(-\pi r^2 \lambda_k\right).
\tag{A2}
$$

Hence, given an arbitrary coefficient value of n, the probability of $r_p > n r_m$ can be given as:

$$
\begin{aligned}
\text{Prob}\,(r_p > n r_m) &= \text{Prob}\,(no\ BS\ closer\ than\ n r_m | r_m) \\
&= \text{Prob}\,(r_p > n r_m, r_m) \\
&= \int_0^\infty \text{Prob}\,(r_p > n r_m, r_m) f_{r_m}(r) dr \\
&= \int_0^\infty \left[\int_{nr}^\infty f_{r_p}(r)\, dr\right] f_{r_m}(r) dr \\
&= \int_0^\infty \left[\int_{nr}^\infty 2\pi r \lambda_p \exp\left(-\pi r^2 \lambda_p\right) dr\right] f_{r_m}(r)\, dr \\
&= \int_0^\infty \left[-\exp\left(-\pi r^2 n^2 \lambda_p\right)\right] 2\pi r \lambda_m \exp\left(-\pi r^2 \lambda_m\right) dr \\
&= \frac{\lambda_m}{\lambda_m + n^2 \lambda_p}.
\end{aligned}
\tag{A3}
$$

Further, given two arbitrary coefficient values of n_a and n_b, the probability of $n_a r_m < r_p < n_b r_m$ can be expressed as:

$$
\begin{aligned}
\text{Prob}\,(n_a r_m < r_p < n_b r_m) &= \text{Prob}\,(r_p > n_a r_m) - \text{Prob}\,(r_p > n_b r_m) \\
&= \frac{\lambda_m}{\lambda_m + n_a^2 \lambda_p} - \frac{\lambda_m}{\lambda_m + n_b^2 \lambda_p}.
\end{aligned}
\tag{A4}
$$

Combing Equation (A4) with Equation (2), the association probabilities of the typical UE can be obtained as Equation (5).

Appendix B. Proof of Lemma 2

According to the Bayes rule, we can obtain the conditional probability as

$$\text{Prob}\left(r_m|r_p > nr_m\right) = \frac{\text{Prob}\left(r_m < r, r_p > nr_m\right)}{\text{Prob}\left(r_p > nr_m\right)}, \tag{A5}$$

where $\text{Prob}\left(r_p > nr_m\right)$ is given in Equation (A3). After taking partial derivation with respect to the variable r, the PDF of distance r between a typical UE to its serving BS under the condition of $r_p > nr_m$ is

$$
\begin{aligned}
f_{r_m|r_p>nr_m}(r) &= \frac{d}{dr}\text{Prob}\left(r_m|r_p > nr_m\right) \\
&= \frac{1}{\text{Prob}\left(r_p > nr_m\right)}\frac{d}{dr}\text{Prob}\left(r_m < r, r_p > nr_m\right) \\
&= \frac{2\pi\lambda_m}{\text{Prob}\left(r_p > nr_m\right)}\frac{d}{dr}\int_0^r \text{Prob}\left(r_p > nr_m|r_m = r\right) f_{r_m}(r)\,dr \\
&= \frac{2\pi\lambda_m}{\text{Prob}\left(r_p > nr_m\right)}\frac{d}{dr}\int_0^r r\exp\left(-\pi r^2 n^2\lambda_p\right)\exp\left(\pi r^2\lambda_m\right)dr \\
&= \frac{\frac{\lambda_m}{\lambda_m + n^2\lambda_p}}{\text{Prob}\left(r_p > nr_m\right)}\frac{d}{dr}\left\{1 - \exp\left[-\pi r^2\left(\lambda_m + n^2\lambda_p\right)\right]\right\} \\
&= \frac{2\pi r\lambda_m}{\text{Prob}\left(r_p > nr_m\right)}\exp\left[-\pi r^2\left(\lambda_m + n^2\lambda_p\right)\right].
\end{aligned} \tag{A6}
$$

Further, we can obtain as:

$$
\begin{aligned}
f_{r_m|n_a r_m < r_p < n_b r_m}(r) &= \frac{d}{dr}\text{Prob}\left(r_m|n_a r_m < r_p < n_b r_m\right) \\
&= \frac{d}{dr}\frac{\text{Prob}\left(r_m < r, n_a r_m < r_p < n_b r_m\right)}{\text{Prob}\left(n_a r_m < r_p < n_b r_m\right)} \\
&= \frac{1}{\text{Prob}\left(n_a r_m < r_p < n_b r_m\right)}\frac{d}{dr}\left[\text{Prob}\left(r_m < r, r_p > n_a r_m\right)\right. \\
&\quad \left.-\text{Prob}\left(r_m < r, r_p > n_b r_m\right)\right] \\
&= \frac{2\pi r\lambda_m}{\text{Prob}\left(n_a r_m < r_p < n_b r_m\right)}\left\{\exp\left[-\pi r^2\left(\lambda_m + n_a^2\lambda_p\right)\right]\right. \\
&\quad \left.- \exp\left[-\pi r^2\left(\lambda_m + n_b^2\lambda_p\right)\right]\right\}.
\end{aligned} \tag{A7}
$$

According to Equations (5) and (A7), we can get the PDFs of distance r between a typical UE l to its serving BS in Equation (6).

Appendix C. Proof of Lemma 3

Due to W_l and N_l can be seen as constant and $f_l(r)$ can be obtained based on Equation (6), we can deduce the closed-form expression of the average achievable downlink rate of UE l as:

$$
\begin{aligned}
R_l &= \frac{W_l}{N_l}E\left[\log_2\left(1 + \gamma_l\right)\right] \\
&= \frac{W_l}{N_l}\int_0^\infty E\left[\log_2\left(1 + \frac{\rho_l P_l h r_l^{-\alpha}}{\rho'_l I_m + I_p + \sigma^2}\right)\right] f_l(r)\,dr,
\end{aligned} \tag{A8}
$$

where P_l is full transmission power of serving BS of user type l, r_l represents the distance between UE l and its serving BS. ρ_l is determined as $\rho_l = \rho$ when $l \in \{pm\}$ and ρ_l' is determined as $\rho_l' = \rho$ when

$l \in \{pm, pp\}$. Otherwise $\rho_l = \rho_l' = 1$. Considering the complexity of Equation (A8), we can deduce $E\left[\log_2 (1 + \gamma_l)\right]$ first as Equation (A9).

$$
\begin{aligned}
E\left[\log_2\left(1 + \frac{\rho_l P_l h r_l^{-\alpha}}{\rho'_l I_m + I_p + \sigma^2}\right)\right] &\overset{(a)}{=} \int_0^\infty \text{Prob}\left[\log_2\left(1 + \frac{\rho_l P_l h r_l^{-\alpha}}{\rho'_l I_m + I_p + \sigma^2}\right) > t\right] dt \\
&= \int_0^\infty \text{Prob}\left[h > (2^t - 1)\left(\rho'_l I_m + I_p + \sigma^2\right) r_l^\alpha \rho_l^{-1} P_l^{-1}\right] dt \\
&\overset{(b)}{=} \int_0^\infty \exp\left[-(2^t - 1)\rho'_l I_m r_l^\alpha \rho_l^{-1} P_l^{-1}\right] \exp\left[-(2^t - 1) I_p r_l^\alpha \rho_l^{-1} P_l^{-1}\right] \\
&\quad \cdot \exp\left[-(2^t - 1)\sigma^2 r_l^\alpha \rho_l^{-1} P_l^{-1}\right] dt \\
&\overset{(c)}{=} \int_0^\infty \mathcal{L}_{I_m}\left(\tau r_l^\alpha \rho'_l \rho_l^{-1} P_l^{-1}\right) \mathcal{L}_{I_p}\left(\tau r_l^\alpha \rho_l^{-1} P_l^{-1}\right) \\
&\quad \cdot \exp\left[-\tau \sigma^2 r_l^\alpha \rho_l^{-1} P_l^{-1}\right] dt,
\end{aligned}
\tag{A9}
$$

where (a) is derived according to $E(X) = \int_0^\infty P[X > x] dx$, (b) is obtained referring to $h \sim \exp(1)$, (c) is obtained through the Laplace transform of I_m and I_p by setting $\tau = 2^t - 1$. Referring to [6], the Laplace transform formulations can be expressed as:

$$
\mathcal{L}_{I_m}\left(\tau r_l^\alpha \rho'_l \rho_l^{-1} P_l^{-1}\right) = \exp\left\{-\pi \lambda_m \left(\frac{\rho'_l P_m}{\rho_l P_l}\right)^{2/\alpha} r^2 \tau^{2/\alpha} \int_{\left[B_m/\left(\rho'_l \rho_l^{-1} B_l \tau\right)\right]^{2/\alpha}}^\infty \frac{1}{1 + x^{\alpha/2}} dx\right\}
\tag{A10}
$$

$$
\mathcal{L}_{I_p}\left(\tau r_l^\alpha \rho_l^{-1} P_l^{-1}\right) = \exp\left\{-\pi \lambda_p \left(\frac{P_p}{\rho_l P_l}\right)^{2/\alpha} r^2 \tau^{2/\alpha} \int_{\left[B_{p,l}/(B_l \tau)\right]^{2/\alpha}}^\infty \frac{1}{1 + x^{\alpha/2}} dx\right\},
\tag{A11}
$$

where B_l denotes the association bias of the typical UE l to select its serving BS. When $l = pp$, $B_l = B_p$; otherwise $B_l = 1$. In order to simplify the analysis, we can define variable $\beta = B_{k,l}/B_l$, $k \in \{m, p\}$, where $B_{m,l} = B_m / \left(\rho'_l \rho_l^{-1}\right)$ and $B_{p,l} = \begin{cases} 1, \text{if } l = up \\ B_p, \text{ otherwise} \end{cases}$.

Substituting Equations (A10) and (A11) into Equation (A9), and further substituting Equation (A9) into Equation (A8), we can obtain the closed-form expression of the average achievable downlink rate of a typical UE l as Equation (9).

References

1. Zhang, S.Q.; Wu, Q.Q.; Xu, S.G.; Li, G.Y. Fundamental green tradeoffs: Progresses, challenges, and impacts on 5G networks. *IEEE Commun. Surv. Tutor.* **2017**, *19*, 33–56. [CrossRef]
2. Gandotra, P.; Jha, R.K.; Jain, S. Green Communication in Next Generation Cellular Networks: A Survey. *IEEE Access.* **2017**, *5*, 11727–11758.
3. Hu, R.Q.; Qian, Y.; Kota, S.; Giambene, G. HetNet-a new paradigm for increasing cellular capacity and coverage. *IEEE Wirel. Commun.* **2011**, *18*, 8–9.
4. Kamel, M.; Hamouda, W.; Youssef, A. Ultra-Dense Networks: A survey. *IEEE Commun. Surv. Tutor.* **2016**, *18*, 2522–2545.
5. Singh, S.; Dhillon, H.S.; Andrews, J.G. Offloading in Heterogeneous Networks: Modeling, Analysis, and Design Insights. *IEEE Trans. Wirel. Commun.* **2013**, *12*, 2484–2497.
6. Tang, H.; Peng, J.L.; Hong, P.L.; Xue, K.P. Offloading performance of range expansion in picocell networks: A stochastic geometry analysis. *IEEE Wirel. Commun. Lett.* **2013**, *2*, 511–514.
7. Merwaday, A.; Mukherjee, S.; Guvenc, I. HetNet capacity with reduced power subframes. In Proceedings of the 2014 IEEE Wireless Communications & Networking Conference (WCNC), Istanbul, Turkey, 6–9 April 2014.
8. Hu, H.; Weng, J.; Zhang, J. Coverage Performance Analysis of FeICIC Low-Power Subframes. *IEEE Trans. Wirel. Commun.* **2016**, *15*, 5603–5614.

9. Bhushan, N.; Li, J.Y.; Malladi, D.; Gilmore, R.; Brenner, D.; Damnjanovic, A.; Sukhavasi, R.T.; Patel, C.; Geirhofer, S. Network densification: The dominant theme for wireless evolution into 5G. *IEEE Commun. Mag.* **2014**, *52*, 82–89.

10. Lopez-Perez, D.; Ding, M.; Claussen, H.; Jafari, A.H. Towards 1 Gbps/UE in Cellular Systems: Understanding Ultra-Dense Small Cell Deployments. *IEEE Commun. Surv. Tutor.* **2015**, *17*, 2078–2101.

11. Kuang, Q.; Utschick, W. Energy Management in Heterogeneous Networks with Cell Activation, User Association, and Interference Coordination. *IEEE Trans. Wirel. Commun.* **2016**, *15*, 3868–3879.

12. Lorincz, J.; Matijevic, T. Energy-efficiency analyses of heterogeneous macro and micro base station sites. *Comput. Electr. Eng.* **2014**, *40*, 330–349.

13. Wei, Y.; Zhang, Z.; Yu, F.R.; Han, Z. Power allocation in hetnets with hybrid energy supply using actor-critic reinforcement learning. In Proceedings of the GLOBECOM 2017—2017 IEEE Global Communications Conference, Singapore, 4–8 December 2017; pp. 1–5.

14. Wei, Y.; Yu, F.R.; Song, M.; Han, Z. User scheduling and resource allocation in HetNets with hybrid energy supply: An actor-critic reinforcement learning approach. *IEEE Trans. Wirel. Commun.* **2018**, *17*, 680–692.

15. Yamamoto, K.; Ohtsuki, T. Parameter optimization using local search for CRE and eICIC in heterogeneous network. In Proceedings of the IEEE International Symposium on Personal, Indoor, and Mobile Radio Communication (PIMRC), Washington, DC, USA, 2–5 September 2014; pp. 1536–1540.

16. Singh, S.; Andrews, J.G. Joint Resource Partitioning and Offloading in Heterogeneous Cellular Networks. *IEEE Trans. Wirel. Commun.* **2014**, *13*, 888–901.

17. Budura, G.; Balint, C.; Otesteanu, M.; Petrut, I. HetNet performance analysis from the eICIC parameters perspective. In Proceedings of the IEEE International Symposium on Electronics and Telecommunications (ISETC), Timisoara, Romania, 27–28 October 2016.

18. Sun, Y.Z.; Wang, Z.J.; Wang, T.; Wu, T.Y.; Fang, Y. Joint Optimization of Parameter for enhanced inter cell interference coordination (eICIC) in LTE-A HetNets. *IEICE Trans. Commun.* **2017**, *100*, 799–807.

19. Yang, K.; Martin, S.; Quadri, D.; Wu, J.S.; Feng, G. Energy-Efficient Downlink Resource Allocation in Heterogeneous OFDMA Networks. *IEEE Trans. Veh. Technol.* **2017**, *66*, 5086–5098.

20. Wang, X.M.; Zheng, F.C.; Zhu, P.C.; You, X.H. Energy-Efficient Resource Allocation in Coordinated Downlink Multicell OFDMA Systems. *IEEE Trans. Veh. Technol.* **2016**, *65*, 1395–1408.

21. Mach, P.; Becvar, Z. Centralized dynamic resource allocation scheme for femtocells exploiting graph theory approach. In Proceedings of the 2014 IEEE Wireless Communications & Networking Conference (WCNC), Istanbul, Turkey, 6–9 April 2014; pp. 1479–1484.

22. Al-Zahrani, A.Y.; Yu, F.R. An Energy-Efficient Resource Allocation and Interference Management Scheme in Green Heterogeneous Networks Using Game Theory. *IEEE Trans. Veh. Technol.* **2016**, *65*, 5384–5396.

23. Xu, L.; Mao, Y.M.; Leng, S.P.; Qiao, G.H.; Zhao, Q.X. Energy-efficient resource allocation strategy in ultra dense small-cell networks: A Stackelberg game approach. In Proceedings of the IEEE International Conference on Communications (ICC), Paris, France, 21–25 May 2017; pp. 1–6.

24. Parmar, A.; Panchal, M.; Shrivastava, A. Time resource partition among users in HetNet using reduced power subframe. In Proceedings of the International Conference on Micro-Electronics and Telecommunication Engineering (ICMETE), Ghaziabad, India, 22–23 September 2016; pp. 517–522.

25. Merwaday, A.; Guvenc, I. Optimization of FeICIC for energy efficiency and spectrum efficiency in LTE-advanced HetNets. *Electron. Lett.* **2016**, *52*, 982–984.

26. Zhang, Y.H.; Cui, Z.Y.; Cui, Q.M.; Yu, X.L.; Liu, Y.J.; Xie, W.L.; Zhao, Y. Energy efficiency analysis of heterogeneous cellular networks with extra cell range expansion. *IEEE Access* **2017**, *5*, 11003–11014.

27. Nie, X.F.; Wang, Y.; Zhang, J.L.; Ding, L.Q. Performance analysis of FeICIC and adaptive spectrum allocation in heterogeneous networks. In Proceedings of the International Conference on Telecommunications (ICT), Limassol, Cyprus, 3–5 May 2017; pp. 1–6.

28. Kumbhar, A.; Güvenç, I.; Singh, S.; Tuncer, A. Exploiting LTE-Advanced HetNets and FeICIC for UAV-assisted public safety communications. *IEEE Access* **2018**, *6*, 783–796.

29. Thakur, R.; Sengupta, A.; Murthy, C.S.R. Improving capacity and energy efficiency of femtocell based cellular network through cell biasing. In Proceedings of the 2013 11th International Symposium and Workshops on Modeling and Optimization in Mobile, Ad Hoc and Wireless Networks (WiOpt), Tsukuba Science City, Japan, 13–17 May 2013; pp. 436–443.

30. Merwaday, A.; Mukherjee, S.; Güvenç, I. Capacity analysis of LTE-Advanced HetNets with reduced power subframes and range expansion. *EURASIP J. Wirel. Commun. Netw.* **2014**, *2014*, 189.

31. Liu, Y.; Chen, C.S.; Sung, C.W.; Singh, C. A game theoretic distributed algorithm for FeICIC optimization in LTE-A HetNets. *IEEE/ACM Trans. Netw. (TON)* **2017**, *25*, 3500–3513. [CrossRef]

32. Ge, X.H.; Tu, S.; Mao, G.Q.; Wang, C.X.; Han, T. 5G Ultra-Dense Cellular Networks. *IEEE Wirel. Commun.* **2016**, *23*, 72–79.

33. Lin, X.; Wang, S. Joint user association and base station switching on/off for green heterogeneous cellular networks. In Proceedings of the IEEE International Conference on Communications (ICC), Paris, France, 21–25 May 2017; pp. 1–6.

34. Sun, Y.Z.; Xu, L.; Wu, Y.T.; Wang, T.; Fang, W.D.; Shan, L.H.; Fang, Y. Energy efficient small cell density optimization based on stochastic geometry in ultra-dense HetNets. In Proceedings of the IEEE 9th International Conference on Communication Software and Networks (ICCSN), GuangZhou, China, 6–8 May 2017; pp. 249–254.

35. Zhang, T.; Zhao, J.; An, L.; Liu, D. Energy efficiency of base station deployment in ultra dense HetNets: A stochastic geometry analysis. *IEEE Wirel. Commun. Lett.* **2016**, *5*, 184–187.

36. Li, L.; Peng, M.; Yang, C.; Wu, Y. Optimization of Base-Station Density for High Energy-Efficient Cellular Networks With Sleeping Strategies. *IEEE Trans. Veh. Technol.* **2016**, *65*, 7501–7514.

37. Renzo, M.D.; Zappone, A.; Lam, T.T.; Debbah, M. System-level modeling and optimization of the energy efficiency in cellular networks—A stochastic geometry framework. *IEEE Trans. Wirel. Commun.* **2018**, *17*, 2539–2556.

38. Lorincz, J.; Matijevic, T.; Petrovic, G. On interdependence among transmit and consumed power of macro base station technologies. *Comput. Commun.* **2014**, *50*, 10–28.

39. Lorincz, J.; Capone, A.; Begusic, D. Impact of service rates and base station switching granularity on energy consumption of cellular networks. *EURASIP J. Wirel. Commun. Netw.* **2012**, *2012*, 342.

40. Andrews, J.G.; Baccelli, F.; Ganti, R.K. A Tractable Approach to Coverage and Rate in Cellular Networks. *IEEE Trans. Commun.* **2011**, *59*, 3122–3134.

41. Jo, H.S.; Sang, Y.J.; Xia, P.; Andrews, J.G. Heterogeneous Cellular Networks with Flexible Cell Association: A Comprehensive Downlink SINR Analysis. *IEEE Trans. Wirel. Commun.* **2012**, *11*, 3484–3495.

42. Singh, S.; Andrews, J.G. Rate distribution in heterogeneous cellular networks with resource partitioning and offloading. In Proceedings of the IEEE Global Communications Conference (GLOBECOM), Atlanta, GA, USA, 9–13 December 2013; pp. 3796–3801.

Article

Clustering Based Physical-Layer Authentication in Edge Computing Systems with Asymmetric Resources

Yi Chen [1], Hong Wen [2,*], Jinsong Wu [3,*], Huanhuan Song [2], Aidong Xu [4], Yixin Jiang [4], Tengyue Zhang [2] and Zhen Wang [2]

[1]   National Key Laboratory of Science and Technology on Communications, University of Electronic Science and Technology of China, Chengdu 611731, China; chenyi1309@126.com
[2]   School of Aeronautics and Astronautics, University of Electronic Science and Technology of China, Chengdu 611731, China; huanhuansong@126.com (H.S.); uestczty@163.com (T.Z.); hswinston716@163.com (Z.W.)
[3]   Department of Electrical Engineering, Universidad de Chile, Santiago 833-0072, Chile
[4]   EPRI, China Southern Power Grid Co., Ltd., Guangzhou 510080, China; xuad@csg.cn (A.X.); jiangyx@csg.cn (Y.J.)
*   Correspondence: sunlike@uestc.edu.cn (H.W.); wujs@ieee.org (J.W.)

Received: 15 March 2019; Accepted: 20 April 2019; Published: 24 April 2019

Abstract: In this paper, we propose a clustering based physical-layer authentication scheme (CPAS) to overcome the drawback of traditional cipher-based authentication schemes that suffer from heavy costs and are limited by energy-constrained intelligent devices. CPAS is a novel cross-layer secure authentication approach for edge computing system with asymmetric resources. The CPAS scheme combines clustering and lightweight symmetric cipher with physical-layer channel state information to provide two-way authentication between terminals and edge devices. By taking advantage of temporal and spatial uniqueness in physical layer channel responses, the non-cryptographic physical layer authentication techniques can achieve fast authentication. The lightweight symmetric cipher initiates user authentication at the start of a session to establish the trust connection. Based on theoretical analysis, the CPAS scheme is secure and simple, but there is no trusted party, while it can also resist small integer attacks, replay attacks, and spoofing attacks. Besides, experimental results show that the proposed scheme can boost the total success rate of access authentication and decrease the data frame loss rate, without notable increase in authentication latencies.

Keywords: edge computing; clustering; physical-layer authentication; lightweight cipher; channel state information; lightweight authentication

---

1. Introduction

With the rapid development of Internet of things (IoT) technologies, various intelligent terminals (devices) have penetrated into our daily lives and works. As is well known, the traditional cloud computing system has some inherent limitations, namely real-time control incompetence [1], heavy network traffic, cloud data privacy insecurity, and so on. Luckily for us, the edge computing paradigm can also meet the key industrial requirements (such as instant links, real-time business, low latency and jitter, data security and privacy protection, and so on) by building small edge data centers [2]. As shown in Figure 1, the edge computing system consists of edge devices (edge servers) who are usually specific high-end servers with powerful central processing unit (CPU), larger memory and storage, and various terminals that usually have limited resources [3] (such as limited computation power, battery, memory, and bandwidth) due to cost constraints. Thus, it is vulnerable for IoT devices to be attacked by hackers or illegal users (such as replay, impersonation, eavesdropping, tampering, and so on) due to asymmetric resources. Identity authentication for communication participants (edge devices and terminals) is the basis and key to information security and privacy protection. Once the authentication system crashes, the whole system will be insecure. Traditional cryptographic ciphers can be divided into two categories, symmetric and asymmetric ciphers. Some of conventional symmetric ciphers are AES, DES or 3DES, and so on. RSA (Rivest, Shamir, and Adleman) and ECC (Elliptic Curve Cryptography) are the common asymmetric algorithms. They have one thing in common, namely large key size, which makes encryption or decryption slow and increases the complexity [4]. However, resource-constrained terminals often fail to satisfy the large memory requirements to store the large key size. Due to the limited resources about terminals, it is not suitable to use traditional complex encryption algorithms to implement access authentication. Therefore, it is necessary to design a lightweight identity authentication program for edge computing systems with asymmetric resources.

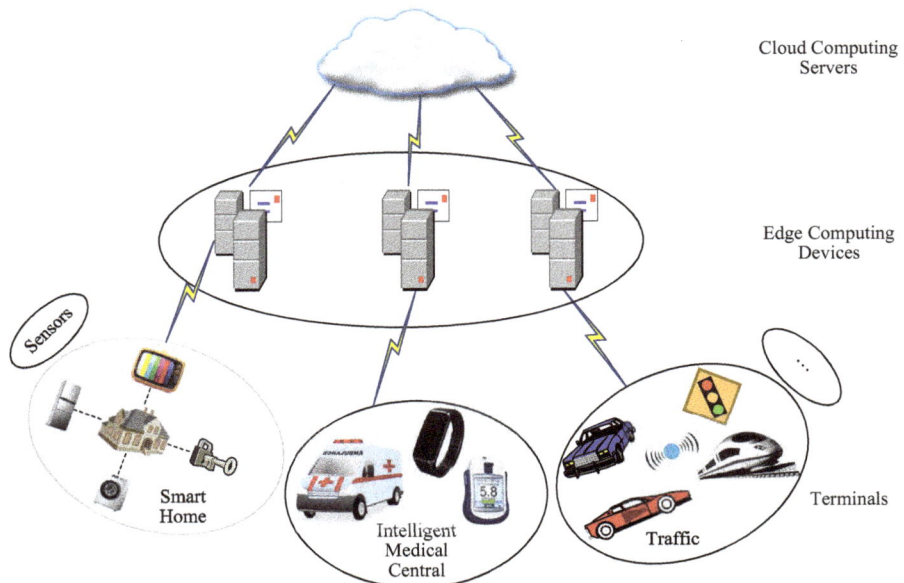

Figure 1. A simplified model of edge computing system.The edge computing system consists of edge computing devices who are usually specific high-end servers with powerful central processing unit, larger memory and storage, and various terminals that usually have limited resources [3] (such as limited computation power, battery, memory and bandwidth).

To provide security for resource-constrained devices, many lightweight symmetric ciphers have been proposed, such as MCRYPTON, HIGHT, PRESENT, MIBS, Piccolo, KLEIN, and so on [5]. They are secure and relatively fast but with low costs, and usually use the same key for both encryption and decryption of data [4]. Additionally, non-cryptographic authentication mechanisms based on physical-layer characteristics have been proposed for information security and privacy protection of devices in recent years [6–9], which have higher levels of security [10]. The authentication technique of physical layer based on channel state information (CSI) is one of the non-cryptographic authentication mechanisms [11], which can augment traditional network security [12]. It is carried out via comparing the similarity of CSI [13,14], which has the physical-layer channel characteristics of spatial-temporal uniqueness and can be extracted from the received data frames. In recent years, there have been many physical layer authentication methods based on machine learning (ML) [15–20]. However, the ML based physical layer authentication approach needs a large number of samples to train the network, which is unrealistic for real-time application. For the authentication technique of physical layer CSI, many research results have also been obtained [12–14,21–25]. However, the authentication rates of these methods need to be improved for their applications. The authentication rate mainly relies on the accuracy of CSI and the determination of test threshold. Finding suitable method to set the threshold according to environment is the most important to get high authentication rate, especially dynamically setting the threshold. Therefore, this paper present a clustering based physical-layer authentication scheme (CPAS). The proposed approach is a tradeoff between the traditional schemes [12–14,22] and machine learning based methods [15–20] for complexity and authentication rate. The advantage of the CPAS scheme is that the proposed method can adjust the decision threshold adaptively by updating the physical-layer channel authentication model and can be performed under limited data frames in the beginning, which can support the fast access.

Clustering is the unsupervised classification of data items into clusters [26]. Cluster analysis with little or no prior knowledge includes advanced techniques across various fields [27]. It plays a significant role in many disciplines [28]. Many researchers have proposed clustering algorithms [29,30]. However, there is little research on physical-layer security using clustering techniques. Considering the idea of clustering, in this research paper, we propose a clustering based physical-layer authentication scheme (CPAS), which is a novel cross-layer secure authentication approach for edge computing system with asymmetric resources. The CPAS scheme combines clustering technique and lightweight symmetric cipher with physical-layer channel state information to achieve two-way authentication between edge devices and terminals. The edge device does not drop data frames directly when physical-layer channel authentication fails, but to activate upper layer authentication to verify the legality of the data frames, which can resist losing legitimate data frames but lead to some processing delay. Moreover, multiple channel state information are used to establish a physical layer channel authentication model in the CPAS scheme, which magnify the differences between the multiple channel state information, but no effect on the performance of authentication. Experimental results show that our proposed scheme can effectively improve the success rate of physical-layer channel authentication, total success rate of access authentication and decrease the data frame loss rate without significantly increasing processing time. It is not only secure but also simple and flexible, especially independent of a third party. In addition, our scheme could resist spoofing attacks, replay attacks and small integer attacks. It can significantly reduce the access authentication complexity and achieve greater security for the edge computing system with asymmetric resources.

We summarize our main contributions as follows.

- We propose the first CPAS scheme, which combines clustering and lightweight symmetric cipher with physical-layer channel state information firstly and can be employed to authenticate mutually between terminals and edge devices. We also show the detailed implementing procedures of the proposed scheme.
- We analyze the security of the proposed scheme and prove that it can resist small integer attacks, replay attacks, and spoofing attacks.
- The CPAS scheme is implemented in a real world environment based on MIMO-OFDM systems. We also show the impacts of adjusting parameters of clusters on the success rate of physical-layer channel authentication, the data frame loss rate, the total success rate of access authentication, and the time cost through experimental results demonstration.

The rest of this paper is organized as follows. Section 2 introduces the basic principles of physical layer channel authentication. The system model and proposed CPAS scheme are presented in Section 3. The security of the proposed scheme is analyzed in Section 4. In Section 5, the experiment results indicate that the proposed CPAS scheme is effective for authentication. We conclude this paper in Section 6.

2. Basic Principles of Physical Layer Channel Authentication

In this section, we briefly present the basic principles of physical-layer channel authentication and show the shortcomings of some authentication schemes.

Xiao et al. designed a physical-layer authentication scheme via exploiting the spatial variability of the radio channel response [13]. However, the proposed scheme in [13] has the disadvantage of authenticating the initial data frame that is usually assumed to be valid. In their scheme, the receivers need to estimate the radio channel response, shown below

$$\underline{\mathbf{H}}_k = [H_k(f_1), \cdots, H_k(f_i), \cdots, H_k(f_M)]^T, \tag{1}$$

where k denotes the data frame index, $f_i = f_0 + \left(\frac{i}{M} - \frac{1}{2}\right)W$, $i = 1, 2, \cdots, M$, f_0 is the center measurement frequency, W is the measurement bandwidth, and M is the number of measurement frequency over the measurement bandwidth.

The receiver utilizes channel state information in two consecutive data frames, $\mathbf{H}_{k-1}$ and $\mathbf{H}_k$, and hypothesis testing to determine whether they come from the same sender or not. Hypothesis testing is the task of deciding which of the two hypotheses, $\mathcal{H}_0$ or $\mathcal{H}_1$, is true, when one is given the value of a random variable [22]. $\mathbf{H}_{k-1}$ and $\mathbf{H}_k$ can be estimated by ILS channel estimation method [23–25]. In the null hypothesis, $\mathcal{H}_0$, the claimant user is the initial sender. The base station accepts this hypothesis if the test statistic T is below some threshold Γ. Otherwise, in the alternative hypothesis, $\mathcal{H}_1$, the claimant is someone else. The notation "$\sim$" is used to indicate accurate values without measurement errors, and thus have

$$\begin{aligned} \mathcal{H}_0 &: \widetilde{\underline{\mathbf{H}}}_k = \widetilde{\underline{\mathbf{H}}}_{k-1} \\ \mathcal{H}_1 &: \widetilde{\underline{\mathbf{H}}}_k \neq \widetilde{\underline{\mathbf{H}}}_{k-1} \end{aligned}. \tag{2}$$

The inherent physical parameters of the multi-path fading channels were exploited to support continuous mutual authentication between wireless terminals by He et al. [22]. He et al. [22] used the information of both amplitude and phase in the channel signature to enhance the communication security. They employed three statistical channel signature information to strengthen physical security. However, in reality, the noisy power is unknown. Thus, the test statistic of channel responses is normalized as follows

$$\Lambda_i = \frac{K_{co}||(\mathbf{H}_{k-i+1}(i) - \mathbf{H}_{k-i}(i)e^{j\varphi})||^2}{||\mathbf{H}_{k-i}(i)||^2}, \tag{3}$$

where "i" is an index, $i = 1, 2, \cdots, S$, "S" is a positive integer, and $S \geqslant 1$. Then, the cumulative summation of the log-likelihood ratio Λ is calculated as

$$\Lambda = K_{co\_S} \sum_{i=1}^{S} \Lambda_i \underset{<\mathcal{H}_0}{\overset{>\mathcal{H}_1}{\gtrless}} \Gamma, \tag{4}$$

where K_{co_S} denotes the normalization factor to let the threshold value $\Gamma \in [0, 1]$. When $S > 1$, it is sequential probability ratio test (SPRT). A SPRT could compare $\widetilde{\underline{\mathbf{H}}}_k$ with all past records ($\widetilde{\underline{\mathbf{H}}}_i$), where $i < k$ in some way. When $S = 1$, it is a likelihood ratio test (LRT). The LRT only compares the estimation in the kth data frame ($\widetilde{\underline{\mathbf{H}}}_k$) with that in the $(k-1)$th data frame ($\widetilde{\underline{\mathbf{H}}}_{k-1}$).

3. System Model and Proposed Scheme

We consider the edge computing scenario shown in Figure 2, which consists of various terminals (T_E), also called Alice, and edge computing devices (ED), also called Bob. They want to exchange messages across a wireless link. It must be assured that the received data frames are all coming from the correct communication pair. Compared with the terminals with limited resources, edge devices are usually specific high-end servers with powerful CPUs, larger memory and storage units. Alice and Bob can perform authentication with each other via exchanging messages in the edge computing system with asymmetric resources. Their evil adversary, Eve, will play the part of an active opponent that injects undesirable messages into the medium in the expectations of spoofing Bob.

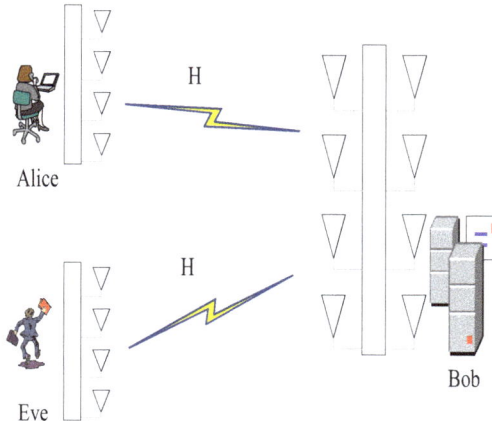

Figure 2. Scenario with Alice (T_E), Bob (ED), and Eve.

The proposed authentication scheme is divided into secret key sharing, initial authentication, physical-layer channel modeling, physical-layer channel authentication, lightweight cryptographic authentication, and model update of physical-layer channel authentication.

3.1. Secret Key Sharing

A secret key named *Key* is shared between Alice and Bob over a secure channel. This is not the essence of this article, thus we omit it here.

3.2. Initial Authentication

The initial authentication between the terminal and the edge computing device is completed through a lightweight cryptographic algorithm by using the same secret key. As shown in Figure 3, the initial full authentication phases are as follows:

(1)   Alice generates a pseudorandom number $PS1$, and encrypts $PS1$ with a lightweight cryptographic algorithm to obtain ciphertext $Y_1 = E_{(key)}(PS1)$, where $E_{(key)}(PS1)$ means that encrypting message, such as the random number $PS1$ in the parentheses by using a lightweight cryptographic algorithm and a secret key. Then, the terminal generates a login request message M_1 and sends it to the edge computing device, where the request message M_1 includes the ciphertext Y_1.

(2)   Bob extracts the channel state information H_1 from the received signal sent by Alice, and then gets the ciphertext Y_1' from decoding data and the plaintext $PS1'$ via decrypting Y_1' with the same lightweight cryptographic algorithm and secret key, where $PS1' = D_{(key)}(Y_1')$, $D_{(key)}(Y_1')$ means that decrypting message, such as Y_1' in the parentheses via using a lightweight cryptographic algorithm and a secret key, and the channel information H_1 is a complex matrix of m rows and n columns.

(3)   Bob generates two pseudorandom numbers $PS2$ and $PS3$, and calculates the ciphertext $Y_2 = E_{(key)}(PS1' \parallel PS2 \parallel PS3)$. Then, Bob sends a response message M_2 to Alice, where M_2 contains the ciphertext Y_2.

(4)　Alice verifies the legitimacy of Bob. When Alice receives the response message M_2', it decodes M_2' to obtain the ciphertext Y_2', and then decrypts Y_2' to obtain the plaintext $(PS1' \parallel PS2' \parallel PS3')$ $= D_{(key)}(Y_2')$. If the $PS1'$ is not equal to $PS1$, Bob is an illegal edge device and Alice cancels the login; otherwise, Alice considers Bob to be a legitimate edge computing device, calculates two response messages M_3 and M_4, and continuously sends them to the edge computing device, where M_3 includes ciphertext $Y_3 = E_{(key)}(PS2')$, and M_4 contains ciphertext $Y_4 = E_{(key)}(PS3')$.

(5)　Bob verifies the legitimacy of Alice. Bob extracts the channel information H_2 and H_3 from the received response messages M_3 and M_4 sent from Alice, and then gets the ciphertext Y_3' and Y_4' from decoding and the plaintext $PS2'$ and $PS3'$ by decrypting Y_3' and Y_4' with the same lightweight cryptographic algorithm and secret key, where $PS2' = D_{(key)}(Y_3')$, $PS3' = D_{(key)}(Y_4')$, the channel information H_2 extracted by Bob from M_3 and H_3 from M_4, H_2 and H_3 are complex matrices of m rows and n columns. If $PS2'$ is equal to $PS2$ and $PS3'$ is matching to $PS3$, Bob considers Alice as a legitimate terminal, and the initial authentication ends; otherwise, Alice is an illegal terminal and Bob cancels the login.

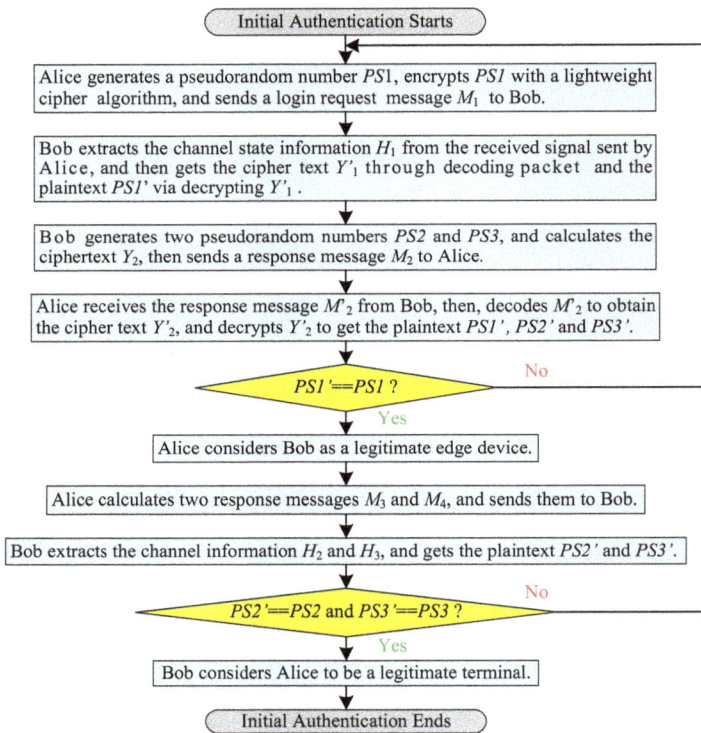

Figure 3. Process flowchart of initial authentication.

3.3. Physical-Layer Channel Modeling

Bob uses the channel state information, detected and estimated within the correlated time, for the physical-layer channel modeling. We consider the idea of clustering that is the task of organizing a set of objects into groups whose members are more similar to each other than to those in other groups (clusters). Bob needs at least three data frames to model the physical layer channel (organize a cluster of similar data frames). As shown in Figure 4, the physical-layer channel model consists of four parts: preprocessing channel state information, locating central position of cluster (channel model), estimating coverage radius of cluster, and clustering physical-layer channel model. Figure 5 is the detailed modeling principle of physical-layer channel.

(1)   Preprocessing channel state information

The channel information H_1, H_2, and H_3, which are extracted during the initial full authentication phase, are complex matrices of m rows and n columns, where m denotes the number of carriers, and n indicates the number of antennas. To obtain the statistical characteristics of channel information, we accumulate the absolute value of the real part and the imaginary part about the complex matrices, respectively. The statistical coordinates of channel information are named as $H_1'(x_1, y_1)$, $H_2'(x_2, y_2)$, and $H_3'(x_3, y_3)$, which are coordinate pairs on the complex plane.

(2)   Locating central position of cluster

After completing the previous sub-step, preprocessing channel information, the central position of cluster (channel model), named as $W(x, y)$, is estimated by

$$\begin{cases} x = \dfrac{min\{x_1, x_2, x_3\} + max\{x_1, x_2, x_3\}}{2} \\ y = \dfrac{min\{y_1, y_2, y_3\} + max\{y_1, y_2, y_3\}}{2} \end{cases}, \tag{5}$$

where $min\{\cdot\}$ represents minimum value, while $max\{\cdot\}$ implies maximum value.

(3)   Estimating coverage radius of cluster

The Euclidean distances between the central position $W(x, y)$ and the statistical position of channel information $H_1'(x_1, y_1)$, $H_2'(x_2, y_2)$, and $H_3'(x_3, y_3)$ are given by

$$\left\| WH_n' \right\| = \sqrt{(x_W - x_{H_n'})^2 + (y_W - y_{H_n'})^2}, \tag{6}$$

where $\|WH_n'\|$ $(n = 1, 2, 3)$ denotes the Euclidean distances between $W(x, y)$ and H_1', H_2', and H_3', respectively. Then, the maximum Euclidean distance is taken as the radius (R) of cluster.

$$R = max\left\{ \left\| WH_1' \right\|, \left\| WH_2' \right\|, \left\| WH_3' \right\| \right\}, \tag{7}$$

where R denotes the radius of cluster. Further, the coverage radius of channel model is obtained by

$$dist = R + \theta, \tag{8}$$

where θ is the adjusting parameter of the coverage radius of channel model.

(4)   Clustering physical-layer channel model

When the central position and the coverage radius of channel model are determined, the categories of physical-layer channel model are defined as

$$C_i = \{W_i, dist_i\}, \tag{9}$$

where i indicates the index of terminal, and different C_i is specified for a different cluster, i.e., a different terminal.

The physical-layer channel modeling is completed.

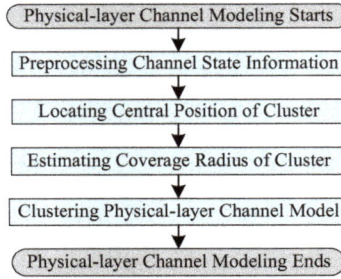

Figure 4. Process flowchart of physical-layer channel modeling.

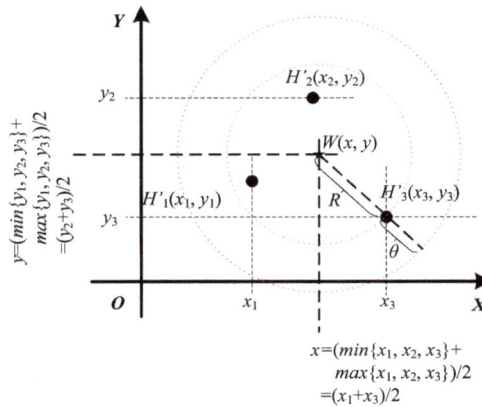

Figure 5. Principle map of physical-layer channel modeling.

3.4. Physical-Layer Channel Authentication

When Bob receives a new data frame, it can directly verify the legality of the data frame according to the established physical-layer channel model. Figure 6 is the process flowchart of physical-layer channel authentication. The detailed authentication principle map of physical-layer channel is exhibited in Figure 7.

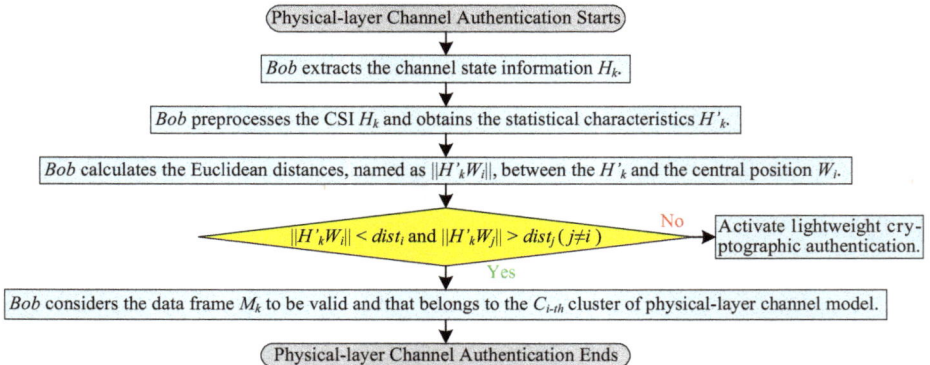

Figure 6. Process flowchart of physical-layer channel authentication.

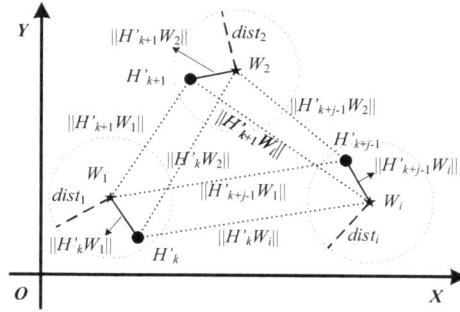

Figure 7. Principle map of physical-layer channel authentication.

(1) Bob extracts the channel information H_k from the received data frame M_k sent from Alice, where, the channel information H_k is a complex matrix of m rows and n columns, the data frame M_k contains the cipher text $Y'_k = E_{(key)}(PS2'_i \oplus PS3'_i)$, "$\oplus$" means XOR function, and the k indicates the index of data frame.

(2) Bob preprocesses the channel information H_k. To obtain the statistical characteristics $H'_k(x_k, y_k)$ of channel information, Bob accumulates the absolute value of the real part and the imaginary part of H_k, respectively. The statistical characteristics $H'_k(x_k, y_k)$ denote the coordinate pairs on the complex plane.

(3) Bob checks the validity of the data frame M_k. Firstly, Bob calculates the Euclidean distances, named as $||H'_k W_i||$, between the H'_k and the central position W_i of the cluster, respectively. Then, Bob compares the sizes of $||H'_k W_i||$ and $dist_i$: when $||H'_k W_i|| < dist_i$ $(i \in S = \{1, 2, \cdots\})$ and $||H'_k W_j|| > dist_j$ $(\forall j \in \{j | j \in S, j \neq i\})$, Bob considers the data frame M_k to be valid and that belongs to the $C_{i\text{-th}}$ cluster (physical-layer channel model); otherwise, Bob activates lightweight cryptographic authentication.

3.5. Lightweight Cryptographic Authentication

During the non-initial authentication phase, if Bob cannot check the validity of the data frame M_k coming from terminal through the physical-layer channel authentication, the lightweight cryptographic authentication will be activated. The process flowchart of lightweight cryptographic authentication is shown in Figure 8.

(1) Bob gains the ciphertext, Y'_k, and the number of data frame, PS_k, which is also a pseudorandom number, via decoding the data frame M_k sent from Alice, where $Y'_k = E_{(key)}(PS2'_i \oplus PS3'_i)$, and the length of the random number is determined according to the actual application scenario. If PS_k matches the previous number of data frame, ED considers M_k as a replayed packet and throws it away; otherwise, Bob goes to next step.

(2) Bob decrypts the ciphertext Y'_k to get the plaintext $(PS2'_i \oplus PS3'_i) = D_{(key)}(Y'_k)$.

(3) Bob checks the validity of the data frame M_k. If $(PS2'_i \oplus PS3'_i)$ does not match $(PS2_i \oplus PS3_i)$, the data frame M_k is illegal and Bob discards it. If $(PS2'_i \oplus PS3'_i)$ is equal to $(PS2_i \oplus PS3_i)$, Bob considers M_k as a valid data frame, and then extracts and records its channel information H_k. When Bob receives j data frames $\{M'_k, M'_{k+1}, \cdots, M'_{k+(j-1)}\}$, namely lightweight cryptographic authentication being activated j times continuously, the model update of physical-layer channel authentication will be activated.

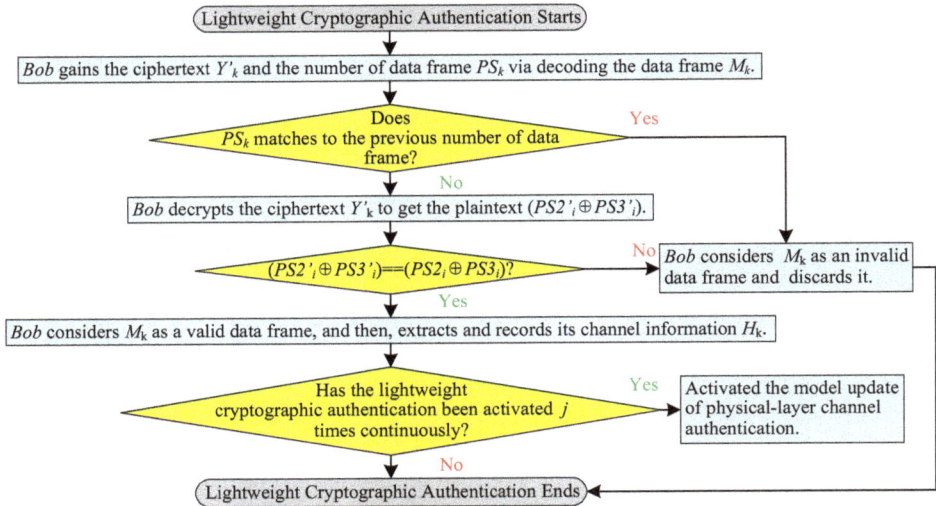

Figure 8. Process flowchart of lightweight cryptographic authentication.

3.6. Model Update of Physical-Layer Channel Authentication

When lightweight cryptographic authentication is activated continuously j times to verify the validity of data frames $\{M'_k, M'_{k+1}, \cdots, M'_{k+(j-1)}\}$, Bob needs to update the physical-layer channel model for a renewed physical-layer authentication, where $j \geqslant 3$. Figure 9 presents the process flowchart of model update of physical-layer channel authentication, which similar to the physical-layer channel modeling also contains four parts: preprocessing the new channel information, locating the new central position of the cluster, estimating the new coverage radius of the cluster, and re-clustering the physical-layer channel model. The detailed model update principle map of the physical-layer channel is displayed in Figure 10.

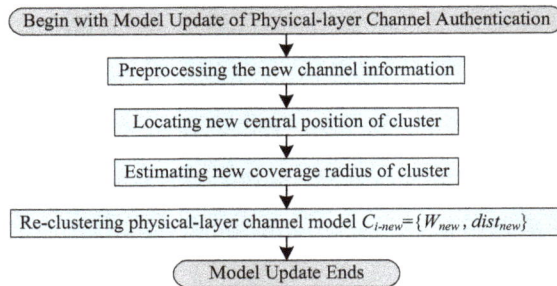

Figure 9. Process flowchart of model update of physical-layer channel authentication.

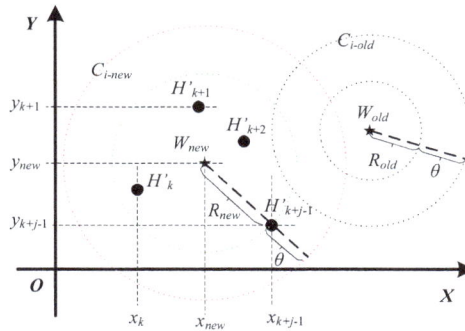

Figure 10. Principle map of model update of physical-layer channel authentication.

(1)   Preprocessing the new channel information

The sequences of channel information H_k, H_{k+1}, $\cdots$, $H_{k+(j-1)}$, which are extracted during the lightweight cryptographic authentication phase, are complex matrices of m rows and n columns. To obtain the statistical characteristics of channel information, we accumulate the absolute values of the real part and the imaginary part about the complex matrices, respectively. The statistical sequences of channel information are named as $H_k'(x_k, y_k)$, $H_{k+1}'(x_{k+1}, y_{k+1})$, $\cdots$, $H_{k+j-1}'(x_{k+j-1}, y_{k+j-1})$, which are coordinate pairs on the complex plane.

(2)   Locating new central position of cluster

After completing the previous sub-step, preprocessing the new channel information, the new central positions of physical-layer channel model, named as $W_{new}(x_{new}, y_{new})$, are estimated by Equation (10).

$$\begin{cases} x_{new} = \frac{min\{x_k, x_{k+1}, \cdots, x_{k+j-1}\} + max\{x_k, x_{k+1}, \cdots, x_{k+j-1}\}}{2} \\ y_{new} = \frac{min\{y_k, y_{k+1}, \cdots, y_{k+j-1}\} + max\{y_k, y_{k+1}, \cdots, y_{k+j-1}\}}{2} \end{cases}. \tag{10}$$

(3)   Estimating new coverage radius of cluster

The Euclidean distances between the new central position $W_{new}(x_{new}, y_{new})$ and the statistical sequences of channel information $H_k'(x_k, y_k)$, $H_{k+1}'(x_{k+1}, y_{k+1})$, $\cdots$, $H_{k+j-1}'(x_{k+j-1}, y_{k+j-1})$, are given by

$$\left\| W_{new} H_n' \right\| = \sqrt{\left(x_{W_{new}} - x_{H_n'}\right)^2 + \left(y_{W_{new}} - y_{H_n'}\right)^2}, \tag{11}$$

where $\|W_{new} H_n'\|$ ($n = k$, $k + 1$, $\cdots$, $k + j - 1$) denote the Euclidean distances. Then, the maximum Euclidean distance is taken as the new radius (R_{new}) of cluster.

$$R_{new} = max\{\|W_{new} H_k'\|, \|W_{new} H_{k+1}'\|, \cdots, \|W_{new} H_{k+j-1}'\|\}, \tag{12}$$

where R_{new} denotes the new radius of channel model. Further, the new coverage radius of cluster is obtained by

$$dist_{new} = R_{new} + \theta, \tag{13}$$

where θ indicates the adjusting parameter of the coverage radius of channel model.

(4)   Re-clustering physical-layer channel model

When obtaining the new central position and the new coverage radius of channel model, the new cluster of physical-layer channel model is redefined as

$$C_{i\text{-}new} = \{W_{i\text{-}new}, dist_{i\text{-}new}\}. \tag{14}$$

The model update of physical-layer channel authentication is completed.

4. Security Analysis

In this section, the proposed CPAS scheme is analyzed with respect to the security.

The proposed CPAS scheme can be used to authenticate mutually between terminals (Alice) and edge devices (Bob) for the edge computing system with asymmetric resources, despite the presence of Eve. In the CPAS scheme, the following security measures are adopted.

Firstly, the lightweight cipher algorithm is one of the security measures. A different lightweight cipher has a different security intensity. CPAS scheme can choose different lightweight cipher flexibly to encrypt data. Bob is usually a specific high-end server. He has the ability to withstand complex computations for different cryptographic algorithms. However, the appropriateness of Alice's ciphers depend on her resources. Besides, there is no trusted party involved in the authentication process. Thus, the strategy is feasible for resource-constrained terminals, if lightweight cipher just keep them safe in a certain time, according to the requirement of application.

The second security measure is the use of pseudorandom number. The replay attacks and small integer attacks cannot be successful since the authentication messages are not the same every time. This is due to the use of dynamic authentication messages combined with a different pseudorandom number in every communication session and every data frame. In other words, the authentication packets generated in different valid phases are different, and the current authentication messages are valid only for the current authentication phase, since the pseudorandom number cannot be enumerated and the valid authentication messages cannot be generated in a period of data transmission. Thus far, researchers have proposed a lot of pseudorandom number generators [31–34]. The periods of different pseudorandom generators are different. For example, the Mersenne Twister MT19937 is a pseudorandom number generator and it has a large period of $2^{19937-1}$ [34]. Bob could still bear its computational complexity. In practical applications, users can choose the appropriate pseudorandom number generator according to their own needs. Thus, the exhaustive attacks and guessing attacks are also impossible, since the authentication messages are not the same every time.

In addition, physical-layer channel state information recognition technique is another security measure. It depends on the spatiotemporal uniqueness of physical-layer channel characteristics, which can be estimated from the received data frames. This can assist CPAS scheme to resist the spoofing attacks. Eve could not convince Bob that she is Alice.

Therefore, the proposed CPAS scheme not only can implement bidirectional authentication between Alice and Bob, but also can withstand replay attacks, small integer attacks, and spoofing attacks.

5. Performance

To examine the performances of the proposed CPAS scheme, we firstly simulated it in MATLAB under different signal-to-noise ratios (SNRs). In the simulations, we set the maximum Doppler shift of 15 Hz, the bandwidth of 1 MHz, the digital modulation method of QPSK, the number of subcarrier 128, the number of multi-paths 5, and 1000 times test.

Detection rate and false alarm rate of physical-layer channel authentication are two critical measurements. Detection rate of physical-layer channel authentication indicates the probability of illegal data frames detection and false alarm rate of physical-layer channel authentication denotes the probability of legitimate data frames detected as illegitimate. When the false alarm rate is smaller and detection rate is bigger, the authentication performance is better, where the false alarm rate of 0 and the detection rate of 1 are the ideal performances. Figure 11 depicts the diagram of detection rate and false alarm rate of physical layer channel authentication for different adjusting parameter θ. The proposed scheme was compared with the LRT and SPRT schemes. The performances of these schemes upgraded gradually with the increase of SNR, while the performance of CPAS was better than those of the other schemes under the same SNR.

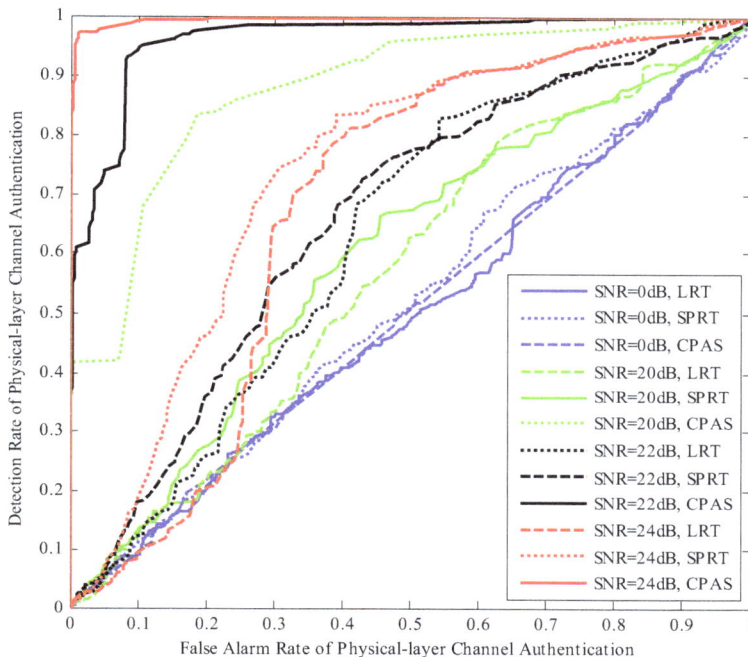

Figure 11. Comparisons of detection rate and false alarm rate under different SNRs.

The simulations in MATLAB demonstrated the advantages of the CPAS scheme, which was also implemented over universal software radio peripheral (USRP) platform [35–37]. Experiments were performed in an office room, which is 8 m long, 7.5 m wide, and 3 m high. Edge computing device was equipped with an 8×8 MIMO system. Terminal was equipped with a 2×2 MIMO system. They worked on the center frequency 3.5 GHz with the sub-bandwidth 2 MHz, the number of subcarrier 128, and the interval of sub-carriers 15.625 kHz. The wavelength of the transmission signal was about 0.086 m. The maximal transmitting power was 15 dBm and transmission gain 20 dB. The communication scheme was based on MIMO-OFDM (Multiple Input and Multiple Output—Orthogonal Frequency Division Multiplexing) and ILS (Improved-scaled Least Squares) was adopted to estimate channels. In our experiments, we employed RC4 algorithm to act a lightweight cryptographic algorithm, which is not the focus of this paper.

We considered the following performance metrics to evaluate the proposed scheme: success rate of physical-layer channel authentication, data frame loss rate, total success rate of authentication, and time cost. Success rate of physical-layer channel authentication indicates the probability of success in physical-layer channel authentication. Data frame loss rate means the ratio of the data frames lost to the data frames received by the receiver. Total success rate of authentication contains the success rate of physical-layer channel authentication and lightweight cryptographic authentication. Time cost represents the time required to authenticate data frames in simulation work, which consists of the time overhead of RC4 key initialization, physical-layer channel authentication (physical-layer channel modeling and model update also included in CPAS scheme), data demodulation, and upper layer cipher authentication. The comparative results are shown in Figures 12–16. The values in the figures are all statistics in 1000 trials.

Figure 12 plots the success rate of physical-layer channel authentication at a given $j = 3$ for varying threshold values or adjusting parameter θ. The success rates of physical-layer channel authentication gradually increased with the increasing adjusting parameter θ. When the adjusting parameter θ was high, greater than 1, the LRT, SPRT, and CPAS schemes contributed to high success rates of physical-layer channel authentication. When θ was less than 1, the success rate of physical-layer channel authentication decreased with the decreasing adjusting parameter θ. This decrease was, however, more significant in the case of the LRT and SPRT schemes. Especially, the LRT and SPRT schemes had near zero success rate of physical-layer channel authentication for θ close to zero due to each data frame received by the edge device being different, but the proposed CPAS scheme had a higher success rate due to three data frames being used to establish a physical-layer channel authentication model. Thus, the proposed scheme had a higher success rate of physical-layer channel authentication when θ was small.

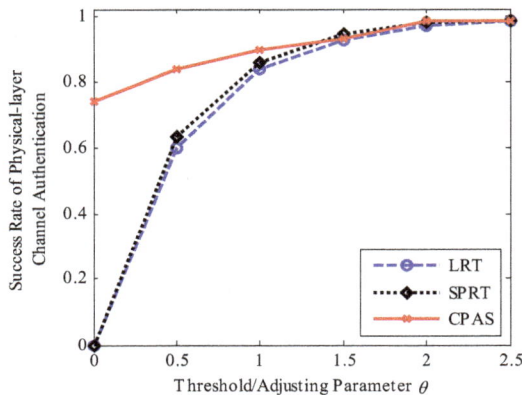

Figure 12. Success rate of physical-layer channel authentication versus threshold values or adjusting parameter θ. It shows the success rate of physical-layer channel authentication of different schemes at different θ.

Figure 13 demonstrates the comparisons among these schemes in terms of data frame loss rate. The data frame loss rate of LRT and SPRT gradually decreased with the increase of the adjusting parameter θ, while the data frame loss rate of the proposed scheme was always close to 0. It is worth noting that LRT scheme had 50% data frame loss rate and SPRT scheme had 33.3% data frame loss rate but the proposed scheme had near zero data frame loss rate when $\theta = 0$. The reason was that Bob dropped the data frame directly when the physical-layer channel authentication failed and upper layer authentication was required before each physical-layer channel authentication in the LRT and SPRT schemes. Our scheme did not discard data frames directly but activated upper layer authentication to check the validity of the data frames. Thus, no matter the value of parameter θ, the data frame loss rate of our proposed scheme was close to zero, as long as the data frame was legitimate.

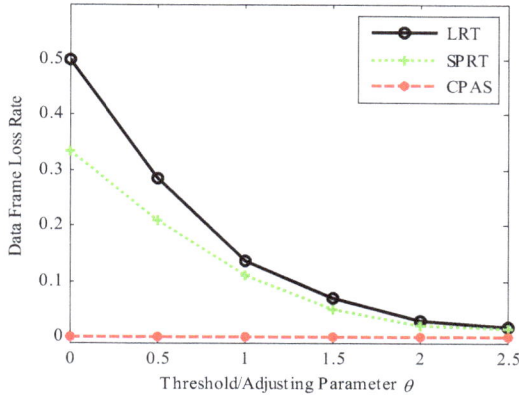

Figure 13. Data frame loss rate.

Figure 14 shows the comparisons among the LRT, SPRT, and CPAS schemes in terms of total success rates of authentication, assumed to be free of attack. The total success rates of physical-layer channel authentication gradually increased with the increase of the threshold value in the LRT and SPRT schemes, while it was always close to 100% with the increase of adjusting parameter θ in the proposed scheme. The reason was that the edge device did not drop data frames directly, when physical-layer channel authentication failed, but activated upper layer authentication to verify the legality of the data frames in the CPAS scheme. This resisted losing legitimate data frames when physical-layer channel authentication failed. However, this led to some processing delay.

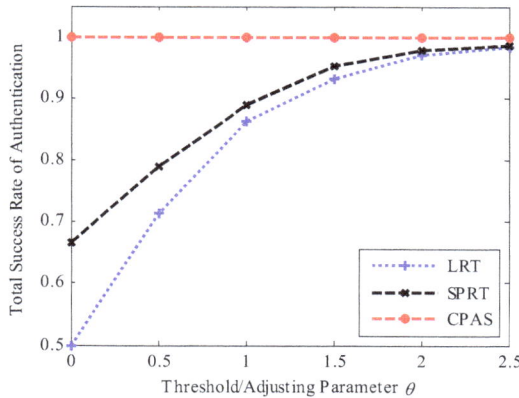

Figure 14. Total authentication success rate of different authentication scheme.

Figures 15 and 16 plot the time costs of data frames authentication in different authentication schemes. The time costs of the LRT, SPRT, and CPAS schemes increased with the increase of the number of data frames on the whole, but decreased with the increase of threshold value. In many experiments, the time cost of traditional cipher authentication scheme (TCAS) also increased linearly with the increase of the number of data frames.

However, as evident from the results, the SPRT scheme needed more time costs than LRT and CPAS schemes when $\theta = 0$, especially with the increase of data frames. The reason was that the data frames must be demodulated before upper layer authentication. That is to say, data demodulation took more time cost before upper layer authentication, which was also a pivotal reason. In the LRT and

SPRT schemes, Bob dropped the data packet directly when the physical-layer channel authentication failed and upper layer authentication was required before each physical-layer channel authentication. In the TCAS scheme, upper layer cipher authentication, which was after demodulation, was needed to verify the validity of each data frame. In the CPAS scheme, Bob did not discard data frames directly, when physical-layer channel authentication failed, but activated upper layer cipher authentication. The low time cost indicates that the CPAS scheme activated the upper layer authentication less frequently, because it had a higher successful rate of physical-layer channel authentication, when $\theta = 0$. The proposed scheme employed j ($j = 3$, in our experiments) data frames to establish a physical-layer channel authentication model, which was more meaningful for practical application, and upper layer authentication to verify the legality of the data frames when physical-layer channel authentication failed.

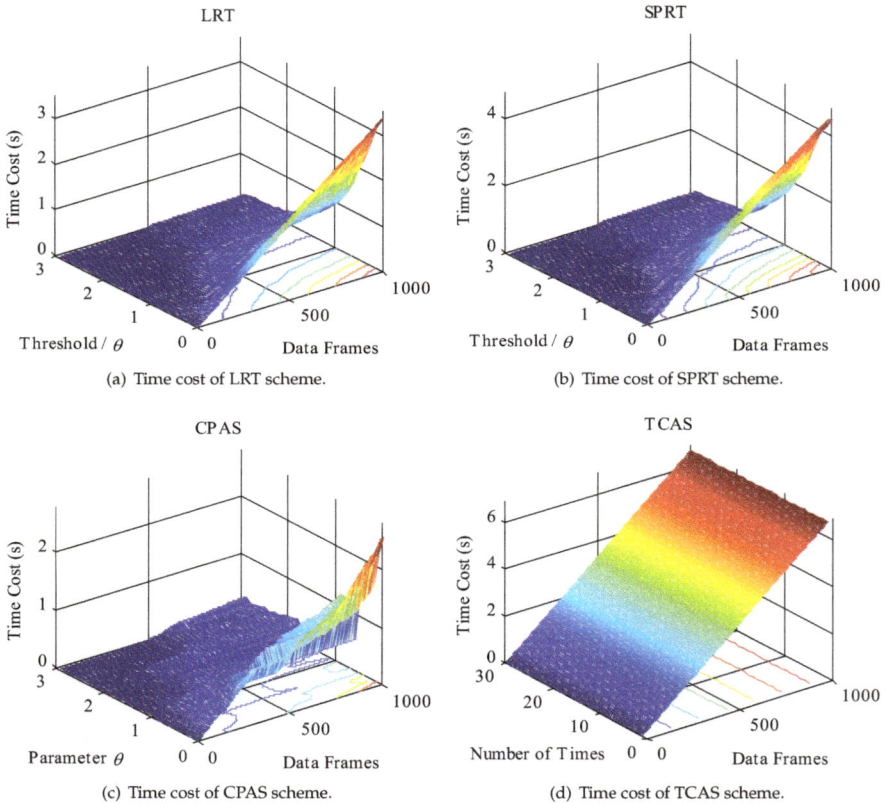

Figure 15. Three dimensional plots of time cost. Note that RC4 algorithm was employed to act a lightweight cryptographic algorithm in the experiments: (**a**) the time cost of LRT scheme; (**b**) the time cost of SPRT scheme; (**c**) the time cost of the proposed scheme, CPAS; and (**d**) the time cost of the traditional cipher authentication scheme, TCAS.

In addition, the CPAS scheme needed more time cost than LRT and SPRT schemes with the increase of parameter θ. The low time cost also manifested that the LRT and SPRT schemes had a higher successful rate of physical-layer channel authentication when the adjusting parameter θ was large. It is worth noting that the time cost differences among the LRT, SPRT, and CPAS schemes decreased with the increase of parameter θ. Therefore, it is feasible to satisfy the requirement of the edge computing system with asymmetric resources, as long as the adjusting parameter θ is appropriate.

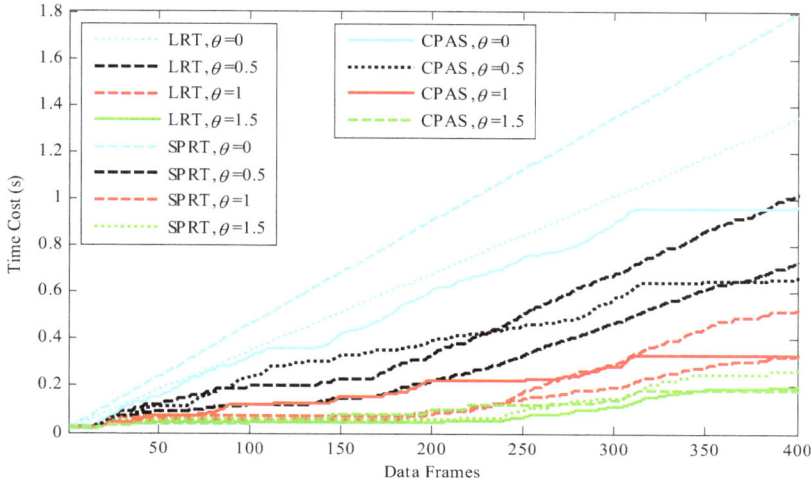

(a) Time cost of LRT, SPRT, and CPAS schemes.

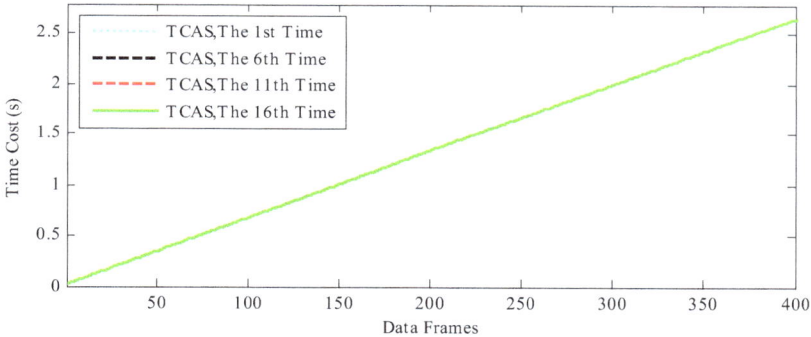

(b) Time cost of TCAS scheme.

Figure 16. Time cost of data frames authentication, where the time cost included the physical layer channel authentication time cost (if any), upper layer cipher authentication time cost, and data demodulation time cost: (**a**) the time cost of LRT, SPRT, and CPAS schemes under different threshold values; and (**b**) the time cost of the TCAS scheme.

6. Conclusions

In this paper, we propose a novel cross-layer secure physical-layer authentication program for edge computing system with asymmetric resources. The proposed scheme combines clustering technology and lightweight symmetric cipher with physical-layer channel state information to achieve mutual authentication between terminals and edge devices. Theoretical analysis and experimental results show that our proposed scheme can effectively boost the total success rate of access authentication and decrease the data frame loss rate but it increases time cost slightly. It is not only secure but also simple and flexible, especially independent of a trusted party. In addition, our scheme could resist spoofing attacks, replay attacks, small integer attacks, exhaustive attacks, and guessing attacks. It can significantly reduce the access authentication complexity and achieve greater security for the edge computing system with asymmetric resources. Therefore, the proposed scheme is very suitable for the resource asymmetric authentication scenario.

Author Contributions: The work was realized with the collaboration of all authors. Conceptualization, Y.C. and H.W.; Data curation, T.Z. and Z.W.; Formal analysis, Y.C., H.W., J.W., H.S. and T.Z.; Funding acquisition, H.W. and J.W.; Investigation, Y.C., A.X., Y.J. and T.Z.; Methodology, Y.C., H.W. and H.S.; Project administration, H.W.; Resources, H.W.; Software, Y.C. and H.S.; Supervision, H.W.; Validation, Y.C. and H.W.; Visualization, Y.C. and H.W.; Writing—original draft, Y.C.; and Writing—review and editing, Y.C., H.W., J.W. and H.S.

Funding: This work was supported by NSFC (No. 61572114), National major R & D program (2018YFB0904900 and 2018YFB0904905); Sichuan sci & tech basic research condition platform project (No. 2018TJPT0041); and Sichuan sci & tech service development project (No. 18KJFWSF0368). This work was also supported in part by Hunan Provincial Nature Science Foundation Project 2018JJ2535, Chile CONICYT FONDECYT Regular Project 1181809.

Conflicts of Interest: The authors declare no conflict of interest.

Abbreviations

The following abbreviations are used in this manuscript:

CPAS	Clustering based physical-layer authentication scheme
CPU	Central processing unit
CSI	Channel state information
ED	Edge computing device
GHz	Giga Hertz
IoT	Internet of things
LRT	Likelihood ratio test
MHz	Mega Hertz
SPRT	Sequential probability ratio test
TCAS	Traditional cipher authentication scheme
USRP	Universal software radio peripheral

References

1. Kawamoto, Y.; Yamada, N.; Nishiyama, H.; Kato, N.; Shimizu, Y.; Zheng, Y. A feedback control-based crowd dynamics management in IoT system. *IEEE Internet Things J.* **2017**, *4*, 1466–1476. [CrossRef]
2. Verma, S.; Kawamoto, Y.; Fadlullah, Z.M.; Nishiyama, H.; Kato, N. A survey on network methodologies for real-time analytics of massive IoT data and open research issues. *IEEE Commun. Surv. Tutor.* **2017**, *19*, 1457–1477. [CrossRef]
3. Rodrigues, T.G.; Suto, K.; Nishiyama, H.; Kato, N. Hybrid method for minimizing service delay in edge cloud computing through VM migration and transmission power control. *IEEE Trans. Comput.* **2017**, *66*, 810–819. [CrossRef]
4. Bhardwaj, I.; Kumar, A.; Bansal, M. A review on lightweight cryptography algorithm for data security and authentication in IoTs. In Proceedings of the 4th International Conference on Signal Processing, Computing and Control (ISPCC 2017), Solan India, 21–23 September 2017; pp. 504–509.
5. Cazorla, M.; Marquet, K.; Minier, M. Survey and benchmark of lightweight block ciphers for wireless sensor networks. In Proceedings of the 10th International Conference on Security and Cryptography (SECRYPT 2013), Reykjavik, Iceland, 29–31 July 2013; pp. 543–548.
6. Wen, H.; Li, S.Q.; Zhu, X.P.; Zhou, L. A framework of the PHY-layer approach to defense against security threats in cognitive radio networks. *IEEE Netw.* **2013**, *27*, 34–39.
7. Wen, H.; Wang, Y.F.; Zhu, X.P.; Li, J.Q.; Zhou, L. Physical layer assist authentication technique for smart meter system. *IET Commun.* **2013**, *7*, 189–197. [CrossRef]
8. Hu, L.; Wen, H.; Wu, B.; Pan, F.; Liao, R.F.; Song, H.H.; Tang, J.; Wang, X.M. Cooperative jamming for physical layer security enhancement in Internet of things. *IEEE Internet Things J.* **2018**, *5*, 219–228. [CrossRef]
9. Xie, F.Y.; Wen, H.; Li, Y.S.; Chen, S.L.; Hu, L.; Chen, Y.; Song, H.H. Optimized coherent integration-based radio frequency fingerprinting in Internet of things. *IEEE Internet Things J.* **2018**, *5*, 3967–3977. [CrossRef]

10. Chen, Y.; Wen, H.; Song, H.; Chen, S.; Xie, F.; Yang, Q.; Hu, L. Lightweight one-time password authentication scheme based on radio-frequency fingerprinting. *IET Commun.* **2018**, *12*, 1477–1484. [CrossRef]

11. Zeng, K.; Govindan, K.; Mohapatra, P. Non-cryptographic authentication and identification in wireless networks. *IEEE Wirel. Commun.* **2010**, *17*, 56–62. [CrossRef]

12. Xiao, L.; Greenstein, L.; Mandayam, N.; Trappe, W. Fingerprints in the Ether: Using the Physical Layer for Wireless Authentication. In Proceedings of the 2007 IEEE International Conference on Communications, Glasgow, UK, 24–28 June 2007; pp. 4646–4651.

13. Xiao, L.; Greenstein, L.J.; Mandayam, N.B.; Trappe, W. Using the physical layer for wireless authentication in time-variant channels. *IEEE Trans. Wirel. Commun.* **2008**, *7*, 2571–2579. [CrossRef]

14. Xiao, L.; Greenstein, L.; Mandayam, N.; Trappe, W. A Physical-Layer Technique to Enhance Authentication for Mobile Terminals. In Proceedings of the 2008 IEEE International Conference on Communications, Beijing, China, 19–23 May 2008; pp. 1520–1524.

15. Wang, N.; Jiang, T.; Lv, S.C.; Xiao, L. Physical-layer authentication based on extreme learning machine. *IEEE Commun. Lett.* **2017**, *21*, 1557–1560. [CrossRef]

16. Pan, F.; Pang, Z.; Luvisotto, M.; Xiao, M.; Wen, H. Physical-Layer Security for Industrial Wireless Control Systems: Basics and Future Directions. *IEEE Ind. Electron. Mag.* **2018**, *12*, 18–27. [CrossRef]

17. Xiao, L.; Wan, X.; Lu, X.; Zhang, Y.; Wu, D. IoT security techniques based on machine learning: How do IoT devices use AI to enhance security? *IEEE Signal Process. Mag.* **2018**, *35*, 41–49. [CrossRef]

18. Xiao, L.; Li, Y.; Han, G.; Liu, G.; Zhuang, W. PHY-Layer spoofing detection with reinforcement learning in wireless networks. *IEEE Trans. Veh. Technol.* **2016**, *65*, 10037–10047. [CrossRef]

19. Xiao, L.; Chen, T.; Han, G.; Zhuang, W.; Sun, L. Game theoretic study on channel-based authentication in MIMO systems. *IEEE Trans. Veh. Technol.* **2017**, *66*, 7474–7484. [CrossRef]

20. Xiao, L.; Wan, X.; Han, Z. PHY-Layer authentication with multiple landmarks with reduced overhead. *IEEE Trans. Wirel. Commun.* **2018**, *17*, 1676–1687. [CrossRef]

21. Pan, F.; Pang, Z.; Xiao, M.; Wen, H.; Liao, R. Clone detection based on physical layer reputation for proximity service. *IEEE Access* **2019**, *7*, 3948–3957. [CrossRef]

22. He, F.; Man, H.; Kivanc, D.; McNair, B. EPSON: Enhanced physical security in OFDM networks. In Proceedings of the IEEE International Conference on Communications (ICC 2009), Dresden, Germany, 14–18 June 2009; pp. 824–828.

23. Li, Y.; Cimini, L.J.; Sollenberger, N.R. Robust channel estimation for OFDM systems with rapid dispersive fading channels. *IEEE Trans. Commun.* **1998**, *46*, 902–915. [CrossRef]

24. Larsson, E.G.; Liu, G.Q.; Li, J.; Giannakis, G.B. Joint symbol timing and channel estimation for OFDM based WLANs. *IEEE Commun. Lett.* **2001**, *5*, 325–327. [CrossRef]

25. Su, W.; Pan, Z. Iterative LS channel estimation for OFDM systems based on transform-domain processing. In Proceedings of the 2007 International Conference on Wireless Communications, Networking and Mobile Computing, Shanghai, China, 21–25 September 2007; pp. 416–419.

26. Jain, A.K.; Murty, M.N.; Flynn, P.J. Data clustering: A review. *ACM Comput. Surv.* **1999**, *31*, 264–323. [CrossRef]

27. Xu, R.; Wunsch, D. Survey of clustering algorithms. *IEEE Trans. Neural Netw.* **2005**, *16*, 645–678. [CrossRef]

28. Singh, D.; Gosain, A. A comparative analysis of distributed clustering algorithms: A survey. In Proceedings of the 2013 International Symposium on Computational and Business Intelligence (ISCBI), New Delhi, India, 24–26 August 2013; pp. 165–169.

29. Fahad, A.; Alshatri, N.; Tari, Z.; Alamri, A.; Khalil, I.; Zomaya, A.Y.; Foufou, S.; Bouras, A. A survey of clustering algorithms for big data: Taxonomy and empirical analysis. *IEEE Trans. Emerg. Top. Comput.* **2014**, *2*, 267–279. [CrossRef]

30. Osman, M.M.A.; Syed-Yusof, S.K.; Abd Malik, N.N.N.; Zubair, S. A survey of clustering algorithms for cognitive radio ad hoc networks. *Wirel. Netw.* **2018**, *24*, 1451–1475. [CrossRef]

31. Alaghi, A.; Hayes, J.P. Survey of stochastic computing. *ACM Trans. Embed. Comput. Syst.* **2013**, *12*, 92:1–92:19. [CrossRef]

32. Han, J.; Chen, H.; Liang, J.H.; Zhu, P.C.; Yang, Z.X.; Lombardi, F. A stochastic computational approach for accurate and efficient reliability evaluation. *IEEE Trans. Comput.* **2014**, *63*, 1336–1350. [CrossRef]

33. Bakiri, M.; Guyeux, C.; Couchot, J.F.; Marangio, L.; Galatolo, S. A hardware and secure pseudorandom generator for constrained devices. *IEEE Trans. Ind. Inform.* **2018**, *14*, 3754–3765. [CrossRef]

34. Harase, S. On the F-2-linear relations of Mersenne Twister pseudorandom number generators. *Math. Comput. Simul.* **2014**, *100*, 103–113. [CrossRef]

35. Borle, K.M.; Chen, B.A.; Du, W.L. Physical layer spectrum usage authentication in cognitive radio: Analysis and implementation. *IEEE Trans. Inf. Forensics Secur.* **2015**, *10*, 2225–2235. [CrossRef]

36. Tong, Z.; Russ, C.; Vanka, S.; Haenggi, M. Prototype of virtual full duplex via rapid on-off-division duplex. *IEEE Trans. Commun.* **2015**, *63*, 3829–3841. [CrossRef]

37. Omar, M.S.; Naqvi, S.A.R.; Kabir, S.H.; Hassan, S.A. An experimental evaluation of a cooperative communication-based smart metering data acquisition system. *IEEE Trans. Ind. Inform.* **2017**, *13*, 399–408. [CrossRef]

sensors

MDPI

Article

Adversarial Samples on Android Malware Detection Systems for IoT Systems

Xiaolei Liu [1], Xiaojiang Du [2], Xiaosong Zhang [1], Qingxin Zhu [1], Hao Wang [3,*] and Mohsen Guizani [4]

[1] School of Information and Software Engineering, University of Electronic Science and Technology of China, Chengdu 610054, China; liuxiaolei@uestc.edu.cn (X.L.); johnsonzxs@uestc.edu.cn (X.Z.); qxzhu@uestc.edu.cn (Q.Z.)

[2] Department of Computer and Information Sciences, Temple University, Philadelphia, PA 19122, USA; dux@temple.edu

[3] Department of Computer Science, Norwegian University of Science and Technology, 7491 Trondheim, Norway

[4] Department of Computer Science and Engineering, Qatar University, Doha 2713, Qatar; mguizani@ieee.org

* Correspondence: hawa@ntnu.no

Received: 22 December 2018; Accepted: 20 February 2019; Published: 25 February 2019

Abstract: Many IoT (Internet of Things) systems run Android systems or Android-like systems. With the continuous development of machine learning algorithms, the learning-based Android malware detection system for IoT devices has gradually increased. However, these learning-based detection models are often vulnerable to adversarial samples. An automated testing framework is needed to help these learning-based malware detection systems for IoT devices perform security analysis. The current methods of generating adversarial samples mostly require training parameters of models and most of the methods are aimed at image data. To solve this problem, we propose a testing framework for learning-based Android malware detection systems (TLAMD) for IoT Devices. The key challenge is how to construct a suitable fitness function to generate an effective adversarial sample without affecting the features of the application. By introducing genetic algorithms and some technical improvements, our test framework can generate adversarial samples for the IoT Android application with a success rate of nearly 100% and can perform black-box testing on the system.

Keywords: Internet of Things; malware detection; adversarial samples; machine learning

1. Introduction

Since many IoT (Internet of Things) devices run Android systems or Android-like systems, with the popularity of IoT devices, Android malware for IoT devices is also increasing. Meanwhile, machine learning has received extensive attention and has gained tremendous application development in many fields, such as financial economics, driverless, medical, and network security. Thus, there are many learning-based Android malware detection systems [1–6].

However, while machine learning brings us great convenience, it also exposes a lot of security problems [7]. Several papers have studied related Android and IoT security issues [8–13]. Scholars in the security field are increasingly concerned about the security issues associated with the lack of fairness and transparency in machine learning algorithms. An attacker can predict certain sensitive information by observing the model, or recover sensitive data in the data set through existing partial data. A current attack method is called a poisoning attack. Biggio B and Zhu attempted to attack the adaptive facial recognition system by a poisoning attack [14–17]. During the model update, they injected malicious data to offset the central value of the recognition feature in the model, so as to achieve the purpose of verifying the attacker's image. Biggio B and Nelson B also attacked the supervised learning algorithm SVM [18]. Experiments show that the test error of the model classifier can be significantly increased during the gradient rise. However, the injected sample data must meet certain constraints in order to deceive the model, and must be the attacker to control the label of the injection point. Yang et al. conducted an experiment on poisoning attacks against neural network learning algorithms [19]. Compared with the direct gradient algorithm, this proposed method can increase the attack sample generation speed by about 240 times.

In fact, although a poisoning attack can make the model go wrong, the attacker has to work hard on how to inject malicious data. Another common method can let models get the wrong result in a short time, that is, adversarial sample attack. Christian Szegedy et al. first proposed the concept of adversarial samples [20]. By deliberately adding minor changes in the dataset, the perturbed samples will cause the model to output a false result with high confidence. Adversarial samples can increase the prediction error of the model, so that the originally correctly classified sample migrates to the other side of the decision area, thereby being classified into another category.

Existing models are vulnerable to adversarial samples [21–24]. For example, in a malware recognition system, by adding a small perturbation to the original software sample, the result of the sample classification can be changed with a high probability, and even the sample can be classified into an arbitrarily designated label according to the attacker's idea. This makes adversarial samples attack a huge hazard to malware recognition systems [25–27].

All of the learning-based Android malware detection systems for IoT devices have the above problems, so a testing framework is needed to test the robustness of these detection systems. To address this challenge, we propose TLAMD, a testing framework for learning-based Android malware detection systems for IoT Devices. When the test results show that the detection system cannot resist the attack of the adversarial samples, it indicates that this detection system has potential safety hazards, and it needs to be reinforced.

Therefore, how to generate effective adversarial samples is the core issue of the entire testing framework. Our approach to generating adversarial samples for the Android IoT malware detection model is based on genetic algorithms. Without the knowledge of the model parameters, the original sample is used as the input of the approach, and finally the adversarial sample of the specific label is generated. The information used is only the probability of the various types of labels output by the model. We hope that this method can be a robust benchmark for the learning-based Android malware detection model for IoT devices. Our contribution is mainly reflected as follows:

1. We migrated the application of adversarial samples from the image recognition domain to the Android malware detection domain of IoT devices. In this process, simply replacing the model's training data from a picture to an Android application is not possible. On the one hand, the data of the binary program is not continuous like the image data. On the other hand, random perturbation of the binary program may lead to the crash of the program, so special processing is required for the Android application to ensure the validity of the adversarial samples. We borrowed the processing method of Kathrin Grosse [28], which realized the disturbance to the Android application by adding the request permission code in the *AndroidManifest.xml* file. The difference is that we have made corresponding analysis and restrictions on the types and quantities of permissions that can be added. This method can ensure that the original function of the app is not affected and can be used normally; and the app can be disturbed in the simplest way to achieve the effect of changing the model detection result;

2. We introduce the genetic algorithm into the adversarial sample generation method and implement the black-box attack against the machine learning model. Without knowing the internal parameters such as the gradient and structure of the target network, it is only necessary to know the probability of various types of labels output by the model. Compared to Kathrin Grosse's approach, our approach not only implements black-box attacks, but also has a higher success rate, almost 100%.

The rest of the paper is organized as follows. Section 2 introduces the related background of our approach. Section 3 presents TLAMD (A Testing Framework for Learning-based Android Malware Detection Systems for IoT Devices). Section 4 presents and discusses our experimental results. Finally, further discussions and conclusions are accomplished in Section 5.

2. Related Background

2.1. Neural Network

The essence of the neural network is a function $y = F(x)$, the input x is an n-dimensional vector, and the output y is an m-dimensional vector. The function F implies the model parameter θ. The purpose of the training network is to calculate the value of the parameter θ from the known partial sample information. After the model is completed, the result of predicting x is to solve the value of y by the function F. In this paper, we mainly study the neural network of the m classifier (that is, the output y is an m-dimensional vector). The output of the last layer of the neural network uses a fully connected layer. The classifier outputs the index with the largest value in the output vector dimension as the result, that is:

$$L(x) = arg \max_{j=1}[F(x)]_i, \tag{1}$$

where $L(x)$ is the category of x.

Define F as a single-layer fully-connected neural network. The output of the $(n-1)$-th layer is the input of the n-th layer, then:

$$y_n = F_n(y_{n-1}). \tag{2}$$

Typical n-layer fully connected neural networks are:

$$F = F_n * F_{n-1} * ... * F_2 * F_1, \tag{3}$$

$$F_n(x) = \sigma(w_n * x + b_n), \tag{4}$$

where σ is a linear or nonlinear activation function. The commonly used activation functions are RELU [29], tanh [30], sigmoid, etc., w is the weight of this layer, and b is the layer offset.

2.2. Genetic Algorithm

The idea of the genetic algorithm is to simulate the biological evolution process of natural selection. Using the thought of evolutionary theory, the process of finding the optimal solution of a certain objective function is simulated into the evolution process of the population. Based on the idea of the population, the algorithm uses a population containing information to perform an optimal search in multiple directions and completes the exchange and reconstruction of information in the search process.

The genetic algorithm can be used to search for the feasible solution space of a problem, and then find the possible optimal solution, which is the uncertainty optimization in the optimization problem. Uncertain optimization is to rely on random variables in the search direction, rather than a certain mathematical expression. Compared with other algorithms, the advantage is that, when the optimization converges to the local extremum, the search result can jump out of an optimal solution and continue to search for a better feasible solution.

By choosing the appropriate objective function, the generation of the adversarial sample can be transformed into a solution to the optimization problem. The process of solving the optimal solution corresponding to the objective function is actually the process of generating the adversarial sample. This shows that genetic algorithms can be effectively applied to machine learning and other fields in terms of parameter optimization and function solving. In terms of parameter optimization, Chen et al. used a parallel genetic algorithm to optimize the parameter selection of Support Vector Machine (SVM) [31]. Experiments show that the proposed method is superior to the grid search in classification accuracy, the number of selected features and running time. Phan et al. proposed a GA-SVM model that can effectively improve classification performance based on genetic algorithm and SVM classifier [32]. Alejandre et al. selected features based on machine learning to detect botnets [33]. A genetic algorithm is used in this method to select the set of features that provide the highest detection rate.

2.3. Adversarial Samples

On many machine learning models, the decision boundary of the classifier has a certain margin of error. That is, when the disturbance satisfies $||\eta||_\infty < \epsilon$, the classifier considers that the perturbed input $x' = x + \eta$ is the same as the original input x. Therefore, when the perturbation value on each feature element is less than ϵ, the classifier cannot discern the difference in the sample. However, changes in input characteristics have a cumulative effect on model predictions. Although the perturbation value on each feature element is small, the accumulated error is sufficient to influence the model prediction result.

On each neuron, the adversarial sample will have the following operations:

$$\omega^T x' = \omega^T (x + \eta) \tag{5}$$

although the adversarial sample has no effect on the classification results of the single-dimensional neuron classifier. However, deep learning has a considerable number of neurons. The weight in each neuron has n dimensions. If the average variation of an element in the weight vector is m, the activation effect will increase by $n * m$. Furthermore, in a high dimensional linear classifier, each individual input feature is normalized. The result is that in the process of deep learning, a small change may not be enough to change the input result, but multiple disturbances to the input will cause the classifier to make a wrong classification result.

Many methods of generating adversarial samples need to know the parameters of the learning model to calculate the perturbation values, but some subsequent studies have shown that without knowing the parameters of the learning model [34–37]. The attacker can interact with the black-box learning model to calculate the samples. Specifically, the attacker can estimate the boundary of the decision region of the model according to the difference of the model output brought by different samples, and then use the estimated boundary as a substitute model. Finally, the adversarial samples are calculated by the parameters of the substitute model. Considering that more and more malicious Android application detection methods based on machine learning, how to evaluate the robustness of these detection methods becomes a new problem. Since most machine learning algorithms are vulnerable to adversarial samples, we have thought of using the generated adversarial samples to test the robustness of these detection methods.

3. Methodology

3.1. Framework

The overview of TLAMD is shown in Figure 1.

When the test results show that the detection system cannot resist the attack against the adversarial sample, it indicates that the system has potential safety hazards and it is necessary to implement such reinforcement measures as distillation defense [38] on the detection system. As we can see, how to generate an adversarial sample is the main challenge of this testing framework. Therefore, we will describe the algorithm in detail for generating an adversarial sample for Android malware.

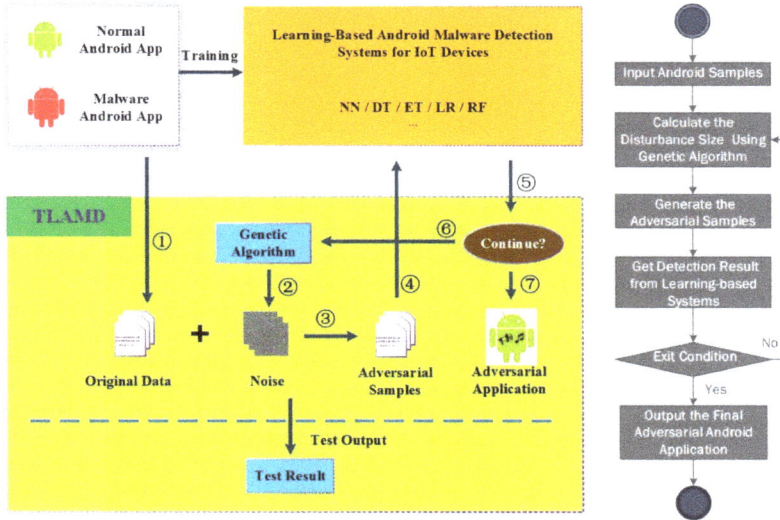

Figure 1. Overview of our testing framework for learning-based Android Malware detection systems for IoT devices. (1) Original Sample Input; (2) Calculate the disturbance size; (3) Generate the adversarial samples; (4) Get detection result from learning-based systems; (5) Determine if the exit condition is met; (6) If not, calculate the new disturbance size using genetic algorithm; (7) If yes, output the final adversarial android application.

3.2. Algorithm

Our goal is to add minor perturbations to the malware without changing the malware functionality, so that the previously trained detection model misidentifies it as normal software. Therefore, our approach generates an adversarial sample by adding permission features to the *AndroidManifest.xml*, and in order not to affect the function of the original malware, the disturbance does not reduce the existing permission features. For a single input sample X, the classifier returns a two-dimensional vector $F(X) = [F_0(X), F_1(X)]$, where $F_0(X)$ indicates the probability that the software is a normal software, $F_1(X)$ indicates the probability that the software is a malware, and satisfies the constraint $F_0(X) + F_1(X) = 1$. We aim to add a perturbation δ to make the classification result $F_1(X + \delta)$ is less than $F_0(X + \delta)$. At the same time, the smaller the δ, the better, that is, the fewer the number of permission features added in the manifest file, the better. For example, for a specific malware x, we use a genetic algorithm to find out which permission features δ are added to x, and finally make x detected as normal software with minimum number of permission added.

From a mathematical point of view, the process of misjudging the detection model by adding the permission features is regarded as a problem to be solved. The feasible solution space of the problem is the disturbance if the detection model is successfully misjudged. The optimal solution is to minimize the disturbance value, that is, add the least permission feature. A genetic algorithm is a type of algorithm that finds the possible optimal solution by searching for a feasible solution space of a problem. Our approach is to use genetic algorithms to search for the minimum perturbation value that causes the detection model to be misjudged. The pseudo code of our approach is shown in Algorithm 1.

Algorithm 1 Generating an adversarial sample.

Require: Popluation Size *pop_size*

$\delta \leftarrow initialization()$

for $i = 0 \rightarrow pop\_size$ **do**

    $P_i \leftarrow Crossover\_Operator()$
    $P_i \leftarrow Mutation\_Operator()$
    Compute $\rightarrow S(\delta)$
    if $F(X + \delta) > 1 - F(X + \delta)$ **then**

        Continue
    else

        Output $\rightarrow \delta$
    end if
end for

The specific steps are as follows:

(1) Randomly generate the population $\delta = P_1, P_2, ..., P_M$. M is the number of individuals, the individual $P_i \in \{0, 1\}^n$ refers to the permission characteristics to be added in the category, and n is the number of permission features in the category. In addition, 1 means to add the corresponding permission; otherwise, 0 means not to add. Our strategy is to only add permissions and not reduce permissions. Therefore, if the original malicious sample has a certain permission feature, the permission cannot be removed, that is, the disturbance is 0.

(2) Determine the fitness function.

$$S(\delta) = \min \; w_1 \cdot F(X + \delta) + w_2 \cdot num(\delta), \tag{6}$$

where w_1 and w_2 represent the two weights, δ is the added small disturbance, $F(X + \delta) \in [0, 1]$ means that the probability of original malicious sample is still detected as a malware, $num(\delta_i)$ indicates the number of permission features added.

When w_1 is much larger than w_2, the sample after the addition of the disturbance must be detected as normal by the detection model to survive, and the individual detected as a malicious sample will be eliminated. The surviving individual must meet the minimum number of added permission features; otherwise, it will also be eliminated. The fitness function defined in this way searches for an optimal solution that can successfully cause the detection model to be misjudged.

(3) Perform mutation operations according to a certain probability to generate new individuals. The mutation refers to adding a disturbance to the corresponding category according to a certain probability, that is, changing the value from 0 to 1, and satisfying the constraint proposed in step (1).

(4) Generate a new generation of the population from the mutation and return to step (2). If the preset number of iterations is reached, the loop is exited.

4. Experiments

4.1. Data Set and Environment

In order to verify the effectiveness of the adversarial sample, we attempt to train five different classifier models, including logistic regression (LR), decision tree (DT), and fully connected neural network (NN) and so on. The hardware environment and software environment of all experiments are shown in Table 1:

All the data we use in the experiments come from the DREBIN dataset [39,40]. The DREBIN dataset has a total of 123,453 sample data for Android applications, including 5560 malicious samples and contains as many as 545,333 behavioral features.

The features of the Android app in this dataset consist of eight categories and are shown in Table 2:

(S1)   Hardware components, which are used to set the hardware permissions required by the software.
(S2)   Requested permissions, which are granted by the user at the time of installation and allow the application software to access the corresponding resources.
(S3)   App components, which include four different types of interfaces: activities, services, content providers and broadcast receivers.
(S4)   Filtered intents, which are used for process communication between different components and applications.
(S5)   Restricted API (Application Programming Interface) calls, access to a series of key API calls.
(S6)   Used permissions, a subset of permissions that are actually used and requested in S5.
(S7)   Suspicious API calls, API calls for allowing access to sensitive data and resources.
(S8)   Network addresses, the IP addresses accessed by the application, including the hostname and URL.

Table 1. The environment of all experiments.

CPU	Inter(R) Core(TM) i5-7400 CPU @ 3.00GHz
Memery	8 GB
Video Card	Inter(R) HD Graphics 630
Operating System	Windows 10
Programming Language	Python 3.6
Development Platform	Jupyter Notebook
Dependence	Tensorflow, Keras, numpy etc.

The first four classes are extracted from the manifest file, and the last four classes are extracted from the disassembly code. Since our method only adds permission requests to the *AndroidManifest.xml* file, we only cover the features in S1 to S4. In Section 4.2.1, we further reduce the feature categories used.

Table 2. Eight features in the DREBIN dataset.

Class	Name	Numbers	Rate (/Total)
S1	Hardware Components	72	0.013%
S2	Requested Permissions	3812	0.704%
S3	App Components	218,951	40.488%
S4	Filtered Intents	6379	1.178%
S5	Restricted API Calls	733	0.136%
S6	Used Permissions	70	0.013%
S7	Suspicious API Calls	315	0.058%
S8	Network Address	310,447	57.4%

4.2. Android Malware Detection Model

First, a detection model is trained to determine whether an Android sample is malware. When the detection model reaches a certain accuracy, our approach is used to generate an adversarial sample for the model.

4.2.1. Feature Extraction

We use a random forest approach to measure the importance of features in the feature extraction phase. The number of features is effectively reduced without affecting the accuracy of detection.

Random forest is an integrated learning in machine learning. It is an integrated classifier composed of multiple sets of decision trees: $h(X, \theta_k), k = 1, 2, ...$, where θ_k is a random variable subject to independent and identical distribution, and k represents the number of decision trees. The principle is to generate multiple decision trees and let them learn independently and make corresponding predictions. Finally, observe which category is selected the most and get the result.

The specific steps are as follows:

(1) Select out of bag (OOB) to calculate the corresponding out-of-bag data deviation $error_1$ for each decision tree.
(2) Add random noise, perturb all samples of OOB, and then calculate the out-of-bag data deviation $error_2$ again.
(3) Define and calculate the importance of the features:

$$I = \sum(error_1 - error_2)/N, \tag{7}$$

where N is the number of forest decision trees.

If $error_2$ is greatly increased after adding random noise, the OOB accuracy rate decreases, indicating that this type of feature has a greater impact on the prediction result, that is, the importance is higher.

The sorting result of feature importance is shown in Figure 2.

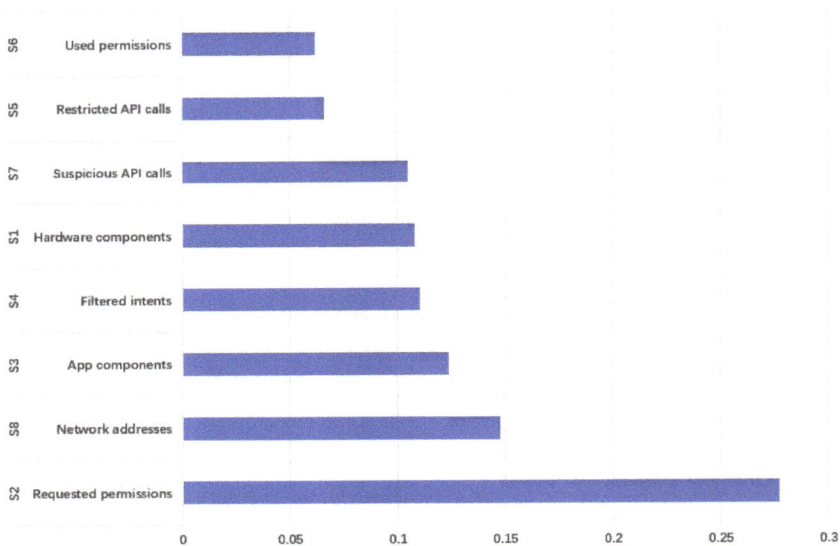

Figure 2. The sorting result of feature importance. The ordinate represents different behavioral feature categories and the abscissa represents the proportion of importance.

As we mentioned before, we only cover the four types of features from S1 to S4. Taking into account the number and importance of various features, we finally choose the two characteristics of S1 and S2.

4.2.2. Training Detection Model

To test the effectiveness of our method for different detection models, we trained five kinds of detection models.

a. Neural Network

Our neural network chooses a two-layer fully connected model with 200 neurons in each connected layer and the activation function is RELU. The output layer of the last layer has two neurons and is the *soft* max activation unit. Furthermore, no dropout operation is performed on each layer. To train our network, we used the gradient descent training method with a batches size of 256. All data was trained five times per iteration.

b. Logistic Regression

Since there are only two types of target predictions, we adopt a two-class logistic regression model. The penalty term selects the L2 paradigm, and the model parameters satisfy the Gaussian distribution, that is, the parameters are constrained so that they do not over-fitting. Considering that the solution problem is not a linear multi-core, and the number of samples is selected to be larger than the number of features, the dual method is not set. Set the condition for stopping the solution is that the loss function is less than or equal to $1 \times e^{-4}$; the category weight defaults to 1. The maximum number of iterations of the algorithm convergence is set to 10.

c. Decision Tree

The decision tree is a tree structure used for classification. The maximum depth of the decision tree is set to 15 to prevent overfitting. The *min_impurity_decrease* is set to 0. The *min_samples_split* is set to 2, indicating the minimum number of samples required for internal node subdivision. The *min_samples_leaf* is set to 10, indicating the minimum number of samples in the leaf node. The *max_leaf_nodes* is set to None, which is expressed as the maximum number of leaf nodes in the decision tree. The *min_weight_fraction_leaf* is set to 0, which represents the minimum value of all sample weights and sums of leaf nodes.

d. Random Forest

Random forest is an integrated learning. Through the bootstrap resampling technique, a number of sample inputs are randomly selected from the original training set with repeated iterations. In this way, a new training set is obtained, and then several decision trees are generated to form a random forest. The *max_feature* is set to auto, that is, a single decision tree can utilize all permission features. The *n_estimators* is set to 20, which means there are 20 decision trees to form the random forest to be trained. The *min_sample_leaf* is set to 20, that is, the minimum number of sample leaves in each decision tree is 20.

e. Extreme Tree

Extra Tree is equivalent to a variant of the random forest. Compared with random forests, the randomness is further calculated when dividing the local best, that is, the selection of the division points is calculated. Most of its parameters are the same as those of random forests, except that *n_estimators* is set to 10 and *max_depth* is set to 50.

Finally, when the five detection models are trained, we test 42,570 samples and the results are shown in Table 3.

Table 3. The detection results of five models.

Models	TP[a]	FP[a]	FN[a]	TN[a]	Accuracy	Precision	Recall
NN (Neural Network)	40770	0	74	1726	99.83%	1	95.95%
LR (Logistic Regression)	40770	0	234	1566	99.45%	1	96.32%
DT (Decision Tree)	40770	0	60	1740	99.86%	1	95.91%
RF (Random Forest)	40770	0	32	1768	99.92%	1	95.85%
ET (Extreme Tree)	40770	0	16	1784	99.96%	1	95.81%

[a] TP = True Positive, FP = False Positive, FN = False Negative, TN = True Negative.

4.3. Simulation Experiments

After getting the trained detection models, we will generate adversarial samples for the five models. The features we add to the *AndroidManifest.xml* file are from S1 or S2. The parameters of the generation algorithm are also different depending on the permission category. The details are as shown in Table 4.

Table 4. The parameters of our approach.

Features	S1: Hardware Components	S2: Requested Permissions
Initialize Probability	1%	0.01%
Mutation Probability	30%	0.5%
Iterations	50	50
Population	150	150
Attacked Samples	1000	1000

The final experimental results are shown in Table 5. In the ten sets of adversarial sample generation experiments for the five detection models, the success rates are above 80%, and most of them are close to 100%. In order to generate these adversarial samples, the average number of permission features added is less than three. On the one hand, it shows that the adversarial sample generated by our method is very effective and our approach is able to be a robust benchmark for the learning-based Android malware detection model for IoT devices; on the other hand, it shows that the existing machine learning algorithms are very vulnerable to the adversarial sample. Our TLAMD test framework is very necessary.

In subsequent experiments, we also performed a reinforcement method for the distillation defense of these models. However, the reinforced model is still unable to resist the attack of adversarial samples, and the success rate of our approach is still close to 100%. This means that, when we want to reinforce existing machine learning models, common methods such as distillation defenses work poorly. We need to find a more effective defense method.

Figure 3 shows the most frequently added permissions in the ten sets of adversarial sample generation experiments for the five kinds of detection models. Compared to other permission features, these permissions are mostly permissions that involve sensitive privacy. In order to verify whether these features have a decisive influence on the model discrimination results, we have conducted further experiments. In the new experiment, we will not allow the algorithm to add the features listed in the figure. However, the success rate of the generated adversarial samples is consistent with the previous one in Table 5, and the number of permission features added is slightly increased. It can be seen that those features that are added more frequently only have greater weight, but have no decisive influence on the results.

Table 5. The results of our approach. [b]

Model	Category	Success Rate	Average of *num* (δ)
NN	S1	1	2.25
	S2	1	2.33
LR	S1	0.998	2.66
	S2	0.995	1.94
DT	S1	0.896	1.05
	S2	0.992	1.68
RF	S1	0.866	2.89
	S2	0.995	9.54
ET	S1	0.833	2.81
	S2	0.945	9.36

[b] Each line of data in the table is the average of the 1000 sample tests results.

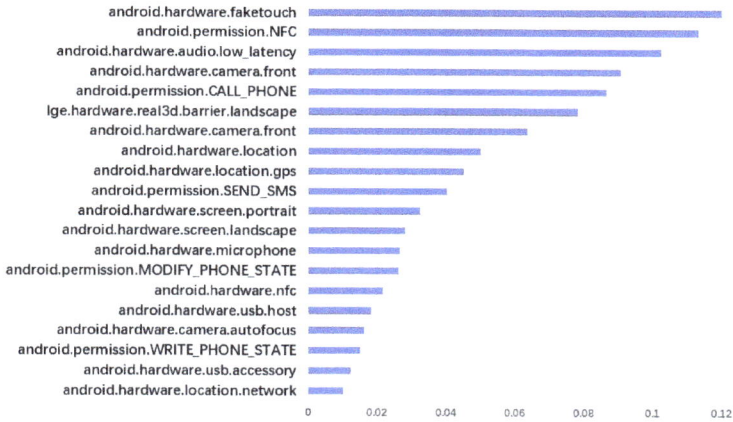

Figure 3. The most frequently added permissions in our adversarial sample generation experiments. The data is the average of $5 \times 2 \times 1000$ samples test results.

Figure 4 is a trend graph of fitness function values as a function of the number of iterations. As the number of iterations increases, the value of the fitness function decreases rapidly. It shows that it is very effective to use the genetic algorithm to solve the problem of generating adversarial samples.

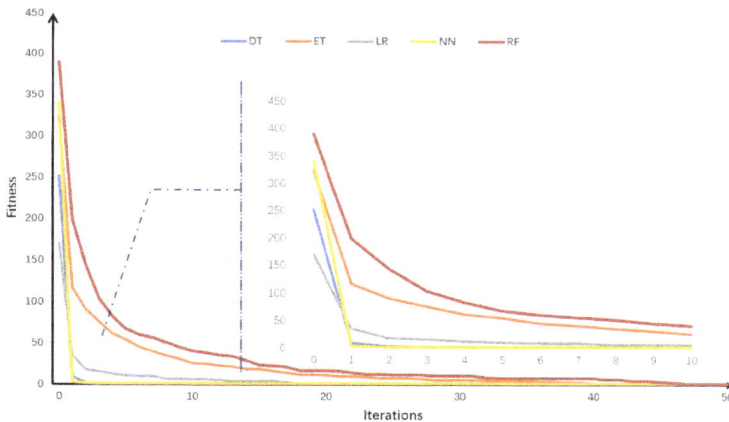

Figure 4. Trend graph of fitness function values with number of iterations.

By comparing the individual models, it can be found that the more complex the detection model, the better the effect of the adversarial samples generated for the model. This phenomenon may be different from what we expected. We believe that one possible reason is that the more complex the model, the more times the feature is processed. This makes small changes in features easily magnified, making the model very sensitive to adversarial samples.

Figure 5 is a box plot of the fitness function values of adversarial samples for five detection models with S1 permission features and Figure 6 is with S2 permission features. As can be seen from the figures, the adversarial samples generated by our approach is very stable. There are only a very small number of divergence points out of 1000 samples. By comparing Figures 5 and 6, the stability of the adversarial sample generated by S2 is better. The reason is that the number of permission features in the S2 list is much larger than the number in the S1 list. This is equivalent to finding the optimal solution of the objective function in a larger space, so there is a greater probability of finding a better solution. Combined with Figure 3, it also provides us with an idea of how to strengthen the learning-based detection model. It is not useful to improve the defense of high-weight permission features. It is necessary to optimize the detection model so that it is not sensitive to small disturbances of all sample features.

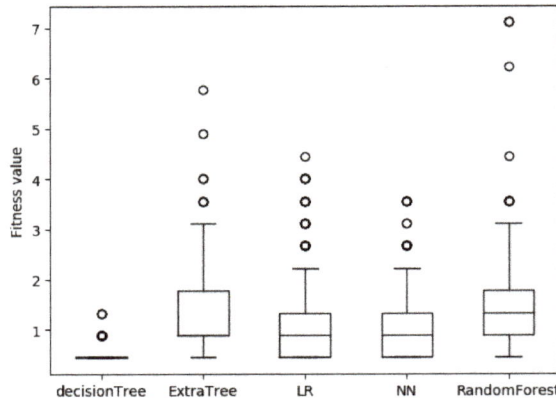

Figure 5. The fitness function values of adversarial samples for five detection models with S1 permission features.

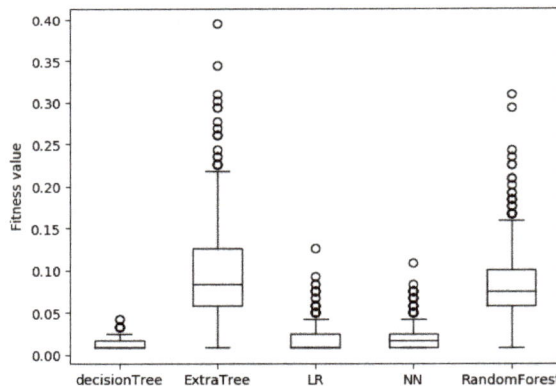

Figure 6. The fitness function values of adversarial samples for five detection models with S2 permission features.

5. Conclusions

To address the challenge of the lack of the testing framework for learning-based Android malware detection systems for IoT devices, we approach TLAMD. Our experimental results show that our approach generates high-quality adversarial samples with a success rate of nearly 100% by adding permission features. In the technical implementation of the TLAMD algorithm, the selection of feature and the range of disturbance are the keys to have a good result. We hope TLAMD can be a benchmark for learning-based IoT Android malware detection model. The limitations of TLAMD is our black-box approach need frequent model requests and our future work includes reducing the requesting times and designing an effective defense approach to reinforce the malware detection model.

Author Contributions: Data curation, H.W.; Formal analysis, X.L.; Investigation, X.L.; Methodology, X.L.; Supervision, Q.Z.; Validation, M.G.; Writing—original draft, X.L.; Writing—review and editing, X.Z. and X.D.

Funding: This research was funded by the National Natural Science Foundation of China under Grant No. 61672170, No. 61871313 and No. 61572115, in part by the National Key R&D Plan under Grant CNS 2016QY06X1205.

Conflicts of Interest: The authors declare no conflict of interest.

References

1. Ham, H.S.; Kim, H.H.; Kim, M.S.; Choi, M.J. Linear SVM-based android malware detection for reliable IoT services. *J. Appl. Math.* **2014**, *2014*, 594501. [CrossRef]
2. McNeil, P.; Shetty, S.; Guntu, D.; Barve, G. SCREDENT: Scalable Real-time Anomalies Detection and Notification of Targeted Malware in Mobile Devices. *Procedia Comput. Sci.* **2016**, *83*, 1219–1225. [CrossRef]
3. Aswini, A.; Vinod, P. Towards the Detection of Android Malware using Ensemble Features. *J. Inf. Assur. Secur.* **2015**, *10*, 9–21.
4. Odusami, M.; Abayomi-Alli, O.; Misra, S.; Shobayo, O.; Damasevicius, R.; Maskeliunas, R. Android Malware Detection: A Survey. In Proceedings of the International Conference on Applied Informatics Applied Informatics, Bogota, Colombia, 1–3 November 2018; Florez, H., Diaz, C., Chavarriaga, J., Eds.; Springer International Publishing: Cham, Switzerland, 2018; pp. 255–266.
5. Misra, S.; Adewumi, A.; Maskeliūnas, R.; Damaševičius, R.; Cafer, F. Unit Testing in Global Software Development Environment. In Proceedings of the International Conference on Recent Developments in Science, Engineering and Technology, Gurgaon, India, 13–14 October 2017; Springer: Berlin, Germany, 2017; pp. 309–317.
6. Alhassan, J.; Misra, S.; Umar, A.; Maskeliūnas, R.; Damaševičius, R.; Adewumi, A. A Fuzzy Classifier-Based Penetration Testing for Web Applications. In Proceedings of the International Conference on Information Theoretic Security, Libertad City, Ecuador, 10–12 January 2018; Springer: Berlin, Germany, 2018; pp. 95–104.
7. Barreno, M.; Nelson, B.; Joseph, A.D.; Tygar, J.D. The security of machine learning. *Mach. Learn.* **2010**, *81*, 121–148. [CrossRef]
8. Wu, L.; Du, X.; Wu, J. Effective defense schemes for phishing attacks on mobile computing platforms. *IEEE Trans. Veh. Technol.* **2016**, *65*, 6678–6691. [CrossRef]
9. Cheng, Y.; Fu, X.; Du, X.; Luo, B.; Guizani, M. A lightweight live memory forensic approach based on hardware virtualization. *Inf. Sci.* **2017**, *379*, 23–41. [CrossRef]
10. Hei, X.; Du, X.; Lin, S.; Lee, I. PIPAC: Patient infusion pattern based access control scheme for wireless insulin pump system. In Proceedings of the 2013 Proceedings IEEE INFOCOM, Turin, Italy, 14–19 April 2013; pp. 3030–3038.
11. Du, X.; Xiao, Y.; Guizani, M.; Chen, H.H. An effective key management scheme for heterogeneous sensor networks. *Ad Hoc Netw.* **2007**, *5*, 24–34. [CrossRef]
12. Du, X.; Guizani, M.; Xiao, Y.; Chen, H.H. Transactions papers a routing-driven Elliptic Curve Cryptography based key management scheme for Heterogeneous Sensor Networks. *IEEE Trans. Wirel. Commun.* **2009**, *8*, 1223–1229. [CrossRef]
13. Du, X.; Chen, H.H. Security in wireless sensor networks. *IEEE Trans. Wirel. Commun.* **2008**, *15*, 60–66.
14. Biggio, B.; Fumera, G.; Roli, F.; Didaci, L. Poisoning adaptive biometric systems. In Proceedings of the Joint IAPR International Workshops on Statistical Techniques in Pattern Recognition (SPR) and Structural and Syntactic Pattern Recognition (SSPR), Hiroshima, Japan, 7–9 Novembe 2012; Springer: Berlin, Germany, 2012; pp. 417–425.
15. Biggio, B.; Didaci, L.; Fumera, G.; Roli, F. Poisoning attacks to compromise face templates. In Proceedings of the 2013 International Conference on Biometrics (ICB), Madrid, Spain, 4–7 June 2013; pp. 1–7.

16. Biggio, B.; Fumera, G.; Roli, F. Pattern recognition systems under attack: Design issues and research challenges. *Int. J. Pattern Recognit. Artif. Intell.* **2014**, *28*, 1460002. [CrossRef]

17. Zhu, X. Super-class Discriminant Analysis: A novel solution for heteroscedasticity. *Pattern Recognit. Lett.* **2013**, *34*, 545–551. [CrossRef]

18. Biggio, B.; Nelson, B.; Laskov, P. Poisoning attacks against support vector machines. *arXiv* **2012**, arXiv:1206.6389.

19. Yang, C.; Wu, Q.; Li, H.; Chen, Y. Generative poisoning attack method against neural networks. *arXiv* **2017**, arXiv:1703.01340.

20. Szegedy, C.; Zaremba, W.; Sutskever, I.; Bruna, J.; Erhan, D.; Goodfellow, I.; Fergus, R. Intriguing properties of neural networks. *arXiv* **2013**, arXiv:1312.6199.

21. Carlini, N.; Wagner, D. Towards evaluating the robustness of neural networks. In Proceedings of the 2017 IEEE Symposium on Security and Privacy (SP), San Jose, CA, USA, 22–26 May 2017; pp. 39–57.

22. Moosavi-Dezfooli, S.M.; Fawzi, A.; Fawzi, O.; Frossard, P. Universal adversarial perturbations. *arXiv* **2017**, arXiv:1610.08401.

23. Warde-Farley, D.; Goodfellow, I. 11 adversarial perturbations of deep neural networks. In *Perturbations, Optimization, and Statistics*; MIT Press: Cambridge, MA, USA, 2016; p. 311.

24. Papernot, N.; McDaniel, P.; Jha, S.; Fredrikson, M.; Celik, Z.B.; Swami, A. The limitations of deep learning in adversarial settings. In Proceedings of the 2016 IEEE European Symposium on Security and Privacy (EuroS&P), Saarbrücken, Germany, 21–24 March 2016; pp. 372–387.

25. Chen, S.; Xue, M.; Fan, L.; Hao, S.; Xu, L.; Zhu, H.; Li, B. Automated poisoning attacks and defenses in malware detection systems: An adversarial machine learning approach. *Comput. Secur.* **2018**, *73*, 326–344. [CrossRef]

26. Chen, L.; Hou, S.; Ye, Y.; Xu, S. Droideye: Fortifying security of learning-based classifier against adversarial android malware attacks. In Proceedings of the 2018 IEEE/ACM International Conference on Advances in Social Networks Analysis and Mining (ASONAM), Barcelona, Spain, 28–31 August 2018; pp. 782–789.

27. Al-Dujaili, A.; Huang, A.; Hemberg, E.; O'Reilly, U.M. Adversarial deep learning for robust detection of binary encoded malware. In Proceedings of the 2018 IEEE Security and Privacy Workshops (SPW), San Francisco, CA, USA, 24 May 2018; pp. 76–82.

28. Grosse, K.; Papernot, N.; Manoharan, P.; Backes, M.; McDaniel, P. Adversarial examples for malware detection. In Proceedings of the European Symposium on Research in Computer Security, Oslo, Norway, 11–15 September 2017; pp. 62–79.

29. Maas, A.L.; Hannun, A.Y.; Ng, A.Y. Rectifier nonlinearities improve neural network acoustic models. In Proceedings of the 30th International Conference on Machine Learning, Atlanta, GA, USA, 16–21 June 2013; Volume 30, p. 3.

30. Mishkin, D.; Matas, J. All you need is a good init. *arXiv* **2015**, arXiv:1511.06422.

31. Chen, Z.; Lin, T.; Tang, N.; Xia, X. A parallel genetic algorithm based feature selection and parameter optimization for support vector machine. *Sci. Progr.* **2016**, *2016*, 2739621. [CrossRef]

32. Phan, A.V.; Le Nguyen, M.; Bui, L.T. Feature weighting and SVM parameters optimization based on genetic algorithms for classification problems. *Appl. Intell.* **2017**, *46*, 455–469. [CrossRef]

33. Alejandre, F.V.; Cortés, N.C.; Anaya, E.A. Feature selection to detect botnets using machine learning algorithms. In Proceedings of the 2017 International Conference on Electronics, Communications and Computers (CONIELECOMP), Cholula, Mexico, 22–24 February 2017; pp. 1–7.

34. Papernot, N.; McDaniel, P.; Goodfellow, I.; Jha, S.; Celik, Z.B.; Swami, A. Practical black-box attacks against machine learning. In Proceedings of the 2017 ACM on Asia Conference on Computer and Communications Security, Abu Dhabi, UAE, 2–6 April 2017; pp. 506–519.

35. Papernot, N.; McDaniel, P.; Goodfellow, I.; Jha, S.; Celik, Z.B.; Swami, A. Practical black-box attacks against deep learning systems using adversarial examples. *arXiv* **2016**, arXiv:1602.02697.

36. Papernot, N.; McDaniel, P.; Goodfellow, I. Transferability in machine learning: from phenomena to black-box attacks using adversarial samples. *arXiv* **2016**, arXiv:1605.07277.

37. Narodytska, N.; Kasiviswanathan, S.P. Simple Black-Box Adversarial Attacks on Deep Neural Networks. In Proceedings of the CVPR Workshops, Honolulu, HI, USA, 21–26 July 2017; pp. 1310–1318.

Sensors **2019**, *19*, 974

38. Papernot, N.; McDaniel, P.; Wu, X.; Jha, S.; Swami, A. Distillation as a defense to adversarial perturbations against deep neural networks. In Proceedings of the 2016 IEEE Symposium on Security and Privacy (SP), San Jose, CA, USA, 23–25 May 2016; pp. 582–597.

39. Arp, D.; Spreitzenbarth, M.; Hubner, M.; Gascon, H.; Rieck, K.; Siemens, C. DREBIN: Effective and Explainable Detection of Android Malware in Your Pocket. In Proceedings of the Network and Distributed System Security Symposium (NDSS), San Diego, CA, USA, 23–26 February 2014; Volume 14, pp. 23–26.

40. Spreitzenbarth, M.; Freiling, F.; Echtler, F.; Schreck, T.; Hoffmann, J. Mobile-sandbox: having a deeper look into android applications. In Proceedings of the 28th Annual ACM Symposium on Applied Computing, Coimbra, Portugal, 18–22 March 2013; pp. 1808–1815.

sensors

MDPI

Article

Measurement Structures of Image Compressive Sensing for Green Internet of Things (IoT)

Ran Li *, Xiaomeng Duan and Yanling Li

School of Computer and Information Technology, Xinyang Normal University, Xinyang 464000, China; dxmLily@163.com (X.D.); liyanling@xynu.edu.cn (Y.L.)
* Correspondence: liran@xynu.edu.cn

Received: 7 November 2018; Accepted: 24 December 2018; Published: 29 December 2018

Abstract: Image compressive sensing (CS) is a potential imaging scheme for green internet of things (IoT). To further make CS-based sensor adaptable to low bandwidth and low power, this paper focuses on finding a good measurement structure, i.e., the organization and storage format of CS measurements. Three potential measurement structures are proposed in this paper, respectively raster structure (RA), patch structure, and layer structure (LA). RA stores CS measurements of each column in an image, and PA packets CS measurements of overlapping patches forming an image. LA enables the measuring of small blocks and recovery of large blocks. All of the three structures avoid high computation complexity and huge memory in the process of measuring and recovery, and efficiently suppress the annoying blocking artifacts which often occur in traditional block structures. Experimental results show that RA, PA, and LA can efficiently reduce blocking artifacts, and produce comforting visual qualities. LA, especially, presents both good time-distortion and rate-distortion performance. By this paper, it is proved that LA is a suitable measurement structure for green IoT.

Keywords: image compressive sensing (CS); green internet of things (IoT); measurement structure; random structural matrices; linear recovery

---

1. Introduction

1.1. Motivation and Objective

Up to this day, compressive sensing (CS) [1,2] has already become a commonplace technology, and it is widely used in imaging applications to sample various signals, e.g., magnetic resonance imaging (MRI) [3], multispectral imaging [4], synthetic aperture radar (SAR) imaging [5], etc. Moreover, in the field of microwave tomography and antenna synthesis, CS also becomes a popular tool, e.g., [6,7] use CS to solve the non-linear inverse scattering problems, and [8,9] use CS to solve power synthesis of maximally-sparse arrays. These applications reflect the low cost of CS in capturing invisible light which brings a cheap and portable sensor. Many researchers recognize that CS is an energy-efficient sampling method which provides a good imaging quality at the same time, so they have turned their attention to the application of CS to green internet of things (green IoT) [10,11].

To achieve green IoT, some frameworks [12–14] of image CS are reviewed and newly developed to meet the demand for these energy-hungry devices. These works focus on how to design sampling and recovery algorithms to get better imaging quality in low-rate and low-power cases, but they have not considered how to organize and store CS measurements to improve the efficiency of sensors in performing CS. In green IoT, measurement structure, i.e., the organization and storage format of CS measurements, is important for sensors to produce a compact bitstream that could reduce the energy consumption of transmission. Therefore, the objective of this paper is to find a good measurement structure to make CS-based sensors energy-efficient.

1.2. Related Work

In existing works, two kinds of measurement structures are used to organize CS measurements, i.e., whole structure (WH) and block structure (BL). WH regards whole image as a column vector, and measures it by using a random matrix. The early works [15–18] on image CS extensively applied WH in CS measuring, and they often spent much time in recovering a small size image by some complex numeric iterations. The limitation of these works results from the difficulty in constructing some excellent matrices, e.g., i.i.d. Gaussian matrix, under WH. When enlarging the image size, WH leads to high computation complexity for measuring and huge memory for storage, e.g., for an image of 512×512 in size, 512 gigabytes storage is required to construct a random matrix with entries being 64-bit floating points. This is impractical especially for low-power sensors in green IoT. To make WH suitable for large-scale image, some memory-friendly structural matrices were proposed, in which the pioneering work is structurally random matrix (SRM) proposed by Do et al. [19]. SRM replaces the matrix entries with operators of scrambling, fast transforming and sub-sampling, and it enables WH to fast encode a large-scale image while providing a good recovery quality. Based on SRM, some works added some special operators to construct better structural matrices, e.g., Zhang et al. [20] used a unit-norm tight frame as a part of SRM, and Hsieh et al. [21] used sparse FFT as the core of SRM. Two defects still exist in SRM-based WH. First, CS measurements are only encoded by scalar quantization (SQ) which, however, does not perform well in rate-distortion performance due to the randomness of measurements. Second, SRM enforces the recovery algorithm to perform matrix-vector product in the form of function handle which, however, are not supported by some popular recovery algorithms. These defects limit the wide use of WH in practice, so some researchers turn their interest to the structure BL.

BL splits a whole image into non-overlapping blocks and then regards a block as a column vector and finally, measures it by using a random matrix. Because blocks are small in size, the unstructured matrix like i.i.d. Gaussian matrix can be used, without the worry of high computation complexity and huge memory. This makes BL practical especially when sensors have limited power and computation resource in green IoT. Many works focus on BL-based sampling, quantization, and recovery methods, e.g., Gan [22] and Mun et al. [23] proposed the smoothed projected Landweber (SPL) algorithm to recover blocks; Yu et al. [24] and Zhang et al. [25] proposed to adaptively measure blocks according to their features; Mun et al. [26], Wang et al. [27], Dinh et al. [28], and Gao et al. [29] proposed various predictive schemes to quantize block measurements. These works win BL more popularity than WH in image CS. However, blocks are different in sparse degree in a fixed space, thus they vary in recovery quality when the same algorithm is performed, which leads to blocking artifacts. The challenge for BL is to suppress the annoying blocking artifacts. The existing works try to overcome this defect by designing sampling and recovery algorithms, but they neglect the fact that BL is the root cause of blocking artifacts. Therefore, we need to find other potential structures to make up for this deficiency of BL.

1.3. Main Contribution

This paper presents three potential measurement structures—raster structure (RA), patch structure (PA), and layered structure (LA)—and they can effectively reduce the blocking artifacts in BL. Compared with WH, these structures have lower power and rate, especially for RA and LA, so they are suitable to be used in sensors of green IoT. We carefully analyze the time-distortion and rate-distortion performance of image CS when deploying various structures, and conclude that LA is the most potential structure among them. Combining LA with linear recovery, CS-based sensor will be more suitable for green IoT.

The rest of this paper is organized as follows. Section 2 briefly describes the two traditional structures: WH and BL. Section 3 presents the three potential structures: RA, PA, and LA. Experimental results are presented in Section 4, and we conclude this paper in Section 5. For easy understanding, the acronyms and notations in this paper are listed in Table 1.

Table 1. List of acronyms and notations.

CS	Compressive Sensing
Green IoT	Green Internet of Things
WH	Whole Structure
BL	Block Structure
RA	Raster Structure
PA	Patch Structure
LA	Layered Structure
SQ	Scalar Quantization
DPCM	Differential Pulse Code Modulation
SRM	Structurally Random Matrix
DCT	Discrete Cosine Transform
FFT	Fast Fourier Transform
GPSR	Gradient Projection for Sparse Reconstruction
PSNR	Peak Signal-to-Noise Ratio

2. Traditional Structures

2.1. Whole Structure

As shown in Figure 1, WH first transfers the 2-D image $X \in \mathbb{R}^{I_c \times I_r}$ ($N = I_c \times I_r$) to 1-D vector $x \in \mathbb{R}^{N \times 1}$ through raster scanning, i.e.,

$$x = \text{Raster}(X) \tag{1}$$

in which Raster($\cdot$) is an operator of raster scanning. Then, the measurement matrix $\Phi \in \mathbb{R}^{M \times N}$ is constructed, and we get the CS measurements $y \in \mathbb{R}^{M \times 1}$ of x as follows,

$$y = \Phi \cdot x \tag{2}$$

in which M is the number of CS measurements, and the subrate S is defined as M/N. These CS measurements y are quantized by SQ to bits, and finally, these bits are packaged, and stored in the output buffering of sensor.

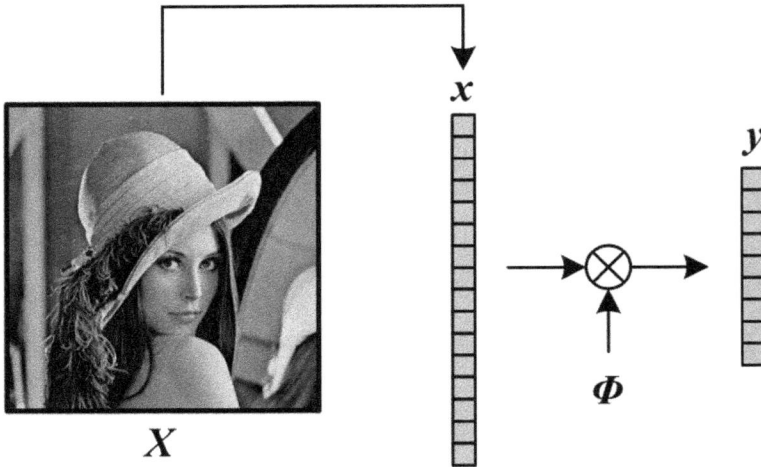

Figure 1. Illustration on WH format.

For a large-size image, limited by the memory size of sensor, it is impossible to construct the measurement matrix $\boldsymbol{\Phi}$, so we cannot perform Equation (2). However, by SRM, we can get an operator equivalent to Equation (2) as follows,

$$y = \boldsymbol{\Phi}(x) \tag{3}$$

in which $\boldsymbol{\Phi}(\cdot)$ is a function handle of SRM. SRM is defined as a product of three matrices, i.e.,

$$\boldsymbol{\Phi} = \sqrt{\frac{N}{M}} DFR \tag{4}$$

in which R is a uniform random permutation matrix, F is an orthonormal matrix selected from some popular fast computable transforms, e.g., fast Fourier transform (FFT), discrete cosine transform (DCT), etc. D is a sub-sampling matrix, and $\sqrt{\frac{N}{M}}$ is to normalize the transform so that the energy of the output vector is almost similar to that of the input vector [19]. The three matrices have all the equivalent operators corresponding to their matrix-vector products, so a function handle of SRM can be designed to replace Equation (2).

WH introduces a challenge for image recovery: the recovery algorithm must support the function handles of measurement and transform matrices. The WH-based recovery model is listed as

$$\hat{\alpha} = \underset{\alpha}{\mathrm{argmin}} \|\alpha\|_0 \text{ s.t. } y = \boldsymbol{\Phi}(\boldsymbol{\Psi}(\alpha)) \tag{5}$$

$$\hat{x} = \boldsymbol{\Psi}(\hat{\alpha}) \tag{6}$$

in which $\|\cdot\|$ is l_0 norm, $\boldsymbol{\Phi}(\cdot)$ is a handle of SRM, and $\boldsymbol{\Psi}(\cdot)$ is a fast transform operator, e.g., DCT, FFT, etc. The existence of $\boldsymbol{\Phi}(\cdot)$ and $\boldsymbol{\Psi}(\cdot)$ requires that $\boldsymbol{\Phi} \cdot x$, $\boldsymbol{\Phi}^T \cdot x$, $\boldsymbol{\Psi} \cdot x$, and $\boldsymbol{\Psi}^T \cdot x$ have the equivalent function handles (the superscript T represents transposition). From the above, it can be seen that WH requires the recovery algorithm to support function handle, so we only select some special algorithms, e.g., gradient projection for sparse reconstruction (GPSR) [15], to recover an image once from WH-based measurements.

2.2. Block Structure

As shown in Figure 2, BL first splits the 2-D image X into L non-overlapping blocks $X_{bi} \in \mathrm{R}^{B \times B}$ ($N_b = B \times B$) and transfer these blocks to 1-D vector $x_{bi} \in \mathbb{R}^{N_b \times 1}$ through raster scanning, i.e.,

$$\{X_{bi} | i = 1, 2, \cdots, L\} = \mathrm{Split}(X) \tag{7}$$

$$x_{bi} = \mathrm{Raster}(X_{bi}) \tag{8}$$

in which Split$(\cdot)$ is an operator of splitting image into non-overlapping blocks. Then, by constructing the block measurement matrix $\boldsymbol{\Phi}_{bi} \in \mathbb{R}^{M_b \times N_b}$, the CS measurements $y_{bi} \in \mathbb{R}^{M_b \times 1}$ of x_{bi} are generated as

$$y_{bi} = \boldsymbol{\Phi}_{bi} \cdot x_{bi} \tag{9}$$

in which M_b is the number of CS measurements for each block, and the subrate S can be computed by M_b/N_b. Due to the small size of block, it requires a small memory size to construct the block measurement matrix $\boldsymbol{\Phi}_{bi}$, and the less computations are invested when performing Equation (9) block-by-block. These block measurements can be quantized to bits by the predictive quantization, e.g., DPCM [26], and finally, these bits are packaged block by block, and progressively stored in the output buffering of sensor. The predictive quantization shows better rate-distortion performance than SQ, so BL has a compact packet, which reduces the size of buffering in sensor.

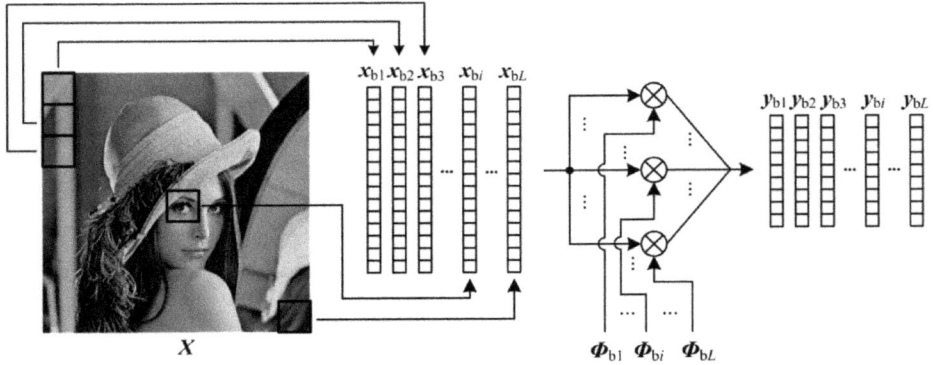

Figure 2. Illustration on BL format.

BL has no special requirement on recovery algorithm. Each block can be reconstructed independently by any recovery algorithm, and especially, different from WH, a linear recovery method can be used to reconstruct blocks, i.e.,

$$\hat{x}_{bi} = \left\{ R \cdot \boldsymbol{\Phi}_{bi}^{\mathsf{T}} \cdot \left[\boldsymbol{\Phi}_{bi} \cdot R \cdot \boldsymbol{\Phi}_{bi}^{\mathsf{T}} \right]^{-1} \right\} \cdot y_{bi} \tag{10}$$

in which R is statistic auto-correlation matrix of x_{bi}, and its element $R_{m,n}$ is approximated as follows,

$$R_{m,n} = (0.95)^{\delta_{m,n}} \tag{11}$$

in which $\delta_{m,n}$ is the spatial distance between two pixels in x_{bi}, $x_{bi,m}$ and $x_{bi,n}$. Compared with conventional recovery algorithms, the linear recovery costs less computation while presenting a good recovery quality. Therefore, BL can both reduce the energy consumption of encoder and decoder, which is more suitable for green IoT. However, there are annoying blocking artifacts in BL-based recovery, so we need to explore potential structures to suppress blocking artifacts while preserving the advantages of BL at the same time.

3. Three Potential Structures

3.1. Raster Structure

A key reason for blocking artifacts in BL is that the block is the basic unit of measuring and recovering. Blocking artifacts are gone once we irregularly spilt an image into sub-areas in other shapes. A simple way is to write the 2-D image in column form as

$$X = \begin{bmatrix} | & | & & | \\ x_{c1} & x_{c2} & \cdots & x_{cl_c} \\ | & | & & | \end{bmatrix} \tag{12}$$

in which $x_{ci} \in R^{l_r \times 1}$ represents the i-th column vector. We can view these column vectors as the raster of image, so this segmentation is called raster structure (RA). As shown in Figure 3, these column vectors can be measured by a small-size matrix $\boldsymbol{\Phi}_{ci} \in R^{l_r \times l_r}$, and the corresponding CS measurements $y_{ci} \in \mathbb{R}^{M_c \times 1}$ can be achieved as

$$y_{ci} = \boldsymbol{\Phi}_{ci} \cdot x_{ci} \tag{13}$$

in which M_c is the number of CS measurements for each column, and the subrate S can be computed by M_c/I_r. Because of high similarity between columns, the predictive quantization shows better rate-distortion performance for RA. Similar to BL, by replacing $\boldsymbol{\Phi}_{bi}$ and y_{bi} in Equation (6) with $\boldsymbol{\Phi}_{ci}$

and y_{ci}, each column vector can be recovered by the linear algorithm, so RA retains the advantages of BL.

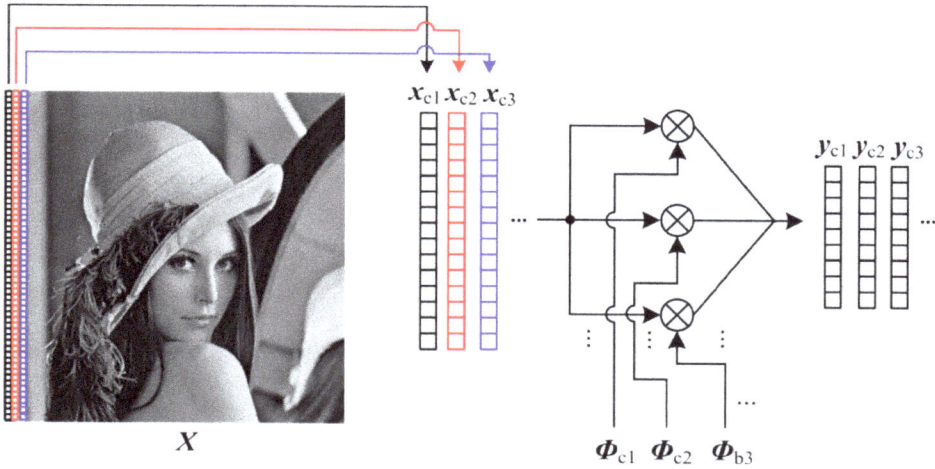

Figure 3. Illustration on RA format.

3.2. Patch Structure

BL performs the recovery algorithm block by block, any two blocks are different in recovery quality because the statistical characteristic of block is non-stationary. Blocking artifacts are usually perceived in the regions where one block has significantly different statistics from its neighbors. By extending BL, we present a patch structure (PA) which split the 2-D image into overlapping patches. As shown in Figure 4, PA first splits the 2-D image X into K overlapping blocks $X_{pi} \in \mathbb{R}^{P \times P}$ ($N_p = P \times P$), and transfer these patches to 1-D vector $x_{pi} \in \mathbb{R}^{N_p \times 1}$ through raster scanning. The size of overlap region between two neighboring patches is $P/2$, i.e., we slide a $P \times P$ patch in an image with the step size being $P/2$. By constructing the block measurement matrix $\Phi_{pi} \in \mathbb{R}^{M_p \times N_p}$, the CS measurements $y_{pi} \in \mathbb{R}^{M_p \times 1}$ of x_{pi} are generated as

$$y_{pi} = \Phi_{pi} \cdot x_{pi} \tag{14}$$

in which M_p is the number of CS measurements for each patch, and the subrate S can be computed by M_p/N_p. The storage of Φ_{pi} is moderate, so PA cannot introduce for sensors excessive burdens of computing and storing. By the predictive quantizing, these measurements are transferred to bits, and transmitted to decoder. By replacing Φ_{bi} and y_{bi} in Equation (6) with Φ_{pi} and y_{pi}, the linear recovery algorithm is used to recover each patch, and these patches are spliced together into a whole image. PA smooths the boundary between two blocks, so it shows efficiency of reducing blocking artifacts.

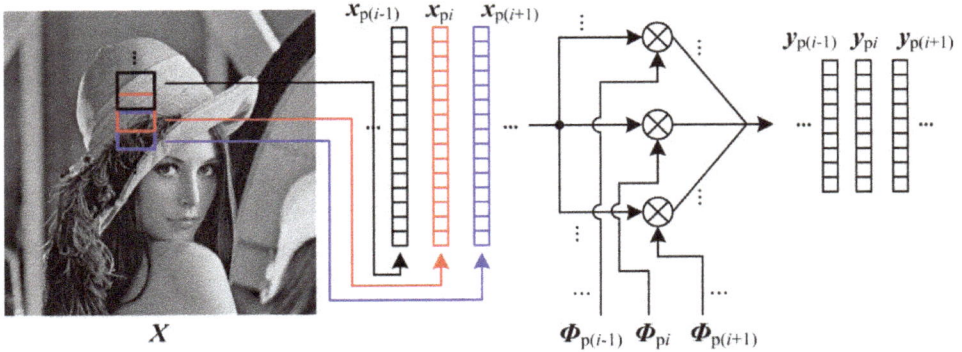

Figure 4. Illustration on PA format.

3.3. Layered Structure

Blocking artifacts results from the differences of neighboring blocks in recovery quality. The smaller the block size is, the more serious blocking artifacts are. The recovery quality is known to be better with a large block [22], so blocking artifacts can be effectively suppressed when we set a large block size in BL. However, considering the light burden of sensor, a small block size is more desired. Especially for predictive quantization, a high spatial correlation exists among small-size blocks, which would increase predicting efficiency of quantization. In view of that, as shown in Figure 5, we present the layered structure (LA) which can splice CS measurements of small blocks into those of large blocks, thus enabling the measuring of small blocks and the recovery of large blocks.

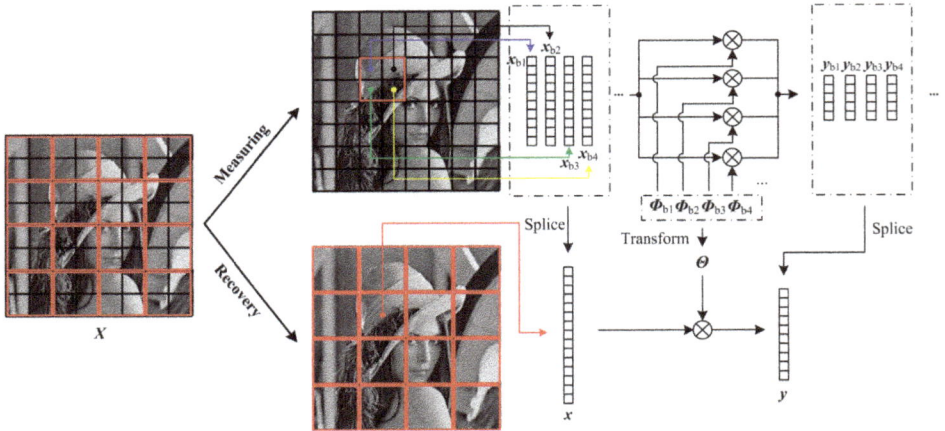

Figure 5. Illustration on LA format. The block-size pair is (B_1, B_2), and B_2/B_1 is 2.

In LA, we set a block-size pair (B_1, B_2), in which B_1 and B_2 are the block size for measuring and recovery, respectively. To achieve measuring small block and recovering large block, B_1 is smaller than B_2, and they satisfy that,

$$B_2 = 2^l \cdot B_1 \quad l = 1, 2, \cdots \tag{15}$$

where l is a positive integer. When performing CS measuring, as the same with BL, LA splits the 2-D image X into L non-overlapping blocks $X_{bi} \in \mathbb{R}^{B_1 \times B_1}$ ($N_b = B_1 \times B_1$) and transfer these blocks to 1-D vector $x_{bi} \in \mathbb{R}^{N_b \times 1}$ through raster scanning. By constructing the block measurement matrix $\Phi_{bi} \in \mathbb{R}^{M_b \times N_b}$, the CS measurements $y_{bi} \in \mathbb{R}^{M_b \times 1}$ of these small blocks are produced, and we gather these CS measurements into the structure at the low layer. By re-organizing the low-layer structure, these CS measurements of small blocks are converted into those of large blocks which form the structure at high layer. The following describes the process of converting structure from low layer to high layer when l is set to be 1. As shown in Figure 5, each red block contains four black blocks which are measured as

$$y_{bi} = \Phi_{bi} \cdot x_{bi} \quad i = 1, 2, 3, 4 \tag{16}$$

and splice y_{bi} into y in rows, i.e.,

$$y = \begin{bmatrix} y_{b1} \\ y_{b2} \\ y_{b3} \\ y_{b4} \end{bmatrix} = \begin{bmatrix} \Phi_{b1} & & & \\ & \Phi_{b2} & & \\ & & \Phi_{b3} & \\ & & & \Phi_{b4} \end{bmatrix} \begin{bmatrix} x_{b1} \\ x_{b2} \\ x_{b3} \\ x_{b4} \end{bmatrix} = \Phi_\Lambda \cdot \begin{bmatrix} x_{b1} \\ x_{b2} \\ x_{b3} \\ x_{b4} \end{bmatrix} \tag{17}$$

in which Φ is the diagonal matrix composed of the block measurement matrices of four black blocks. These CS measurements of red blocks are gathered into the high-layer structure. However, from Equation (17), it can be seen that the column vector y does not correspond to the column vector x of the red block, so Φ_Λ needs to be transformed into a proper matrix. By an elementary column transformation, it is easy to transform $[x_{b1}; x_{b2}; x_{b3}; x_{b4}]$ into x, which can be represented as

$$\begin{bmatrix} x_{b1} \\ x_{b2} \\ x_{b3} \\ x_{b4} \end{bmatrix} = E \cdot x \tag{18}$$

in which E is an elementary column transformation matrix. Plugging Equation (18) into Equation (17), we can get

$$y = \begin{bmatrix} y_{b1} \\ y_{b2} \\ y_{b3} \\ y_{b4} \end{bmatrix} = \Phi_\Lambda \cdot \begin{bmatrix} x_{b1} \\ x_{b2} \\ x_{b3} \\ x_{b4} \end{bmatrix} = \Phi_\Lambda E \cdot x = \Theta \cdot x \tag{19}$$

in which $\Theta = \Phi_\Lambda E$. According to Equation (19), we construct CS measuring formula of the red block, so we can use the CS measurements of black blocks to linearly recover the red ones by Equation (6). When l is set to be larger than 1, the construction of LA can be done in the manner similar to the above.

4. Experimental Results

In this section, we subjectively and objectively evaluate the reconstruction quality of some 512×512 test images using different measurement structures in image CS. These test images include *Lenna*, *Barbara*, *Peppers*, *Goldhill*, and *Mandrill*. In all experiments, the subrate S is set to be in the range of $[0.1, 0.5]$. For WH, DCT-based SRM [19] is used to construct the measurement matrix, SQ is used to quantize CS measurements, and GPSR algorithm is used to recover images. For BL, RA, PA, and LA, to ensure a fair comparison, i.i.d. Gaussian matrix is selected to be measurement matrix, DPCM is used to quantize CS measurements, and images are recovered linearly. The block size is set to be 16 in BL, and LA uses different block-size pairs in different experiments. Subjective evaluation shows the reconstructed images at different subrates using different structures. In objective evaluation, we measure the recovery quality in terms of peak signal-to-noise ratio (PSNR), bitrate, and encoding time. The variation of PSNR with bitrate is called rate-distortion, and the variation of PSNR with encoding time is called time-distortion. Bitrate indicates the size of data transmitted by sensor, and encoding time gives an indication of the amount of energy expended while encoding, so rate-distortion reflects the transmission efficiency, and time-distortion the energy efficiency. All experiments are conducted under the following computer configuration: Intel(R) Core (TM) i7@3.30 GHz CPU, 8 GB RAM, Microsoft Windows 764 bits, and MATLAB Version 7.6.0.324 (R2008a).

4.1. Subjective Evaluation

Figure 6 presents the visual recovery results of *Lenna* by various measurement structures at different subrates. When subrate S is 0.1, due to insufficient CS measurements, the reconstructed image by any structure lose structural details. It can be seen that, WH presented a disappointing result due to serious blurs, and BL bring lots of blocking artifacts. Although RA and PA can remove the blocking artifacts, they introduce other problems—e.g., obvious stripes in RA, global fuzzy in PA. When setting the block-size pair (8, 16), LA has a good visual perception when compared with BL, though many blocking artifacts still exist. When subrate S is increased to be 0.3, bad effects can be efficiently suppressed in WH, PA, and RA, and blocking artifacts almost disappear in BL and LA. When subrate S is 0.5, all structures provide satisfying visual results. we can see from the above that, none of the structures can guarantee good visual results when subrate S is 0.1, therefore, we expect to improve the visual quality at a low subrate. As shown in Figure 7, with a large l, LA can satisfy our expectation. When l is set to be 2, i.e., the block-size pair is (8, 32), blocking artifacts are reduced efficiently at subrate $S = 0.1$, and even when the block-size pair is (8, 64), blocking artifacts are almost imperceptible. When setting a small block size, as shown in Figure 7d–f, LA can still ensure a high efficiency for reducing blocking artifacts. Given the above, LA has greater potential than other structures from subjective view.

Figure 6. *Cont.*

Figure 6. Visual quality comparison on test image Lenna for various measurement structures. Note: the block-size pair of LA is (8, 16).

(**a**) (8, 16)　　　　　　(**b**) (8, 32)　　　　　　(**c**) (8, 64)

Figure 7. *Cont.*

(**d**) (4, 16)   (**e**) (4, 32)   (**f**) (4, 64)

Figure 7. Visual quality comparison on test image Lenna for LA structures with various block-size pairs when the subrate is 0.1.

4.2. Objective Evaluation

Table 2 presents the average encoding time of various structures on all test images when setting different subrates. WH and PA have a high execution time at any subrate, and their average time is respectively 31.74 and 36.05 ms on all subrates. BL, RA, and LA have a low execution time, among which RA does the best with only 10.86 ms on all subrates. Table 3 presents the average decoding time of various structures on all test images at different subrates. We can see that, the decoding time of WH is much longer than those of other structures, indicating that the non-linear recovery brings a heavy burden for WH. Due to linear recovery, BL, RA, PA, and LA all have a low decoding time, especially for LA, with only 6.03 ms at all subrates. Tables 2 and 3 both considered, we can see that LA is a good structure for green IoT because it guarantees a low complexity at both encoder and decoder. Figure 8 shows average time-distortion and rate-distortion curves on test images for various measurement structures. It can be seen that, WH and PA have a poor time-distortion performance due to high encoding time, the time-distortion performance of BL is close to that of LA, and RA works the best among all structures. As for rate-distortion performance, WH and PA perform poorly, and LA works the best among all structures. Although RA has a good time-distortion performance, it produces more bits than LA. Figure 9 shows average time-distortion and rate-distortion curves on test images for LA structures with various block-size pairs. It can be seen that the rate-distortion performance can be significantly improved when block size B_1 of measuring is much smaller than block size B_2 of recovery, but different block-size pairs have little impact on time-distortion performance. To sum up, LA can save more bits while keeping both a low encoding time and a good recovery quality. Given the positive relation between encoding time and energy consumption, LA is the most suitable one among all structures for Green IoT from the objective view.

Table 2. Execution time (ms) of encoder for various measurement structures.

Subrate	0.1	0.2	0.3	0.4	0.5	Average
WH	30.68	31.17	31.57	31.97	33.28	31.74
BL	16.09	16.63	17.46	18.57	19.52	17.65
RA	6.82	8.57	10.69	12.91	15.31	10.86
PA	33.01	35.72	36.06	37.36	38.12	36.05
LA	16.66	17.37	18.46	19.31	20.41	18.44

Note: the block-size pair of LA is (8, 16).

Table 3. Execution time (ms) of decoder for various measurement structures.

Subrate	0.1	0.2	0.3	0.4	0.5	Average
WH	1.22×10^5	1.00×10^5	0.65×10^5	0.52×10^5	0.31×10^5	0.74×10^5
BL	10.12	10.21	10.16	10.14	10.33	10.19
RA	5.67	7.92	8.66	10.05	12.88	9.04
PA	21.18	21.71	21.92	21.72	21.59	21.62
LA	5.39	5.69	6.03	6.51	6.54	6.03

Note: the block-size pair of LA is (8, 16).

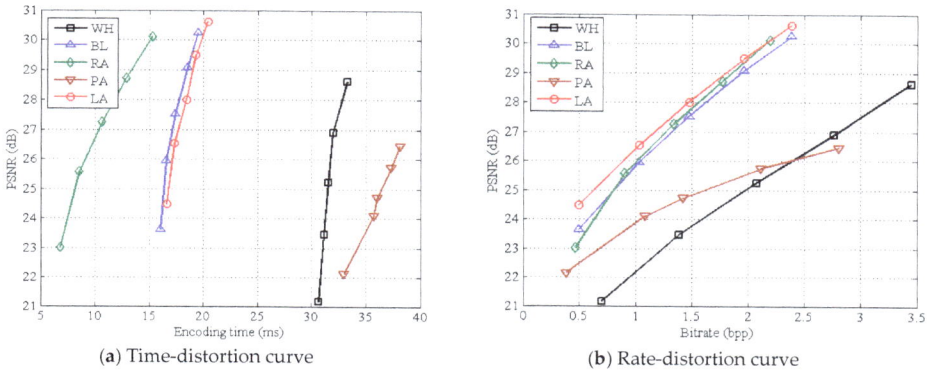

(**a**) Time-distortion curve

(**b**) Rate-distortion curve

Figure 8. Average time-distortion and rate-distortion curves on test images for various measurement structures. Note: the block-size pair of LA is (8, 16).

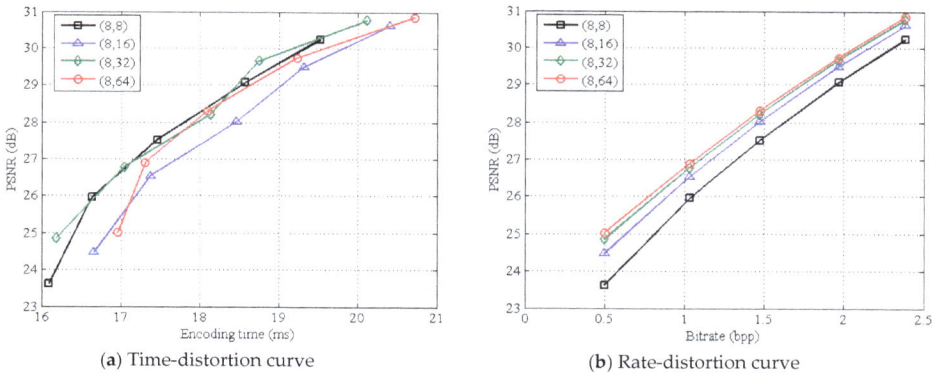

(**a**) Time-distortion curve

(**b**) Rate-distortion curve

Figure 9. Average time-distortion and rate-distortion curves on test images for LA structures with various block-size pairs.

4.3. Effects of Measurement Matrices

In this section, we present the effects of different measurement matrices on the performance of BL, RA, PA, and LA. We select two classic measurement matrices, Gaussian matrix and permuted fast Fourier transform (PFFT) matrix [30], and two hardware-friendly matrices, ±Bernoulli matrix [31] and DCT-based SRM [19]. To ensure a fair comparison, all structures use linear algorithm to reconstruct test images. Table 4 presents the average encoding time of various structures on all test images at all subrates using different measurement matrices. It can be seen that, no matter which matrix is used, RA costs less encoding time than other structures, PA requires the most time among all structures, and BL and LA have a moderate encoding complexity. For any structure, the encoding time varies slightly even when different matrices are used. Table 5 presents the average PSNR values of various structures on all test images at all subrates using different measurement matrices. It can be seen that LA gets higher PSNR values than others regardless of matrix used. For any structure, Gaussian matrix provides higher PSNR values than others, and ±Bernoulli and SRM matrices have a little PSNR degradation compared with Gaussian matrix. We can find from the above that both encoding time and imaging quality considered, LA provides a good balance for any measurement matrix, and ±Bernoulli and SRM matrices have little effect on the performance of different structures compared with Gaussian matrix. In order to make LA hardware-friendly, therefore, Gaussian matrix can be replaced with ±Bernoulli and SRM matrices.

Table 4. Average execution time (ms) of encoder for different structures using different measurement matrices.

	Gaussian	PFFT	±Bernoulli	SRM
BL	17.65	18.35	18.14	18.17
RA	10.86	11.13	10.67	10.66
PA	36.05	34.23	33.11	33.28
LA	18.44	18.71	18.14	18.61

Note: the block-size pair of LA is (8, 16).

Table 5. Average PSNR values (dB) for different structures using different measurement matrices.

	Gaussian	PFFT	±Bernoulli	SRM
BL	27.28	25.70	26.88	26.69
RA	26.94	16.22	26.92	26.94
PA	24.63	24.22	24.38	24.24
LA	27.82	26.09	27.53	27.48

Note: the block-size pair of LA is (8, 16).

4.4. Effects of Recovery Algorithms

In this section, we present the effects of different recovery algorithms on the performance of different structures. All structures use DCT-based SRM to measure test images. The four recovery algorithms, namely GPSR, orthogonal matching pursuit (OMP) [32], Bayesian recovery [33], and linear recovery, are used to reconstruct test images. To ensure good performance of non-linear algorithms, the block size is set to be 32 for BL and PA, and the block-size pair of LA is set to be (32, 64). Table 6 presents the average decoding time of different structures on all test images at all subrates using different recovery algorithms. We can see that non-linear algorithms cost far more decoding time than linear algorithm for any structure, and LA costs more decoding time than BL, RA, PA, for linear algorithm, indicating that linear algorithm has a light computational burden compared with non-linear algorithms, and LA increases the decoding computation. Table 7 presents the average PSNR values of different structures on all test images at all subrates using different recovery algorithms. It can be seen that linear algorithm provides higher PSNR values than non-linear algorithms for any structure, especially that LA cooperated with linear recovery achieves the highest PSNR value. From the above, we can find that linear algorithm is more suitable for BL, RA, PA, and LA in terms of either computational complexity or objective quality, and LA can improve the recovery quality at the cost of some computational burden increase.

Table 6. Average execution time (ms) of decoder for different measurement structures using different recovery algorithms.

	GPSR	OMP	Bayesian	Linear
WH	7.44×10^4	–	–	–
BL	9.34×10^4	5.57×10^3	6.26×10^4	114.36
RA	9.00×10^4	5.56×10^3	6.64×10^4	127.11
PA	1.14×10^5	2.99×10^3	7.73×10^4	71.97
LA	5.32×10^5	1.70×10^5	2.14×10^5	3.94×10^3

Note: due to the unavailability of function handle, OMP, Bayesian and linear algorithms cannot be used for WH.

Table 7. Average PSNR values (dB) for different measurement structures using different recovery algorithms.

	GPSR	OMP	Bayesian	Linear
WH	25.08	–	–	–
BL	22.93	23.07	24.13	28.01
RA	19.80	20.63	21.63	26.94
PA	21.03	21.61	21.57	25.44
LA	23.04	23.32	24.47	28.12

Note: due to the unavailability of function handle, OMP, Bayesian and linear algorithms cannot be used for WH.

5. Conclusions

In this paper, we have reviewed two traditional measurement structures of image CS, i.e., WH and BL, and propose three potential structures for Green IoT, i.e., RA, PA, and LA. As a straightforward structure, WH stores CS measurements of a whole image at a time, but must use SRM to measure and recover image in order to avoid the huge storage and computations. BL organizes CS measurements block by block, with high energy efficiency and low storage, but it produces serious blocking artifacts at the low subrate. By replacing non-overlapping block in BL with raster and overlapping block, RA and PA can remove blocking artifacts, though bringing other problems. LA splits image into small blocks, and re-organizes these measurements of small blocks into those of large ones. Since small block improves the efficiency of predictive quantization, and large block enhances the recovery quality, LA can efficiently suppress blocking artifacts. We perform several experiments to evaluate the subjective and objective performances of all structures. At a low subrate, by setting a proper block-size pair, LA almost eliminates blocking artifacts, and provides better visual quality than other structures. It shows good time-distortion and rate-distortion performances, especially for rate-distortion performance, it can significantly improve it by setting a large block size of recovering. Both subjective and objective evaluation considered, it can be seen that LA is the most suitable for green IoT among structures.

This paper only presents an exploratory research, and there are many intriguing questions that future work should consider. For instance, for PA, a patch overlaps its neighbors, so we can design more efficient recovery algorithm by using the correlation among patches. Also, parallel computing can be adopted to speed up the linear recovery for LA.

Author Contributions: R.L. contributed to the concept of the research, performed the data analysis and wrote the manuscript; X.D. helped perform the data analysis with constructive discussions and helped performed the experiment. Y.L. revised the manuscript, and provided the necessary support for the research.

Funding: This work was supported in part by the National Natural Science Foundation of China, under grant nos. 61501393 and 61572417, in part by Nanhu Scholars Program for Young Scholars of Xinyang Normal University, and in part by Innovation Team Support Plan of University Science and Technology of Henan Province (no. 19IRTSTHN014).

Conflicts of Interest: The authors declare no conflict of interest.

References

1. Baraniuk, R.G.; Goldstein, T.; Sankaranarayanan, A.C.; Studer, C.; Veeraraghavan, A.; Wakin, M.B. Compressive video sensing: Algorithms, architectures, and applications. *IEEE Signal Process. Mag.* **2017**, *34*, 52–66. [CrossRef]
2. Candè, E.J.; Wakin, M.B. An introduction to compressive sampling. *IEEE Signal Process. Mag.* **2008**, *25*, 21–30. [CrossRef]
3. Mun, S.; Fowler, J.E. Motion-compensated compressed-sensing reconstruction for dynamic MRI. In Proceedings of the 2013 IEEE International Conference on Image Processing, Melbourne, Australia, 15–18 September 2013; pp. 1006–1010.
4. Li, C.; Sun, T.; Kelly, K.F.; Zhang, Y. A compressive sensing and unmixing scheme for hyperspectral data processing. *IEEE Trans. Image Process.* **2012**, *21*, 1200–1210. [PubMed]
5. Wu, J.; Liu, F.; Jiao, L.C.; Wang, X. Compressive sensing SAR image reconstruction based on Bayesian framework and evolutionary computation. *IEEE Trans. Image Process.* **2011**, *20*, 1904–1911. [CrossRef] [PubMed]
6. Bevacqua, M.T.; Crocco, L.; Donato, L.D.; Isernia, T. Non-linear inverse scattering via sparsity regularized contrast source inversion. *IEEE Trans. Comput. Imaging* **2017**, *3*, 296–304. [CrossRef]
7. Bevacqua, M.T.; Isernia, T. Boundary indicator for aspect limited sensing of hidden dielectric objects. *IEEE Geosci. Remote Sens. Lett.* **2018**, *15*, 838–842. [CrossRef]
8. Morabito, A.F. Synthesis of maximum-efficiency beam arrays via convex programming and compressive sensing. *IEEE Antennas Wirel. Propag. Lett.* **2017**, *16*, 2404–2407. [CrossRef]
9. Morabito, A.F.; Rocca, P. Reducing the number of elements in phase-only reconfigurable arrays generating sum and difference patterns. *IEEE Antennas Wirel. Propag. Lett.* **2015**, *14*, 1338–1341. [CrossRef]
10. Arshad, R.; Zahoor, S.; Shah, M.A.; Wahid, A.; Yu, H. Green IoT: An investigation on energy saving practices for 2020 and beyond. *IEEE Access* **2017**, *5*, 15667–15681. [CrossRef]
11. Hu, J.; Luo, J.; Zheng, Y.; Li, K. Graphene-grid deployment in energy harvesting cooperative wireless sensor networks for Green IoT. *IEEE Trans. Ind. Inform.* **2018**. [CrossRef]
12. Colonnese, S.; Biagi, M.; Cattai, T.; Cusani, R.; De Vico Fallani, F.; Scarano, G. Green compressive sampling reconstruction in IoT networks. *Sensors* **2018**, *18*, 2735. [CrossRef] [PubMed]
13. Chen, P.; Ahammad, P.; Boyer, C.; Huang, S.I.; Lin, L.; Lobaton, E.; Meingast, M.; Oh, S.; Wang, S.; Yan, P.; et al. CITRIC: A low-bandwidth wireless camera network platform. In Proceedings of the ACM/IEEE International Conference on Distributed Smart Cameras, Stanford, CA, USA, 7–11 September 2008; pp. 1–10.
14. Asif, M.S.; Fernandes, F.; Romberg, J. Low-complexity video compression and compressive sensing. In Proceedings of the 2013 Asilomar Conference on Signals, Systems and Computers, Pacific Grove, CA, USA, 3–6 November 2013; pp. 579–583.
15. Figueiredo, M.A.T.; Nowak, R.D.; Wright, S.J. Gradient projection for sparse reconstruction: Application to compressed sensing and other inverse problems. *IEEE J. Sel. Top. Signal Process.* **2007**, *1*, 586–597. [CrossRef]
16. Bioucas-Dias, J.M.; Figueiredo, M.A.T. A new TwIST: Two-step iterative shrinkage/thresholding algorithms for image restoration. *IEEE Trans. Image Process.* **2007**, *16*, 2992–3004. [CrossRef]
17. Becker, S.; Bobin, J.; Candès, E.J. NESTA: A fast and accurate first-order method for sparse recovery. *SIAM J. Imaging Sci.* **2011**, *4*, 1–39. [CrossRef]
18. Beck, A.; Teboulle, M. A fast iterative shrinkage-thresholding algorithm for linear inverse problems. *SIAM J. Imaging Sci.* **2009**, *2*, 183–202. [CrossRef]
19. Do, T.T.; Gan, L.; Nguyen, N.H.; Tran, T.D. Fast and efficient compressive sensing using structurally random matrices. *IEEE Trans. Signal Process.* **2012**, *60*, 139–154. [CrossRef]
20. Zhang, P.; Gan, L.; Sun, S.; Ling, C. Modulated unit-norm tight frames for compressed sensing. *IEEE Trans. Signal Process.* **2015**, *63*, 3974–3985. [CrossRef]
21. Hsieh, S.H.; Lu, C.S.; Pei, S.C. Compressive sensing matrix design for fast encoding and decoding via sparse FFT. *IEEE Signal Process. Lett.* **2018**, *25*, 591–595. [CrossRef]
22. Gan, L. Block compressed sensing of natural images. In Proceedings of the 2007 15th International Conference on Digital Signal Processing, Cardiff, UK, 1–4 July 2007; pp. 403–406.
23. Mun, S.; Fowler, J.E. Block compressed sensing of images using directional transforms. In Proceedings of the 2009 16th IEEE International Conference on Image Processing, Cairo, Egypt, 7–10 November 2009; pp. 3021–3024.

24. Yu, Y.; Wang, B.; Zhang, L. Saliency-based compressive sampling for image signals. *IEEE Signal Process. Lett.* **2010**, *17*, 973–976.
25. Zhang, J.; Xiang, Q.; Yin, Y.; Chen, C.; Luo, X. Adaptive compressed sensing for wireless image sensor networks. Multimedia Tools and Applications. *Multimed. Tools Appl.* **2017**, *76*, 4227–4242. [CrossRef]
26. Mun, S.; Fowler, J.E. DPCM for quantized block-based compressed sensing of images. In Proceedings of the 2012 Proceedings of the 20th European Signal Processing Conference, Bucharest, Romania, 27–31 August 2012; pp. 1424–1428.
27. Wang, L.; Wu, X.; Shi, G. Binned progressive quantization for compressive sensing. *IEEE Trans. Image Process.* **2012**, *21*, 2980–2990. [CrossRef] [PubMed]
28. Dinh, K.Q.; Shim, H.J.; Jeon, B. Measurement coding for compressive imaging using a structural measuremnet matrix. In Proceedings of the 2013 IEEE International Conference on Image Processing, Melbourne, Australia, 15–18 September 2013; pp. 10–13.
29. Gao, X.; Zhang, J.; Che, W.; Fan, X.; Zhao, D. Block-based compressive sensing coding of natural images by local structural measurement matrix. In Proceedings of the 2015 Data Compression Conference, Snowbird, UT, USA, 7–9 April 2015; pp. 133–142.
30. Duarte, M.; Wakin, M.; Baraniuk, R. Fast reconstruction of piecewise smooth signals from incoherent projections. Presented at the Workshop on Signal Processing with Adaptive Sparse Structured Representations, Rennes, France, 16–18 November 2005; pp. 1–4.
31. Baraniuk, R.; Davenport, M.; DeVore, R. A simple proof of the restricted isometry property for random matrices. *Constr. Approx.* **2008**, *28*, 253–263. [CrossRef]
32. Tropp, J.A.; Gilbert, A.C. Signal recovery from random measurements via orthogonal matching pursuit. *IEEE Trans. Inf. Theory* **2007**, *53*, 4655–4666. [CrossRef]
33. Ji, S.; Xue, Y.; Carin, L. Bayesian compressive sensing. *IEEE Trans. Signal Process.* **2008**, *56*, 2346–2356. [CrossRef]

sensors

MDPI

Article

Matching SDN and Legacy Networking Hardware for Energy Efficiency and Bounded Delay †

Pablo Fondo-Ferreiro *,‡, Miguel Rodríguez-Pérez ‡, Manuel Fernández-Veiga ‡ and Sergio Herrería-Alonso ‡

atlanTTic Research Center, University of Vigo, 36310 Vigo, Spain; miguel@det.uvigo.gal (M.R.-P.); mveiga@det.uvigo.es (M.F.-V.); sha@det.uvigo.es (S.H.-A.)
* Correspondence: pfondo@det.uvigo.es; Tel.: +34-986-818-684
† This paper is an extended version of our paper published in Fondo-Ferreiro, P.; Rodríguez-Pérez, M.; Fernández-Veiga, M. Implementing Energy Saving Algorithms for Ethernet Link Aggregates with ONOS. In Proceedings of the 2018 Fifth International Conference on Software Defined Systems (SDS), Barcelona, Spain, 23–26 April 2018 and Fondo Ferreiro, P.; Rodríguez Pérez, M.; Fernández Veiga, M. QoS-aware Energy-Efficient Algorithms for Ethernet Link Aggregates in Software-Defined Networks. In Proceedings of the 9th Symposium on Green Networking and Computing (SGNC 2018), Split-Supetar, Croatia, 13–15 September 2018.
‡ These authors contributed equally to this work.

Received: 3 October 2018; Accepted: 8 November 2018; Published: 13 November 2018

Abstract: Both economic and environmental costs are driving much research in the area of the energy efficiency of networking equipment. This research has produced a great amount of proposals. However, the majority of them remain unimplemented due to the lack of flexibility of current hardware devices and a certain lack of enthusiasm from commercial vendors. At the same time, Software-Defined Networking (SDN) has allowed customers to control switching decisions with a flexibility and precision previously unheard of. This paper explores the potential convergence between the two aforementioned trends and presents a promising power saving algorithm that can be implemented using standard SDN capabilities of current switches, reducing operation costs on both data centers and wired access networks. In particular, we focus on minimizing the energy consumption in bundles of energy-efficient Ethernet links leveraging SDN. For this, we build on an existing theoretical algorithm and adapt it for implementing with an SDN solution. We study several approaches and compare the resulting algorithms not only according to their energy efficiency, but also taking into account additional QoS metrics. The results show that the resulting algorithm is able to closely match the theoretical results, even when taking into account the requirements of delay-sensitive traffic.

Keywords: energy-efficient Ethernet; QoS; SDN; real-time traffic; ONOS

1. Introduction

Nowadays, public concern about energy consumption of networking equipment is increasing due to not only environmental reasons, but also economic ones. Inside data centers, the network consumes up to 20% of its total power [1]. If we focus on wireless networks, the reduction of energy consumption is also a Key Performance Indicator (KPI) according to the 5G Infrastructure Public Private Partnership (5G-PPP) [2]. As a result, a wide range of solutions has been proposed to reduce the energy consumption of networking equipment. However, many of these new techniques remain unimplemented due to the lack of flexibility in current networks. For instance, some of the recent proposals in the literature require changes to basic protocols, whereas others require changes to the networking equipment [3–6].

In parallel, the Software-Defined Networking (SDN) paradigm is being embraced by the networking community. The SDN paradigm moves the forwarding logic from the devices themselves to the SDN applications, which run on top of the logically-centralized controller. In these networks, switches are just pure forwarding fabrics instructed by the SDN controller, and the network policies are programmed through software applications, which run on top of the SDN controller. The flexibility that this paradigm introduces has led to its extensive adoption in data centers. SDN is also deemed as a key enabler technology for 5G networks [7]. The adoption of SDN in these networks is seen as an opportunity to improve the energy efficiency of the communications infrastructure, overcoming the limitations of current networks by virtue of its programmability and flexibility.

Energy-Efficient Ethernet (EEE) [8,9] is the standard for saving energy in Ethernet interfaces. Despite the large savings in energy consumption achievable with EEE over single Ethernet links, the overall consumption in Ethernet link aggregates is not proportional to the offered load and depends largely on the actual traffic share among the links. In this paper, we adapt efficiently an analytical solution to the problem of minimizing energy in bundles of EEE links by leveraging the operational principles of SDN networks. Therefore, combining analysis with SDN capabilities, legacy switches equipped with EEE line cards can run energy-aware traffic distribution algorithms even if the vendors do not build support for them in the hardware/firmware. Specifically, we design, build and analyze three energy-efficient SDN-compatible flow allocation algorithms from the point of view of energy consumption, packet loss rate and transmission latency, both through simulations and also with an implementation on top of the Open Network Operating System (ONOS) SDN controller. Since the energy saving nature of the algorithms can make latency increase, we also consider traffic with different QoS requirements. Subsequently, two different solutions are proposed to handle the low-latency traffic while at the same time reducing the energy consumption. Our solutions are validated both through extensive simulation experiments and by implementing the algorithms in ONOS.

This paper extends our previous work presented in [10,11] by discussing further the related work, describing the solutions in detail, studying a new mechanism for estimating the rate of each flow and providing a thorough analysis of the algorithms. The rest of the paper is organized as follows. Section 2 introduces the related work. We describe our proposal for minimizing energy consumption in Section 3. Section 4 shows the QoS-aware algorithms. Results are discussed in Section 5. Finally, we draw some conclusions in Section 6.

2. Related Work

The advantages offered by SDN networks for advanced traffic management have been the subject of study of prior works that helped to understand the best way to use SDN for spending less energy. In [12], the authors carried out a survey on energy efficiency, identifying which components involved in the SDN networks can be configured in a dynamic way in order to reduce energy usage. Most of the solutions analyzed rely on re-routing the flows in the network so that the number of active switches is minimized. Thus, these devices can be put in a low-energy state or eventually turned off, reducing the power consumption of the network. These proposals are termed traffic-aware, since they need to know the traffic load that is currently passing through the network. The use of the centralized view of the topology provided by the SDN controller is a key assumption in this approach.

GreenSDN [13] is an emulation environment built using Mininet and POX, which is a Python-based SDN controller. The authors summarized the difficulties they found implementing an SDN environment with green capabilities. They presented an integration of three energy-saving protocols operating at different layers: adaptive link rate operating at the chip level, synchronized coalescing working at the node level and the sustainability-oriented network management system, which considers the whole topology to maintain the balance between QoS and energy savings. Particularly relevant to this paper is the mechanism proposed at the node level, which also exploits the Low Power Idle (LPI) state defined in the IEEE 802.3az standard. Nevertheless, GreenSDN does not consider setting individual ports in low power idle mode when the traffic traverses an aggregate between switches.

The energy-efficient flow routing problem is formulated in [14] as an optimization problem and solved with a heuristic—named the Energy Monitoring and Management Application (EMMA)—which aims to concentrate the traffic on the smallest possible set of nodes in the whole topology, so that the number of idle switches is maximized. EMMA is implemented as an ONOS application and evaluated in a network emulated through Mininet. The solution requires that the flows have previously declared their demanded rates. Thereby, when a new flow arises, EMMA tries to allocate all the active flows in the subset of the network topology that is currently active. If the active topology cannot support the flows, a new allocation for the whole network topology is computed. Analogously, when a flow is removed, EMMA attempts to re-route existing flows so that energy consumption of the network diminishes.

ElasticTree [1] and ECODANE [15] present solutions focused on data center networks, exploiting the redundancy in their internal switching networks and the variability in the workload that the data center must support over time. ElasticTree is a heuristic algorithm that adjusts the set of active devices to gauge the changes in the traffic load. The heuristic was validated over a testbed composed of production OpenFlow switches, using real traffic traces from an e-commerce website, saving up to 50% of energy. In ECODANE, an emulation framework, is built around Mininet and NOX, composed of five modules: the optimizer, which is in charge of determining the minimum subset of the topology that needs to be active, the power control module that manages the power states of the switches, the forwarding module that manages the flow rules installed in the switches to forward the traffic, the traffic generator, which generates the traffic to perform simulations, and the data center network itself. Their results obtained between a 10% and 35% energy reduction, depending on the source and destination of the traffic. However, none of these two solutions considers the characteristics of EEE links to reduce their energy consumption or explore the particular case of link aggregates between switches.

There is a clear line of work dedicated to exploring the interactions between resource activation/deactivation, routing decisions and energy savings. Early works focused on powering down redundant resources, e.g., switches and links, to reduce energy consumption during periods with low load. Some examples of these early works are [3,16–18], where the authors studied the energy savings obtained by link aggregation in metropolitan optical networks. All of these proposed different formulations to obtain the proper set of nodes to keep active, but only under the assumption of on-off power profiles in the devices. There is also research considering other more advanced link power profiles. For instance, transmission links with super-linear cost functions were studied in [19] with the goal of calculating the maximum power savings. New insights were provided in [20], after it was demonstrated that traffic consolidation can increase energy consumption for certain power profiles. The main problem with all these proposals was their high complexity, since the energy minimization problem is NP-complete [5]. Accordingly, a number of proposals in the literature resorted to heuristics for solving the problems. For instance, reference [21] took advantage of genetic algorithms to get close to the optimum, and [22] provided a heuristic based on particle swarm optimization. In general, these works overlook the problems derived from the practical implementation of the algorithmic results. Finally, only [23] reformulated the energy-saving problem considering an SDN-capable network and extended the problem to consider the usage of network function virtualization, so that computing tasks can also be moved across the network to enable greater energy savings.

A common feature of the prior works is that the underlying algorithms operate in relatively long time frames, and so, their response can be slow. Moreover, they address a network flow allocation problem, globally. In contrast, we focus on flow allocation in a single-link aggregate between two switches (see [6], which is the theoretical basis of this paper), and our techniques work in much shorter timescales, in the order of a single frame transmission. It was found therein that the optimum minimum-energy allocation of flows into a bundle of EEE links turns out to follow a water filling policy for typical consumption profiles, as explained in [6]. That is, a new port will only be used to transmit a packet if the packet cannot be transmitted by any of the ports already being used, since they are operating at its full capacity. This result holds for the main classes of mechanisms used to manage the power state of the IEEE 802.3az ports, i.e., frame transmission and burst transmission modes [24]. In addition, reference [6] presented an algorithm to achieve these results, which operates on a per-packet basis, deciding the port that will be used for each packet based on the occupation of the ports. Following a naive water filling algorithm and only diverting traffic to a new idle port when the previous ones are completely used at their full capacity will lead to an unbounded delay. Hence, a simple modification is proposed to maintain the average delay bounded to a target value, by using the average delay of the already queued packets to determine the output port.

3. SDN Application Design

In this section, we address the energy-efficient allocation of traffic flowing between two switches through an aggregate of EEE links, from the point of view of a software-defined network.

3.1. Background and Problem Statement

We will consider a link aggregate composed of L IEEE 802.3az links, of identical capacity C. The traffic traversing that bundle is represented by F flows. Let $x_i \in [0, C)$ denote the estimated rate of flow i, and let $p_i \in \{1, \dots, L\}$ be the port where that flow is allocated. According to [25], the normalized individual energy consumption of a single EEE interface for any uncorrelated incoming traffic distribution is:

$$E(\rho) = 1 - (1 - \sigma_{\text{off}})(1 - \rho)\frac{T_{\text{off}}(\rho)}{T_{\text{off}}(\rho) + T_S + T_W},\tag{1}$$

where ρ is the traffic load, σ_{off} is the relative energy consumption of the EEE idle mode and T_S and T_W are the time needed to enter and exit the LPI mode, respectively. They are constant parameters dependent on the physical interface characteristics. Finally, $T_{\text{off}}(\cdot)$ is the average length of the idle periods, which depends on both the algorithm governing the idle mode and the actual traffic load. Let $\vec{\rho} = [\rho_1, \dots, \rho_L]$ be the load allocation to the links forming the bundle. Then,

$$E_B(\vec{\rho}) = \frac{1}{L}\sum_{i=1}^{L} E(\rho_i)\tag{2}$$

is the normalized energy consumption of the whole bundle for a given load allocation. In [6], it was proven that, for any arbitrary algorithm governing the idle mode, $E_B(\cdot)$ is minimized iff:

$$x_i = \min\left\{C, X - \sum_{j<i} x_j\right\}, \quad i = 1, \dots, L,\tag{3}$$

where X denotes the total traffic allocation and x_i is the traffic allocation to the i-th port, i.e., $\rho_i = x_i/C$. That is, the minimum energy consumption is obtained when a water filling algorithm is used to assign the traffic among the links in the bundle. This is because the typical energy consumption profile of a single Ethernet link is a concave function of the traffic load (see Figure 1). In addition to the theoretical solution, a practical algorithm is also provided by [6], but assuming that the switches operate on a packet-by-packet basis. For two reasons, this algorithm cannot be directly implemented as an SDN application:

- The ideal algorithm considers that the switch individually decides for each packet which port will be used to forward it, based on the instantaneous occupation of the ports, a packet-level operation. SDN does not allow forwarding each packet individually, since its data plane works at the flow level, applying the same actions to the packets of a flow once a matching rule is found in its flow table (i.e., forwarding the packets to the same port). In addition to the action prescribed by the flow rule, the counters associated with the port are updated.

- The current queue occupation of each port is used to determine the forwarding port. Unfortunately, this state variable is not usually provided by SDN switches (e.g., it is not considered in OpenFlow).

Throughout the rest of this section, we will present the main architecture of the SDN application and the new flow allocation algorithms that realize the solution presented in [6].

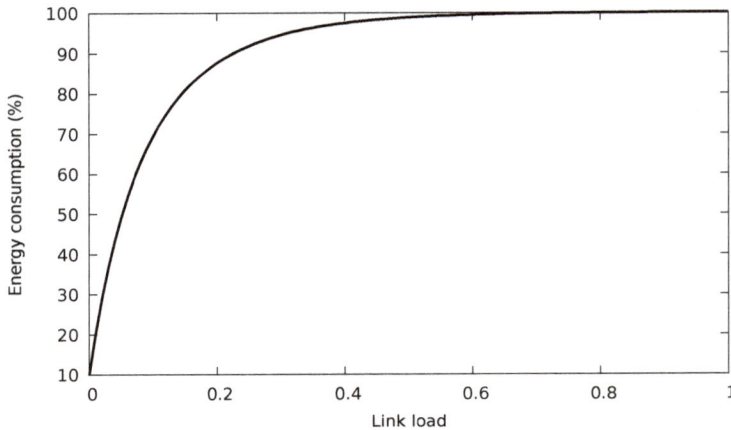

Figure 1. Consumption of a 10 Gb/s IEEE 802.3az port using frame transmission. © 2018 IEEE. Reprinted, with permission, from [10].

3.2. Designing the SDN Application

We devised a reactive forwarding behavior. That is, a low-priority rule will be installed in the switches to send the packets to the SDN controller, so the packets that do not match any flow rule with higher priority will be sent to the SDN controller, which has a centralized view of the topology and will act in response. The controller will transfer this packet to our application, which runs on top of the controller. Next, the application will determine which port the packet should be forwarded to and install a medium-priority flow rule in the switch. Future packets classified in this same flow will be directly forwarded by the switch at the line rate.

The medium-priority flows installed are defined by the destination MAC address of the packets and also the first eight bits of the destination IP address. Eight bits attain a good trade-off for the granularity of flows: enough to spread the traffic among the ports of the bundle and to keep the flow tables of the switches small to avoid performance degradation.

Since the controller maintains a full view of the network topology, it can compute shortest paths to the packet destination. For unknown destinations, the controller floods the packet out of all ports except the input port and using only one port of each bundle, without installing a flow rule for this packet yet. Therefore, when a packet is received, the application determines the next hop switch to which the packet should be forwarded. When the next hop is behind a bundle of links, our application selects at random a port of the bundle to forward the packet and installs the flow in the switch accordingly. The random selection is performed since the application does not have prior information about the transmission rate of this new flow. Allocating the flow to the highest loaded port would cause excessive losses if the flow demands a high rate. On the contrary, using an idle port would activate its hardware, drastically increasing the energy consumption if the flow demands a rate that can be handled by ports already active. In any case, it is important to note that this is just an initial transient when a new flow appears on the network.

The above description is the part of the SDN application that manages the packet forwarding in the transport infrastructure. It uses a customized flow definition, but spreads flows at random in the bundles between switches since there exists no a priori information about the flow rates, thus not performing any energy-aware optimization. Now, we proceed to describe how to optimize the energy consumption in the bundles. Figure 2 displays the flow diagram of the control application. The application performs the following tasks, some of them periodically:

1. Retrieve the list of switches.
2. For each switch, identify the neighbors of the switch (i.e., the switches that a link to it).
3. For each neighbor, retrieve the ports in the switch that are connected with the neighbor.
4. If there is more than one port (i.e., there is a bundle between the two switches), retrieve the flows installed in the switch that forward packets to a port of this bundle.
5. Predict the rate of each flow; that is to say, the amount of traffic that the flow will transmit in the next interval.
6. Compute a new allocation for these flows to the ports of the bundle in a way that energy consumption is minimized.
7. Instruct the switch to modify the flow rules that have changed their allocation.

Two of these deserve further discussion: rate prediction and flow allocation.

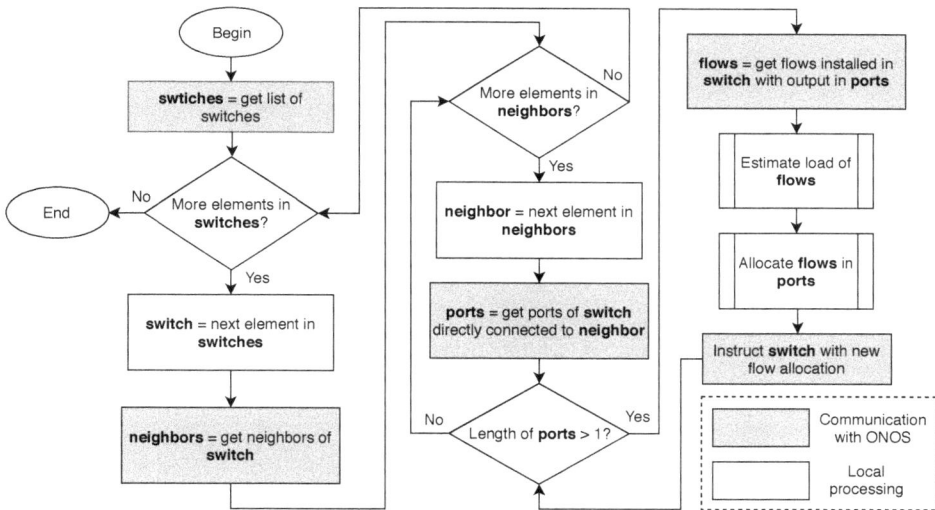

Figure 2. High level application logic.

3.3. Flow Rate Prediction

The rate demanded by a flow in the following measurement interval is estimated leveraging the counters associated with the flow rules. These counters include the number of packets and bytes that have matched with this flow along with the duration of the flow (i.e., the time that the flow has been active). Thus, the bytes transmitted in the current measurement period are the difference between two counter queries by the controller. The estimated rate is simply the number of transmitted bytes divided by the sampling period. If the flow was not present in the previous interval, we use the value of the duration of the flow instead of the sampling period, for an accurate measure. For the prediction of the flow rate in the next measurement period, we tested two simple estimators:

1. The measured value in the previous interval.
2. An exponentially-weighted moving average (EWMA) with the measured rates of the flow in past intervals. The estimated rate is:

$$R_n = \begin{cases} M_n, & n = 0 \\ \alpha \cdot M_n + (1 - \alpha) \cdot R_{n-1}, & n > 0 \end{cases} \tag{4}$$

where R_n is the value of the EWMA at time n (thus, the rate estimated for the interval $n + 1$), M_n is the measured rate in the interval n and the constant $\alpha \in (0, 1]$ is a parameter that tunes the relevance of the samples as time goes by.

We have analyzed the quality of the estimation with the two methods, calculating the error in the estimation as the absolute value of the difference between the estimated value and the real value, for each flow in each interval. Just for a reference, the results for a 32.5 Gb/s trace are shown in Figure 3 (Although we have not included them for space reasons, results for several other traffic traces have been produced with similar results. The traffic traces are the same as those used in Section 5, where their exact characteristics are described).

Clearly, using directly the measurements performs very similar to the usage of the EWMA for the different values of α and the sampling periods studied. We can also notice large errors in the estimated rate for sampling periods lower than 0.1 s. This is due to the high variability of the rates for such a small time window. Therefore, we will directly use the rate of each flow in the previous interval to forecast its rate in the next interval, since it is a simpler method than the EWMA one and provides almost the same accuracy.

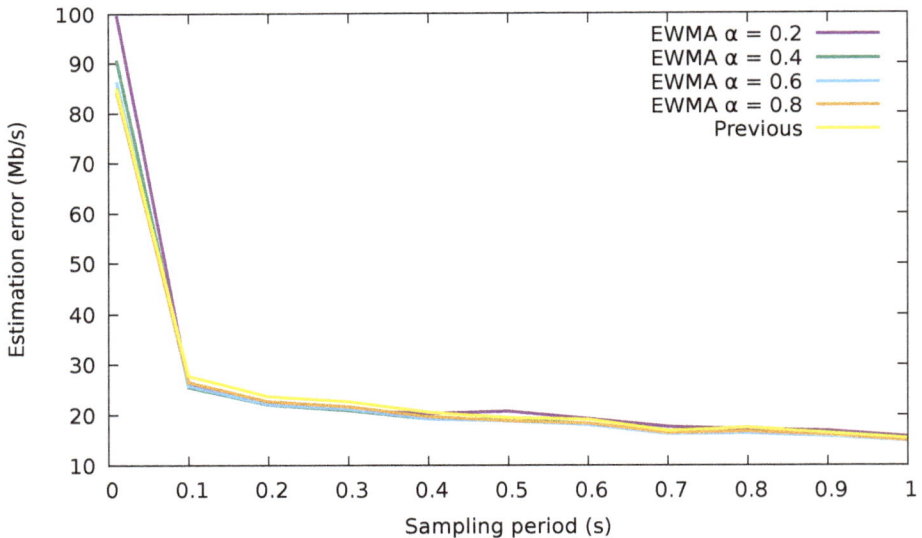

Figure 3. Average error in the estimation of the flow rate for different periods. EWMA, exponentially-weighted moving average.

3.4. Flow Allocation Algorithm

In this section, we describe three new flow allocation algorithms implemented to reduce the energy consumption. The purpose of these algorithms is to select the flows according to their traffic rate and assign them to the available links according to the optimum flow allocation algorithm [6].

The input of this algorithm will be the set of flows to be allocated along with their estimated rates, and also the set of ports that make up the aggregate.

3.4.1. Greedy Algorithm

GA attempts to fill the ports to the maximum of their capacity, only allocating a flow in an empty port if it does not fit in any of the already active ones. Therefore, GA attempts to use only the minimum number of ports, filling them up to their nominal capacity and leaving the maximum number of ports empty. The latter problem is actually a combinatorial NP-hard problem (bin packing), so we propose using the first-fit decreasing (FFD) [26] algorithm as a heuristic suboptimal solution. Note that finding the optimum one would require evaluating $|ports|^{|flows|}$ combinations and is not scalable. Instead of this, our GA algorithm will consider the flows in a decreasing order based on their estimated rates and will allocate them to the highest loaded port. Note that although similar to a water filling approach, GA does not operate at the packet level, but at the flow level.

The pseudocode of the GA algorithm is shown in Figure 4, where the bound is always set to zero. Firstly, the flows are sorted decreasingly on their estimated rate. Next, the flows are allocated in a sequential order so that the port occupation is maximized: the ports are evaluated in a predefined order (e.g., by port identifier), and the flow is allocated to the first port it fits. We consider that a flow fits in a port when the sum of all the flow rates assigned to the port is less than the capacity of the port. This algorithm is akin to the classical FFD heuristic solution of the bin packing problem [26]. We expect GA to yield low values of energy consumption. Nevertheless, since it pushes to use the full capacity of the ports, packet delay can grow quickly, as pointed out in [6]. Furthermore, since rate estimations are noisy, the estimation errors could lead to a non-negligible level of packet losses for almost any buffer size.

```
1   allocate_greedy(flows, ports, bound=0) {
2       // Hold assigned port for each flow
3       flow_allocation[1..|flows|] = ∅
4
5       // Sort flows by decreasing load value
6       ordered_flows = sort(load(flows), DECREASING)
7
8       // Initialize occupation of the ports to 0
9       port_load[1..|ports|] = 0
10      port_flows[1..|ports|] = 0
11
12      for flow ∈ ordered_flows {
13          for port ∈ ports {
14              if ((port_flows[port] == 0) ||
15                  (port_load[port] + load(flows)[flow]
16                   ≤ 1 - bound/port_flows[port])) {
17                  // Update port with the load of this flow
18                  port_load[port] += load(flows)[flow]
19                  port_flows[port] += 1
20                  flow_allocation[flow] = port
21                  break
22              }
23          }
24      }
25
26      return flow_allocation
27  }
```

Figure 4. Pseudocode for the greedy algorithms. Note that the greedy algorithm (GA) is identical to bounded GA (BGA) when setting the variable bound to zero. © 2018 IEEE. Reprinted, with permission, from [10].

3.4.2. Bounded Greedy Algorithm

BGA is a straightforward modification of GA, in that it attempts to bound the packet delay and also to reduce the packet losses in GA. One reason for the losses is that using the ports very close to their capacity leads easily to buffer overload if the rate is not accurately estimated. Thus, BGA avoids using links close to their capacity just by setting a threshold in the maximum allowable load allocated on a port. Specifically, we limit the fraction of the port capacity that can be used to an increasing function in the number of flows already allocated to each port p, with the following function:

$$\rho_{max}^{p} = 1 - B/F_p, \tag{5}$$

where F_p is the number of flows already allocated to the port and B denotes the fraction of space that cannot be used in a port when there is only one flow allocated to it. The idea is that rate prediction errors between the different flows should compensate the global estimation error, and thus, the higher the number of flows, the higher the link occupation that is safe to attain. For the rest, the algorithm operates in the same way as GA. The pseudocode is also shown in Figure 4, where *bound* is the fraction of space that cannot be used in a port when there is only one flow allocated to it, i.e., *bound* = B.

3.4.3. Conservative Algorithm

Despite the effort of the BGA to mitigate packet losses and control the delay of the packets, the results might not be acceptable yet, as we will later show. Hence, we designed a better algorithm that does not only minimize energy consumption, but also reduces packet losses. The idea behind this algorithm is to first compute the minimum number of ports necessary for the next interval. This value is lower bounded by:

$$L_{used} = \left\lceil \frac{\sum_{i=1}^{F} x_i}{C} \right\rceil. \tag{6}$$

Then, the flows are distributed among the L_{used} ports in a way that tries to spread the load evenly. Although this minimizes the individual occupation of each link, it does not follow a water filling approach. However, as we will show later, this does not really degrade energy consumption. The reason is that the individual energy consumption rises very quickly with the occupation of the link, as Figure 1 illustrates.

As a consequence, once a port reaches an occupation higher than 20%, it makes little difference for the energy consumption the actual traffic load assigned to it. Thus, the conservative algorithm (CA) prefers to have its used ports with a balanced traffic occupation avoiding the need for using the ports at full capacity in many situations, like frequently happens with the GA and BGA. Figure 5 shows normalized power profiles of an eight-link bundle with the ideal water filling algorithm and the power profile expected for the conservative algorithm.

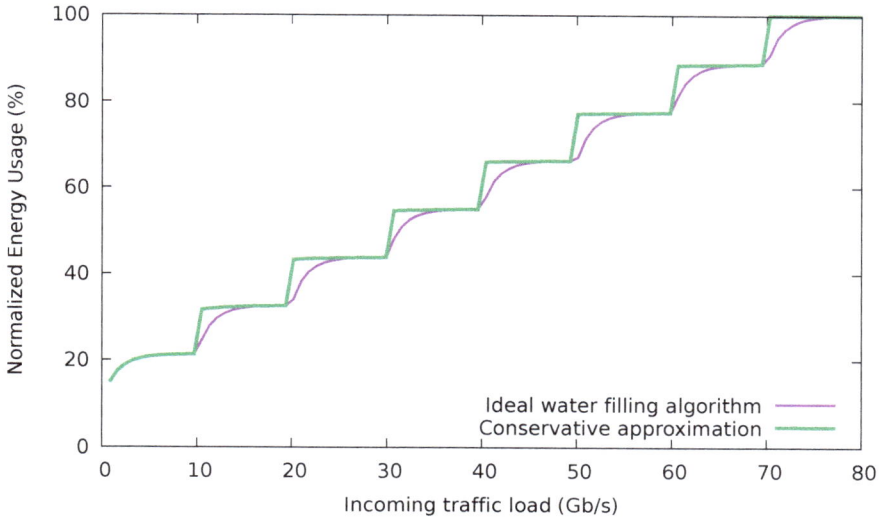

Figure 5. Comparison between the normalized power profiles of an eight-link bundle when using the water filling and the conservative algorithm.

For further reduction of the likelihood of packet losses, rather than using the total estimated load to compute the number of ports, we add to this value a safety margin, M, which we empirically set to 20%. The number of ports to be used is calculated therefore as:

$$L_{\text{used}} = \left\lceil \frac{\sum_{i=1}^{N} x_i}{C} + M \right\rceil . \tag{7}$$

After determining the L_{used}, the CA proceeds with the allocation of the flows attempting to achieve a balanced distribution of the flows to ports, both in terms of rate and number of flows. To accomplish this, the algorithm performs a minimization of the occupation of the ports. The pseudocode is shown in Figure 6, which first sorts the flows in decreasing order of their estimated rate and then sequentially allocates the flows to the port with the lowest occupation among those that will be used in the interval. Note that this algorithm is also capable of maintaining some ports idle, reducing energy consumption.

```
1   safety_margin = 20%
2
3   allocate_conservative(flows, ports) {
4       // Hold assigned port for each flow
5       flow_allocation [1..|flows|] = ∅
6
7       expected_load = sum(load(flows)) + safety_margin
8       minimum_ports = ceil(expected_load)
9
10      // Only use the minimum number of ports
11      used_ports = ports [1..minimum_ports]
12
13      // Sort flows by decreasing load value
14      ordered_flows = sort(load(flows), DECREASING)
15
16      // Initialize occupation of the ports to 0
17      port_occupation [1..|used_ports|] = 0
18
19      for flow ∈ ordered_flows {
20          port = get_port_min_occupation(port_occupation)
21          // Update port with the load of this flow
22          port_occupation [port] += load(flows)[flow]
23          flow_allocation [flow] = port
24      }
25
26      return flow_allocation
27
28  }
```

Figure 6. Pseudocode for the conservative algorithm. © 2018 IEEE. Reprinted, with permission, from [10].

4. Energy-Efficient Algorithms with Bounded Delay

In this section, we consider flows with different quality of service (QoS) requirements in terms of latency. We will introduce two modifications to the energy-efficient algorithms described in Section 3 in order to consider the demands of low-latency flows. The specific mechanism used to identify low-latency flows is not relevant for this work since the method actually employed does not affect the algorithm at all. Thus, we will assume that the low-latency flows are tagged with a well-known differentiated services code point (DSCP), which is carried in the IP header. As a result, we will work with two types of flows: low-latency flows and best-effort flows.

These modifications are generic enough to be applied to any of the different energy-efficient algorithms presented above. However, since CA outperforms the other two algorithms in packet losses and delay with a minimum penalty in energy savings, as will be later shown in Section 5.1, during the rest of this paper, we will only use CA as the energy-efficient algorithm, for the sake of simplicity.

4.1. Spare Port Algorithm

The spare port algorithm (SPA), shown in Figure 7, exploits the fact that, when the traffic load is moderate, the energy-efficient algorithms concentrate the traffic and leave some spare ports. Therefore, this new algorithm trades an increase of energy consumption in unused ports so as to provide expedited service to low-latency flows. SPA works in two phases, first allocating best-effort flows and then the low-latency ones.

1. In the first phase, the energy-saving algorithm is directly applied without modifications, but only to the best-effort flows.
2. In the second phase, the remaining low-latency flows are assigned to the least occupied port among those in the bundle.

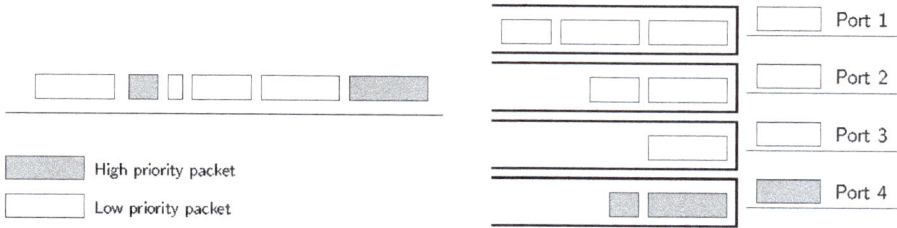

Figure 7. Conceptual diagram of the spare port algorithm in a four-link bundle.

SPA can perform well under the assumptions that low-latency traffic represents a small fraction of both the total traffic and the spare port capacity and that best-effort flows do not exhaust the capacity of the bundle. Then, in the next step, one of the unused ports will be used to forward low-latency traffic, without increasing the delay of the best-effort traffic. However, we need to discuss the limitations of this algorithm when some of the assumptions do not hold:

1. If the traffic demand is so high such that all the ports in the bundle must be dedicated to best-effort flows, low-latency traffic will not be forwarded through a single port. As a result, both low-latency and best-effort traffic will be treated in the same way, without meeting the needs of premium traffic.
2. If the amount of low-latency traffic is significant, the energy consumption of the spare port can drastically increase because of the energy profile of an EEE link, which rises very quickly with the port occupation (cf. Figure 1).

4.2. Two Queues Algorithm

The spare port algorithm may increase the energy consumption if the amount of delay-sensitive traffic is significant. More importantly, SPA will not be able to satisfy the demands of flows with low-latency requirements when there is a high load due to best-effort traffic. To solve this, we leverage the ability of most SDN switches to have multiple queues attached to a physical port. These queues can be defined with different priorities. In fact, this is the standard way of providing QoS in SDN devices as stated in the OpenFlow (OF) specification [27]. Although this capability is not required, it is provided by most of the devices, such as Open vSwitch, which is presumably the most widely-used OF-enabled switch.

In the two queues algorithm (TQA), we define two queues with different priorities inside each physical port of the switches, as shown in Figure 8: the queue with the highest priority will only serve low-latency traffic, and the other will forward best-effort traffic. The algorithm operates in two phases, determining first the port and then the queue inside each port:

1. The first phase consists of directly applying the unmodified energy-efficient algorithm described in Section 3 to the whole set of flows, both including low-latency and best-effort, treated equally. The whole set of flows is allocated in a few ports.
2. The second phase sets the queue inside the assigned port for every flow. Low-latency flows are assigned to the high-priority queue of the ports, whilst best-effort flows are assigned to the low-priority queue.

Clearly, the allocation of flows to ports is actually given by the energy-efficient algorithm, but thanks to the introduction of multiple first-in first-out (FIFO) queues inside the port, we prioritize flows with stringent QoS requirements in terms of latency, thus providing an expedited service. The decision of the next packet to be served by a port is straightforward: each time the port ends the transmission of a packet, it will pick the next packet to be transmitted from the non-empty queue with the highest priority. In other words, delay-sensitive packets have non-preemptive priority over the rest.

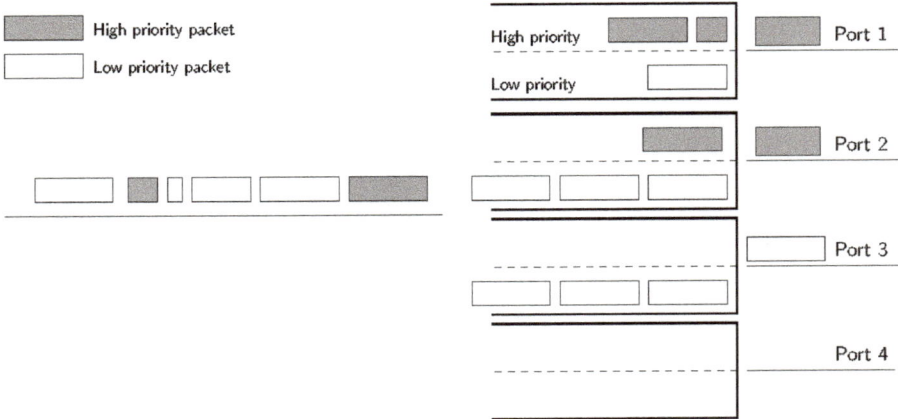

Figure 8. Conceptual diagram of the two queues algorithm in a four-link bundle.

Using two queues per port overcomes the limitations of SPA. Firstly, the service received by the high-priority traffic is independent of the amount of best-effort traffic load. Secondly, the expedited service does not lead to increased energy consumption, since the joint allocation phase still uses the energy-efficient algorithm. Thus, the energy consumption will be equal to the original energy-efficient algorithm described in Section 3. The main drawback is that using two queues increases the delay of best-effort traffic, and this becomes more noticeable as the share of traffic demanding expedited service is higher. However, since the use of the priority queues only implies a reordering of the packets in a port, the average delay of all the packets will not change. Thus, the maximum delay of best-effort packets is still bounded.

5. Results

Throughout this section, we analyze the performance of the proposed algorithms to then assess the correct behavior of the proper ONOS application.

The algorithms have been analyzed in a scenario composed of bundle of five 10 Gb/s copper based Ethernet links (10 GBASE-T) aggregated between two switches with two hosts attached to them, one of them serving as the source for the traffic, and the other one acting as the traffic sink. Figure 9 shows a diagram of the setup.

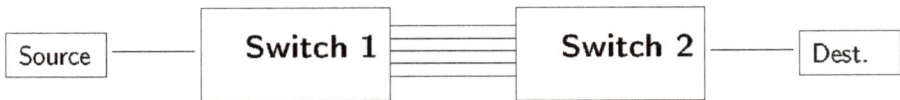

Figure 9. Basic experimental topology.

The experiments have been carried out with an in-house developed simulator available for download [28]. The simulator shares most of the code with the ONOS implementation, thus reducing the developing time and helping in the validation. As for the traffic itself, we have employed real traffic traces from the public CAIDA (Center for Applied Internet Data Analysis) dataset [29]. As the original traces have been captured in 10 Gb/s Ethernet links, they have a relative low average throughput, so we have constructed new traffic traces, reducing the inter-arrival times by a constant factor, producing new 6.5, 13, 19.5, 26 and 32.5 Gb/s traces.

5.1. Flow Allocation Algorithms

The first experiment evaluated the variation of the energy consumption with the duration of the sampling period of the algorithms for a rate of 32.5 Gb/s. Figure 10a shows the results of the three flow allocation algorithms for a buffer size of 10,000 packets. There was a fourth algorithm, named equitable, that distributed the flows uniformly among all the ports without regards to energy efficiency, serving as a baseline for comparison. The energy consumption attained by the three energy-saving algorithms was practically the same. Besides, we can also observe from Figure 10a that low sampling periods (e.g., lower than 0.1 s) presented higher consumption than those greater than 0.1 s. This probably corresponds to mispredictions of the flow characteristics, as already shown in Figure 3. Finally, note that the obtained energy consumption was very close to the optimum. According to (3), the best allocation was obtained when the rate allocation vector was $\vec{x} = [10\,\text{Gb/s}, 10\,\text{Gb/s}, 10\,\text{Gb/s}, 2.5\,\text{Gb/s}, 0\,\text{Gb/s}]$; in other words, this minimum consumption was achieved when the 32.5 Gb/s load was distributed in the bundle in the following way: three ports fully utilized carrying 10 Gb/s, one transmitting 2.5 Gb/s and the last one with no traffic. In that case, and for the usual EEE parameters ($\sigma_{\text{off}} = 0.1$, $T_W = 4.48\,\mu\text{s}$, $T_S = 2.88\,\mu\text{s}$ for 10 Gb/s links and $T_{\text{off}}(\lambda) = \lambda^{-1}e^{-\lambda T_S}$ for the frame transmission mode [25]), the bundle consumption was $\frac{1}{5}(1 + 1 + 1 + 0.83 + 0.1) = 0.78$ using (2). This is just a little less than the 79% energy consumption obtained by CA.

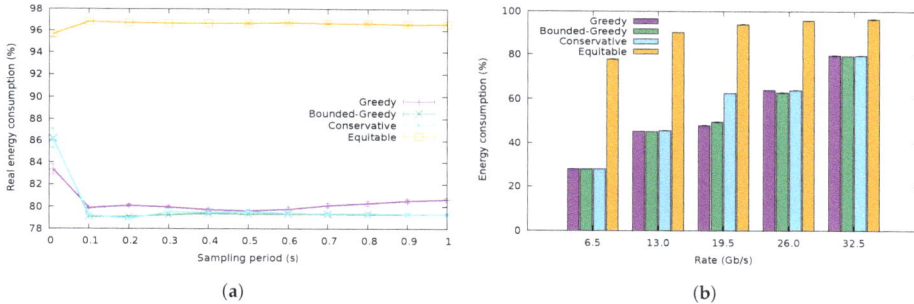

Figure 10. Energy consumption. (**a**) Energy consumption for different sampling periods, (**b**) Energy consumption for different traffic traces.

The energy consumption for the different traffic traces is shown in Figure 10b for a sampling period of 0.5 s and a buffer size of 10,000 packets. We can observe that the energy consumption was almost identical for the three proposed algorithms and that it was considerably lower than that of the non-energy-efficient equitable algorithm. There was just a slight difference in the case of the 19.5 Gb/s, where CA consumed a bit more than the greedy algorithms. This was because its safety margin (M) made it use three ports, while the greedy algorithms would try to allocate the flows using just two ports. Nevertheless, the consumption attained by CA was indeed much lower than that of the equitable one.

Figure 11a presents the variation of the packet loss rate with the sampling period for a 10,000-packet buffer size, while Figure 11b explores the packet losses introduced for different buffer sizes, using a sampling period of 0.5 s. GA was the one with the highest losses, followed by BGA, then CA and finally the equitable algorithm. These results confirm that the greedy algorithms can lead to high loss rates when the flow rates are underestimated. The conservative algorithm, however, was able to trade a small increment in energy consumption for an acceptable loss rate for buffer sizes from 1000 packets onward. Furthermore observe how for the highest sampling rates, packet losses diminished, as the algorithm adapted faster to rate variations; although, as seen before, energy usage also incremented.

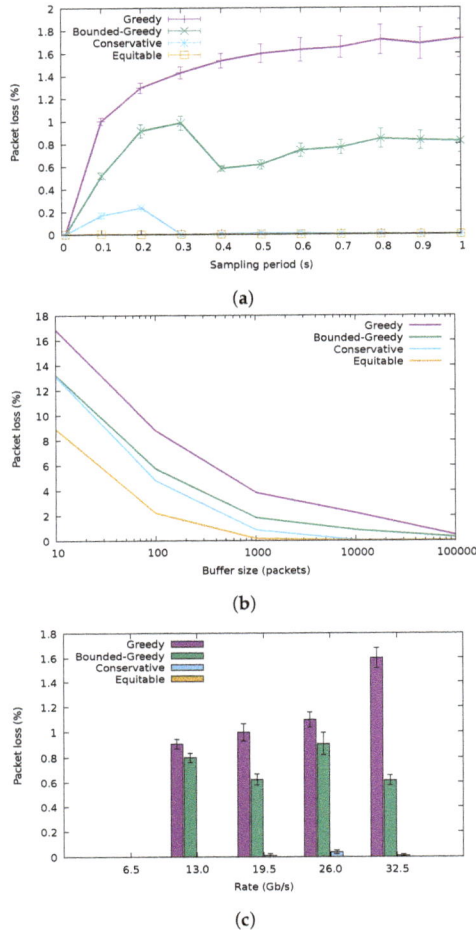

Figure 11. Packet loss rate. (**a**) Packet loss percentage variation with the duration of the sampling period. (**b**) Packet loss percentage variation with the buffer size. © 2018 IEEE. Reprinted, with permission, from [10]. (**c**) Packet loss for different traffic traces.

The packet loss rates for the different traffic traces are shown in Figure 11c, where the sampling period is set to 0.5 s and the buffer size to 10,000 packets. As expected, GA and BGA were the ones having the highest losses in every case, with CA and equitable algorithms showing negligible losses. In the case of the 6.5 Gb/s trace, losses were not recorded, as expected.

The results for packet transmission delay are depicted in Figure 12. In particular, Figure 12a shows average packet delay variation versus sampling period for a 10,000-packet buffer size. The average delay for GA was about 4 ms, which is considerably higher than that of the other algorithms, whereas the delay for BGA was about 1.5 ms, which is still a high value. The delay for CA was, however, an order of magnitude lower, about 250 μs. For reference, the delay of the equitable algorithm sat around 50 μs, being, as expected, the lowest one. Figure 12b shows the average packet delay experienced by the different traffic traces with the different algorithms. For the 6.5 Gb/s trace, the three energy-saving algorithms behaved identically, using just one port for all the traffic. Furthermore, for CA, the delay of the packets using the 26 Gb/s trace was higher than that using the 32.5 Gb/s one, as, in the latter case, there was one more link in use, but with lower load. For the rest of the traces, the results were in accordance with those shown in Figure 12a.

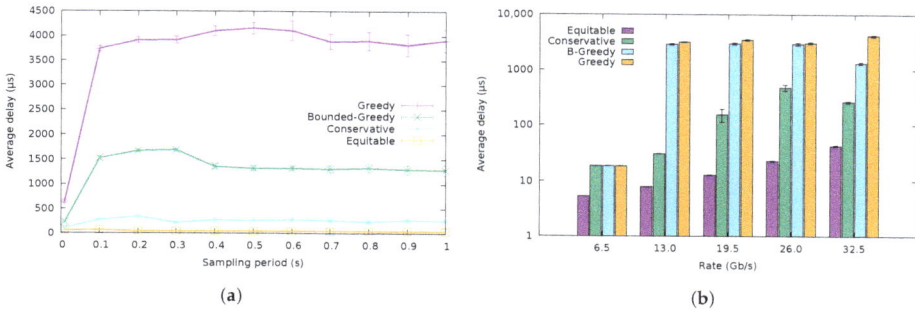

Figure 12. Packet delay. (**a**) Packet delay for different sampling periods, (**b**) Average packet delay for the different traffic traces.

5.2. QoS-Aware Algorithms

To test the performance of the two proposed QoS scheduling algorithms, we have created an additional traffic trace of 45.5 Gb/s reducing again the inter-arrival times of the original CAIDA trace. Additionally, we have added a source of low-latency traffic, consisting of a synthetic traffic trace made of relatively small packets (100 and 200 bytes) and deterministic inter-arrival times corresponding to the desired final average rate. We have used CA as the flow allocation algorithm using a sampling period of 0.5 s and a buffer size limited to 10,000 packets to provide negligible (below 0.05%) packet losses, as per the results of the previous section.

Figure 13a shows the average delay of the packets with low-latency requirements using the QoS-aware algorithms and that obtained using the baseline conservative one. The results in the figure correspond to the best-effort traffic trace of 32.5 Gb/s, while we varied the rate of the low-latency traffic (We have omitted the results using lower rates for the best-effort traffic for the sake of brevity, since the results are analogous). The unmodified CA yielded considerably worse results than the QoS-aware algorithms, producing a delay higher than 100 μs. The fluctuations for the different rates of the low-latency traffic come from the fact that the low-latency flows would be allocated to a different port in each case, being forced to compete with a different amount of normal traffic.

Both QoS-aware algorithms significantly reduced the average delay. The SPA delay stayed around 5 μs, while TQA added less than 2 μs for all the tested transmission rates. The main delay contribution for SPA was the time needed to wake up the interface ($TW = 4.48$ μs), which would be usually idle at the arrival of a low-latency packet. This was not the case for TQA, as low-latency traffic shared the port with best-effort traffic.

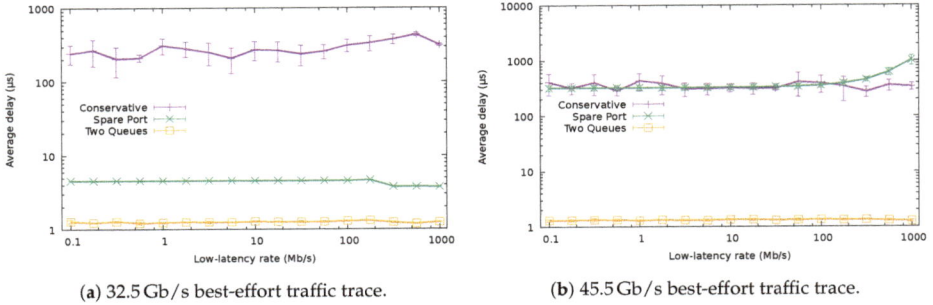

(**a**) 32.5 Gb/s best-effort traffic trace.   (**b**) 45.5 Gb/s best-effort traffic trace.

Figure 13. Average delay of the low-latency traffic.

Figure 13b shows the results when the system is already experiencing a very high load due to best-effort traffic (45.5 Gb/s). Both SPA and the non-QoS-aware CA experimented with an average delay higher than 200 µs, fluctuating up to 1000 µs, depending on the actual low-latency rate. On the other hand, TQA maintained the latency lower than 2 µs. These results confirm that SPA was not capable of providing a low latency service in high load scenarios, since all the ports were already busy forwarding best-effort traffic.

Figure 14 compares the average delay of the best-effort packets using the QoS-aware algorithms with the average delay of these packets when using CA for the 32.5 Gb/s best-effort traffic trace and varying low-latency traffic. When the rate of low-latency packets was very low (e.g., lower than 100 Mb/s), the average delay of best-effort packets was identical for the two QoS algorithms and CA, being around 264 µs. Nevertheless, as this rate increased, the delay exhibited by CA and TQA rose. On the other hand, the delay of the SPA remained unaffected by the rate of the low-latency traffic, since it was being forwarded through a different port than the best-effort traffic.

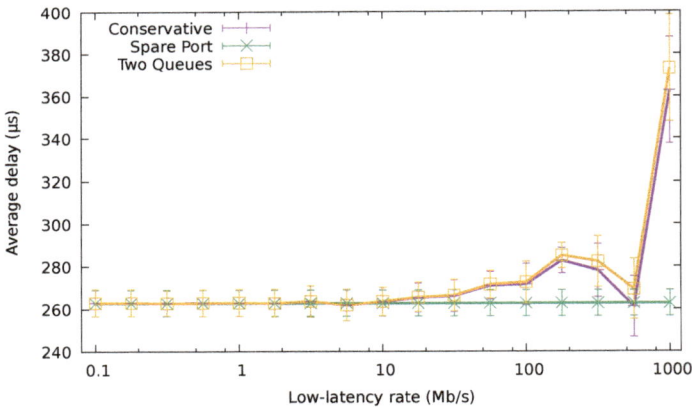

Figure 14. Average delay of the normal packets for the 32.5 Gb/s best-effort traffic trace.

Finally, Figure 15 shows the average energy consumption of the bundle using the different QoS-aware algorithms and also CA in the same traffic conditions. Again, while the amount of high-priority traffic was negligible (i.e., lower than 10 Mb/s), the three algorithms drew the same amount of energy. As expected, TQA achieved exactly the same consumption as CA irrespective of the low-latency traffic rate. However, for values higher than 10 Mb/s, the energy usage increased rapidly for SPA, reaching nearly 100% for rates above 100 Mb/s. This confirms that energy consumption can rise quickly in SPA as soon as the amount of high-priority traffic forwarded in the spare port becomes significant.

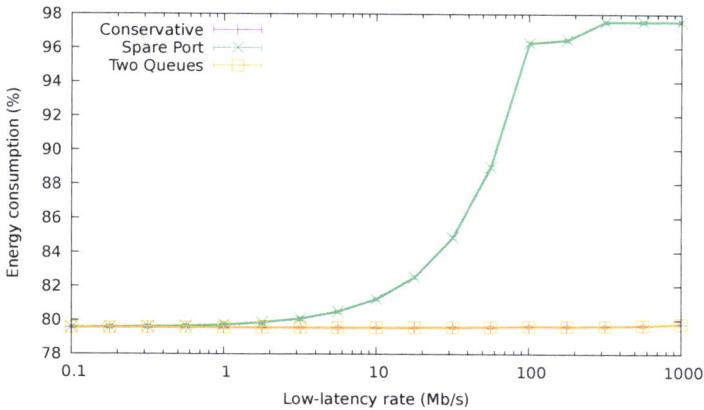

Figure 15. Normalized energy consumption for the 32.5 Gb/s best-effort traffic trace.

5.3. ONOS Application Results

The previous sections have measured the efficiency of the proposed algorithms via a simulation study. We also tested the correctness and feasibility of the proposal with an actual implementation of the application. To this end, we implemented the proposed SDN application on top of the ONOS, emulating the experimental topology with Mininet in order to evaluate the proper operation of the application. The Open vSwitch switches employed by Mininet have an OpenFlow API accessible by ONOS, but it cannot reproduce exactly the EEE capabilities, so we measured the average occupation of each outgoing link as a proxy for the corresponding energy consumption.

We evaluated our application with the 32.5 Gb/s traffic trace used in the previous experiments (Results for the other traffic traces, namely the 6.5, 13, 19 and 26 Gb/s ones, have been omitted for the sake of brevity, but otherwise show consistent results).

We used `tcpreplay` to transmit it, but at a rate of just about 330 Mb/s, since the computer used for the experiments was not capable of transmitting this traffic trace at higher rates (We have used an Intel® Core™ i7-4710HQ (4th Generation) at 2.5 GHz).

Accordingly, the nominal capacity of the interfaces of the bundle had been scaled to 100 Mb/s, and we have also scaled the sampling period to 10 s. The occupation of each port of the bundle averaged throughout twelve intervals in ten independent executions is shown in Table 1. Despite the fact that the actual consumption of 100 Mb/s interfaces would be different, this experiment allowed us to validate the behavior of the algorithm.

Table 1. Average port occupation of the ports of the bundle.

Algorithm	Occupation (%)					
	Port 1	Port 2	Port 3	Port 4	Port 5	Average
Greedy	92.57	97.83	97.05	30.36	0.02	63.57
Bounded-Greedy	83.46	81.16	95.08	61.27	0.02	64.20
Conservative	84.17	83.60	80.78	79.76	0.02	65.67
Equitable	83.89	80.52	54.13	53.63	57.23	65.88

The results of the real implementation matched the simulation results. We see that GA used three ports to more than 90% of their nominal capacity, one to about 30%, and left the other one unused.

These values describe the behavior of a water filling algorithm, as desired per design. BGA avoids having three ports so close to their nominal capacity, although one port still presented an occupation higher than 95%. As the flows were assigned in decreasing rate order, less flows were allocated to the first ports. In fact, in this case, 1.56 were allocated on average to the first port, 6.56 to the second and 96.91 to the third one. The high number of flows allocated to the third port explains why its occupation was so high. CA behaved exactly as desired, using four ports around 80% occupation and leaving the last one empty. The equitable algorithm spread the traffic evenly among all the ports of the bundle. Note that the 0.02% usage of the last port in the three energy-efficient algorithms was due to the flows being assigned randomly during the first interval. The small average occupation differences were mostly due to packet losses, which occurred whenever more than 100 Mb/s were assigned to a port during an interval.

Table 2 collects the average energy consumption averaged throughout the intervals for the same ten independent executions. As we can observe, the differences in the energy consumption among the three energy-efficient algorithms were minimal, and all of them consumed about 18% less than the baseline equitable algorithm. They only differed in the consumption in port 4, which consumed about 7% less with GA than with BGA and CA. This is in accordance to the simulations.

Table 2. Average energy consumption of the ports of the bundle.

Algorithm	Energy Consumption (%)					
	Port 1	Port 2	Port 3	Port 4	Port 5	Average
Greedy	99.89	99.99	99.99	92.36	10.24	80.49
Bounded-Greedy	99.80	99.90	99.98	99.38	10.24	81.86
Conservative	99.77	99.92	99.88	99.89	10.24	81.94
Equitable	99.78	99.90	99.04	98.97	99.27	99.39

We have also validated the QoS algorithms with the ONOS application using the setup depicted in Figure 16. This time, the setup consisted of three switches (numbered from 1 to 3) and eight hosts (numbered from 1 to 8). Hosts 1 to 4 were connected to Switch 1 and Hosts 5 to 8 are connected to Switch 3. These edge switches were connected to an inner switch by their respective four-link bundles. All the interfaces in this scenario had a nominal capacity of 1 Gb/s.

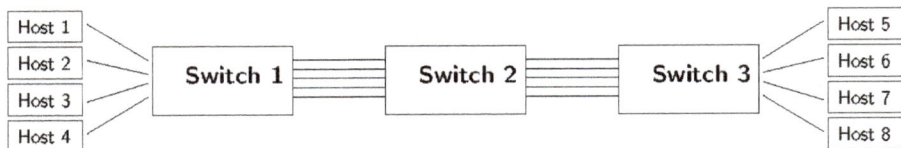

Figure 16. Experimental topology used for QoS-aware algorithms' validation.

In this network, three UDP flows without latency requirements were originated in Hosts 1, 2 and 5, with respective destinations in Hosts 5, 6 and 7. These flows have been created with the `iperf3` tool. The first two clients send traffic at 700 Mb/s, while the third one at 600 Mb/s. This way we force the flows to be allocated on the first three ports of each bundle. Then, we added three lightweight flows from Host 4 to Host 8 tagged with a predefined differentiated services code point (DSCP)value, so that they can be identified as low-latency by our ONOS application. The purpose of these lightweight flows is to measure the latency suffered by the low-latency packets, using the different scheduling algorithms.

Figure 17 shows box and whisker plots with the round-trip time (RTT) of 10,000 packets of the lightweight flows using the different algorithms. The whiskers show that 95% of the samples and outliers have been removed for the sake of clarity. We can see that traffic without real-time requirements suffered a substantial latency in this scenario, around 50 ms. This performance is expected, since the flow was allocated in the same port and queue as the 600 Mb/s big flow. As a result, the packets of the small flows have to contend with the packets of the big flow, yielding considerable waiting times in the queue of the port, which are indeed the main contributions to this large RTT.

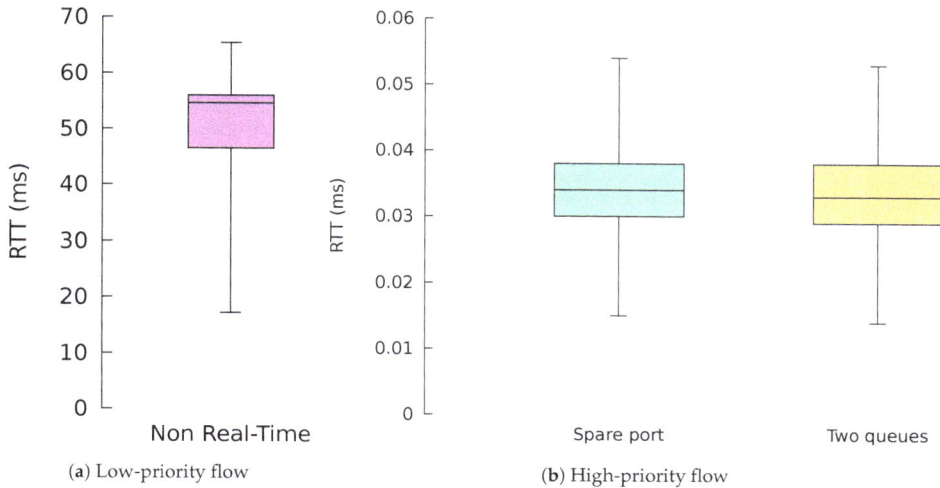

(a) Low-priority flow   (b) High-priority flow

Figure 17. Round-trip time (RTT) for the different algorithms.

Regarding the QoS-aware algorithms, both of them managed to decrease the round-trip time of low-latency traffic by three orders of magnitude in this scenario. The SPA algorithm was using the low-priority queue of the port that does not contain any big flow, thus providing low latency. On the other hand, the TQA algorithm was using the same port as the 600 Mb/s flow, but using the high-priority queue for the lightweight flow rather than the low-priority one as in the case of the big flow. Additionally, despite the algorithms using different ports and queues, their performance in terms of latency was really solid and stable, without major fluctuations, as desired.

5.4. Discussion

Simulation results exhibit the existing trade-off between energy consumption, traffic delay and packet losses. Certainly, the analysis shows that CA is the best algorithm. Although the greedy algorithms can be slightly more energy efficient than CA in some scenarios, they could lead to unacceptable packet delays and losses. On the other hand, the computational complexity of the three algorithms is roughly the same. However, CA allows tuning in a fine-grained manner the trade-off between packet delay and energy savings through the safety margin. Increasing the safety margin will contribute to reducing the packet delay and the likelihood of losses, at the cost of a slight increment in the energy consumption in some situations.

It is also important to ponder about the adequate value for the sampling period. Although the use of low values of the sampling period (e.g., 0.01 s) exhibits low delays and packet losses (with the subsequent increase in energy consumption), values lower than 0.5 s are hardly implementable in practice, since they result in a huge (and unmanageable) overhead of control traffic. Moreover, for low sampling periods, this frequent rerouting can harm the performance of TCP, as studied in [30].

6. Conclusions

The main focus of this paper has been the minimization of the energy consumption in networking equipment with SDN capabilities when the traffic traverses an aggregate of links between two switches. We elaborated a solution in the form of an SDN application that efficiently concentrates the traffic flows in a few ports of the bundle, dynamically adapting to the variations in the traffic demand. We firstly proposed three allocation algorithms and analyzed them in terms of energy consumption, packet delay and packet losses. We validated the algorithms using real traffic traces through simulation and also in a real implementation on top of the ONOS SDN controller. The obtained results confirm the expected operation of the algorithms, showing that the SDN capabilities of networking equipment can be used to reduce energy consumption in bundles of EEE links up to 50%, without the need for modifying the firmware of the devices.

We also proposed two modifications to the previous algorithms to offer a low-latency service to traffic with stringent QoS requirements while keeping the energy consumption reduced: the SPA algorithm uses the last port of the bundle to transmit high-priority traffic, while the TQA algorithm sets up a low priority and a high priority queue in each output port and transmits low-latency traffic on the high-priority queue. The results showed that the algorithms are able to provide a low-delay service to time-sensitive traffic, achieving a reduction of some orders of magnitude. Moreover, even under the situation of normal traffic congestion, one of the proposals manages to continue offering an accelerated service.

This work could be extended in some lines. First, the centralized view of the whole topology that the SDN controller provides to the applications can be harnessed so that inner switches can reuse the flow allocations performed in the edge switches, in the case where multiple link bundles are present in the network. It will also be interesting to test our solutions in a testbed composed of hardware OF-enabled devices with IEEE 802.3az ports controlled by our ONOS application. Finally, we see a research opportunity in reducing the control plane traffic required, since our work has been focused just on data plane traffic. Minimizing control plane traffic will contribute to the overall energy reduction.

Author Contributions: Conceptualization, M.R.-P. and S.H.-A.; methodology, M.R.-P.; software, P.F.-F.; validation, P.F.-F. and S.H.-A.; formal analysis, M.F.-V.; investigation, P.F.-F.; resources, M.R.-P.; writing, original draft preparation, P.F.-F.; writing, review and editing, M.R.-P., M.F.-V. and S.H.-A.; supervision, M.R.-P. and S.H.-A.; project administration, M.F.-V.; funding acquisition, M.R.-P. and M.F.-V.

Funding: This research was funded by the "Ministerio de Economía, Industria y Competitividad" through the project TEC2017-85587-R of the "Programa Estatal de Investigación, Desarrollo e Innovación Orientada a los Retos de la Sociedad" (partly financed with FEDER funds).

Conflicts of Interest: The authors declare no conflict of interest.

References

1. Heller, B.; Seetharaman, S.; Mahadevan, P.; Yiakoumis, Y.; Sharma, P.; Banerjee, S.; McKeown, N. Elastictree: Saving energy in data center networks. *NSDI* **2010**, *10*, 249–264.
2. The 5G Infrastructure Public Private Partnership KPIs. Available online: https://5g-ppp.eu/kpis/ (accessed on 10 November 2018).
3. Chiaraviglio, L.; Mellia, M.; Neri, F. Minimizing ISP Network Energy Cost: Formulation and Solutions. *IEEE/ACM Trans. Netw.* **2012**, *20*, 463–476. [CrossRef]
4. Jung, D.; Kim, R.; Lim, H. Power-saving strategy for balancing energy and delay performance in WLANs. *Comput. Commun.* **2014**, *50*, 3–9. [CrossRef]
5. Kim, Y.M.; Lee, E.J.; Park, H.S.; Choi, J.K.; Park, H.S. Ant colony based self-adaptive energy saving routing for energy efficient Internet. *Comput. Netw.* **2012**, *56*, 2343–2354. [CrossRef]
6. Rodríguez Pérez, M.; Fernández Veiga, M.; Herrería Alonso, S.; Hmila, M.; López García, C. Optimum Traffic Allocation in Bundled Energy-Efficient Ethernet Links. *IEEE Syste. J.* **2018**, *12*, 593–603. [CrossRef]
7. Agiwal, M.; Roy, A.; Saxena, N. Next generation 5G wireless networks: A comprehensive survey. *IEEE Commun. Surv. Tutor.* **2016**, *18*, 1617–1655. [CrossRef]
8. *IEEE Standard for Information Technology–Local and Metropolitan Area Networks–Specific Requirements–Part 3: CSMA/CD Access Method and Physical Layer Specifications Amendment 5: Media Access Control Parameters, Physical Layers, and Management Parameters for Energy-Efficient Ethernet*; IEEE Std 802.3az-2010 (Amendment to IEEE Std 802.3-2008); IEEE: Piscataway, NJ, USA, 2010; pp. 1–302. [CrossRef]

9.   Christensen, K.; Reviriego, P.; Nordman, B.; Bennett, M.; Mostowfi, M.; Maestro, J.A. IEEE 802.3az: The road to Energy Efficient Ethernet. *IEEE Commun. Mag.* **2010**, *48*, 50–56. [CrossRef]

10. Fondo-Ferreiro, P.; Rodríguez-Pérez, M.; Fernández-Veiga, M. Implementing Energy Saving Algorithms for Ethernet Link Aggregates with ONOS. In Proceedings of the 2018 Fifth International Conference on Software Defined Systems (SDS), Barcelona, Spain, 23–26 April 2018; pp. 118–125. [CrossRef]

11. Fondo Ferreiro, P.; Rodríguez Pérez, M.; Fernández Veiga, M. QoS-aware Energy-Efficient Algorithms for Ethernet Link Aggregates in Software-Defined Networks. In Proceedings of the 9th Symposium on Green Networking and Computing (SGNC 2018), Split-Supetar, Croatia, 13–15 September 2018; In press.

12. Tuysuz, M.F.; Ankarali, Z.K.; Gözüpek, D. A Survey on Energy Efficiency in Software Defined Networks. *Comput. Netw.* **2017**, *113*, 188–204. [CrossRef]

13. Rodrigues, B.B.; Riekstin, A.C.; Januário, G.C.; Nascimento, V.T.; Carvalho, T.C.M.B.; Meirosu, C. GreenSDN: Bringing energy efficiency to an SDN emulation environment. In Proceedings of the 2015 IFIP/IEEE International Symposium Integrated Network Managment, Ottawa, ON, Canada, 11–15 May 2015; pp. 948–953. [CrossRef]

14. Tadesse, S.S.; Casetti, C.; Chiasserini, C.F.; Landi, G. Energy-efficient traffic allocation in SDN-based backhaul networks: Theory and implementation. In Proceedings of the 2017 14th IEEE Annual, IEEE, Consumer Communications & Networking Conference (CCNC), Las Vegas, NV, USA, 8–11 January 2017; pp. 209–215.

15. Huong, T.; Schlosser, D.; Nam, P.; Jarschel, M.; Thanh, N.; Pries, R. ECODANE—Reducing energy consumption in data center networks based on traffic engineering. In Proceedings of the 11th Würzburg Workshop on IP: Joint ITG and Euro-NF Workshop Visions of Future Generation Networks (EuroView2011), Wurzburg, Germany, 1–3 August 2011.

16. Yang, Y.; Xu, M.; Li, Q. Towards fast rerouting-based energy efficient routing. *Comput. Netw.* **2014**, *70*, 1–15. [CrossRef]

17. Cianfrani, A.; Eramo, V.; Listanti, M.; Polverini, M.; Vasilakos, A.V. An OSPF-integrated routing strategy for QoS-aware energy saving in IP backbone networks. *IEEE Trans. Netw. Serv. Manag.* **2012**, *9*, 254–267. [CrossRef]

18. Lorincz, J.; Mujaric, E.; Begusic, D. Energy consumption analysis of real metro-optical network. In Proceedings of the 2015 38th International Convention on Information and Communication Technology, Electronics and Microelectronics (MIPRO), Opatija, Croatia, 25–29 May 2015; pp. 550–555. [CrossRef]

19. Chiaraviglio, L.; Ciullo, D.; Mellia, M.; Meo, M. Modeling sleep mode gains in energy-aware networks. *Comput. Netw.* **2013**, *57*, 3051–3066. [CrossRef]

20. Garroppo, R.; Nencioni, G.; Tavanti, L.; Scutella, M.G. Does Traffic Consolidation Always Lead to Network Energy Savings? *IEEE Commun. Lett.* **2013**, *17*, 1852–1855. [CrossRef]

21. Lu, T.; Zhu, J. Genetic Algorithm for Energy-Efficient QoS Multicast Routing. *IEEE Commun. Lett.* **2013**, *17*, 31–34. [CrossRef]

22. Galán-Jiménez, J.; Gazo-Cervero, A. Using bio-inspired algorithms for energy levels assessment in energy efficient wired communication networks. *J. Netw. Comput. Appl.* **2014**, *37*, 171–185. [CrossRef]

23. Huin, N. Energy Efficient Software Defined Networks. Ph.D. Thesis, Université Côte d'Azur, Nice, France, 2017.

24. Herrería-Alonso, S.; Rodríguez-Pérez, M.; Fernández-Veiga, M.; López-García, C. How efficient is energy-efficient Ethernet? In Proceedings of the 2011 3rd International Congress on IEEE, Ultra Modern Telecommunications and Control Systems and Workshops (ICUMT), Budapest, Hungary, 5–7 October 2011; pp. 1–7.

25. Herrería Alonso, S.; Rodríguez Pérez, M.; Fernández Veiga, M.; López García, C. A GI/G/1 Model for 10 Gb/s Energy Efficient Ethernet Links. *IEEE Trans. Commun.* **2012**, *60*, 3386–3395. [CrossRef]

26. Johnson, D.S. Near-Optimal Bin Packing Algorithms. Ph.D. Thesis, Massachusetts Institute of Technology, Cambridge, MA, USA, 1973.

27. ONF. OpenFlow® Switch Specification, Version 1.5.1. Available online: http://sdn.ifmo.ru/Members/shkrebets/sdn_4115/sdn_theory/openflow-switch-v1.5.1.pdf/view (accessed on 10 November 2018).

28. Fondo-Ferreiro, P. SDN Bundle Network Simulator. Available online: https://pfondo.github.io/sdn-bundle-simulator/ (accessed on 10 November 2018).

29. The CAIDA UCSD Anonymized Internet Traces 2016. 6 April 2016 13:03:00 UTC. Available online: https: //www.caida.org/data/passive/passive_2016_dataset.xml (accessed on 10 November 2018).
30. Carpa, R.; de Assuncao, M.D.; Glück, O.; Lefèvre, L.; Mignot, J.C. Evaluating the Impact of SDN-Induced Frequent Route Changes on TCP Flows. In Proceedings of the 2017 13th International Conference on Network and Service Management (CNSM), Tokyo, Japan, 26–30 November 2017; pp. 1–9. [CrossRef]

![sensors logo] *sensors*

MDPI

Article

Impact of Node Speed on Energy-Constrained Opportunistic Internet-of-Things with Wireless Power Transfer

Seung-Woo Ko [1] and Seong-Lyun Kim [2,*]

1   Department of Electrical and Electronic Engineering, The University of Hong Kong, Pok Fu Lam, Hong Kong, China; swko@eee.hku.hk
2   School of Electrical and Electronic Engineering, Yonsei University, 50 Yonsei-Ro, Seodaemun-Gu, Seoul 03722, Korea
*   Correspondence: slkim@yonsei.ac.kr; Tel.: +82-2-2123-5862

Received: 15 June 2018; Accepted: 16 July 2018; Published: 23 July 2018

Abstract: *Wireless power transfer* (WPT) is a promising technology to realize the vision of *Internet-of-Things* (IoT) by powering energy-hungry IoT nodes by electromagnetic waves, overcoming the difficulty in battery recharging for massive numbers of nodes. Specifically, *wireless charging stations* (WCS) are deployed to transfer energy wirelessly to IoT nodes in the charging coverage. However, the coverage is restricted due to the limited hardware capability and safety issue, making mobile nodes have different battery charging patterns depending on their moving speeds. For example, slow moving nodes outside the coverage resort to waiting for energy charging from WCSs for a long time while those inside the coverage consistently recharge their batteries. On the other hand, fast moving nodes are able to receive energy within a relatively short waiting time. This paper investigates the above impact of node speed on energy provision and the resultant throughput of energy-constrained opportunistic IoT networks when data exchange between nodes are constrained by their intermittent connections as well as the levels of remaining energy. To this end, we design a two-dimensional Markov chain of which the state dimensions represent remaining energy and distance to the nearest WCS normalized by node speed, respectively. Solving this enables providing the following three insights. First, faster node speed makes the inter-meeting time between a node and a WCS shorter, leading to more frequent energy supply and higher throughput. Second, the above effect of node speed becomes marginal as the battery capacity increases. Finally, as nodes are more densely deployed, the throughput becomes scaling with the density ratio between mobiles and WCSs but independent of node speed, meaning that the throughput improvement from node speed disappears in dense networks. The results provide useful guidelines for IoT network provisioning and planning to achieve the maximum throughput performance given mobile environments.

Keywords: internet-of-things; opportunistic networks; wireless power transfer; inter-meeting time; Markov chain; node speed; battery capacity; node density

1. Introduction

Wireless mobile devices are currently pervasive, and the number of the devices is expected to be ever-growing when *Internet-of-Things* (IoT) and smart cities emerge in the near future [1]. This tendency makes their energy supply required not only huge but also so frequent that the existing wired charging technologies cannot cope with them. Faced with the energy supply problem, *wireless power transfer* (WPT) is fast becoming recognized as a viable solution [2] enabling the recharging of batteries without plugs and wires if there is an apparatus to perform WPT, known as a *wireless charging station* (WCS). However, due to the limited capability of the state-of-art WPT technique and concerns about human safety [3], it is impractical to radiate electromagnetic waves with higher power from a WCS, making the charging coverage restricted. Multiple numbers of WCSs can be installed to cover the entire network area, but excessive deployment cost is required.

This paper addresses the energy provisioning issue of *energy-constrained opportunistic IoT networks* (This is originated from an opportunistic IoT network where IoT nodes exchange information via *device-to-device* (D2D) communications based on their interaction [4,5]). These have been extensively studied in a wide range of fields, e.g., healthcare, logistics, and car navigation. We add the term "energy-efficient" to highlight the energy provision problem where nodes' opportunistic connections to other nodes and WCSs lead data transmission and energy charging, respectively. These features of energy supply and consumption yield the following energy dynamics, which is the main theme of this work. Particularly, we pay attention to the random mobility of an IoT node followed by its moving pattern, making it possible to supply energy in spite of WCSs' limited charging coverage. In other words, a WCS can transfer energy to nodes who get into the charging coverage, which is referred to as "meeting" throughout the paper. The resultant energy provision of the IoT node thus depends on its meeting pattern, dominantly affected by moving speed as shown in Figure 1. When a node moves slowly, as an example, it remains in the charging coverage of the WCS and can receive energy continuously. Once out of the coverage, on the other hand, it may take an extremely long time to receive energy again. Consequently, such an irregular energy provision occurs where some devices consistently receive energy while others suffer from the lack of energy. As a node moves faster, this energy-starving duration is likely to be shorter, leading to a more regular pattern of energy supply. This difference motivates us to investigate the relation between speed, energy provision and the resultant throughput.

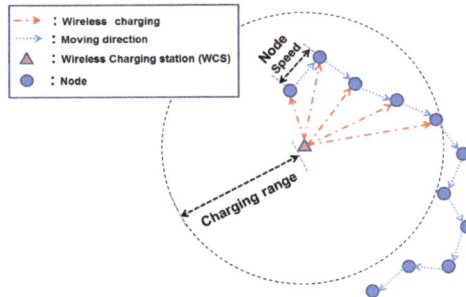

Figure 1. The pattern of wireless charging when node speed is slow. During the period that a node is in the charging coverage of the WCS, it receives energy from WCS continuously. Once a node is out of the charging range, on the other hand, it takes a long time to receive energy from WCS again.

1.1. Wireless Power Transfer

WPT is a key enabler to realize the vision of next-generation mobile networks, e.g., IoT and smart cities by overcoming the challenge with battery charging. With the aid of WPT, it is possible to deploy thousands of IoT sensors at a low cost. In addition, WPT along with energy harvesting enables the facilitating of designing green networks (see, e.g., [6,7]). Due to its promising potential and the interdisciplinary nature, many new research issues arise in the area of WPT and are widely studied in different fields. Recent advancements in the area can be found in numerous surveys such as [8].

The most widely-used WPT method is the magnetic inductive coupling that electric power is delivered by means of an induced magnetic field. The drawback of this method is its power transfer efficiency that diminishes significantly unless the transmitter and the receiver are close in contact. Recently, there have been efforts to develop WPT technology of which the efficiency remains high within the range from several to tens of meters. In [9], Kurs et al. suggest a novel method called magnetic resonant coupling where electric power is transferred from one to the other with high efficiency when two devices are tuned to the same resonant frequency. However, its high efficiency requirement is so tight that it is vulnerable to the misalignment between a transmitter and a receiver especially when the distance between the two becomes larger. Some sophisticated tracking and alignment techniques are proposed for practical use, e.g., frequency matching [10], impedance matching [11] and resonant isolation [12], but the use of magnetic resonant coupling for long-range battery charging in harsh mobile environments like vehicular scenarios is still doubtful. Another approach is microwave power transfer where *radio-frequency* (RF) waves are delivered to recharge the battery by using advanced techniques of wireless communications, e.g., directional beamforming [13], backscatter communication [14,15], and full duplex communication [16]. Recent studies on WPT consider practical factors to design and optimize realistic systems, e.g., imperfect channel state information [17], nonlinear energy harvesting efficiency [18,19], and waveform design [20]. Microwave power transfer theoretically enlarges the charging coverage more than tens of meters, but controversial issues on health impairments caused by RF exposure have not been resolved yet, making it difficult to use commercially.

1.2. Applying Wireless Power Transfer to Wireless Networks

There have been several studies incorporating WPT into the design of energy-constrained wireless networks. One research thrust focuses on the design of the efficient recharging protocol to make every node always active. For example, a *wireless charging vehicle* (WCV) is suggested in [21] that visits all nodes to recharge their batteries. The optimal travel path is derived to avoid the battery depletion of each node. The work of the optimal WCV routing is generalized in [22] such that the battery charging is enabled in every place in the entire network under the consideration of the trade-off between charging efficiency and distance. In [23], the optimal routing for safe charging problem is studied where no location in the networks has electromagnetic radiation exceeding a given threshold. A prototype testbed of the routing platform is constructed by using off-the-shelf RF energy transfer hardware equipment in [24] to demonstrate the performance of wireless sensor networks powered by RF energy transfer. A distributed recharging protocol is proposed in [25] where multiple WCVs wirelessly provide energy to nodes given the limited network information. The concept of Qi-ferry is introduced in [26], which is similar to WCV except the fact that Qi-ferry consumes its own residual energy when it is moving. In other words, longer travel distance of Qi-ferry visits more nodes but accelerates its energy depletion. They optimize its travel path reflecting the above trade-off. Nevertheless, these papers [21–26] are based on the assumption that WPT-enabled devices have knowledge of full or limited geographical information for all rechargeable nodes, hardly estimated in mobile environments.

Recently, there have been some trials to exploit node mobility for battery charging in WPT-aided mobile networks. In [27], the energy provision based on the node mobility is introduced where nodes can harvest excessive energy in a power-rich area and store it for later use in a power-deficient area. The number of necessary WCSs for continuous operation of every node is analyzed, but some practical aspects like node speed and battery capacity are ignored. In [28], the performance of energy-constrained mobile networks is analyzed using stochastic geometry assuming that the energy arrival process of each node as an *independent and identically distributed* (i.i.d.) sequence, which is reasonable only when the speed of each node is extremely fast. Delay-limited and delay-tolerant communications with WPT are respectively studied in [29,30], where a node can move to a few rechargeable points according to predetermined transition probabilities. In [31], an intentional mobility to WPT-enabling locations for battery charging, called a spatial attraction, is modeled as a Markov chain and analyzed to show the improvement of the coverage rate by the optimally controlled power and charging range. These papers [27–31] do not consider node speed in spite of its significant effects on the energy arrival process. In [32], a *quality of energy provisioning* (QoEP) is defined as the expected portion of time a node sustains its operation when mobiles are moving within a given range of node speed. It is shown that QoEP converges to one as battery capacity or node speed increases. The analytical results are based on the continuous transmission model where a node keeps transmitting data whenever it has energy. In IoT networks, on the other hand, data is transmitted discontinuously according to a few specific conditions, making it more challenging to analyze.

1.3. Contributions and Organization

In this work, we study the performance of energy-constrained opportunistic IoT networks where the opportunistic behaviors of mobile nodes affect the patterns of data transmission and energy charging. Specifically, data delivery is enabled between nodes when (1) they are intermittently connected and (2) a transmitting node receives energy from WCSs enough to deliver data. To reflect the above energy dynamics, we design a two-dimensional Markov chain of which the horizontal and vertical state dimensions represent the remaining energy and the distance to the nearest WCS, respectively. We derive its steady-state probabilities and aim at explaining the effect on throughput. The main contributions of this paper are summarized below.

- **Inter-meeting time vs. Throughput**: Higher node speed reduces the frequency of lengthy inter-meeting times between a node and a WCS and eventually improves the throughput. The inter-meeting time is interpreted as an energy-starving duration. We explain the phenomenon through the stochastic distribution of the inter-meeting time in Proposition 1.
- **Node speed vs. battery capacity**: A slow-moving node stays in the charging coverage for a long time. It saves enough energy to endure a lengthy inter-meeting time if its battery capacity, the maximum amount of energy stored in the battery, is large enough. In Proposition 2, we show that a fast-moving node achieves the same throughput when the battery capacity becomes infinite.
- **Throughput scaling law**: In Proposition 3, we prove that the throughput scaling is given as $\Theta\left(\min\left(1, \frac{m}{n}\right) c^{\min\left(1, \frac{m}{n}\right)}\right)$ (We recall that the following notation: (i) $f(n) = O(g(n))$ means that there exists a constant c and integer N such that $f(n) \leq cg(n)$ for $n > N$. (ii) $f(n) = \Theta(f(n))$ means that $f(n) = O(g(n))$ and $g(n) = O(f(n))$.) where n and m denote the number of nodes and WCSs respectively, and c is a constant ($0 < c < 1$). As the network becomes denser, the throughput depends on the ratio $\frac{m}{n}$ and becomes independent of node speed.

Note that the approach in this work is similar to that of our previous work [33] as both apply a Markov chain to model an energy-constrained mobile network. In [33], it is assumed that nodes follow the i.i.d. mobility model, which allows us to include only the residual energy status as a Markov chain state. On the other hand, our current work focuses on finite node speed, which limits node movement within a restricted area. In other words, the current node location depends on the previous one. Therefore, we should take into account not only the residual energy, but also the location information of a node when designing a Markov chain model. Our paper illustrates that the throughput under the i.i.d. mobility model in [33] can be understood as an upper bound, which is achievable when (i) node speed becomes faster; (ii) battery capacity becomes larger or (iii) node density increases.

The rest of this paper is organized as follows: In Section 2, we explain our models and metrics. In Section 3, we introduce a two-dimensional Markov chain design and derive its steady state probabilities. In Section 4, we verify how the node speed effect is affected by battery capacity and node density. Finally, we conclude this paper in Section 5.

2. Models and Metrics

2.1. Network Description

Consider an energy-constrained IoT network where n nodes and m WCSs are randomly distributed in a torus area (A torus area refers to finite and boundary-less region such that one side's edge is connected to the opposite one. In this model, the boundary effect disappears, enabling the analysis of the performance tractably from the viewpoint of one typical node.) of size $\sqrt{S} \times \sqrt{S}$

(in meter^2). Time is slotted and one slot is large enough to transmit a single packet. A node is assumed to change its direction randomly at every slot with constant speed of v (meter/slot), namely, we have:

$$\|X_\ell(t+1) - X_\ell(t)\| = v, \tag{1}$$

where $X_\ell(t)$ is the location of node ℓ at slot t and $\|\cdot\|$ means the Euclidian distance. The assumption makes sense because an IoT node requires much longer latency than conventional cellular networks to transmit data, say up to a few seconds [34].

A node enables transmitting its packet to one of its neighbors within r (in meters) defined as the transmission range. For an interference model, we use the well-known protocol model [35] where the packet transmission is successful only when the other transmitting nodes are no less than r. Too large r leads to frequent transmission failures because there are many interfering nodes. To avoid excessive interference, the transmission range r is set to the average distance to the nearest node in the area.

2.2. Two-Phase Routing

A pair of source and destination nodes is given randomly. Unless there is the corresponding destination node of a source node within the range r, its packet should be delivered via a relay node. This paper adopts the *two-phase routing* [35] as follows:

- *Mode switch.* In the beginning of each slot, a node becomes a transmitter or a receiver with probability q or $1-q$, respectively. Without loss of generality, we set $q = 0.5$.
- *Phase 1.* In odd slots, let us consider node ℓ becomes a transmitter. If there is at least one receiver within transmission range r, node ℓ forwards its packet to one of them. This receiver node can be the destination of node ℓ.
- *Phase 2.* In even slots, let us consider node ℓ becoming a receiver. If there is at least one transmitter within transmission range r and one of them has a packet whose destination is node ℓ, it forwards the packet to node ℓ. This transmitter can be the source of node ℓ.

In [35], the throughput of the two-phase routing is defined as follows:

Definition 1. *(Throughput) Let $M_\ell(t)$ be the number of node ℓ's packets that its corresponding destination node receives during t slots. We say that the throughput of Λ is feasible for every S-D pair if:*

$$\liminf_{t \to \infty} \frac{1}{t} M_\ell(t) \geq \Lambda. \tag{2}$$

When a node transmits a packet, a constant amount of energy is consumed defined as one *unit of energy* (It is implicitly assumed that a *modulation and coding scheme* (MCS) is fixed and constant power is required to deliver a packet within the transmission range. It is interesting to adjust the level of MCS to improve the energy efficiency, which is outside the scope of current work.). A node is called *active* when it has at least one unit of energy. Otherwise, the node is *inactive*. Let p_{on} denote the probability that a node is active, defined as an *active probability*. In [33], the throughput Λ is given as follows:

$$\Lambda = \frac{1}{2} q p_{\text{on}} \exp\left(-\frac{\pi}{4} q p_{\text{on}}\right) \left(1 - \exp\left(\frac{\pi}{4}(-1+q)\right)\right). \tag{3}$$

It is shown that the throughput Λ (3) depends on p_{on}, which is determined by the process of energy recharging introduced in the sequel.

2.3. Recharging Mechanism by Wireless Charging Stations

Inactive nodes are unable to transmit packets in their buffers. To supply energy to them, m WCSs are deployed in the network. WCSs recharge nodes via WPT. No interference between data transmission and energy transfer exists because each of them use a separated band.

315

The energy transferred to a mobile is given by the product of the maximum deliverable units of energy E and the energy transfer efficiency $\tau(x)$, where x is the distance to a WCS. Let R_y denote the maximum distance that a node can receive y units of energy. Without loss of generality, the efficiency $\tau(x)$ is a monotone decreasing function of x, and the charging range R_y is determined by finding the value of x that $E \cdot \tau(x)$ becomes y, such that $R_y = \{x : E \cdot \tau(x) = y\}$. Let $Y_s(t)$ denote the location of WCS s at slot t. The distance between node ℓ and WCS s is given as $\|X_\ell(t) - Y_s(t)\|$, and the amount of recharged energy v (in units of energy) is

$$v(\|X_\ell(t) - Y_s(t)\|) = \begin{cases} E, & \text{if } \|X_\ell(t) - Y_s(t)\| \leq R_E, \\ k, & \text{if } R_{k+1} < \|X_\ell(t) - Y_s(t)\| \leq R_k, k = 1, \cdots, E-1, \\ 0, & \text{otherwise,} \end{cases} \tag{4}$$

where $R_E < R_{E-1} < \cdots < R_1$. Define a *charging range* as the maximum distance that a node receives at least one unit of energy from the connected WCS. Given the above recharging mechanism, the charging range is equivalent to R_1. The time required to transfer energy from a WCS to a node is extremely short compared to one slot. This means that the contact duration is long enough to deliver up to E units of energy under finite speed. It is a reasonable assumption because the maximum power transfer rate of magnetic resonance coupling is 12 (in Watts) [36], whereas that of an IoT device is 23 (in dBm), approximately 0.2 (in Watts).

The battery of each node is recharged by one of WCSs. When a node is in the coverage of multiple WCSs, it is assumed to receive energy from one of them due to the practical alignment technique limitation. The maximum battery capacity of each node is set to L units of energy. If the sum of residual and recharged energy are larger than L, a node stores up to L units of energy, and the remaining is thrown out. A WCS can recharge up to u nodes within one slot using the technique of tracking resonance frequencies. For example, it is experimentally shown in [37,38] that up to two devices can be charged by using the technique of the said resonant frequency splitting and load balancing, respectively. When there are more than u nodes within the coverage, the WCS randomly selects u nodes among them.

Each WCS always monitors its own remaining energy. If the remaining energy is below a certain level, it communicates with an operator station by using its communication module. The operator station then sends the charging vehicle, which recharges the WCS before its battery runs out. This means that all WCSs always have sufficient energy.

3. Stochastic Modeling of Energy-Efficient Opportunistic Internet-of-Things

In this section, we design a two-dimensional Markov chain in which the horizontal and vertical state dimensions represent the residual energy and the distance to the nearest WCS, respectively. We first outline our Markov chain design, and then derive the steady state probabilities to determine the active probability P_{on} (3).

3.1. Two-Dimensional Markov Chain

The state space of the proposed two-dimensional Markov chain Ψ is given as follows:

$$\Psi = \{(e, d) : 0 \leq e \leq L, 0 \leq d \leq M\}, \tag{5}$$

where parameter e is the number of remaining units of energy, and d is a discrete number indicating the distance to the nearest WCS by the following rule:

$$
d = \begin{cases}
0, & \text{if } \min_s \|X_\ell(t) - Y_s(t)\| \le R_1, \\
1, & \text{else if } \min_s \|X_\ell(t) - Y_s(t)\| \le R_1 + v, \\
\vdots & \\
k, & \text{else if } \min_s \|X_\ell(t) - Y_s(t)\| \le R_1 + kv, \\
\vdots & \\
M, & \text{otherwise,}
\end{cases}
\tag{6}
$$

where $\min_s \|X_\ell(t) - Y_s(t)\|$ represents the distance from node ℓ to the nearest WCS (in meters), and the charging coverage R_1 and node speed v are specified in (1) and (4), respectively. The number M in (6) can be interpreted as the resolution of the Markov chain in the sense that larger M is able to express the trajectory of node ℓ more accurately. The number d is defined as a *relative distance* meaning that a physical distance (in meters) is normalized by node speed v.

Figure 2 represents an example of the two-dimensional Markov chain when WCS can deliver up to two units of energy to a node within one slot ($E = 2$). The state transitions are explained as follows:

- **State transition by node mobility**: The state transitions to the up or down arise when the relative distance d (6) becomes shorter or longer, respectively. Let $P_{i,j}$ denote the probability that the relative distance d is changed from a to b, i.e.,

$$
P_{i,j}(t) = \Pr[d = j \text{ at slot } t + 1 \,|\, d = i \text{ at slot } t].
\tag{7}
$$

The mobility model follows a time-invariant Markov process of which the transition probabilities are constant regardless of slot t, and $P_{i,j}(t)$ can be simply expressed as $P_{i,j}$ by omitting the index t. The exact form of $P_{i,j}$ is in Appendix A.1 with its derivation. All transition probabilities $P_{i,j}$ are constant regardless of the residual energy status.

- **State transition by data transmission**: The state transition to the left happens when node ℓ transmits a packet to one of neighbors nodes. Let p_t denote a probability that an active node can transmit its packet as

$$
p_t = q \cdot \left[1 - \left\{ 1 - (1 - q)\frac{\pi r^2}{S} \right\}^{n-1} \right].
\tag{8}
$$

The detailed derivation is in [33]. Unless its residual energy e is zero, the transmission probability p_c is constant regardless of the relative distance d (6).

- **State transition by energy charging**: The state transition to the right arises when the node is recharged by a WCS. This event only happens when the node is selected by one of WCSs is in the charging coverage, and these are only stipulated on the lowest state transition ($d = 0$). Recall that each WCS can charge up to u nodes in a given slot. We define a charging probability p_c as the probability that node ℓ becomes one of u selected nodes, i.e.,

$$
p_c = \frac{1 - \gamma(u,n)^m}{1 - \left(1 - \frac{\pi R_1^2}{S}\right)^m},
\tag{9}
$$

where $\gamma(u,n) = 1 - \frac{\pi R_1^2}{S} F(u-2; n-1, \frac{\pi R_1^2}{S}) - \frac{u}{n}\left(1 - F(u-1; n, \frac{\pi R_1^2}{S})\right)$ and $F(k; n, p) = \sum_{i=0}^{k} \binom{n}{i} p^i (1-p)^{n-i}$ is the *cumulative distribution function* (CDF) of the binomial distribution with parameters k, n and p. The derivation is given in Appendix A.2. The number of recharged units of energy depends on the distance to its associated WCS. Let $\beta(k)$ denote a probability a node receives k units of energy as follows:

$$\beta(k) = \begin{cases} \frac{R_k^2 - R_{k+1}^2}{R_1^2}, & \text{if } k = 1, \cdots, E-1, \\ \frac{R_k^2}{R_1^2}, & \text{if } k = E. \end{cases}$$

A node in the charging coverage thus receives k units of energy with probability $p_c \beta(k)$.

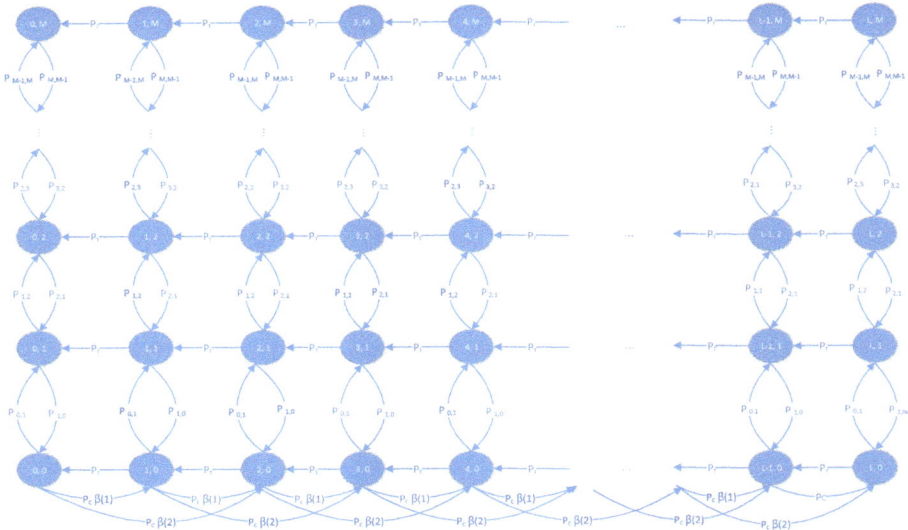

Figure 2. Two-dimensional Markov chain of which the horizontal and vertical state dimensions represent the number of remaining units of energy and the relative distance to the nearest WCS normalized by node speed, respectively.

3.2. Steady State Probability and Throughput

Let $\pi_{e,d}$ denote the steady state probability when the residual energy of node ℓ is e units of energy and the relative distance is d. Then, we make the following steady state vector $\pi = \begin{bmatrix} \pi_{0,0}, \cdots \pi_{0,M}, & \pi_{1,0}, \cdots \pi_{1,M}, \cdots & \pi_{L,0}, \cdots \pi_{L,M} \end{bmatrix}$, which is partitioned according to the number of remaining units of energy, i.e., $\pi = \begin{bmatrix} \pi_0 & \pi_1 & \cdots & \pi_L \end{bmatrix}$, where

$$\pi_e = \begin{bmatrix} \pi_{e,0} & \pi_{e,1} & \cdots & \pi_{e,M} \end{bmatrix}. \tag{10}$$

In order to derive π, we make the following balance equation:

$$\pi Q = 0, \quad \pi \mathbf{1} = 1, \tag{11}$$

where $\mathbf{1}$ is the column vector where every entity is one, and Q is the generating matrix of the corresponding Markov chain:

$$Q = \begin{pmatrix} B_0 & A_2 & A_3 & 0 & 0 & \cdots & 0 & 0 & 0 \\ A_0 & A_1 & A_2 & A_3 & 0 & \cdots & 0 & 0 & 0 \\ 0 & A_0 & A_1 & A_2 & A_3 & \cdots & 0 & 0 & 0 \\ \vdots & \vdots & \vdots & \vdots & \vdots & \ddots & \vdots & \vdots & \\ 0 & 0 & 0 & 0 & 0 & \cdots & A_1 & A_2 & A_3 \\ 0 & 0 & 0 & 0 & 0 & \cdots & A_0 & A_1 & A_2 + A_3 \\ 0 & 0 & 0 & 0 & 0 & \cdots & 0 & A_0 & A_1 + A_2 + A_3 \end{pmatrix}. \tag{12}$$

Its sub-matrices B_0, A_0, A_1, A_2 and A_3 are expressed as follows:

$$\mathbf{B_0} = \begin{pmatrix} -P_{0,1} - p_c & P_{0,1} & 0 & \cdots & 0 \\ P_{1,0} & -P_{1,0} - P_{1,2} & P_{1,2} & \cdots & 0 \\ 0 & P_{2,1} & -P_{2,1} - P_{2,3} & \cdots & 0 \\ \vdots & \vdots & \vdots & \ddots & \vdots \\ 0 & 0 & 0 & \cdots & -P_{M,M-1} \end{pmatrix},$$

$$\mathbf{A_0} = \begin{pmatrix} p_t & \cdots & 0 \\ \vdots & \ddots & \vdots \\ 0 & \cdots & p_t \end{pmatrix} = p_t \mathbf{I}, \quad \mathbf{A_1} = \mathbf{B_0} - \mathbf{A_0}, \mathbf{A_2} = \begin{pmatrix} p_c \beta(1) & \cdots & 0 \\ \vdots & \ddots & \vdots \\ 0 & \cdots & 0 \end{pmatrix}, \quad \mathbf{A_3} = \begin{pmatrix} p_c \beta(2) & \cdots & 0 \\ \vdots & \ddots & \vdots \\ 0 & \cdots & 0 \end{pmatrix}.$$

After solving the balance equation of (11), we can caculate the active probability P_{on} as

$$P_{\text{on}} = \sum_{e=1}^{L} \pi_e \mathbf{1} = 1 - \pi_0 \mathbf{1}. \tag{13}$$

With (3), the throughput Λ is given as

$$\Lambda = \frac{1}{2} q \left(1 - \pi_0 \mathbf{1} \right) \exp \left(-\frac{\pi}{4} q \left(1 - \pi_0 \mathbf{1} \right) \right) \left(1 - \exp \left(\frac{\pi}{4} (-1 + q) \right) \right). \tag{14}$$

4. Performance Evaluation of Energy-Efficient Opportunistic Internet-of-Things

Based on the aforementioned mathematical framework, this section attempts to analyze the effects of node speed on throughput in terms of inter-meeting time, battery capacity, and node density, each of which is verified by comparing Monte-Carlo simulations.

4.1. Inter-Meeting Time and Throughput

This subsection aims at analyzing the effect of node speed v on throughput Λ using the inter-meeting time defined as follows:

Definition 2. *(Inter-meeting time) Consider that there are node ℓ and WCS s in the network. The inter-meeting time T_I is the interval between adjacent meeting events between node ℓ and WCS s:*

$$T_I = \inf \{ t \geq 0 : Z_{t+k} = 1 \mid Z_k = 1 \}, \tag{15}$$

where Z_t is an indicator to check whether a meeting event occurs between node ℓ and WCS s at time t. If $\|X_\ell(t) - Y_s(t)\| \leq R_1$, we set Z_t to one. Otherwise, $Z_t = 0$.

The inter-meeting time T_I is related to the energy starving period of the node because the node has no opportunity to receive energy until it meets one of WCSs. The stochastic features of T_I is thus related to an energy provision process of an arbitrary node. Let P denote an M by M matrix of which the elements represents the transition probability $P_{i,j}$ (7) ($1 \leq a, b \leq M$):

$$
P = \begin{pmatrix} p_1 \\ p_2 \\ \vdots \\ p_M \end{pmatrix} = \begin{pmatrix} P_{1,1} & P_{1,2} & \cdots & P_{1,M-1} & P_{1,M} \\ P_{2,1} & P_{2,2} & \cdots & P_{1,M-1} & P_{2,M} \\ \vdots & \vdots & \ddots & \vdots & \vdots \\ P_{M,1} & P_{M,2} & \cdots & P_{M-1,M} & P_{M,M} \end{pmatrix},
\tag{16}
$$

where $p_d = \begin{pmatrix} P_{d,1} & P_{d,2} \cdots & P_{d,M-1} & P_{d,M} \end{pmatrix}$. From P (16), we derive the stochastic distribution of inter-meeting time T_I in the following Proposition:

Proposition 1. *The complementary cumulative distribution function (CCDF) of the inter-meeting time T_I is*

$$
\Pr\left[T_I > t\right] = \sum_{i=1}^{M} \gamma_i \lambda_i^{t-1},
\tag{17}
$$

where λ_i is the ith eigenvalue of P (16) ($1 > \lambda_1 > \cdots > \lambda_M > 0$). The coefficient γ_i is

$$
\gamma_i = p_0 a_i b_i^T,
$$

where vectors a_i and b_i are the right-hand and left-hand eigenvectors of λ_i such that $P a_i = \lambda_i a_i$ and $b_i^* P = \lambda_i b_i^*$, respectively.

Proof. see Appendix A.3. □

Figure 3a shows the CCDFs of inter-meeting time T_I. We use the energy transfer efficiency function in [39], i.e., $\tau(x) = -0.0958x^2 - 0.0377x + 1.0$, which is obtained through the curve fitting of the experimental results of [40]. We numerically measure the inter-meeting time T_I by changing the node speed as $v = 1, 2, 3$ and 6 (meters/slot). When the length of one slot is set to a second, the concerned sets of speed represent the cases of stationary, walking, slow vehicle and fast vehicle, respectively [41]. It is shown that higher node speed v reduces the number of lengthy inter-meeting times. A node with faster speed can reach the charging coverage of the WCS within a few slots, reducing the occurrence of lengthy inter-meeting times. A node with higher speed enables to move a new location far away from its previous one. In other words, the event of meeting WCS depends on the ratio of the charging coverage to the network area, i.e., $\frac{1}{\mu} = \frac{\pi R_1^2}{S^2} \approx 0.053$ as does the i.i.d. mobility model. With increased node speed, the distribution converges to that of the i.i.d. mobility model following the exponential distribution with parameter $\mu \approx 18.7174$. It is verifed by simulation that the analytic result in Proposition 1 follows similar tendencies of practical mobility models e.g., Brownian motion and random waypoint (See Appendix A.4).

The CCDF of T_I of (17) is the sum of powered eigenvalues with the exponent t. As t becomes larger, it is simplified by the largest eigenvalue λ_1 because other terms decay faster:

$$
\Pr\left[T_I > t\right] \approx \lambda_1^t.
\tag{18}
$$

The eigenvalue λ_1 is called the *spectral radius* of matrix P (16). As the spectral radius becomes smaller, the approximated CCDF (18) decreases much faster in the regime of large t. This indicates that lengthy inter-meeting times are infrequent when λ_1 is small. In Table 1, we summarize this spectral radius λ_1 as a function of node speed v and show that λ_1 is a non-increasing function of node speed *v*. Consequently, higher node speed decreases spectral radius λ_1 and produces fewer occurrences of lengthy inter-meeting times.

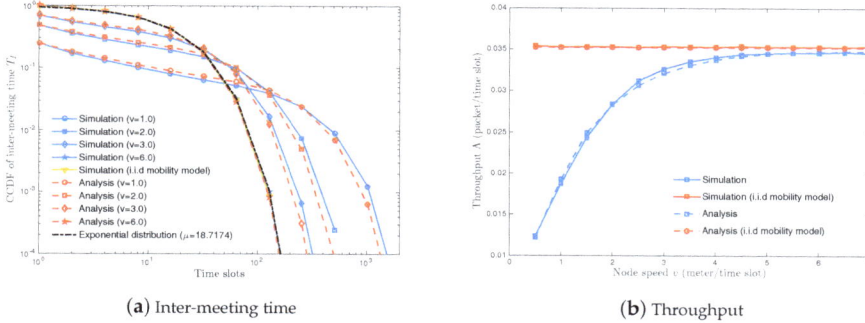

(**a**) Inter-meeting time   (**b**) Throughput

Figure 3. (**a**) The CCDF of inter-meeting time under different node speed *v* (meter/slot); (**b**) expected throughput as a function of node speed *v* (meter/slot). Parameters: Network size $S = 400$ (in meter^2), battery capacity $L = 10$ (units of energy), the maximum number of simultaneous transferable nodes $u = 1$, the maximum transferable energy per slot $E = 3$ (in units of energy), the number of nodes $n = 10$, and the number of WCSs $m = 1$.

Table 1. Spectral radius λ_1 as a function of node speed *v* (meters/slot). The same parameter setting is used as Figure 3.

	$v = 0.5$	$v = 1.0$	$v = 1.5$	$v = 2.0$	$v = 2.5$	$v = 3.0$	$v = 3.5$	$v = 4.0$	$v = 4.5$	$v = 5.0$	$v = 5.5$	$v = 6.0$
λ_1	0.9985	0.9953	0.9903	0.9845	0.9780	0.9714	0.9649	0.9585	0.9534	0.9492	0.9471	0.9457

The above feature of the inter-meeting time affects the energy provision process as shown in Figure 3b. When node speed *v* is 0.5 (meters/slot), the throughput Λ is nearly one-third of that of the i.i.d. mobility model. A node is unable to receive energy for a long time due to frequent lengthy inter-meeting time and remains in an inactive state. This results in the decrease in throughput. As *v* increases, on the other hand, the inter-meeting time decreases. This leads to the reduction in energy-starving period and the improvement of throughput.

4.2. Battery Capacity and Throughput

Consider a slow-moving node who stays in the charging coverage for a long duration. The node can receive energy continuously from the connected WCS. Nevertheless, the node is unable to save more than *L* units of energy due to the battery capacity constraint. In other words, the node can remain active longer with larger battery capacity. To explain the phenomenon, we make the following proposition.

Proposition 2. *When the battery capacity L becomes infinite, the throughput of an energy-constrained network Λ becomes independent of node speed v as*

$$\Lambda = \frac{q}{2}\rho \left(1 - \exp\left(\frac{\pi(q-1)}{4}\right)\right) \cdot \exp\left(-\frac{\pi}{4}q \cdot \rho\right), \tag{19}$$

where

$$\rho = \min\left[1, \frac{p_c}{p_t}\left\{1 - \left(1 - \frac{\pi R_1^2}{S}\right)^m\right\}\left(\sum_{i=k}^{E} k\beta(k)\right)\right]. \tag{20}$$

Proof. See Appendix A.5. □

Figure 4 represents the throughput Λ as a function of battery capacity L. As L increases, Λ increases and converges as mentioned in Proposition 2 (see the black dotted line). A noticeable point is that Proposition 2 is achievable under a finite battery capacity. If a node can store enough energy to sustain the inter-meeting time, it remains in an active state and achieves the throughput in Proposition 2. We calculate the mean of the inter-meeting time $E[T_I]$ from (18) and the spectral radius λ_1 in Table 1:

$$E[T_I] = \sum_{t=0}^{\infty} \Pr[T_I > t] \approx \sum_{t=0}^{\infty} (\lambda_1)^t = \frac{1}{1 - \lambda_1}. \tag{21}$$

When the battery capacity L is no less than $E[T_I]$, the throughput Λ becomes the same as that in Proposition 2 (19). For example, when node speed v is 0.5 or 1.5 (meters/slot), its spectral radius λ_1 is 0.9985 or 0.9903 (see Table 1) and its corresponding $E[T_I]$ becomes 666.67 or 103.09, respectively. As a result, a battery capacity larger than $E[T_I]$ is understood as a necessary condition to achieve the upper bound in Proposition 2.

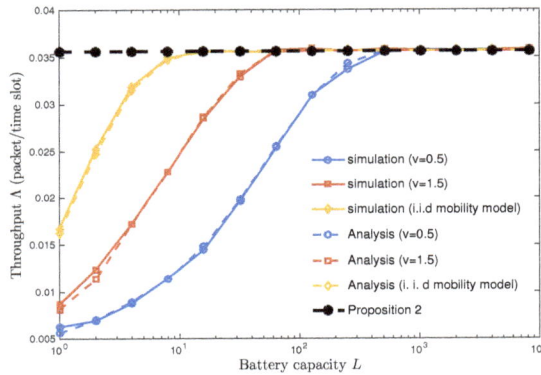

Figure 4. Throughput vs. battery capacity L. The same parameter setting as in Figure 3 is used unless specified.

4.3. Node Density and Throughput

Since the seminal work by Grossglauser and Tse [35], investigating the relationship between throughput Λ and node density n has been the most fundamental issue with mobile networks; therefore, the impact of irregular energy provision due to low node speed has not yet been studied. In this subsection, we investigate this effect through some numerical evaluations and the following throughput scaling law.

Proposition 3. *The scaling law of the throughput Λ is:*

$$\Lambda = \Theta\left(\min\left(1, \frac{m}{n}\right) c^{\min\left(1, \frac{m}{n}\right)}\right), \tag{22}$$

where $0 < e^{-\frac{\pi \cdot u}{4 \cdot a}} < c \leq e^{-\frac{\pi \cdot u}{4 \cdot a}\left(\sum_{k=1}^{E} k\beta(k)\right)} < 1$, *and* $a = 1 - e^{-\frac{\pi}{4}(1-q)}$.

Proof. See Appendix A.6. □

Proposition 3 indicates that the throughput Λ is a function of the ratio of the number of WCSs m and the number of nodes n, and independent of node speed v. A node with low speed receives energy from WCSs irregularly, yielding the decrease of throughput. Compared with fast-moving one, it needs more WCSs to maintain the same throughput. As the network becomes denser, however, the penalty due to slow speed disappears and we only consider the ratio $\frac{m}{n}$ when installing WCSs. In order to achieve the constant throughput of $\Theta(1)$ as in [35], for example, $\Theta(n)$ WCSs is required regardless of node speed.

Note that the scaling law in Proposition 3 of (22) is the same as that of the i.i.d. mobility model in [33]. Figure 5 shows that the throughput Λ always converges to that of the i.i.d. mobility model as the number of nodes n increases. This implies that a high node density makes nodes look as if they are moving at a fast speed in the sense that the i.i.d. mobility model allows a node to increase moving speed v up to the network size. When calculating the throughput of a dense mobile network with WPT, it is a reasonable assumption that nodes move according to the i.i.d. mobility model.

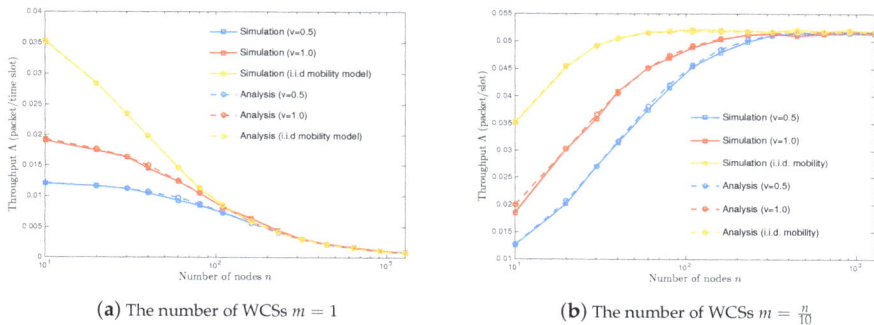

(**a**) The number of WCSs $m = 1$

(**b**) The number of WCSs $m = \frac{n}{10}$

Figure 5. Throughput vs. node density n. The same parameter setting as in Figure 3 is used unless specified.

5. Conclusions

In this paper, we have investigated the throughput of an energy-constrained opportunistic IoT network where WCSs are deployed to recharge IoT nodes when they are in the charging coverage. Given the network architecture, the energy provision pattern follows the speed of the corresponding node. Namely, a slow-moving node outside a WCS's charging coverage waits a long time for energy supply from WCSs, whereas a fast-moving one can receive energy from a WCS within a short interval. The analytical and numerical results have shown that this distinct energy provisioning difference leads to the throughput gap between slow- and fast-moving nodes especially when the battery capacity is finite and IoT nodes are sparsely deployed. This finding provides useful guidelines for designing energy-efficient opportunistic IoT networks. First, a WCS should prioritize the charging opportunity of nodes depending on its speed to improve energy provision efficiency. Second, the battery capacity of an IoT node can be minimized by predicting its speed. Finally, the deployment strategy of WCSs should be different depending on node speed such that a relatively less number of WCSs per node is enough to support the whole nodes in area of high mobility, e.g., motorway, while more WCSs per node are required to guarantee the same throughput in area of slow mobility, e.g., pedestrian way.

The current work can be extended in several directions. In this work, we consider the simple mobility model where each node moves without preference. In practical, on the other hand, people are likely to visit some popular places frequently, making it very challenging to supply enough energy due to the relatively high node density in the area. Next, considering vehicular scenarios such that safety-information is disseminated based on *vehicle-to-everything* (V2X) communication is aligned with the recent trend of wireless communications. Moreover, considering the economic aspect of WCSs is another interesting avenue for future research.

Author Contributions: The contributions of S.-W.K. were to analyze the main results and verify them by simulation, while S.-L.K. suggested the key motivation and advised the idea about the mathematical framework for tractable analysis.

Funding: This research was supported by a grant to Bio-Mimetic Robot Research Center Funded by Defense Acquisition Program Administration, and by Agency for Defense Development (UD160027ID).

Conflicts of Interest: The authors declare no conflict of interest.

Abbreviations

The following abbreviations are used in this manuscript:

WPT	Wireless power transfer
IoT	Internet-of-things
WCS	Wireless charging station
D2D	Device-to-device
RF	Radio-frequency
WCV	Wireless charging vehicle
i.i.d.	Independent and identically distributed
QoEP	Quality of energy provisioning
MCS	Modulation and coding scheme
CDF	Cumulative distribution function
CCDF	Complementary cumulative distribution function
V2X	Vehicular-to-everything
BM	Brownian motion
RWP	Random way point
BMAP	Batch Markovian arrival process
QBD	Quasi-birth-death

Appendix A

Appendix A.1. Derivation of Transition Probability $P_{i,j}$ (7)

Let D_t denote the distance between a node and a WCS at time t. Since nodes and WCSs are uniformly distributed in a torus area, the conditional probability that D_{t+1} is smaller than or equal to x_2 given $D_t = x_1$ is

$$\Pr\left[D_{t+1} \leq x_2 | D_t = x_1\right] = \begin{cases} 0, & \text{if } v + x_2 < x_1 \text{ or } v - x_2 > x_1, \\ \dfrac{\arccos\left(\frac{v^2 + x_1^2 - x_2^2}{2vx_1}\right)}{\pi}, & \text{if } |v - x_2| \leq x_1 < v + x_2, \\ 1, & \text{if } x_2 - v > x_1, \end{cases} \tag{A1}$$

which is based on the fact that nodes and WCSs are uniformly distributed in a torus area. From the conditional probability (A1), we derive the joint cumulative distribution function (CDF) that D_t is smaller than or equal to x_1, and D_{t+1} is smaller than or equal to x_1:

$$\Pr\left[D_t \leq x_1, D_{t+1} \leq x_2\right] = \int_0^{x_1} \Pr\left[D_{t+1} \leq x_2 | D_t = x\right] f_{D_t}(x) dx. \tag{A2}$$

Using (A2), we calculate the following joint probability:

$$\begin{aligned} cc \Pr\left[x_1 \leq D_t \leq x_2, x_3 \leq D_{t+1} \leq x_4\right] &= \Pr\left[D_t \leq x_2, D_{t+1} \leq x_4\right] \\ &- \left[D_t \leq x_1, D_{t+1} \leq x_4\right] - \Pr\left[D_t \leq x_2, D_{t+1} \leq x_3\right] + \Pr\left[D_t \leq x_1, D_{t+1} \leq x_3\right]. \end{aligned} \tag{A3}$$

By inserting the boundary values of the i and j states in (6) into x_1, x_2, x_3 and x_4 in (A3), we can derive the joint probability $\alpha_{i,j}$ that relative distances d of (6) in slot t and $t+1$ are i and j, respectively. For example, to calculate $\alpha_{0,1}$, we set $x_1 = 0$, $x_2 = R_1$, $x_3 = R_1$ and $x_4 = R_1 + v$.

Define $A_{a,b}$ as the joint CCDF that distances d_t and d_{t+1} are respectively no less than a and b when the number of WCSs m is one, which is expressed as the sum of $\alpha_{i,j}$, i.e., $A_{a,b} = \sum_{i=a}^{M} \sum_{j=b}^{M} \alpha_{i,j}$. Noting that each location of WCSs is independent, we derive $P_{i,j}$ in terms of $A_{a,b}$ as follows:

- If $i = 0$,

$$
P_{0,j} = \begin{cases}
1 - \dfrac{A_{0,1}{}^m - A_{1,1}{}^m}{1 - \left(1 - \frac{\pi R_1{}^2}{S}\right)^m}, & \text{if } j = 0, \\[3ex]
\dfrac{A_{0,1}{}^m - A_{1,1}{}^m}{1 - \left(1 - \frac{\pi R_1{}^2}{S}\right)^m}, & \text{if } j = 1, \\[3ex]
0, & \text{otherwise.}
\end{cases}
$$

- If $0 < i < M$,

$$
P_{i,j} = \begin{cases}
1 - \dfrac{A_{i,i}{}^m - A_{i+1,i}{}^m}{\left\{1 - \frac{\pi(R_1 + (i-1)v)^2}{S}\right\}^m - \left\{1 - \frac{\pi(R_1 + iv)^2}{S}\right\}^m}, & \text{if } j = i - 1, \\[4ex]
\dfrac{A_{i,i}{}^m - A_{i+1,i}{}^m - A_{i+1,i}{}^m + A_{i+1,i+1}{}^m}{\left\{1 - \frac{\pi(R_1 + (i-1)v)^2}{S}\right\}^m - \left\{1 - \frac{\pi(R_1 + iv)^2}{S}\right\}^m}, & \text{if } j = i, \\[4ex]
\dfrac{A_{i,i+1}{}^m - A_{i+1,i+1}{}^m}{\left\{1 - \frac{\pi(R_1 + (i-1)v)^2}{S}\right\}^m - \left\{1 - \frac{\pi(R_1 + iv)^2}{S}\right\}^m}, & \text{if } j = i + 1, \\[4ex]
0, & \text{otherwise.}
\end{cases}
$$

- If $i = M$,

$$
P_{M,j} = \begin{cases}
\dfrac{A_{M,M-1}{}^m - A_{M,M}{}^m}{\left(1 - \frac{\pi(R_1 + Mv)^2}{S}\right)^m}, & \text{if } j = M - 1, \\[3ex]
\dfrac{A_{M,M}{}^m}{\left(1 - \frac{\pi(R_1 + Mv)^2}{S}\right)^m}, & \text{if } j = M, \\[3ex]
0, & \text{otherwise.}
\end{cases}
$$

Appendix A.2. Derivation of Charging Probability p_c (9)

Given that there are h WCSs within R_1 from a node, the probability that the node is charged by one of the WCSs $p_c(h)$ is

$$
p_c(h) = 1 - \left[1 - \sum_{\ell=0}^{n-1} \min\left[1, \frac{u}{\ell+1}\right] f\left(\ell; n-1, \frac{\pi R_1{}^2}{S}\right) \right]^h = 1 - \Gamma^h, \tag{A4}
$$

where $f(n; k, p)$ is the probability density function of the binomial distribution with parameters n, k and p, and

$$
\Gamma = 1 - F\left(u - 2; n - 1, \frac{\pi R_1{}^2}{S}\right) - \frac{u\left(1 - F\left(u - 1; n, \frac{\pi R_1{}^2}{S}\right)\right)}{n\pi R^2}. \tag{A5}
$$

The probability that there are h WCSs within R_1 from a node is $f\left(h; m, \frac{\pi R_1^2}{S}\right)$. Therefore, the charging probability p_c is

$$p_c = \frac{\sum_{h=1}^{m} p_c(h) f\left(h; m, \frac{\pi R_1^2}{S}\right)}{1 - \left(1 - \frac{\pi R_1^2}{S}\right)^m}. \tag{A6}$$

The denominator of (A6) represents the probability that the node is in one of the WCS's charging coverage. After substituting (A4) into (A6), the charging probability p_c becomes

$$
ccp_c = 1 - \frac{\sum_{h=1}^{m} \binom{m}{h} \left(A \frac{\pi R_1^2}{S}\right)^h \left(1 - \frac{\pi R_1^2}{S}\right)^{m-h}}{1 - \left(1 - \frac{\pi R_1^2}{S}\right)^m} = \frac{1 - \sum_{h=0}^{m} \binom{m}{h} \left(A \frac{\pi R_1^2}{S}\right)^h \left(1 - \frac{\pi R_1^2}{S}\right)^{m-h}}{1 - \left(1 - \frac{\pi R_1^2}{S}\right)^m}
$$
$$
= \frac{1 - \left(A \frac{\pi R_1^2}{S} + 1 - \frac{\pi R_1^2}{S}\right)^m}{1 - \left(1 - \frac{\pi R_1^2}{S}\right)^m}. \tag{A7}
$$

Plugging (A5) into (A7) completes the proof.

Appendix A.3. Proof of Proposition 1

According to [42], the CCDF of T_I is $\Pr\left[T_I > t\right] = p_0 P^{t-1} \mathbf{1}$. Assume that matrix P (16) is invertible (It is a reasonable assumption that the transition probability, expressed as a row vector in P, is independent of the current location status d (6) unless speed is infinite, and P is likely to be a full rank matrix guaranteeing the existence of M eigenvalues. It is also verified numerically under numerous combinations of parameter settings.), and it can be diagonalized as follows:

$$
P = VDV^{-1} = \begin{pmatrix} a_1 & a_2 & \cdots & a_M \end{pmatrix} \begin{pmatrix} \lambda_1 & 0 & \cdots & 0 \\ 0 & \lambda_2 & \cdots & 0 \\ \vdots & \vdots & \ddots & \vdots \\ 0 & 0 & \cdots & \lambda_M \end{pmatrix} \begin{pmatrix} b_1^T \\ b_2^T \\ \vdots \\ b_M^T \end{pmatrix}. \tag{A8}
$$

Therefore, P^{t-1} is

$$
P^{t-1} = V(D)^{t-1} V^{-1} = \sum_{i=1}^{M} G_i \cdot (\lambda_i)^{t-1}, \tag{A9}
$$

where $G_i = a_i b_i^T$ are $M \times M$ matrices of which the sum is an identity matrix $\left(\sum_{i=1}^{M} G_i = I\right)$. From (A9), $\Pr\left[T_I > t\right]$ is given as

$$
\Pr\left[T_I > t\right] = \sum_{i=1}^{M} p_0 G_i \mathbf{1} \lambda_i^{t-1} = \sum_{i=1}^{M} \gamma_i \cdot \lambda_i^{t-1}.
$$

Matrix P (16) is called a sub-stochastic matrix because every row sum is 1 except the first one due to a strictly positive transition probability $P_{1,0}$. Noting that every eigenvalue of an irreducible sub-stochastic matrix is less than 1, λ_i should be smaller than one for every i.

Appendix A.4. Comparison with Practical Mobility Models

We measure the inter-meeting time of *Brownian motion* (BM) and *random way point* (RWP), which is shown in Figure A1 that the analytic result tends to overestimate the frequency of long inter-meeting than the practical models due to its constant moving speed. Nevertheless, the overall tendencies are quite similar especially when speed becomes high.

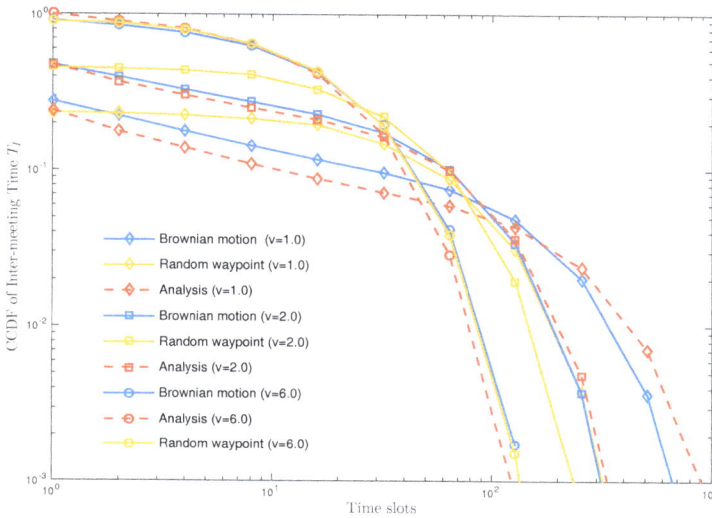

Figure A1. The CCDF of inter-meeting time of practical mobility models. The same parameter setting as in Figure 3 is used unless specified.

Appendix A.5. Proof of Proposition 2

As $L \to \infty$, this Markov process (12) becomes *batch Markovian arrival process* (BMAP), of which the stochastic process can be described by the mean steady-state arrival rate $\bar{\lambda}$ derived as follows. First, an infinite generator D is given as

$$
D = B_0 + A_2 + A_3 + \cdots + A_{E-1} = \begin{pmatrix}
-P_{0,1} & P_{0,1} & 0 & \cdots & 0 \\
P_{1,0} & -P_{1,0} - P_{1,2} & P_{1,2} & \cdots & 0 \\
0 & P_{2,1} & -P_{2,1} - P_{2,3} & \cdots & 0 \\
\vdots & \vdots & \vdots & \ddots & \vdots \\
0 & 0 & 0 & \cdots & -P_{M,M-1}
\end{pmatrix}.
\tag{A10}
$$

Let ϕ_k denote steady state probability that the relative distance d is k. We make the following row vector ϕ as

$$
\phi = \left[\phi_0, \phi_1, \phi_2, \cdots, \phi_M \right] = \left[\sum_{j=0}^{L} \pi_{0,j}, \sum_{j=0}^{L} \pi_{1,j}, \sum_{j=0}^{L} \pi_{2,j}, \cdots, \sum_{j=0}^{L} \pi_{M,j} \right],
$$

which can be derived by solving the following equations.

$$
\phi D = 0, \quad \phi 1 = 1.
\tag{A11}
$$

From (A11), we have

$$\phi_k = \begin{cases} 1-\left(1-\frac{\pi R_1^2}{S}\right)^m, & \text{if } k = 0, \\ \left(1-\frac{\pi(R_1+Mv)^2}{S}\right)^m, & \text{if } k = M, \\ \left\{1-\frac{\pi(R_1+(k-1)v)^2}{S}\right\}^m - \left\{1-\frac{\pi(R_1+kv)^2}{S}\right\}^m, & \text{otherwise.} \end{cases} \tag{A12}$$

Finally, we derive the mean steady-state arrival rate $\bar{\lambda}$ from (A12),

$$\bar{\lambda} = \phi\left(\sum_{k=1}^{E} kA_{k+1}\right)\mathbf{1} = p_c\left\{1-\left(1-\frac{\pi R_1^2}{S}\right)^m\right\}\left(\sum_{k=1}^{E} k\beta(k)\right).$$

If $\bar{\lambda} < p_t$, the active probability P_{on} is

$$P_{\text{on}} = \frac{\bar{\lambda}}{p_t} = \frac{p_c}{p_t}\left\{1-\left(1-\frac{\pi R_1^2}{S}\right)^m\right\}\left(\sum_{k=1}^{E} k\beta(k)\right). \tag{A13}$$

Otherwise, $P_{\text{on}} = 1$. After inserting (A13) into (3), we complete the proof.

Appendix A.6. Proof of Proposition 3

For the first step, we check the ratio $\frac{p_c}{p_t}$ as the number n increases,

$$\frac{p_c}{p_t} \approx \frac{1-\left(1-\frac{u}{n}\right)^m}{q\left(1-(1-\frac{\pi R_1^2}{S})^m\right)\left(1-e^{-\frac{\pi}{4}(1-q)}\right)} = \begin{cases} \frac{m}{n}\frac{u}{q\left(1-(1-\frac{\pi R_1^2}{S})^m\right)\left(1-e^{-\frac{\pi}{4}(1-q)}\right)} & \text{if } m = O(n), \\ \frac{1}{q\left(1-(1-\frac{\pi R_1^2}{S})^m\right)\left(1-e^{-\frac{\pi}{4}(1-q)}\right)}, & \text{otherwise.} \end{cases} \tag{A14}$$

We already proved the upper bound in Proposition 2 (19) as follows:

$$\Lambda \le \Lambda_{\text{upper}} = \Theta\left(\min\left(1,\frac{m}{n}\right)c_1^{\min\left(1,\frac{m}{n}\right)}\right), \tag{A15}$$

where $c_1 = e^{-\frac{\pi \cdot u}{4 \cdot a}\left(\sum_{k=1}^{E} k\beta(k)\right)}$.

In order to derive the lower bound, consider that each WCS only delivers one unit of energy to a node in a slot ($E = 1$). Since the submatrices A_2 and A_3 in the generating matrix Q (12) is null matrices, the Markov process becomes finite *Quasi-Birth-Death* (QBD) Process. In [43], the authors showed that the steady state probability vector π_k (10) of finite QBD can be expressed in a matrix geometric form:

$$\pi_k = v_1 R_1^k + v_2 R_2^{L-k}. \tag{A16}$$

Here, matrices R_1 and R_2 are

$$R_1 = -A_2(A_1+\eta A_0)^{-1}, \tag{A17}$$
$$R_2 = -A_0(A_1+A_0 G)^{-1}, \tag{A18}$$

where η is the spectral radius of R_1, and G is the square matrix that the every element of the first column is one and the others are zero. Detailed derivations of R_1 and R_2 are in [44]. Row vectors v_1 and v_2 satisfy the following conditions:

$$
\begin{bmatrix} v_1 \ v_2 \end{bmatrix}
\begin{bmatrix}
B_1+R_1A_0 & R_1{}^{L-1}(A_0+R_1(A_0+B_0)) \\
R_2{}^{L-1}(R_2B_0+A_0) & A_0+A_1+R_2A_0
\end{bmatrix} = 0,
\tag{A19}
$$

$$
\left(v_1 \sum_{i=0}^{L} R_1{}^{i} + v_1 \sum_{i=0}^{L} R_1{}^{i} \right) \mathbf{1} = 1.
\tag{A20}
$$

The boundary condition (A19) is derived by inserting (A16) into the first and last columns of the balance Equation (11), and condition (A20) means the summation of the entire steady state probabilities is one:

$$
X_1 = \begin{vmatrix}
\frac{p_t}{p_c}\frac{P_{0,1}+p_c}{P_{1,0}}+1 & \frac{p_t}{p_c}\frac{P_{0,1}+p_c}{P_{1,0}}\frac{P_{1,2}}{P_{2,1}} \\
\frac{p_t}{p_c}\frac{P_{0,1}+p_c}{P_{1,0}}+1 & \frac{p_t}{p_c}\frac{P_{0,1}+p_c}{P_{1,0}}\frac{P_{1,2}}{P_{2,1}}+\frac{p_t}{p_{2,1}}+1
\end{vmatrix}, \quad
X_2 = \begin{vmatrix}
\frac{p_t}{p_c}\frac{P_{0,1}}{P_{1,0}} & \frac{p_t}{p_c}\frac{P_{0,1}}{P_{1,0}}\frac{P_{1,2}}{P_{2,1}} \\
\frac{p_t}{p_c}\frac{P_{0,1}+p_c}{P_{1,0}}+1 & \frac{p_t}{p_c}\frac{P_{0,1}+p_c}{P_{1,0}}\frac{P_{1,2}}{P_{2,1}}+\frac{p_c}{P_{2,1}}+1
\end{vmatrix},
$$

$$
X_3 = \begin{vmatrix}
\frac{p_t}{p_c}\frac{P_{0,1}}{P_{1,0}} & \frac{p_t}{p_c}\frac{P_{0,1}}{P_{1,0}}\frac{P_{1,2}}{P_{2,1}} \\
\frac{p_t}{p_c}\frac{P_{0,1}+p_c}{P_{1,0}}+1 & \frac{p_t}{p_c}\frac{P_{0,1}+p_c}{P_{1,0}}\frac{P_{1,2}}{P_{2,1}}
\end{vmatrix},
\tag{A21}
$$

where $\begin{vmatrix} a & b \\ c & d \end{vmatrix} = ad-bc$.

From the Equation (A16), the active probability P_{on} (13) is rewritten as follows:

$$
P_{\mathrm{on}} = 1-(v_1+v_2R_2{}^{L})\mathbf{1}.
\tag{A22}
$$

Since the active probability P_{on} is a non-decreasing function of the battery capacity L, we make the following inequality condition:

$$
P_{\mathrm{on}} \geq 1-(v_1+v_2R_2)\mathbf{1} = \left(v_1 \sum_{i=0}^{1} R_1{}^{i} + v_1 \sum_{i=0}^{1} R_1{}^{i} \right)\mathbf{1} - (v_1+v_2R_2)\mathbf{1} = (v_1R_1+v_2)\mathbf{1}.
\tag{A23}
$$

All elements in matrix R_1 is zero except the first row, and the first element of v_1 is zero. Therefore, v_1R_1 becomes zero and the above inequality (A23) becomes

$$
P_{\mathrm{on}} \geq v_2\mathbf{1} = \sum_{i=0}^{M} v_{2,i} \geq v_{2,0}.
\tag{A24}
$$

From the condition (A19), we make the following relations between v_1 and v_2,

$$
v_1 = -v_2\left(R_2+p_tB_0{}^{-1}\right),
\tag{A25}
$$

$$
v_1+v_2(I+R_2) = \phi.
\tag{A26}
$$

After inserting (A25) into (A26), we derive v_2 as

$$
v_2 = \phi\left(I-p_tB_0{}^{-1}\right)^{-1}.
\tag{A27}
$$

According to numerical verifications, it is checked that $v_{2,0}$ is an increasing function of the resolution factor M. When $M = 3$, the vector v_2 becomes:

$$v_{2,0} = \phi \frac{1}{\left(\frac{p_t}{p_c}+1\right)X_1 - \frac{p_t}{p_c}X_2 + \frac{p_t}{p_c}X_3} \begin{pmatrix} X_1 \\ -X_2 \\ X_3 \end{pmatrix} = \frac{\phi_1 X_1 - \phi_2 X_2 + \phi_3 X_3}{\left(\frac{p_t}{p_c}+1\right)X_1 - \frac{p_t}{p_c}X_2 + \frac{p_t}{p_c}X_3} > \frac{p_c}{p_t}\frac{\left(\phi_1 - \phi_2 \frac{X_2}{X_1} + \phi_3 \frac{X_3}{X_1}\right)}{2}, \quad (A28)$$

where X_1, X_2 and X_3 are described in (A21). As n increases, the ratios $\frac{X_2}{X_1}$ and $\frac{X_2}{X_1}$ reduce to zero. From the inequalities (A24) and (A28), the active probability P_{on} is

$$P_{\text{on}} > \frac{p_c}{p_t}\frac{\phi_1}{2} = \frac{p_c}{p_t}\frac{\left\{1 - \left(1 - \frac{\pi R_1^2}{S}\right)^m\right\}}{2}. \quad (A29)$$

We can derive the lower bound of the throughput Λ as

$$\Lambda > \Lambda_{\text{lower}} = \Theta\left(\min\left(1, \frac{m}{n}\right)c_2^{\min\left(1, \frac{m}{n}\right)}\right), \quad (A30)$$

where $c_2 = e^{-\frac{\pi \cdot u}{8 \cdot a}}$. From the upper bound (A15) and lower bound (A30), we complete the proof.

References

1. Zanella, A.; Bui, N.; Castellani, A.L.V.; Zorzi, M. Internet of Things for Smart Cities. *IEEE Inter. Things J.* **2014**, *1*, 22–32. [CrossRef]
2. Lin, J.C. Wireless power transfer for mobile applications and health effect. *IEEE Antennas Propag. Mag.* **2013**, *55*, 250–253. [CrossRef]
3. Huang, K.; Zhou, X. Cutting the last wires for mobile communications by microwave power transfer. *IEEE Commun. Mag.* **2015**, *53*, 86–93. [CrossRef]
4. Guo, B.; Zhang, D.; Wang, Z.; Yu, Z.; Zhou, X. Opportunistic IoT: Exploring the harmonious interaction between human and the internet of things. *J. Netw. Comput. Appl.* **2013**, *36*, 1531–1539. [CrossRef]
5. Atzori, L.; Iera, A.; Morabito, G. From "smart objects" to "social objects": The next evolutionary step of the internet of things. *IEEE Commun. Mag.* **2014**, *52*, 97–105. [CrossRef]
6. Tran, H.V.; Kaddoum, G. Green Cell-less Design for RF-Wireless Power Transfer Networks. In Proceedings of the Wireless Communications and Networking Conference (WCNC), Barcelona, Spain, 16–18 April 2018.
7. Tran, H.V.; Kaddoum, G. RF wireless power transfer: Regreening future networks. *IEEE Potential* **2018**, *2*, 35–41. [CrossRef]
8. Zeng, Y.; Clerckx, B.; Zhang, R. Communications and signal design for wireless power transmission. *IEEE Trans. Commun.* **2017**, *65*, 2264–2290. [CrossRef]
9. Kurs, A.; Karalis, A.; Moffatt, R.; Joannopoulos, J.D.; Fisher, P.; Soljacic, M. Wireless power transfer via strongly coupled magnetic resonances. *Science* **2007**, *317*, 83–86. [CrossRef] [PubMed]
10. Sample, A.P.; Meyer, D.A.; Smith, J.R. Analysis, experimental results, and range adaption of magnetically coupled resonators for wireless power transfer. *IEEE Trans. Ind. Electron.* **2011**, *58*, 544–554. [CrossRef]
11. Huang, Y.; Shinohara, N.; Mitani, T. Impedance Matching in Wireless Power Transfer. *IEEE Trans. Microw. Theory Tech.* **2017**, *65*, 582–590. [CrossRef]
12. Kim, S.J.; Kwon, U.K.; Yoon, S.K.; Tarokh, V. Near Field Resonator Isolation System: Theory to Implementation. *IEEE Trans. Circuits Syst. I Reg. Pap.* **2013**, *60*, 1175–1187. [CrossRef]
13. Huang, K.; Lau, V.K.N. Enabling Wireless Power Transfer in Cellular Networks: Architecture, Modeling and Deployment. *IEEE Trans. Wirel. Commun.* **2014**, *13*, 902–912. [CrossRef]
14. Zhu, G.; Ko, S.-W.; Huang, K. Inference from randomized transmissions by many backscatter sensors. *IEEE Trans. Wirel. Commun.* **2018**, *17*, 3111–3127. [CrossRef]
15. Lyu, B.; Yang, Z.; Gui, G.; Sari, H. Optimal Time Allocation in Backscatter Assisted Wireless Powered Communication Networks. *Sensors* **2017**, *17*, 1258.

16. Kang, X.; Ho, C.K.; Sun, S. Full-Duplex Wireless-Powered Communication Network With Energy Causality. *IEEE Trans. Wirel. Commun.* **2015**, *14*, 5539–5551. [CrossRef]
17. Tran, H.-V.; Kaddoum, G.; Shin, O. Joint Channel Resources Allocation and Beamforming in Energy Harvesting Systems. *IEEE Wirel. Commun. Lett.* **2018**. [CrossRef]
18. Clerckx, B.; Bayhuzina, E.; Yates, D.; Mitcheson, P.D. Waveform Optimization for wireless power transfer with nonlinear energy harvester modeling. In Proceedings of the 2015 International Symposium on Wireless Communication Systems (ISWCS), Brussels, Belgium, 25–28 August 2015.
19. Tran, H.-V.; Kaddoum, G.; Truong, K.T. Resource allocation in SWIPT networks under a non-linear energy harvesting model: Power efficiency, user fairness, and channel non-reciprocity. *arXiv* **2018**, arXiv:1806.07926.
20. Clerckx, B.; Bayguzina, E. A low-complexity adaptive multisine waveform design for wireless power transfer. *IEEE Antennas Wirel. Propag. Lett.* **2017**, *16*, 2207–2210. [CrossRef]
21. Xie, L.; Shi, Y.; Hou, Y.T.; Sherali, H.D. Making sensor networks immoral: An energy-renewal approach with wireless power transfer. *IEEE/ACM Trans. Netw.* **2012**, *20*, 1748–1761. [CrossRef]
22. Fu, L.; Cheng, P.; Gu, Y.; Chen, J.; He, T. Optimal charging in wireless rechargeable sensor networks. *IEEE Trans. Veh. Technol.* **2016**, *65*, 278–291. [CrossRef]
23. Dai, H.; Liu, Y.; Chen, G.; Wu, X.; He, T.; Liu, A.X.; Ma, H. Safe charging for wireless power transfer. *Proc. IEEE* **2017**, *25*, 3531–3544. [CrossRef]
24. Sangare, F.; Xiao, Y.; Niyato, D.; Han, Z. Mobile Charging in Wireless-Powered Sensor Networks: Optimal Scheduling and Experimental Implementation. *IEEE Trans. Veh. Technol.* **2017**, *66*, 7400–7410. [CrossRef]
25. Madhja, A.; Nikoletseas, S.; Raptis, T.P. Distributed wireless power transfer in sensor networks with multiple mobile chargers. *Comput. Netw.* **2015**, *80*, 89–108. [CrossRef]
26. Li, K.; Luan, H.; Shen, C.-C. Qi-Ferry: Energy-constrained wireless charging in wireless sensor networks. In Proceedings of the 2012 IEEE Wireless Communications and Networking Conference (WCNC), Paris, France, 14 April 2012; pp. 573–577.
27. He, S.; Chen, J.; Jiang, F.; Yau, D.K.Y.; Xing, G.; Sun, Y. Energy provisioning in wireless rechargeable sensor networks. *IEEE Trans. Mob. Comput.* **2013**, *12*, 1931–1942. [CrossRef]
28. Huang, K. Spatial throughput of mobile ad hoc network with energy harvesting. *IEEE Trans. Inf. Theory* **2013**, *59*, 7597–7612. [CrossRef]
29. Niyato, D.; Pink, T.H.; Saad, W.; Kim, D. Cooperation in delay tolerant networks with wireless energy transfer: Performance analysis and optimization. *IEEE Trans. Veh. Technol.* **2015**, *64*, 3740–3754. [CrossRef]
30. Niyato, D.; Wang, P. Delay-limited communications on mobile node with wireless energy harvesting: Performance analysis and optimization. *IEEE Trans. Veh. Technol.* **2014**, *63*, 1870–1885. [CrossRef]
31. Kim, J.; Park, J.; Ko, S.-W.; Kim, S.-L. User attraction via wireless charging in downlink cellular networks. In Proceedings of the 2016 14th International Symposium on Modeling and Optimization in Mobile, Ad Hoc, and Wireless Networks (WiOpt), Tempe, AZ, USA, 9–13 May 2016.
32. Dai, H.; Chen, G.; Wang, C.; Wang, S.; Wu, X.; Wu, F. Quality of energy provisioning for wireless power transfer. *IEEE Trans. Parallel Distrib. Syst.* **2015**, *26*, 527–537. [CrossRef]
33. Ko, S.-W.; Yu, S.M.; Kim, S.-L. The capacity of energy-constrained mobile networks with wireless power transfer. *IEEE Commun. Lett.* **2013**, *17*, 529–532. [CrossRef]
34. Landström, S; Bergström, J.; Westerberg, E.; Hammarwall, D. NB-IoT: A sustainable technology for connecting billions of devices. *Ericsson Technol. Rev.* **2016**, *4*, 2–11.
35. Grossglauser, M.; Tse, D.N.C. Mobility increases the capacity of ad hoc wireless networks. *IEEE/ACM Trans. Netw.* **2012**, *10*, 477–486. [CrossRef]
36. Kesler, M. *Highly Resonant Wireless Power Transfer: Safe, Efficient, and Over Distance*; Witricity: Water Town, MA, USA, 2013.
37. Cannon, B.; Hoburg, J.; Stancil, D.; Goldstein, S. Magnetic resonant coupling as a potential means for wireless power transfer to multiple small receivers. *IEEE Trans. Power Electron.* **2009**, *27*, 1819–1825. [CrossRef]
38. Fu, M.; Zhang, T.; Ma, C.; Zhu, X. Efficiency and optimal loads analysis for multiple-receiver wireless power transfer systems. *IEEE Trans. Microw. Theory Tech.* **2015**, *63*, 801–812. [CrossRef]
39. Xie, L.; Shi, Y.; Hou, Y.T.; Sherali, H.D.; Midkiff, S.F. On renewable sensor networks with wireless energy transfer: The multi-node case. In Proceedings of the 2012 9th annual IEEE communications society conference on Sensor, mesh and ad hoc communications and networks (SECON), Seoul, Korea, 18–21 August 2012.

40. Kurs, A.; Moffatt, R.; Soljacic, M. Simultaneous mid-range power transfer to multiple devices. *Appl. Phys. Lett.* **2010**, *96*, 044102-1–044102-3. [CrossRef]

41. Shim, T.; Park, J.; Ko, S.-W.; Kim, S.-L.; Lee, B.; Choi, J. Traffic convexity aware cellular networks: A vehicular heavy user perspective. *IEEE Wirel. Commun.* **2016**, *23*, 88–94. [CrossRef]

42. Meyer, C.D. *Matrix Analysis and Applied Linear Algebra*; Society for Industrial and Applied Mathematics: Philadelphia, CA, USA, 2000.

43. Akar, N.; Sohraby, K. Finite and infinite QBD chains: A simple and unifying algorithm approach. In Proceedings of the 1997 IEEE International Performance, Computing and Communications Conference, Scottsdale/Phoenix, AZ, USA, 5–7 February 1997; pp. 1105–1113.

44. Ramswami, V.; Latouche, G. A general class of Markov process with explicit matrix-geometric solutions. *Oper. Res. Spektrum* **1986**, *9*, 209–218. [CrossRef]

![sensors logo] *sensors*

MDPI

Article

Closed-Form Expression for the Symbol Error Probability in Full-Duplex Spatial Modulation Relay System and Its Application in Optimal Power Allocation

Le Van Nguyen [1], Ba Cao Nguyen [1], Xuan Nam Tran [1] and Le The Dung [2,3,*,†]

[1] Advanced Wireless Communications Group, Le Quy Don Technical University, Hanoi 11917, Vietnam; nguyenlevan2211@gmail.com (L.V.N.); bacao.sqtt@gmail.com (B.C.N.); namtx@mta.edu.vn (X.N.T.)
[2] Division of Computational Physics, Institute for Computational Science, Ton Duc Thang University, Ho Chi Minh City 758307, Vietnam
[3] Faculty of Electrical and Electronics Engineering, Ton Duc Thang University, Ho Chi Minh City 758307, Vietnam
* Correspondence: lethedung@tdtu.edu.vn
† Current address: Institute for Computational Science, Ton Duc Thang University, 19 Nguyen Huu Tho Street, Tan Phong Ward, District 7, Ho Chi Minh City 758307, Vietnam.

Received: 26 October 2019; Accepted: 3 December 2019; Published: 6 December 2019

Abstract: Full-duplex (FD) communication and spatial modulation (SM) are two promising techniques to achieve high spectral efficiency. Recent works in the literature have investigated the possibility of combining the FD mode with SM in the relay system to benefit their advantages. In this paper, we analyze the performance of the FD-SM decode-and-forward (DF) relay system and derive the closed-form expression for the symbol error probability (SEP). To tackle the residual self-interference (RSI) due to the FD mode at the relay, we propose a simple yet effective power allocation algorithm to compensate for the RSI impact and improve the system SEP performance. Both numerical and simulation results confirm the accuracy of the derived SEP expression and the efficacy of the proposed optimal power allocation.

Keywords: spatial modulation; multiple-input multiple-output; full-duplex; self-interference cancellation; symbol error probability

1. Introduction

Full-duplex (FD) communication and spatial modulation (SM) are two promising techniques to increase the spectral efficiency of wireless systems [1–3]. Theoretically, an FD communication system can double the spectral efficiency as its transceivers can transmit and receive signals at the same time and on the same frequency [4,5]. Besides increasing the channel capacity, the FD communication systems can also reduce the end-to-end and feedback delay, improve network security and solve the hidden terminal problem. Therefore, the FD mode has found its applications in various wireless systems such as sensor networks, massive MIMO , relaying networks and possibly future wireless networks such as the fifth-generation (5G) and beyond. Unfortunately, the residual self-interference (RSI) due to imperfect self-interference cancellation (SIC) limits the capacity and performance of the FD communication systems [6]. In the literature, numerous solutions such as relay selection scheme and adaptive transmission [7,8] and optimal power allocation [9–11] were proposed to reduce the impact of the RSI and improve the performance of the FD relay communication system. These solutions significantly improved the capacity, outage performance and energy efficiency of the single-input single-output (SISO) systems.

Meanwhile, the multiple-input multiple-output (MIMO) transmission is another solution that can also achieve high capacity and better performance over the SISO systems [12]. However, the hardware of the MIMO system is more complex, that is, more radio frequency (RF) chains, as both transmitter and receiver user multiple antennas. The MIMO receiver also requires a high-complexity detector to deal with the inter-channel interference (ICI) due to simultaneous transmissions on the same frequency from different antennas. In that context, SM can tackle these issues correctly as it activates only one transmit antenna for transmission and requires only low-complexity iterative maximal-ratio combining (MRC) at the receiver for signal estimation [12–15]. Therefore, the combined SM and FD system achieves higher spectral efficiency while reducing the complexity requirement at the transceiver, particularly of the relay node.

In the literature, using FD relay in SM systems has attracted great interest because this integrated FD-SM system has the advantages of both spectral efficiency improvement and performance enhancement. Numerous works focused on analyzing the system performance in terms of the outage probability (OP) [2], bit error rate (BER) [3], average symbol error probability (SEP), ergodic capacity [16] and average BER [17,18]. Specifically, in Reference [2], the lower and upper bounds of the OP of the SM-MIMO system with decode-and-forward (DF) FD relay were derived over cascaded $\alpha - \mu$ fading channels. It also demonstrated that the RSI had a strong impact on the OP performance of the system. The work in Reference [3] considered the SM-MIMO system with amplify-and-forward (AF) FD/half-duplex (HD) relaying. It successfully derived a new unified tight upper-bound for the system BER. The results of the paper indicate that the SM-MIMO-FD relay system can improve the BER and the spectral efficiency if suitable SIC techniques are applied. Under the same assumption of the RSI, References [16–18] investigated the SM-MIMO-FD relay systems which can exploit the benefits of the FD transmission mode. The approximate expressions of SEP ([16]) and BER ([17,18]) were also derived for performance evaluation.

Although the previous works conducted various performance analyses, their results were limited to either upper and lower bounds or approximate expressions but not the exact closed-form expressions of SEP and BER. Therefore, it is required to have exact mathematical expressions for the performance evaluation rather than the upper bound or approximate ones for better understanding the system behaviors. Moreover, since the FD mode significantly degrades the system performance, besides effective SIC techniques, there should be other solutions such as power allocation to compensate for this degradation. Motivated by the previous works, in this paper, we aim to derive the exact closed-form expression of the SEP of the SM-FD relay system with DF protocol applied at the relay to enhance the system performance. Based on the derived expression of SEP, we can determine the optimal power allocation for the FD relay to reduce the RSI impact on the SEP performance. So far, this is the first work that successfully derives the exact closed-form expression of SEP and use it for optimizing the power allocation for SM-MIMO-FD relay systems. The main contributions of this paper can be summarized as follows:

- We analyze the SM-MIMO-FD relay system where SM is used at the source and the relay nodes under the impact of the RSI caused by the imperfect SIC. Unlike previous works, we derive the exact closed-form expression of SEP for the system over the Rayleigh fading channel.
- We propose an algorithm to calculate the optimal transmission power of the FD relay. Based on this algorithm, we obtain the optimal power allocation for the considered system. The proposed optimal transmission power algorithm significantly improves the SEP performance, especially in the low SNR region. Additionally, using the derived expression of SEP, we can also examine the influences of the number of transmitting/receiving antennas and the RSI on the system performance in the case with and without optimal power allocation.

The rest of this paper is organized as follows. Section 2 presents the system model. Section 3 provides the detailed derivations of the closed-form expression of SEP. The optimal power allocation algorithm for the FD relay is developed in Section 3. Section 4 presents the analytical and simulation results for performance evaluation. Finally, Section 5 concludes the paper.

2. System Model

We consider a single-user single-carrier SM-MIMO-FD relay system with a source node S, which transmits its signal to a destination node D via a relay node R as shown in Figure 1. Particularly, S and D operate in the half-duplex (HD) mode with N_S transmitting antennas and N_D receiving antennas, respectively. The relay node operates in the FD mode with N_{tR} transmitting antennas and N_{rR} receiving antennas.

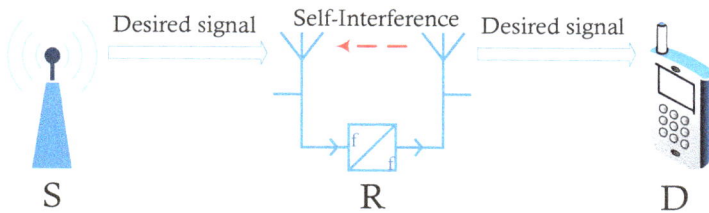

Figure 1. System model of the considered SM-MIMO-FD relay system.

At time slot t, the received signal at R can be calculated as

$$\mathbf{y}_R(t) = \sqrt{P_S}\mathbf{h}_i^R x_S(t) + \sqrt{P_R}\mathbf{h}_j^R x_R(t) + \mathbf{z}_R(t), \tag{1}$$

where $\mathbf{h}_i^R$ and $\mathbf{h}_j^R$ are respectively the channel vector from the ith active antenna of S to N_{rR} receiving antennas of R and from the jth active antenna of R to all reception antennas of R. These channels are assumed to undergo flat Rayleigh fading, which can be modeled by independent and identically distributed complex Gaussian random variables with zero mean and unit variance. x_S and x_R are the transmitted signals at the ith antenna of S and the jth antenna of R, respectively; P_S and P_R are the average transmission power at S and R, respectively; $\mathbf{z}_R$ is the noise vector whose elements follow a complex Gaussian distribution with variance σ^2.

At the FD relay, we assume that the transmitting and receiving antennas are both directional, thus there will be no direct link which causes the self-interference (SI) from the transmit to the receive antenna. This SI is mainly due to reflections caused by multipath propagation. We also assume that the system can use all SIC techniques in the three domains, that is, propagation, analog and digital, to remove the SI [19,20]. Specifically, R can use all available isolation techniques to suppress SI. It can use the cross-polarization transmission to isolate the transmitting and receiving antennas [19,21]. In the analog domain, thanks to the SI awareness of analog circuits, the transmitted signal via the transmit antenna is collected and then subtracted from the received signal. The RSI is then converted to the digital domain for further SIC via digital signal processing. As R knows its transmitted signal, it can subtract the SI from the received signal by using SI channel estimation [20,22]. Thanks to all these SIC techniques, the relay node can achieve up to 110 dB SI suppression [23]. Moreover, since the SI is canceled from the received signal in the analog and digital domain by reconstructing the SI signal, the RSI is in effect the resulted errors due to the imperfect reconstruction or more correctly, the imperfect SI channel estimation. Moreover, as the digital-domain cancellation is done after a quantization operation, RSI at the relay $\mathbf{r}_{SI}$ can be modeled using complex Gaussian random variable [4,17,20,23] with zero mean and variance of σ_{RSI}^2, that is $\sigma_{RSI}^2 = \tilde{\Omega}P_R$ where $\tilde{\Omega}$ denotes the SIC capability at the relay. Therefore, the received signal at R can be rewritten from (1) as

$$\mathbf{y}_R(t) = \sqrt{P_S}\mathbf{h}_i^R x_S(t) + \mathbf{r}_{SI}(t) + \mathbf{z}_R(t),\tag{2}$$

and the received signal at the destination D is then given by

$$\mathbf{y}_D(t) = \sqrt{P_R}\mathbf{h}_j^D x_R(t) + \mathbf{z}_D(t),\tag{3}$$

where $\mathbf{h}_j^D$ is the channel vector from the jth antenna of R to N_D receiving antennas of D; $\mathbf{z}_D$ is the noise vector at D. Both S and R are assumed to have the same N_t transmitting antennas for the same expected spectral efficiency. In the SM system, to estimate the transmitted bits, the receiver needs to use joint ML detection for both the activated transmit antenna and the M-ary modulated symbols. This joint detector is computationally complex, especially for the SM system with a large number of antennas. Near-ML low-complexity detectors such as those summarized in Reference [24] or that for the index modulation in the frequency domain [25] are more favorable for practical implementation. In this paper, as we are interested in analyzing the effect of the RSI due to the FD mode on the system performance, we assume that the receivers of both R and D can perfectly estimate the transmitted antenna index of the respective transmitters for the ML detection.

From (2) and (3), we can calculate the instantaneous signal-to-interference-plus-noise-ratios (SINRs) of $S - R$ and $R - D$ links as follows

$$\gamma_R = \frac{P_S\|\mathbf{h}_i^R\|^2}{\sigma_{RSI}^2 + \sigma^2} = \|\mathbf{h}_i^R\|^2\bar{\gamma}_R,\tag{4}$$

$$\gamma_D = \frac{P_R\|\mathbf{h}_j^D\|^2}{\sigma^2} = \|\mathbf{h}_j^D\|^2\bar{\gamma}_D,\tag{5}$$

where $\bar{\gamma}_R = \frac{P_S}{\sigma_{RSI}^2 + \sigma^2}$ and $\bar{\gamma}_D = \frac{P_R}{\sigma^2}$ denote the average SINR at R and the average signal-to-noise-ratio (SNR) at D, respectively.

Since the relay node uses the DF protocol, the instantaneous end-to-end SINR of the considered system is defined as

$$\gamma_{e2e} = \min(\gamma_R, \gamma_D).\tag{6}$$

where γ_R and γ_D are respectively the instantaneous SINRs at R and D.

3. Optimal Power Allocation for FD Mode

To find the optimal power allocation for the FD relay, we first calculate the SEP of the considered system using the definition given in Reference [26] as follows

$$SEP = a\mathbb{E}\{Q(\sqrt{b\gamma_{e2e}})\} = \frac{a}{\sqrt{2\pi}}\int_0^\infty F_{\gamma_{e2e}}\left(\frac{t^2}{b}\right)e^{-\frac{t^2}{2}}\,dt,\tag{7}$$

where a and b are constants whose values depend on the modulation types, for example, $a = 1, b = 2$ for the binary phase-shift keying (BPSK) modulation [26]. The values of a and b are determined using Table 6.1 of Reference [26]; $Q(x)$ is the Gaussian function; γ_{e2e} is the instantaneous end-to-end SINR of the considered system which is determined in (6); $F_{\gamma_{e2e}}(.)$ is cumulative distribution function (CDF) of γ_{e2e} [21,27]. After some mathematical manipulations, (7) becomes

$$\text{SEP} = \frac{a\sqrt{b}}{2\sqrt{2\pi}} \int_0^\infty \frac{e^{-bx/2}}{\sqrt{x}} F_{\gamma_{e2e}}(x)dx. \tag{8}$$

To obtain the closed-form expression of (8), we calculate the CDF, $F_{\gamma_{e2e}}(x)$, of the probability that the instantaneous end-to-end SINR falls below a defined threshold. Mathematically, $F_{\gamma_{e2e}}(x)$ is expressed as

$$\begin{aligned} F_{\gamma_{e2e}}(x) &= \Pr\left\{\log_2(N_t) + \log_2(1 + \gamma_{e2e}) < \mathcal{R}\right\} \\ &= \Pr\left\{\gamma_{e2e} < 2^{\mathcal{R} - \log_2(N_t)} - 1\right\}, \end{aligned} \tag{9}$$

where $\mathcal{R}$ is the minimum data transmission rate of the considered system; the term $\log_2(N_t)$ denotes the number of bits which is used for activating the transmit antenna at the transmitters of (S or R).

Using the probability law of two independent variables $\mathcal{A}$ and $\mathcal{B}$ [28], that is $\Pr\{\mathcal{A} \cup \mathcal{B}\} = \Pr\{\mathcal{A}\} + \Pr\{\mathcal{B}\} - \Pr\{\mathcal{A}\}\Pr\{\mathcal{B}\}$, we have

$$\begin{aligned} F_{\gamma_{e2e}}(x) &= \Pr\left\{\gamma_{e2e} < 2^{\mathcal{R} - \log_2(N_t)} - 1\right\} = \Pr\{\gamma_{e2e} < x\} \\ &= \Pr\{\gamma_R < x\} + \Pr\{\gamma_D < x\} \\ &\quad - \Pr\{\gamma_R < x\}\Pr\{\gamma_D < x\}, \end{aligned} \tag{10}$$

where $x = 2^{\mathcal{R} - \log_2(N_t)} - 1$.

To calculate $F_{\gamma_{e2e}}(x)$ in (10), we first start with the CDF and probability distribution function (PDF) of the channel gain which follows Rayleigh fading distribution, that is,

$$F_{|h|^2}(x) = \Pr\{|h|^2 < x\} = 1 - \exp\left(-\frac{x}{\Omega}\right), x \geqslant 0, \tag{11}$$

$$f_{|h|^2}(x) = \frac{1}{\Omega}\exp\left(-\frac{x}{\Omega}\right), x \geqslant 0, \tag{12}$$

where $\Omega = \mathbb{E}\{|h|^2\}$ is the average channel gain; $\mathbb{E}$ denotes the expectation operator. In this paper, for the ease of presentation, we choose $\Omega = 1$ for all channel gains.

Then, we apply (11) and (12) to compute the probability in (10) as

$$\begin{aligned} \Pr\{\gamma_R < x\} &= \Pr\left\{\|\mathbf{h}_i^R\|^2 \bar{\gamma}_R < x\right\} \\ &= \Pr\left\{\|\mathbf{h}_i^R\|^2 < \frac{x}{\bar{\gamma}_R}\right\}. \end{aligned} \tag{13}$$

Based on the CDF of the summation of channel gains [20], this probability is calculated as

$$\Pr\{\gamma_R < x\} = 1 - e^{-\frac{x}{\bar{\gamma}_R}} \sum_{i=0}^{N_{tR}-1} \frac{1}{i!} \frac{x^i}{\bar{\gamma}_R^i}. \tag{14}$$

Using similar calculations, we can obtain $\Pr\{\gamma_D < x\}$ as

$$\Pr\{\gamma_D < x\} = 1 - e^{-\frac{x}{\bar{\gamma}_D}} \sum_{j=0}^{N_D-1} \frac{1}{j!} \frac{x^j}{\bar{\gamma}_D^j}. \tag{15}$$

Finally, $F_{\gamma_{e2e}}(x)$ in (10) can be given by

$$F_{\gamma_{e2e}}(x) = 1 - e^{-\frac{x}{\tilde{\gamma}_R} - \frac{x}{\tilde{\gamma}_D}} \sum_{i=0}^{N_{tR}-1} \sum_{j=0}^{N_D-1} \frac{1}{i!j!} \frac{x^{i+j}}{\tilde{\gamma}_R^i \tilde{\gamma}_D^j}. \tag{16}$$

Remark: As shown in (16), the active antenna at the transmitter is hidden in the variable x because $x = 2^{R - \log_2(N_t)} - 1$. Moreover, under the assumption that the receiver can estimate the index of the transmitter's activated antenna, it can successfully decode the transmitted bits used to modulate the active antenna. Substituting $F_{\gamma_{e2e}}(x)$ in (16) into (8), we obtain the closed-form expression of SEP as follows:

$$\begin{aligned}
\text{SEP} &= \frac{a\sqrt{b}}{2\sqrt{2\pi}} \left[\int_0^\infty \frac{e^{-bx/2}}{\sqrt{x}} dx - \int_0^\infty \frac{e^{-bx/2}}{\sqrt{x}} e^{-\frac{x}{\tilde{\gamma}_R} - \frac{x}{\tilde{\gamma}_D}} \sum_{i=0}^{N_{tR}-1} \sum_{j=0}^{N_D-1} \frac{1}{i!j!} \frac{x^{i+j}}{\tilde{\gamma}_R^i \tilde{\gamma}_D^j} dx \right] \\
&= \frac{a}{2} - \frac{a\sqrt{b}}{2\sqrt{2\pi}} \sum_{i=0}^{N_{tR}-1} \sum_{j=0}^{N_D-1} \frac{\Gamma(i+j+\frac{1}{2})}{i!j!\tilde{\gamma}_R^i \tilde{\gamma}_D^j (\frac{1}{\tilde{\gamma}_R} + \frac{1}{\tilde{\gamma}_D} + \frac{b}{2})^{i+j+\frac{1}{2}}}.
\end{aligned} \tag{17}$$

It is worth noting that we have used equations ([29] (3.361.2)) and ([29] (3.381.4)) to solve the first and second integrals in (17), respectively.

For the purpose of improving system performance and reducing the impact of the RSI in FD mode, we can calculate the optimal transmission power of R to minimize the system SEP. The optimal transmission power of R for minimizing the system SEP, denoted by P_R^*, is defined as

$$P_R^* = \arg \min_{P_R} \text{SEP}. \tag{18}$$

To explicitly determine the minSEP in (18), we begin with two terms in (17), that is, SEP_1 and SEP_2, as follows

$$\text{SEP} = \underbrace{\frac{a}{2}}_{\text{SEP}_1} - \underbrace{\frac{a\sqrt{b}}{2\sqrt{2\pi}} \sum_{i=0}^{N_{tR}-1} \sum_{j=0}^{N_D-1} \frac{\Gamma(i+j+\frac{1}{2})}{i!j!\tilde{\gamma}_R^i \tilde{\gamma}_D^j (\frac{1}{\tilde{\gamma}_R} + \frac{1}{\tilde{\gamma}_D} + \frac{b}{2})^{i+j+\frac{1}{2}}}}_{\text{SEP}_2}, \tag{19}$$

where

$$\text{SEP}_1 = \frac{a}{2}, \tag{20}$$

and

$$\text{SEP}_2 = \frac{a\sqrt{b}}{2\sqrt{2\pi}} \sum_{i=0}^{N_{tR}-1} \sum_{j=0}^{N_D-1} \frac{\Gamma(i+j+\frac{1}{2})}{i!j!\tilde{\gamma}_R^i \tilde{\gamma}_D^j (\frac{1}{\tilde{\gamma}_R} + \frac{1}{\tilde{\gamma}_D} + \frac{b}{2})^{i+j+\frac{1}{2}}}. \tag{21}$$

As mentioned in Section 3, after (7), since a and b are constants, $\text{SEP}_1 = \frac{a}{2}$ is also constant. Therefore, we have

$$\min \text{SEP} = \min(\text{SEP}_1 - \text{SEP}_2) = \text{SEP}_1 - \max \text{SEP}_2 = \max \text{SEP}_2. \tag{22}$$

Next, (22) can be rewritten as

$$\min \text{SEP} = \max \text{SEP}_2 = \max \left(\frac{a\sqrt{b}}{2\sqrt{2\pi}} \sum_{i=0}^{N_{tR}-1} \sum_{j=0}^{N_D-1} \frac{\Gamma(i+j+\frac{1}{2})}{i!j!\tilde{\gamma}_R^i \tilde{\gamma}_D^j (\frac{1}{\tilde{\gamma}_R} + \frac{1}{\tilde{\gamma}_D} + \frac{b}{2})^{i+j+\frac{1}{2}}} \right), \tag{23}$$

and (23) can be presented as

$$\min \text{SEP} = \max \left(\frac{a\sqrt{b}}{2\sqrt{2\pi}} \sum_{i=0}^{N_{rR}-1} \sum_{j=0}^{N_D-1} \frac{\Gamma(i+j+\frac{1}{2})}{i!j!} \times \frac{1}{\tilde{\gamma}_R^i \tilde{\gamma}_D^j (\frac{1}{\tilde{\gamma}_R} + \frac{1}{\tilde{\gamma}_D} + \frac{b}{2})^{i+j+\frac{1}{2}}} \right). \tag{24}$$

It is noted that, in (24), a, b, N_{rR} and N_D are constants; i and j are antenna indices which do not depend on the transmission power of S and R. Moreover, $\Gamma(i+j+\frac{1}{2})$ is also a constant for the certain values of i and j. Therefore, (24) is maximized when

$$\frac{1}{\tilde{\gamma}_R^i \tilde{\gamma}_D^j (\frac{1}{\tilde{\gamma}_R} + \frac{1}{\tilde{\gamma}_D} + \frac{b}{2})^{i+j+\frac{1}{2}}} \tag{25}$$

is maximized or

$$\tilde{\gamma}_R^i \tilde{\gamma}_D^j (\frac{1}{\tilde{\gamma}_R} + \frac{1}{\tilde{\gamma}_D} + \frac{b}{2})^{i+j+\frac{1}{2}} \tag{26}$$

is minimized.

In summary, the minSEP in (18) is given by

$$\min \text{SEP} = \min \left(\frac{a}{2} - \frac{a\sqrt{b}}{2\sqrt{2\pi}} \sum_{i=0}^{N_{rR}-1} \sum_{j=0}^{N_D-1} \frac{\Gamma(i+j+\frac{1}{2})}{i!j!\tilde{\gamma}_R^i \tilde{\gamma}_D^j (\frac{1}{\tilde{\gamma}_R} + \frac{1}{\tilde{\gamma}_D} + \frac{b}{2})^{i+j+\frac{1}{2}}} \right)$$

$$= \min \left(\tilde{\gamma}_R^i \tilde{\gamma}_D^j (\frac{1}{\tilde{\gamma}_R} + \frac{1}{\tilde{\gamma}_D} + \frac{b}{2})^{i+j+\frac{1}{2}} \right). \tag{27}$$

Denote $P_R = \alpha P_S$ and $f(\alpha) = \tilde{\gamma}_R^i \tilde{\gamma}_D^j (\frac{1}{\tilde{\gamma}_R} + \frac{1}{\tilde{\gamma}_D} + \frac{b}{2})^{i+j+\frac{1}{2}}$. Now we need to find α^*, which is the optimal value of α. Then for this given α^*, we can obtain P_R^*. The procedure for obtaining α^* and P_R^* is summarized in the following Algorithm 1.

Algorithm 1 Calculation of optimal α^* and P_R^*

1: Solve $\frac{\partial f(\alpha)}{\partial \alpha} = 0$ for $\alpha = \alpha_0$;

2: **if** $\begin{cases} \alpha_0 > 0 \\ \dfrac{\partial f(\alpha)}{\partial \alpha} < 0 \text{ for } \alpha < \alpha_0 \\ \dfrac{\partial f(\alpha)}{\partial \alpha} > 0 \text{ for } \alpha > \alpha_0 \end{cases}$

3: **then**
4:    Output optimal value of α
      $\alpha^* = \alpha_0$;
      thus
      $P_R^* = \alpha^* P_S$
5: **else**
6:    Output optimal value $\alpha^* = \varnothing$;
7: **end**

We will explain step-by-step the process of Algorithm 1 as follows.

- **Step 1:** We take the derivative of $\frac{\partial f(\alpha)}{\partial \alpha}$ with respect to α and solve $\frac{\partial f(\alpha)}{\partial \alpha} = 0$ to obtain the stationary point α_0. Specifically, after some basic algebra calculations, we obtain the following equation

$$f'(\alpha) \leq 2\tilde{\Omega}P_S\alpha^2 - 2bP_S\alpha - 9. \tag{28}$$

Then, α_0 can be calculated as

$$\alpha_0 = \frac{bP_S + \sqrt{P_S(b^2 P_S + 18\tilde{\Omega})}}{2\tilde{\Omega}P_S}. \tag{29}$$

- **Step 2:** We check whether $\frac{\partial f(\alpha)}{\partial \alpha}$ is negative or positive in a specific interval to determine maximum or minimum point. If $\frac{\partial f(\alpha)}{\partial \alpha}$ is negative when $\alpha < \alpha_0$ and positive when $\alpha > \alpha_0$, an optimal transmission power P_R^* exits and it is given by

$$P_R^* = \frac{bP_S + \sqrt{P_S(b^2 P_S + 18\tilde{\Omega})}}{2\tilde{\Omega}}. \tag{30}$$

It is worth noting that (30) is the bounded optimal transmission power at the FD relay node and it is often used for the systems with complex mathematical expressions.

- **Step 3:** Otherwise, if $\alpha^* = \varnothing$, depending on whether $\frac{\partial f(\alpha)}{\partial \alpha}$ is less or greater than 0, we can select an appropriate value of α^* to get P_R^*.

4. Numerical Results

In this section, to validate the derived mathematical expressions in the previous sections, we provide analytical results together with the Monte-Carlo simulation results for comparison. For ease of presentation, both S and R use two transmitting antennas, that is $N_t = 2$, while the number of receiving antennas N_{rR} and N_D are set to be equal and varies from 2 to 4 for performance evaluations. The SIC capability used for evaluation is $\tilde{\Omega} = \{-10, -5, 0\}$ dB. For an LTE relay, the typical transmission power ranges from 23 dBm to 30 dBm. Taking 30 dBm for consideration, the RSI levels are $\sigma_{RSI}^2 = \{20, 25, 30\}$ dBm, respectively. In all figures, we define the average SNR for the case without optimization as follows: SNR $= \frac{P_S}{\sigma^2} = \frac{P_R}{\sigma^2}$. In the case with optimization, the average SNR is defined by SNR at R, that is, SNR $= \frac{P_S}{\sigma^2}$. The analytical curves are plotted using Equation (17) while the markers refer to Monte-Carlo simulation results. The simulation results were obtained using 10^6 channel realizations.

Figure 2 plots the SEP of the considered SM-MIMO-FD relay system versus the SNR in dB for two modulation schemes, that is, BPSK ($a = 1$, $b = 2$) and 4-QAM ($a = 2$, $b = 1$). We should remind that a and b are constants, which depend on the types modulation scheme. These values for each type of modulation are given in Table 6.1 of Reference [26]. As shown in Figure 2, similar patterns of SEPs can be observed in both BPSK and 4-QAM modulation schemes. However, the system with 4-QAM modulation has higher SEP than the system with BPSK modulation. Moreover, the benefit of optimization is also reduced as the modulation order increases. Therefore, although our analysis method can be applied for all modulation types, we use the BPSK modulation in the following figures to clearly show the advantage of our proposed optimization algorithm in reducing the SEP of the considered SM-MIMO-FD relay system.

Figure 3 compares the SEPs of the SM-MIMO-FD relay system in two cases, that is $\alpha = 1$ (without optimal power allocation, $P_R = P_S$) and $\alpha = \alpha^*$ (with optimal power allocation, $P_R = P_R^*$) in (30) for different numbers of receiving antennas of R and D. Although we have used $N_{rR} = N_D = 2, 3, 4$ to obtain this figure, it is worth noting that we can use different numbers of receiving antennas at R and D for numerical calculation using the closed-form expression of SEP. The modulation used for evaluation is BPSK with parameters $a = 1, b = 2$. The typical SIC capability of $\tilde{\Omega} = -10$ dB is used for calculation. It is easy to see from the figure that the analytical results perfectly match the simulation ones. Although the SEPs of both the cases with and without optimal power allocation suffer the same error floors in high SNR regime, the SEP with optimal power allocation is significantly lower than that without optimal power allocation in low SNR regime. For example, when SNR = 8 dB, the SEP in the case with α^* is approximately ten times less than the case without α^*.

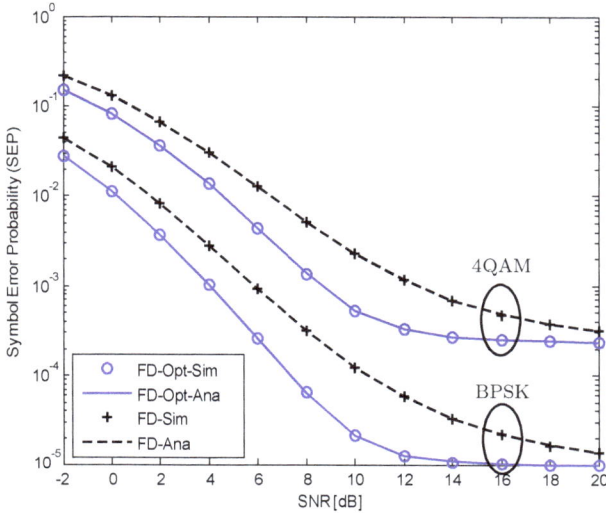

Figure 2. The SEP of the considered SM-MIMO-FD relay system with and without optimal power allocation versus the SNR for different modulation schemes, $N_{rR} = N_D = 4$.

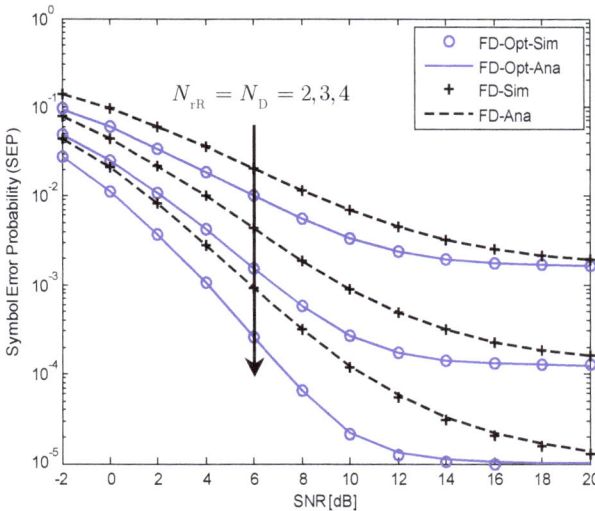

Figure 3. The SEP of the considered SM-MIMO-FD relay system with and without optimal power allocation when BPSK modulation is used, $N_{rR} = N_D = 2, 3, 4; \tilde{\Omega} = -10$ dB.

Figure 4 shows the impact of the RSI on the SEP of the system with and without optimal power allocation for different SIC capabilities, that is, $\tilde{\Omega} = -10, -5, 0$ dB. When the RSI is small ($\tilde{\Omega} = -10$ dB), the difference in SEPs in the cases with and without using α^* is large. However, when the RSI becomes larger ($\tilde{\Omega} = -5$ dB), the benefit of using α^* decreases. Therefore, it is necessary to combine SIC techniques with optimal power allocation to achieve the best performance for this system.

Figures 3 and 4 can be served as the guideline to determine when the optimal power allocation should be used. Specifically, for SNR < 20 dB, we use P_R^* to improve the system performance. For higher SNR, that is SNR > 20 dB, we use $P_R = P_S$ to reduce the signal processing complexity at the FD relay. The advantage of the system with optimization is that it requires only SNR ≥ 8 dB to achieve a reliable voice transmission.

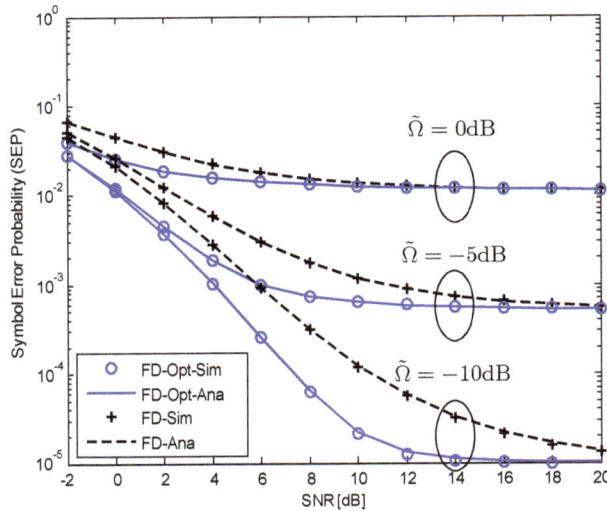

Figure 4. The impact of RSI on the SEP of the considered SM-MIMO-FD relay system with and without optimal power allocation; $\tilde{\Omega} = -10, -5, 0$ dB; $N_{rR} = N_D = 4$.

5. Conclusions

In this paper, we have analyzed the performance of the SM-MIMO-FD DF relay system and derived the closed-form expression of the system SEP. Understanding the importance of the optimal power allocation to the SM-MIMO-FD relay system, especially in the case of imperfect SIC, we have proposed an optimal power allocation algorithm for the FD relay to minimize the system SEP. Both numerical and simulation results showed that the RSI has a substantial impact on the SEP performance. However, the SEP of the system with optimal power allocation is significantly lower compared with that of the system without optimal power allocation. This result confirms the effectiveness of using optimal power allocation for compensating the impact of the RSI in the SM-MIMO-FD DF relay system.

Author Contributions: The main contributions of L.V.N., B.C.N. were to create the main ideas and execute performance evaluation by extensive simulation while X.N.T., L.T.D. worked as the advisors to discuss, create, and advise the main ideas and performance evaluations together.

Funding: This research received no external funding.

Conflicts of Interest: The authors declare no conflict of interest.

References

1. Jiao, B.; Wen, M.; Ma, M.; Poor, H.V. Spatial modulated full duplex. *IEEE Wirel. Commun. Lett.* **2014**, *3*, 641–644. [CrossRef]
2. Bhowal, A.; Kshetrimayum, R.S. Outage probability bound of decode and forward two-way full-duplex relay employing spatial modulation over cascaded α- μ channels. *Int. J. Commun. Syst.* **2019**, *32*, e3876. [CrossRef]

3. Koc, A.; Altunbas, I.; Basar, E. Two-way full-duplex spatial modulation systems with wireless powered af relaying. *IEEE Wirel. Commun. Lett.* **2018**, *7*, 444–447. [CrossRef]

4. Nguyen, B.C.; Tran, X.N.; Hoang, T.M. Performance Analysis of Full-Duplex Vehicle-to-Vehicle Relay System over Double-Rayleigh Fading Channels. *Mob. Netw. Appl.* **2019**, 1–10. [CrossRef]

5. Tran, X.N.; Nguyen, B.C.; Tran, D.T. Outage probability of two-way full-duplex relay system with hardware impairments. In Proceedings of the 3rd International Conference on Recent Advances in Signal Processing, Telecommunications & Computing (SigTelCom), Hanoi, Vietnam, 3–6 July 2019; pp. 135–139.

6. Nguyen, B.C.; Tran, X.N. Performance Analysis of Full-Duplex Amplify-and-Forward Relay System with Hardware Impairments and Imperfect Self-Interference Cancellation. *Wirel. Commun. Mob. Comput.* **2019**, *2019*, 10. [CrossRef]

7. Sofotasios, P.C.; Fikadu, M.K.; Muhaidat, S.; Freear, S.; Karagiannidis, G.K.; Valkama, M. Relay Selection Based Full-Duplex Cooperative Systems Under Adaptive Transmission. *IEEE Wirel. Commun. Lett.* **2017**, *6*, 602–605. [CrossRef]

8. Fikadu, M.K.; Sofotasios, P.C.; Valkama, M.; Muhaidat, S.; Cui, Q.; Karagiannidis, G.K. Outage probability analysis of dual-hop full-duplex decode-and-forward relaying over generalized multipath fading conditions. In Proceedings of the IEEE 11th International Conference on Wireless and Mobile Computing, Networking and Communications (WiMob), Abu Dhabi, United Arab Emirates, 19–21 October 2015; pp. 414–421. [CrossRef]

9. Kim, H.; Lim, S.; Wang, H.; Hong, D. Optimal power allocation and outage analysis for cognitive full duplex relay systems. *IEEE Trans. Wirel. Commun.* **2012**, *11*, 3754–3765. [CrossRef]

10. Sofotasios, P.C.; Fikadu, M.K.; Muhaidat, S.; Cui, Q.; Karagiannidis, G.K.; Valkama, M. Full-Duplex Regenerative Relaying and Energy-Efficiency Optimization Over Generalized Asymmetric Fading Channels. *IEEE Trans. Wirel. Commun.* **2017**, *16*, 3232–3251. [CrossRef]

11. Chen, L.; Meng, W.; Hua, Y. Optimal power allocation for a full-duplex multicarrier decode-forward relay system with or without direct link. *Signal Process.* **2017**, *137*, 177–191. [CrossRef]

12. Le, M.T.; Ngo, V.D.; Mai, H.A.; Tran, X.N.; Di Renzo, M. Spatially modulated orthogonal space-time block codes with non-vanishing determinants. *IEEE Trans. Commun.* **2013**, *62*, 85–99. [CrossRef]

13. Nguyen, T.P.; Le, M.T.; Ngo, V.D.; Tran, X.N.; Choi, H.W. Spatial modulation for high-rate transmission systems. In Proceedings of the IEEE 79th Vehicular Technology Conference (VTC Spring), Seoul, Korea, 18–21 May 2014; pp. 1–5.

14. Kumbhani, B.; Kshetrimayum, R. Outage probability analysis of spatial modulation systems with antenna selection. *Electron. Lett.* **2014**, *50*, 125–126. [CrossRef]

15. Mesleh, R.Y.; Haas, H.; Sinanovic, S.; Ahn, C.W.; Yun, S. Spatial modulation. *IEEE Trans. Veh. Technol.* **2008**, *57*, 2228–2241. [CrossRef]

16. Narayanan, S.; Ahmadi, H.; Flanagan, M.F. On the performance of spatial modulation MIMO for full-duplex relay networks. *IEEE Trans. Wirel. Commun.* **2017**, *16*, 3727–3746. [CrossRef]

17. Raviteja, P.; Hong, Y.; Viterbo, E. Spatial modulation in full-duplex relaying. *IEEE Commun. Lett.* **2016**, *20*, 2111–2114. [CrossRef]

18. Zhang, J.; Li, Q.; Kim, K.J.; Wang, Y.; Ge, X.; Zhang, J. On the performance of full-duplex two-way relay channels with spatial modulation. *IEEE Trans. Commun.* **2016**, *64*, 4966–4982. [CrossRef]

19. Everett, E.; Sahai, A.; Sabharwal, A. Passive self-interference suppression for full-duplex infrastructure nodes. *IEEE Trans. Wirel. Commun.* **2014**, *13*, 680–694. [CrossRef]

20. Nguyen, B.C.; Hoang, T.M.; Tran, P.T. Performance analysis of full-duplex decode-and-forward relay system with energy harvesting over Nakagami-m fading channels. *AEU Int. J. Electron. Commun.* **2019**, *98*, 114–122. [CrossRef]

21. Yang, K.; Cui, H.; Song, L.; Li, Y. Efficient Full-Duplex Relaying With Joint Antenna-Relay Selection and Self-Interference Suppression. *IEEE Trans. Wirel. Commun.* **2015**, *14*, 3991–4005. [CrossRef]

22. Nguyen, B.C.; Tran, X.N.; Tran, D.T. Performance analysis of in-band full-duplex amplify-and-forward relay system with direct link. In Proceedings of the 2nd International Conference on Recent Advances in Signal Processing, Telecommunications & Computing (SigTelCom), Ho Chi Minh, Vietnam, 29–31 January 2018; pp. 192–197.

23. Bharadia, D.; McMilin, E.; Katti, S. Full duplex radios. In *ACM SIGCOMM Computer Communication Review*; ACM: New York, NY, USA, 2013; Volume 43, pp. 375–386.

24. Wen, M.; Zheng, B.; Kim, K.J.; Di Renzo, M.; Tsiftsis, T.A.; Chen, K.C.; Al-Dhahir, N. A Survey on Spatial Modulation in Emerging Wireless Systems: Research Progresses and Applications. *IEEE J. Sel. Areas Commun.* **2019**, *37*, 1949–1972. [CrossRef]

25. Wen, M.; Basar, E.; Li, Q.; Zheng, B.; Zhang, M. Multiple-mode orthogonal frequency division multiplexing with index modulation. *IEEE Trans. Commun.* **2017**, *65*, 3892–3906. [CrossRef]

26. Goldsmith, A. *Wireless Communications*; Cambridge University Press: Cambridge, UK, 2005.

27. Cui, H.; Ma, M.; Song, L.; Jiao, B. Relay selection for two-way full duplex relay networks with amplify-and-forward protocol. *IEEE Trans. Wirel. Commun.* **2014**, *13*, 3768–3777. [CrossRef]

28. Leon-Garcia, A.; Leon-Garcia, A. *Probability, Statistics, and Random Processes for Electrical Engineering*, 3rd ed.; Pearson/Prentice Hall: Upper Saddle River, NJ, USA, 2008.

29. Jeffrey, A.; Zwillinger, D. *Table of Integrals, Series, and Products*; Academic Press: Cambridge, MA, USA, 2007.

sensors

MDPI

Review

A Survey on Recent Trends and Open Issues in Energy Efficiency of 5G

Muhammad Usama and Melike Erol-Kantarci *

School of Electrical Engineering and Computer Science, University of Ottawa, Ottawa, ON K1N 6N5, Canada
*   Correspondence: melike.erolkantarci@uottawa.ca

Received: 13 May 2019; Accepted: 11 July 2019; Published: 15 July 2019

Abstract: The rapidly increasing interest from various verticals for the upcoming 5th generation (5G) networks expect the network to support higher data rates and have an improved quality of service. This demand has been met so far by employing sophisticated transmission techniques including massive Multiple Input Multiple Output (MIMO), millimeter wave (mmWave) bands as well as bringing the computational power closer to the users via advanced baseband processing units at the base stations. Future evolution of the networks has also been assumed to open many new business horizons for the operators and the need of not only a resource efficient but also an energy efficient ecosystem has greatly been felt. The deployment of small cells has been envisioned as a promising answer for handling the massive heterogeneous traffic, but the adverse economic and environmental impacts cannot be neglected. Given that 10% of the world's energy consumption is due to the Information and Communications Technology (ICT) industry, energy-efficiency has thus become one of the key performance indicators (KPI). Various avenues of optimization, game theory and machine learning have been investigated for enhancing power allocation for downlink and uplink channels, as well as other energy consumption/saving approaches. This paper surveys the recent works that address energy efficiency of the radio access as well as the core of wireless networks, and outlines related challenges and open issues.

Keywords: 5G; energy-efficiency; sustainability

1. Introduction

Advances in telecommunication systems around the world have always been pushing the wireless infrastructure to be more resilient and scalable. Ever growing faster data rates and a demand for the highest quality of service has been a strong constraint when energy conservation needs to be considered. Data rates as high as that of 1 Gbps have been foreseen with the advent of 5G. In addition, with an explosive number of heterogeneous devices coming online, including sensors for home security, tablets, and wearable health monitors, the computational power of base stations must increase. An estimated 50% increase in the computing power of baseband units has been predicted to handle this traffic burst [1]. Thus, the focus on energy-efficiency needs to include optimization of computational complexity in addition to optimization of transmission power.

An estimated 75% of the Information and Communications Technology (ICT) industry is supposed to be wireless by 2020 and today 5% of the world's carbon footprint is coming from this industry alone. A consensus between academia and industry dictates that the foreseen $1000\times$ capacity gain must be achieved with either the present energy consumption or lower [2]. Thanks to energy-efficiency efforts world-wide, energy consumption in the 5G realm, in terms of bits/joule, has been considered as an important design parameter. In 4th generation (4G), the concept of small cells has been introduced to increase the coverage and capacity. Therefore, [3] conducted an analysis on energy consumption per unit area for a heterogeneous deployment of cells for fourth generation networks. With 5G, small cells are inevitable in deployments due to their advantage of improved traffic handling within a smaller area as well as the shorter cell ranges that result from the use of higher frequencies. Yet, the increasing number of base stations translate into more energy consumption, although the increase in consumption will not be linear. Small cells, or in other words densification, calls for sophisticated management of resources. Most recently, intelligent resource allocation and control techniques utilizing machine learning algorithms have been suggested to help next generation radios in their autonomous reconfiguration for improving the data rates, energy efficiency and interference mitigation. Overall, the emerging sophistication in both User Equipment (UE) and network side has increased the energy consumption and thus objective functions have been devised to maximize the energy efficiency, harvested energy and energy aware transmission [4]. Many of the existing energy efficiency improvement techniques include the use of green energy sources for base stations, modifying the coverage area of a base station depending upon the load level, putting lightly loaded base stations to sleep and load balancing by handing over the UEs to the macro base station. A survey on these technologies for the 5G Radio Access Network (RAN) can be found in [5].

This survey has been aimed to contribute towards a greener and a sustainable telecommunication's ecosystem by reviewing and bringing together some of the latest ideas and techniques of energy conservation at base station and network level. A high level diagram shows the areas addressed in Figure 1. A few of the prominent examples include the introduction of a newer Radio Resource Control (RRC) state for context signalling and cutting down on the redundant state changes [6]. Utilization of advanced clustering and caching techniques on the RAN side have been highly appreciated for their benefits of improving the latency of getting the data requested by a group of users and possibly eliminating the factor of clogging the network by a huge number of requests for the same content [7,8]. A case study of commercial resource sharing among different operators bears fruitful results in terms of reduced deployment costs and good data rates with minimum interference among them [9]. The upcoming sections introduce the basics of energy efficiency, provide justification for the need of gauging the energy consumption and then present the most recent research works carried out for the optimization at different levels of the architecture. This survey bears its uniqueness in its holistic approach to energy-efficiency by covering radio, core and computing side of 5G. This paper is also different than the surveys in the literature [1–4], as it focuses on works published in the last few years where the majority of the studies focus on concepts specific to the new 5G standard.

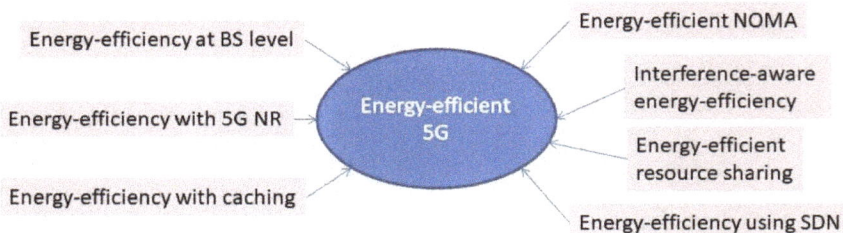

Figure 1. Outline of the energy-efficiency schemes included in this survey.

2. Background on Energy Efficiency

A formal relationship between energy efficiency and Signal to Interference Noise Ratio (SINR) has been presented in [2] using the bit/joule notion. Meanwhile, Reference [4] lays the foundation for energy efficiency in different parts of the network including base stations and the core network. In the literature, energy saving and use of green energy resources have been the two mainstream approaches to offer energy efficiency. Among the energy saving techniques, cell-switch off techniques have been widely exploited. For instance, in the EU FP7 ABSOLUTE project, an energy aware middleware has been proposed that would use the capacity-based thresholds for activation of the base stations [10]. In several other studies, data offloading has been considered as an energy-efficient approach. Furthermore, authors in [11] have put together several techniques for not only reducing the energy consumption from the traditional energy sources but also for surveying newer Energy Efficiency (EE) schemes in the End-to-End (E2E) system. One of the remarkable mentions by the authors includes the implementation of 3rd Generation Partnership Project (3GPP) compliant EE manager that would be responsible for monitoring energy demands in an E2E session and for implementation of the policies needed for catering to the ongoing energy demand.

In addition to energy saving approaches, recently simultaneous wireless energy transfer has been studied. Furthermore, local caching techniques have been proved to be beneficial for relieving the load on the backhaul network by storing the content locally and limiting the re-transmissions, hence reducing energy consumption. Similarly, a cloud based RAN has been envisioned as a possible solution for the computational redistribution in [2,4,12]. Many of the tasks previously performed by a base station (BS) would be taken away to a data center and only decision making for Radio Frequency (RF) chains as well as baseband to RF conversion would be given to base stations. Traffic pattern and demands would then be catered for well before time and redundant BS would be put to sleep mode according to [13]. Furthermore, full duplex Device-to-Device (D2D) communication with uplink channel reuse has been considered to improve SINR and transmission power constraints. A gain of 36% energy efficiency has been demonstrated using the full duplex scheme with enhanced self-interference mitigation mechanism instead of half duplex [14].

As machine learning is penetrating more and more into the operation of wireless networks, Reference [15] suggests that machine learning algorithms would greatly help to predict the hot spots so that other resources could be switched off when not needed.

The concept of energy efficiency being treated as a key performance indicator in the upcoming 5G standard considers it to be a global ambition, but it cannot be declared as a specific actionable item on either the operator or vendor side. Divide and conquer approach has been applied to the entire network and improvements have been targeted at either component level, equipment level or at network level employing newer algorithms at both BS and UE side. This discussion advocates the fact that operators would have the leverage of tuning their network for a balance between quality of service and energy consumption. In the following sections, we introduce the recent works in energy-efficiency in 5G as highlighted in Table 1 preceding to a discussion on open issues and challenges.

Table 1. Summary of surveyed works.

Optimization Scope	Problem Addressed	Citation
	Dissection of a BS and figures for energy consumption	[1]
	Downlink Massive MIMO Systems: Achievable Sum Rates and Energy Efficiency Perspective for Future 5G Systems	[16]
	Energy Efficiency in massive MIMO based 5G networks: Opportunities and Challenges	[17]
EE at the BS level	EE improvement by a Centralized BB processing design	[18]
	Analytical modelling of EE for a heterogeneous network	[19]
	Energy Efficiency Metrics for Heterogeneous Wireless Cellular Networks	[20]
	Incentive based sleeping mechanism for densely deployed femto cells	[21]
	Sector based switching technique	[22]

Table 1. *Cont.*

Optimization Scope	Problem Addressed	Citation
	On interdependence among transmit and consumed power of macro base station technologies	[23]
	Utilization of Nash product for maximizing cooperative EE	[24]
	Energy Efficiency in Wireless Networks via Fractional Programming Theory	[25]
	Energy efficiency maximization oriented resource allocation in 5G ultra-dense network: Centralized and distributed algorithms	[26]
	Comparison of Spectral and Energy Efficiency Metrics Using Measurements in a LTE-A Network	[27]
	Energy Management in LTE Networks	[28]
	Energy-efficient resource allocation scheduler with QoS aware supports for green LTE network	[29]
	Interference-area-based resource allocation for full-duplex communications	[30]
	A resource allocation method for D2D and small cellular users in HetNet	[31]
	Highly Energy-Efficient Resource Allocation in Power Telecommunication Network	[32]
	EE enhancement with RRC Connection Control for 5G New Radio (NR)	[6]
	Proactive caching based on the content popularity on small cells	[7]
	Cooperative Online Caching in Small Cell Networks with Limited Cache Size and Unknown Content Popularity	[33]
	Economical Energy Efficiency: An Advanced Performance Metric for 5G Systems	[34]
	Energy-efficient design for edge-caching wireless networks: When is coded-caching beneficial?	[35]
	Content caching in small cells with optimized UL and caching power	[36]
	An effective cooperative caching scheme for mobile P2P networks	[37]
	EE analysis of heterogeneous cache enabled 5G hyper cellular networks	[8]
EE at the network level	Motivation for infrastructure sharing based on current energy consumption figures	[2,38]
	Energy efficiency in 5G access networks: Small cell densification and high order sectorisation	[39]

Table 1. *Cont.*

Optimization Scope	Problem Addressed	Citation
	Energy-Efficient User Association and Beamforming for 5G Fog Radio Access Networks	[40]
	Global energy and spectral efficiency maximization in a shared noise-limited environment	[9]
	EE Resource Allocation in NOMA	[41]
	Concept and practical considerations of non-orthogonal multiple access (NOMA) for future radio access	[42]
	Optimum received power levels of UL NOMA signals for EE improvement	[43]
	Spectral efficient nonorthogonal multiple access schemes (NOMA vs RAMA)	[44]
	Non-Orthogonal Multiple Access: Achieving Sustainable Future Radio Access	[45]
	Mode Selection Between Index Coding and Superposition Coding in Cache-based NOMA Networks	[46]
EE at the network level	Use case of shared UE side distributed antenna System for indoor usage	[47]
	Optimized Energy Aware 5G Network Function Virtualization	[48]
	Energy Efficient Network Function Virtualization in 5G Networks	[49]
	Network Function Virtualization in 5G	[50]
	A Framework for Energy Efficient NFV in 5G Networks	[51]
	Energy efficient Placement of Baseband Functions and Mobile Edge Computing in 5G Networks	[52]
	Energy Efficiency Benefits of RAN-as-a-Service Concept for a Cloud-Based 5G Mobile Network Infrastructure	[53]
	Dynamic Auto Scaling Algorithm (DASA) for 5G Mobile Networks	[54]
	Design and Analysis of Deadline and Budget Constrained Autoscaling (DBCA) Algorithm for 5G Mobile Networks	[55]
EE using SDN technology	Impact of software defined networking (SDN) paradigm on EE	[56]
	EE gains from the separated control and data planes in a heterogeneous network	[57]
EE using ML techniques	Machine Learning Paradigms for Next-Generation Wireless Networks	[58]
	Switch-on/off policies for energy harvesting small cells through distributed Q-learning	[59]

Table 1. *Cont.*

Optimization Scope	Problem Addressed	Citation
EE using ML techniques	Duty cycle control with joint optimization of delay and energy efficiency for capillary machine-to-machine networks in 5G communication system	[60]
	Distributed power control for two tier femtocell networks with QoS provisioning based on Q-learning	[61]
	Spectrum sensing techniques using both hard and soft decisions	[62]
	EE resource allocation in 5G heterogeneous cloud radio access network	[63]

3. Review of EE Techniques at the Base Station Level

Radio access network (RAN) has been considered as single unit for energy efficiency improvement, and inclusion of these enhancements across the network would have a significant impact on the overall energy efficiency. Metrics for gauging EE in this perspective include the improvements in the architecture and chipset design for the baseband units, cell switch off techniques, incorporation of small cells, interference reduction among the neighboring cells and caching as well as the newer RRC state for UEs for conservation of the battery power.

3.1. Base Station Energy Consumption and Cell Switch Off Techniques

Knowing the accurate energy consumption of a base station constitutes an important part of the understanding of the energy budget of a wireless network. For this purpose, authors in [1] have specifically discussed energy conservation at equipment level by presenting the breakdown of a base station. A typical BS has been presented by dividing it into five parts, namely antenna interface, power amplifier, RF chains, Baseband unit, mains power supply and the DC-DC supply. These modules have been shown in Figure 2. An important claim has been made stating that up to 57% of the power consumption at a base station is experienced at the transmission end, i.e., the power amplifier and antenna interface. Yet, with small cells, the power consumption per base station has been reduced due to shorter distances between the base stations and the users [1,19]. In [19], analytical modelling of the energy efficiency for a heterogeneous network comprising upon macro, pico and femto base stations has been discussed. To a certain extent emphasis has been put on the baseband unit which is specifically in charge of the computing operations and must be sophisticated enough to handle huge bursts of traffic. A baseband unit has been described to be composed of four different logical systems including a baseband system used for evaluating Fast Fourier Transforms (FFT) and wireless channel coding, the control system for resource allocation, the transfer system used for management operations among neighbouring base stations and finally the system for powering up the entire base station site including cooling and monitoring systems. Furthermore, the use of mmWave and massive MIMO would need an even greater push on the computation side of the base station since more and more users are now being accommodated. The study in [16] discusses the achievable sum rates and energy efficiency of a downlink single cell M-MIMO systems under various precoding schemes whereas several design constraints and future opportunities concerning existing and upcoming MIMO technologies have been discussed in [17]. The computation power of base station would increase when number of antennas and the bandwidth increases. In the case of using 128 antennas the computation power would go as high as 3000 W for a macrocell and 800 W for a small cell according to [1].

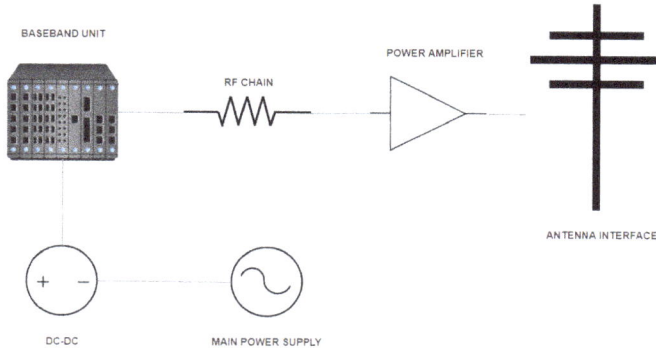

Figure 2. Modules of a typical base station.

Authors in [18] have discussed the utility of taking most of the baseband processing functionality away from the base station towards a central, more powerful and organized unit for supporting higher data rates and traffic density. Users have envisioned experiencing more flexibility using this central RAN since they would be able to get signaling from one BS and get data transfer through another best possible neighboring BS. Visible gains in latency and fronthaul bandwidth have thus been observed by having stronger backhaul links but this research avenue still needs to be formally exploited for devising globally energy efficient mechanisms. The choice of the best suited BS would allow the network to have a lower transmission power thus increasing the energy efficiency. An analysis of throughput as a performance metric has been provided for a two-tier heterogeneous network comprising upon macro and femto cells in [20]. The claimed improvement in throughput originates from a distributed mesh of small cells so that the minimal transmission distance between the end user and the serving base station would be cashed out in terms of reduced antenna's transmission power. Considering these findings on BS energy consumption, cell switch-off techniques have been explored in the literature. An incentive based sleeping mechanism for densely deployed femtocells has been considered in [21] and energy consumption reduction up to 40% has been observed by turning the RF chains off and only keeping the backhaul links alive. The key enabler here would be to have prompt toggling between active and sleep modes for maintaining the quality of service. According to [21], a "sniffer" component installed at these small cells that would be responsible for detecting activity in the network by checking the power in uplink connections, a value surpassing the threshold, would indicate a connection with the macrocell. Mobility Management Entity (MME) has also been suggested to potentially take a lead by sending wake up signals to the respective femtocells and keeping others asleep. In contrast to the usual techniques of handing their users over to the neighbouring base stations and turning that cell off, it would be beneficial to give incentives to users for connecting to a neighbouring cell if they get to have better data rates. Authors in [22] have conducted a thorough study for classification of the switching techniques as well as calculation of the outage probability of UEs, under realistic constraints. Their claim states that the energy consumption of the base station is not directly proportional to its load so an improved switching algorithm was needed that would allow the UEs to maintain the SINR thresholds. They have thus brought forward a sector based switching technique for the first time. Furthermore, their claim favors an offline switching technique instead of a more dynamic online scheme because of practical constraints such as random UE distribution and realistic interference modelling. Authors in [23] discuss influence of the transmit power scaling and on/off switching on instantaneous macro base stations power consumption. The proposed power consumption models have been claimed to be used as generic models for the relationship between transmitted and consumed power for macro base stations of different technologies and generations. In addition to these techniques, recently, machine learning techniques have been used to implement cell switch off which are discussed in Section 6.

3.2. Interference-Aware Energy Efficiency Techniques in 5G Ultra Dense Networks

The advantages of small cell deployment, in terms of increased system capacity and better load balancing capability, have been discussed in the previous sections. Yet, it is important to mention that densification suffers from added system complexity. Therefore, energy efficiency as well as spectral efficiency becomes harder to evaluate. Nash energy efficiency maximization theory has been presented for discussing the relationship between energy and spectral efficiency in [24]. Both are inversely related to each other, increase in one of them demands a natural decrease in the other quantity which usually has been the case of medium to high transmission power. Most of the research conducted in ultra-dense small cell networks has been on coming up with techniques optimizing both energy efficiency (EE) and spectral efficiency (SE). Authors in [24] also brings forth the idea of gaining energy efficiency at the cost of spectral efficiency where the small cells are under the coverage of a macro cell and pose interference issues due to the sharing of bandwidth among them.In such a scenario, all the small cells participate in energy efficiency maximization according to a game theoretic methodology. The suggested game theoretic model has been deemed to be a distributed model and utilizes Nash product for maximizing cooperative energy efficiency. Analysis of the algorithms shows that energy efficiency, although it increases with the increase in the number of small cells, it saturates after about 200 cells and afterwards only experiences a minor increase. Fractional programming has been extensively used in [25] for modelling the energy efficiency ratio for a Point-to-Point (P2P) network as well as for a full scaled communication network using MIMO. EE has been considered as a cost benefit ratio and minimum rate constraints have been put together for modelling real life scenarios. In addition, fairness in resource allocation has been considered a major factor in the overall energy distribution. These two constraints might tend to increase the power consumption in case the minimum thresholds tend to be too high. Adding to the use cases of fractional programming, [26] laid out a robust distributed algorithm for reducing the adverse effects of computational complexity and noise towards resource allocation. Authors in [27], have presented an experimental setup for defining the right kind of key performance indicators when measuring either EE or SE. The setup includes a set of UE(s), three small BS(s) and running iperf traffic using User Datagram Protocol (UDP) and File Transfer Protocol (FTP). Results have indicated that utilization of a higher bandwidth would not increase the power consumption, that throughput must incorporate the traffic density and that the idle power of the equipment needs to be considered for energy consumption calculations. In [28], use of varying transmission power levels by the aid of custom power levels in a two-tier network has been encouraged for the optimization of needed power in Long Term Evolution (LTE). Intelligent switching of control channels in the DL and tuning the power levels according to the UE's feedback have been envisioned to aid in allocation of the resource blocks with an optimum power. Authors in [29], have discussed the opportunities for the less explored domain of user scheduling in LTE. 3GPP has no fixed requirement on scheduling and thus researchers have devised their own mechanisms depending upon their pain points. Authors have proposed the idea of associating Quality of Service (QoS) with scheduling for accommodating cell edge users. Authors in [30] have proposed a resource allocation technique for minimizing the interference at the UE side. Considering a full duplex communication setup, a circular interference area for a DL UE has been demarcated by the BS based upon a predefined threshold. Resource block for this UE has been shared by an UL UE from outside the interference region for keeping the mutual interference to a minimal level. Simulation results claim to improve the overall network throughput based on the efficient pairing of UEs but the throughput might degrade with a large increase in the distance between the paired UEs. A heuristic algorithm presented in [31] improves the system throughput using resource reuse in the three-tier architecture while regulating the interference regions of UEs being served by either macro BS, small BS or in a D2D way. Visible gains in the throughput have been noted with an increased user density for an efficient user selection and having a minimum distance between the UEs being served in a D2D fashion for a stronger link retention. Moreover in [32], authors have constructed objective functions for EE maximization and have thus compared max-min power consumption model against their nonlinear fractional optimization model. Results have been promising for a reduction in the power consumption because of the mutual participation of cells as their number starts to increase.

3.3. Energy Efficiency Enhancement with RRC Connection Control for 5G New Radio (NR)

In [6], the authors discuss the rapid UE battery drainage which is due to the fact that terminals remain in radio resource control's (RRC) ACTIVE state even when they are not interacting with the network. In the 5G networks, the RRC INACTIVE state has greatly been altered where a UE could benefit from the stored context and go through a lower number of state transitions. 5G NR would thus get rid of the constant monitoring of physical downlink control channel (PDCCH) for the incoming transmissions. The proposed improvement brings a 50% less energy consumption at the modem and 18% for the entire device. Referring to the traditional RRC mechanism, only two states were available, namely RRC ACTIVE and RRC IDLE mode. Consumer's usage mainly dictates the time being spent in either of the two states. Typically, when a phone has not been used, the user inactivity timer would expire, putting the UE in IDLE state and as soon as it would go into the IDLE state its context would be removed from the core network. With the new RRC INACTIVE state, the UE context would still be stored when it would stop its communication with the network resulting in a reduced signaling overhead. However, the UE would still need to update eNodeB/gNodeB (evolved NodeB/next generation evolved NodeB) with its context for a valid state change. Figure 3 illustrates the state diagram of the new model. For this state to be widely utilized it should ensure minimum signaling and power consumption. The authors have evaluated the performance of this proposed scheme based on the shorter user inactivity timer achieving quicker state transitions to INACTIVE state and incurring less signaling. Power consumption analysis has been conducted for usage between different applications which validates the claim of authors. Similar analyses have been conducted to eliminate the prolonged connected mode discontinuous reception or better known as the Connected mode DRX (C-DRX) of upto 10 s for short data transfers and avoid the state changes. Signaling overhead also increases with the increase in either UE mobility or shorter user activity timers. However, the worst-case scenario would be to have the UE receive content just after its transition to the INACTIVE state, thus incurring extra RRC signaling. According to the proposed scheme, 5G NR can greatly benefit from this state by having an extended UE life and a lower need for S1 signaling.

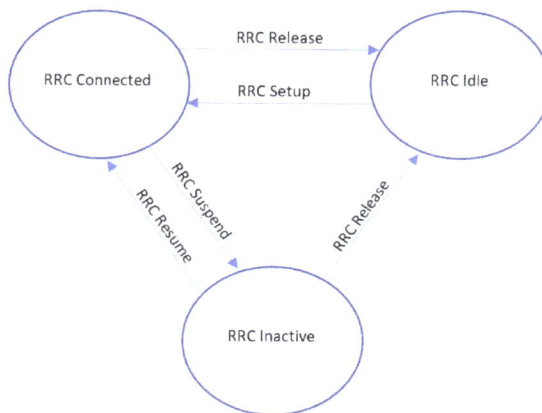

Figure 3. State diagram for radio resource control (RRC) signalling including the 'inactive' state.

3.4. Energy Efficient and Cache-Enabled 5G

In [7], the idea of proactive caching based on the content popularity on small cells has been proposed for improving the energy efficiency. Owing to the abundance of small cells, networks are getting constrained by the overall backhaul link capacity and much of the load is corresponding to transactions of the same requests repeatedly. Energy efficiency has been evaluated with regards to the content placement techniques and more emphasis has been put into organizing the content based on user locations and constantly fine tuning the clusters based on the content popularity distribution instead of spanning the same content across the network. Various topologies are shown in Figure 4. Energy efficiency has been formulated in relation to the small cell density vector. A heterogeneous file popularity distribution has been considered and a popularity vector has been maintained at every user. Users have been grouped into clusters depending upon the similarity in their interests and the cached files are an average of these popularity vectors. Users would usually be allowed to communicate with the base station within a specified distance of their cluster and in case of a cache miss event, the content would then be requested from the core via backhaul links. Spanning the same data across the network tends to sacrifice the information diversity and hence a content-based clustering approach has been brought forward. Simulations have been presented to demonstrate that with the increased base station density, significant energy efficiency gains have been experienced since the allocation problem gets simplified and interference and transmission powers would be reduced. In [34] a unique approach for addressing the energy efficiency challenge has been presented. The proposed E3 ratio thus incorporates a cost factor when calculating the number of UEs being served against the power spent over this operation by the BS. It has been made clear that although the cost factor might not have a direct impact on the spectral efficiency, it would be an important factor when regulating the cost of the entire network. Thus, operators have been addressed to carefully incorporate the features of edge caching and gigabit X-haul links to strike a fair balance between the cost overhead and the need of the feature. Otherwise it would be an overkill which has been meant to be strictly avoided. Mathematical analysis for EE maximization presented in [35] supports the fact that for the cases of low user cache size, non coded schemes should be utilized for a faster delivery system. Highlight of the research work conducted in [33] has been the assumption of a finite cache memory for a more realistic analysis. Delay bounds of an online cooperative caching scheme have been brought forward as compared to offline and a random caching scheme. The cache being periodically updated promises to deliver a tighter user association and aims to have minimum possible latency. The algorithm also aims to accurately cache the data in highest demand with an increased user density. Application of cooperative caching on P2P networks has been discussed in [37], authors have demonstrated the effectiveness of the algorithm by the segmentation of cache memory at the base stations. It would not only keep track of the cached data of the highly demanded information but would also record data paths and the newly requested data. The simulations have illustrated the usefulness of this optimization technique by the reduced number of hops and latency. On the other hand, uplink energy conservation has been considered in the context of dense small cells [36].

In [8], energy efficiency analysis of heterogeneous cache enabled 5G hyper cellular networks was performed. The control and user plane separation is considered to aid in devising enhanced access schemes and retain fairness in service. Furthermore, base station on-off strategy is taken into account to help in cutting down costs spent on redundant small cells [8]. In that scenario, macro cells would be the masters handling mobility, home subscriber and the user admission whereas small cells would be the slave part of the radio resource management scheme. With this increasing growth of the network infrastructure, irregularities in traffic behavior must be taken into account along with the actual user distribution for a realistic scenario. Caching has been sought after as a viable solution for reducing the end to end latency by storing content at the base stations. Small cells would typically involve macro base station in its communication with the UE in a semi sleep mode and ensure that it would always be aware of the UE positioning in the network as well as the cache memory statistics. Macro cell also ensures that the UE would be served by the closest and best possible small cell and would turn off the remaining ones to concentrate on a specified area for improving the throughput. On the other hand, there would be a predefined search radius and content would be fetched from a neighbouring base station within that distance. Otherwise, UE would associate to the macro base station for getting access to the needed content. Expressions for the coverage probability for the UE to get signal to interference (SIR) ratio within the threshold, throughput and power consumption and efficiency have been documented in [8].

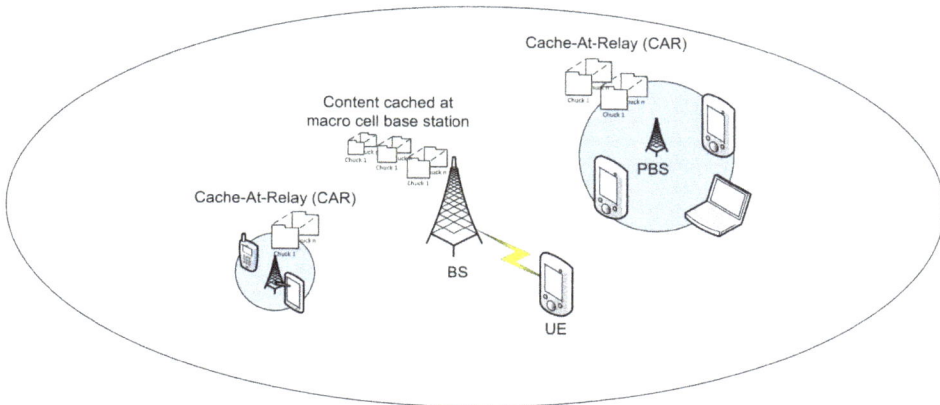

Figure 4. Illustration of different cache topologies.

4. Review of EE Techniques at the Network Level

A collective approach has been adopted for addressing the overall EE challenge considering both access and core network. EE has thus been gauged by the extent of resource sharing among different operators in the urban environment, utilization of efficient resource allocation schemes for fully exploiting the available spectrum, deploying middle ware for coverage enhancement (reduction in the distance between UE and BS would lower the needed transmission power), harnessing maximum computational muscle for accommodating massive incoming user requests yet have the ability to scale instantly (virtualization) and deploying machine learning and Software Defined Network (SDN) technologies for a fine grained control over the resources. An efficient usage of these capabilities would thus lead to the quality of service retention as well as an excellent power management methodology.

4.1. Resource Sharing in 5G with Energy-Efficiency Goal

Spectrum and physical resource sharing needs to be considered for accomplishing the energy efficiency goal of 5G. However, the need of service quality retention with respect to throughput and packet drops must also be addressed. Thoughts on infrastructure sharing have been gaining enough traction owing to several factors, for example, lack of space acquisition for site deployment or utilizing the available resources at their full potential and refraining from any new deployment. This section puts together the studies for bringing improvements in energy efficiency by a mutual sharing of infrastructure. Operators would have the flexibility of resource sharing at either full or partial level naturally emphasizing improved security for their equipment. Additionally, the cost of commissioning every site would lead to a higher expenditure and would minimize the expected revenues. Projects such as EARTH and GREEN TOUCH detail this avenue and brings forth an expectation of a decreased energy consumption by 1000 folds [2,38]. For this level of sophisticated resource sharing, a complete knowledge about the functionality and capacity of the network entities needs to be available which may not be possible in practice. However, the avenue of spectrum sharing still welcomes more discussion and aims to be a potential pathway for gaining solutions to the resource scarcity problem. Details of system level simulations for comparisons drawn between energy consumption and shared infrastructure at different load levels have been documented in [38] where a gain of up to 55% for energy efficiency in the dense areas has been demonstrated. Other significant advantages of resource sharing would include less interference by a planned cell deployment in accordance with the user demands per area. These efforts aim to eliminate the problems of either over provisioning or under-utilization of the deployed network entities. Authors in [40] have discussed the application of an improved resource allocation in a fog RAN. The suggested idea relies upon the fact that the usage of a centralized baseband processing unit, which, while increasing the processing power of the system, remains at risk of getting outdated measurements from the radio heads because of larger transport delays. The suggested algorithm starts off by switching off the redundant access points for conserving the energy and then modifying the beam weights for providing the end user with an optimum signal to leakage and noise ratio. User association is made centrally and then the information gets passed on to the fog access points after being scheduled for users. Following this phase, the proposed greedy algorithm tracks the global as well as the local energy efficiency readings and switches off the access points not needed until the rising trend of global energy efficiency ceases. Simulations have been carried out using a layout of macro and pico cells showing about a three-fold increase in the reported Channel State Information (CSI). Furthermore, authors in [39] have demonstrated the EE gains in a dynamic six-sector BS, capable of operating at either one or a maximum of all the sectors fully functioning, to be up to 75% as compared to the case of an always on approach.

In [9], a case study of infrastructure sharing between different operators has been presented as well. Service level agreement between the participating operators is defined and handled by multi-objective optimization methods. In such a shared environment, QoS should go hand in hand with fair resource utilization. Authors have specifically considered the case of obeying operator specific energy and spectral efficiency criteria along with the global spectral and energy efficiency maximization. The most prominent outcomes of this research are the global energy and spectral efficiency maximization in a shared noise-limited environment and the application of the framework to a network shared by any number of operators each serving different numbers of users and an optimal fulfillment of utility targets. Detailed mathematical analysis has been presented for system modelling with noise and interference constraints. SINR equations, which originally were used as a starting point, were thus gradually modified by incorporating weighting factors for influencing the priorities. This model turns out to be working in a polynomial complexity and maximizes the given objective function. Moreover, maximum and minimum bounds have been enclosed. In the paper, authors have presented the application of the mathematical tools by presenting the case of a base station installed in a crowded place such as an airport or shopping mall where the site owner is the neutral party and the frequency resources are either pooled or one of the operators grants some of his portion to others. Firstly, the case of two operators has been presented when they do not have any global constraints and the multi-objective problem set of noise limited scenario would be used. Secondly, site owner restricts the interference level or the global energy efficiency for both the operators and both of them target a minimum QoS constraint. Thirdly, there would be three operators with the same condition as of the first case. The work has laid the foundation to establish the criterion for the energy-spectral trade off in a single/multi carrier scenario.

4.2. Energy Efficient Resource Allocation in NOMA

In 5G, attempts have been made to possibly explore the area of non-orthogonal multiple access (NOMA), employing power control for saving resources in both time and frequency domain. This concept is highlighted in the following Figure 5. Operators would benefit from this technique by getting to serve the maximum number of users within the same frequency band, thus improving spectral efficiency [41]. This research area has been active for a while now for the reasons of increasing the network capacity and improving the data rates. An intelligent coordination among the base stations must be in place for maximum utilization of the available overall network energy. This corresponds to the fact that the harvested green energy has mostly been volatile, and a constant input source could not be guaranteed. For this reason, a detailed mathematical model has been presented for the power control of the UEs being serviced for minimizing interference as much as possible. A comparison of user association based genetic algorithms against a fixed transmit power was drawn. NOMA based techniques were demonstrated to outperform the conventional techniques for EE improvement for a larger number of nodes. The application was extended to a two-tier RAN having a macro base station covering a region of several pico base stations, being powered by both green and conventional energy sources. The proposed mathematical model uses a ratio of the network's data rate over the entire energy consumption as the network utility. Incorporation of improved user association techniques were suggested in [42] for improvement of user throughput and error containment in NOMA. In [43], authors presented the mathematical feasibility for the utilization of successive interference cancellation at the receiver side. The signal that is being processed considers others to be noise, cancels them out and its iterative nature aims to decode all of them. With an increase in the number of transmitters having a fixed SINR, a linear relationship has been observed. On the other hand, this formulation might lead to a saturation point for the explosive number of IoT devices.

The authors in [44], have taken an interesting approach for a fair comparison of NOMA and a relay-aided multiple access (RAMA) technique and a simulation was carried out for maximization of the sum rate. It was established via mathematical formulation that sum rate is an increasing function of user's transmission power and for the cases of a high data rate demand of the farthest user, NOMA proved to have maximized the sum rate. Distance between the users has been a key figure and with an increased separation between them, NOMA provides maximum rates whereas for the smaller separation relay-based setup provides a good enough sum rate. Authors in [45] have endorsed the advantages of nonorthogonal multiple access (NOMA) for the future radio access networks. Apart from the fact that the technique aids in getting a better spectral efficiency, authors instead have analyzed the feasibility of acquiring a better energy efficiency out of it as well. Considering the example of one base station serving two users, relationships between SE and EE have been observed which reflects that NOMA can potentially regulate the energy within the network by the allocation of more bandwidth to a cell center user in the uplink and more power to the cell edge user in the downlink. Considering the potential of NOMA, the problem was tackled with respect to its deployment scenario for the maximum exploitation. For a single cell deployment, EE mapping against resource allocation was considered as an NP hard problem because each user would be competing for the same radio resource, however, user scheduling and multiple access methods would aid for improving this situation. For the network level NOMA, a joint transmission technique could be beneficial for organizing the traffic load on the radio links and users must be scheduled accordingly when it comes to energy harvesting to keep the users with critical needs prioritized. Lastly, Grant free transmission has been studied for saving the signaling overhead, as soon as the user acquires data in its buffer it should start the uplink transmission and selection of the received data would be based upon its unique multiple access signature. Multiple access signature is deemed to be the basis of this proposal, but the signature pool must be carefully devised with an optimal tradeoff between the pool size and mutual correlation. It would greatly help for collision avoidance and detection. The users remain inactive for cutting down on the grant signaling and hence more energy is typically conserved. The proposed hybrid technique transitions between grant free and scheduled NOMA based on the current traffic load which eventually lowers down the collision probability and improves latency. In contrast with the above works that have discussed the use cases of caching in orthogonal multiple access (OMA), authors in [46] explored index based chaching instead of superposition chaching while adopting a sub optimal user clustering technique for significant reductions in the transmitted power while using NOMA. Owing to the enormous number of users, optimal user clustering was discouraged and user association based upon their differences in terms of link gain and cached data was suggested instead. The iterative power allocation algorithm was demonstrated to converge after several iterations.

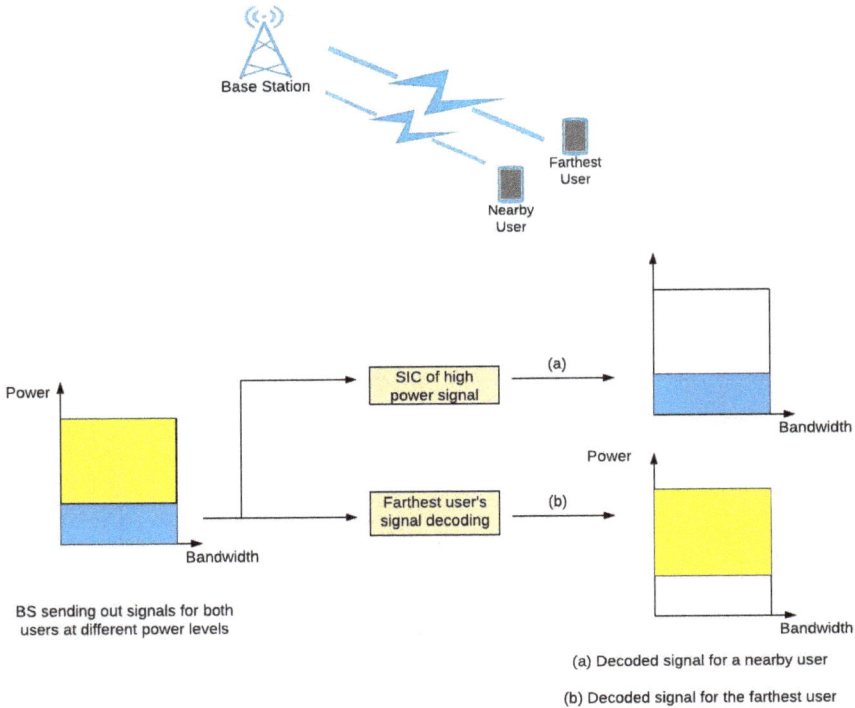

Figure 5. Concept of non-orthogonal multiple access (NOMA) technology.

4.3. Energy Efficient 5G Outdoor-Indoor Communication

The research in [47] discusses a use case of shared UE side distributed antenna system for indoor usage where a combination of distributed antenna and MIMO technology is used for getting enhancements in the coverage area and utilization of unlicensed frequencies for accommodating more users. The use of both licensed as well as unlicensed bands simultaneously needs a redesign of the current resource allocation algorithms [47]. In this work, resource allocation has been considered to be a non-convex optimization for increasing the end to end energy efficiency. The suggested topology demands installation of a shared UE side multiple antenna hardware between a single antenna base station (outdoor) and arbitrary number of single antenna UEs (indoor) which are called shared user equipment (UE)-side distributed antenna system (SUDACs). These SUDACs would be able to communicate the channel information with their neighbouring SUDAC units installed. In contrast with the relaying in the LTE-A system, SUDACs could be installed at different locations by the users and still be able to operate in both licensed and unlicensed bands simultaneously. The problem statement boils down to defining the energy efficiency in terms of the bits exchanged between base station and the UEs via SUDACs per joule of energy. It has been shown in [47] that application of this model exploits the frequency and spatial multiplexing of UEs and increases the system efficiency as compared to the case when SUDACs is not involved.

4.4. Energy Efficient Virtualization in 5G

Virtualization has been a very sought out way of reducing the time to market for the newer mobile technologies but with the emerging technological trends it might be a very useful way forward for reducing the energy consumption. In this case, hardware would serve as a bare metal for running multiple applications simultaneously for saving up on the cost of additional deployments of dedicated hardware and software components [48]. Most of the functions previously deployed on dedicated hardware would now be rolling out as software defined network functions thus promising scalability, performance maximization and mobility with in the cellular network. The virtual network architecture described in [50] lays out the interconnection between several virtual as well as the physical units being interconnected to form a larger system. A generalized 5G architecture incorporating virtualization has been illustrated in Figure 6. The smooth integration of different technologies with virtualized environment thus becomes the key of reaping the expected efficiency outcomes. Resource and operations management plays a vital role in actively regulating the system for a fine tuned state of execution that helps mitigate issues including redundancy and keeping the operating expenses under control. Furthermore, usage of an openflow switch would come in handy for efficient packet traversal within the network. Significant advantage in terms of reduced energy consumption of about 30% have been experienced by incorporating the current architecture with Network Function Virtualization (NFV). Authors have assumed an ideal case scenario that the virtual BBU will not consume any energy when it stays idle and also the advantage of the enormous computational pool in the form of cloud have been used.

Authors in [49] presented the significant energy conservation advantages of having virtual nodes in both access as well as the core network instead of having the physical nodes for executing only a single function. The proposed topology suggests baseband pooling for higher performance in the cloud, a direct gigabit optical connection from the remote radio heads to the core network and an even distribution of the core network nodes. The nearest available core network node would then be the one responsible of serving the incoming requests from the respective radio heads. The proposed architecture boasts the flexibility of resource distribution by having a single node running multiple virtualized access/core network functions e.g., serving gateway, packet gateway, etc. and the readiness of activating these functions wherever needed based on the work load. A visible gain of about 22% was recorded using mixed integer linear programming for modelling the work load across the nodes and both the core and access network were virtualized. Apart from the EE gains, a higher performance would also be achieved because of a reduced distance between the node requesting and the node serving the request. Research in [51] extends the same idea where the EE gains are deemed to be higher with an increased number of virtual function deployments in the access network which typically consumes more energy, about 70% of the entire demand of the end to end network. The suggested topology entails gigabit optical connectivity as the fronthaul technology instead of the Common Public Radio Interface (CPRI) connection between radio and baseband units. This brings out more deployment opportunities for the virtual machines by having more active nodes closer to the user. Authors documented a gain of about 19% with the proposed architecture. According to the authors in [52], existing RAN architecture needs modification for meeting the upcoming traffic demands. Baseband unit has been decomposed into two main parts, namely distributed unit and a central unit. Both units find their optimal placements either close to the users for serving the low latency demands or in remote areas for providing a pool of computational power. Mobile edge computing uses the same concept and NFV proves to be an enabling technology to use it to its full potential. The network layout comprises upon active antenna units and the central office for edge and access computation. Mobile edge computing units were housed along with the distributed and the central units and was the aggregator for the traffic. Both latter functions were virtualized on general purpose processors and finally the electronic switch was responsible for the traffic routing. Simulations conducted on this topology have revealed about 20% power saving as compared to the case of fixed deployment of hardware units. Moreover, Reference [53] also supports the idea of flexible centralization of RAN functions of small cells. Prominent outcomes would comprise upon interference mitigation in a dense deployment and reduced radio access processing. Authors in [54] devised an analytical model for calculating the optimal number of active operator's resources. Dynamic Auto Scaling Algorithm, or DASA, was envisioned to provide a way for operators to better understand their cost vs performance trade off and authors have thus used real life data from Facebook's data center for a realistic estimation. On top of the already established legacy infrastructure comprising mainly upon mobile management entity, serving gateway, packet gateway and the policy & charging function, 3GPP has now proposed specifications for a virtualized packet core providing on demand computational resources for catering to the massive incoming user requests. A comparison was drawn between the consumed power and the response time of the servers for the jobs in a queue by varying different factors including total number of virtual network function (VNF) instances, total number of servers available as well as the rate of the incoming jobs, total system capacity and the virtual machine (VM) setup times. Trends recorded from the plots have signified the saturation point of the system and have paved a way for operators to optimize their infrastructure to be robust without taking in more power than needed. Similarly [55] extends the above mentioned approach by taking into account the rejection of incoming requests in case the saturation point has been reached. A more realistic framework was presented that incorporates either dropping the jobs from the queue or even blocking them out from being registered until some resources could be freed up.

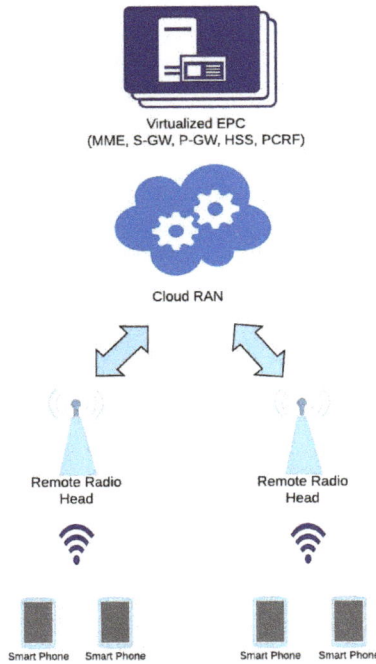

Figure 6. A 'virtualized' 5G architecture.

5. Review of SDN Technology for Enhancing EE

5.1. Energy Monitoring and Management in 5G with Integrated Fronthaul and Backhaul

The impact of software defined networking (SDN) on energy-efficiency was explored in [56]. The tremendous increase in the user density in a given area not only demands an energy efficient hardware but also demands for certain modifications in the control plane. Energy Management and Monitoring Applications (EMMA) were designed for observing the energy consumption in fronthaul as well as the backhaul network constituents. A monitoring layer was implemented over an SDN controller which observes the underlying operational domains including mmWave links and analogue

Radio over Fiber technology (RoF). This topology is shown in Figure 7. The energy management framework was extended to provide analysis on virtual network slices as well by gathering the real time power consumption data of a server by a power meter installed with it and then incorporating it with the respective flows. EMMA is based upon a SDN/NFV integrated transport network using a Beryllium framework and supports features including energy monitoring of the access network and the optimization of power states for the nodes. Furthermore, an analytics module provide statistics on the traffic consumption by the currently ongoing services, Provisioning manager would help in setting up new network connections and dynamic routing of connections for the ongoing sessions based upon the energy aware routing algorithms. Authors have envisioned EMMA as a fronthaul technology for providing coverage for high speed trains. It comprises upon a context information module for collection of data for mobility, a statistics module for storing the contextual data and updating it regularly, and lastly the management module for consuming this data and making real time moves in the network by switching on the nodes as the train approaches and switching them off when it leaves. Significant energy savings ranging between 10 to 60% were demonstrated using the real life data by switching on the nodes exactly when needed and keeping them asleep otherwise [56].

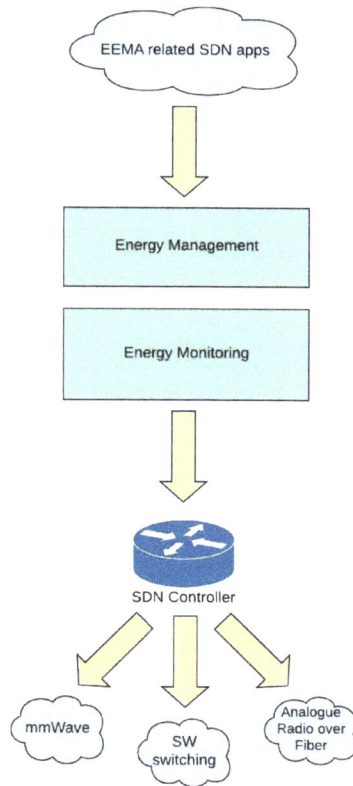

Figure 7. Energy Monitoring and Management via software defined networking (SDN).

5.2. Utility of Sleep Mode Energy Savings

In [57], authors discussed about getting benefited from the separated control and data planes in a heterogeneous network. Since this concept was not used in the previous generation of networks, further exploitation of this feature is expected to yield significant energy reductions. It was proposed that control plane communication would be done via low frequency macro cells and data plane information exchange would take place through high frequency femto cells. Detailed statistics about the daily traffic load and information about the kind of base stations deployed were tabulated. The application of the regular cell switch off technique especially in the off peak hours would yield reduced energy consumption of up to 48%. On the other hand, incorporation of power modulation at the femto cells would keep them operational and would yield energy savings of up to 27%. If the extreme isolation of control and data plane could be relieved then the macro cell would be able to serve the users with not just the control signaling but also with data transfer at low frequencies, resulting in a higher percentage of energy savings in the network. In addition to this concept, Reference [64] discusses the possibility of achieving 50–80% energy savings by incorporation of the energy aware heuristic algorithms.

6. Machine Learning Techniques for Energy-Efficiency in 5G

Recently, machine learning techniques have been employed to various areas of wireless networks including approaches to enhance energy efficiency of the wireless network [58]. A typical example would include a smart transmission point, such as the one shown in Figure 8 that would evolve itself overtime by its observations.

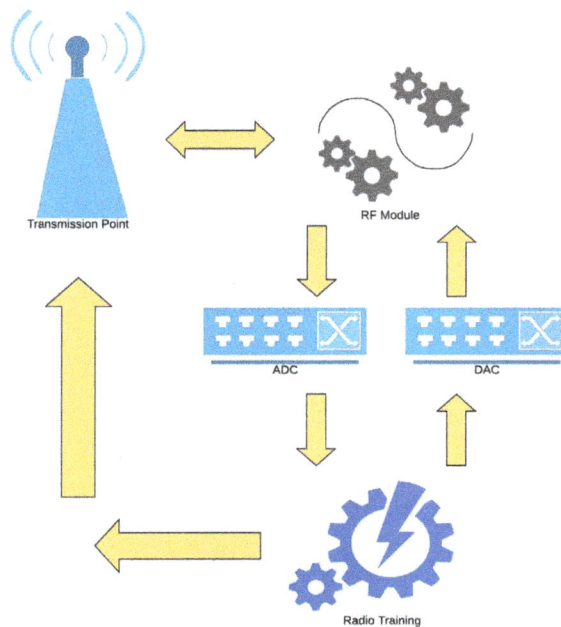

Figure 8. Dissection of a smart antenna.

In [59], the authors proposed switch-on/off policies for energy harvesting small cells through distributed Q-learning. A two tier network architecture was presented for discussion on on-off switching schemes based upon reinforcement learning. It is assumed that small cells are equipped to get their associated macrocell to transfer its load over to them and they themselves would rely upon the harvested energy, for example, solar energy. Application of Q-learning enables them to learn about the incoming traffic requests over time so they could tweak their operation to an optimal level. The proposed scenario includes a macro cell running on electricity and small cells running on solar energy with a distributed Q learning technique being used to gain knowledge about the current radio resource policies. Reward function for the online Q-learning proposes to turn off the small cells if users experience higher drop rates or use the ones that would already be on to take the burden from the macro cell. On the other hand, authors in [60] devised a novel EE and E2E delay duty cycle control scheme for controllers at the gateway of cellular and capillary networks. Formulation of a duty cycle control problem with joint-optimization of energy consumption and E2E delay was addressed followed by the distributed duty cycle control scheme.

In [61], the authors highlighted a distributed power control for two tier femtocell networks with QoS provisioning based on q-learning. Power control in the downlink of the two tier femtocell network was discussed and an effective network capacity measure was introduced for incorporating the statistical delay. Self-organization of small cells was also discussed with the perspective of Q-learning and utilization of a non cooperative game theory [61]. The proposed system model involves a macro base station covering several femtocells in its vicinity, each of them serving their own set of users. Expressions for SINR for both macro and femto cell users were also documented [61]. For the consumer's energy efficiency, Pareto optimization was opted for as compared to the traditional multi-user scenarios, focusing on a system level energy efficiency instead.

Meanwhile in [62], the deployment of macro and pico base stations were made similar to the above scenario. However, the random deployment of femto BS by consumers cause interference problems and cognitive radio technology was put together with these femto BS for an improved spectrum access. Spectrum sensing techniques provide benefits for UL transmission since the femto cells are power limited as compared to the macro cells. Detailed mathematical analysis for spectrum sensing techniques using both hard and soft decisions were demonstrated in [62]. Authors formulated objective functions in such a way that although they are computing optimal power allocation for the users, the whole scheme incorporates constraints for energy efficiency maximization. In [63], the authors also use machine learning techniques for energy-efficient resource allocation in 5G heterogeneous cloud radio access network. Cloud radio access networks are considered as a key enabler in upcoming 5G era by providing higher data rates and lower inter cell interference. It consists of both small cells and macro base stations for accommodating more users, providing them with superior quality of service and for enhancing coverage area respectively where resources are scheduled through a cloud RAN. A resource allocation scheme was put together with the aim of maximizing energy efficiency of UEs served by the radio heads while minimizing inter tier interference [63]. Available spectrum was divided into two resource blocks and assigned to different UE groups depending upon their location and QoS demands. A central controller interfaced with the baseband unit pool gets to learn about the network state through the interfaced macro base station and then take certain actions needed for energy efficiency optimization. Furthermore, compact state representation was utilized for approximating algorithm's convergence. The resource block as well as the power allocation with respect to energy saving in the downlink channel of remote radio heads in accordance with the QoS constraints has also been documented. Since the given model depends upon the prior UE knowledge for it to make transitions for optimization, Q-learning was proposed to practically model the objectives and system specifications. The resource allocation is mainly carried out at the controller in the BBU pool and the control signalling is carried out via the X1 and S1 links. The hierarchy of UEs and RRHs operate under macro base station and convey their states to the controller.

7. Challenges and Open Issues

In accordance with the increase in the computational demand from the base stations, in the upcoming 5G networks, energy efficiency needs to be scaled up by 100–1000 times in contrast with the traditional 4G network [1]. Since the transmission ranges would have been scaled down due the dense small cell deployment, the energy efficiency evaluation will potentially revolve around the computational side as compared to the transmission side previously. Storage functions for local data caching should also be considered in this evaluation, since it would potentially be common in the forthcoming networks. Scheduling schemes should be enhanced to involve an optimal number of antennas and bandwidth for resource allocation. The trade-off between transmission and computational power should be optimized considering the effects of the kind of transmission technology involved. Software Defined Networking might be a potential fix for this issue, yet it needs further exploration. Moreover, authors in [65] proposed the intermediate delays from source to destination to be incorporated in the energy efficiency formulation for an even more realistic estimation.

Most of the ongoing research has been discussing energy efficiency from a lot of different perspectives but so far a unifying approach has not been reached. Green Touch project has taken such an initiative but more exploration is needed for a stronger understanding [2].

With the explosive small cell deployment, 5G network would be interference limited so orthogonal transmission techniques might not be practical. The framework of sequential fractional programming might be extended for energy efficiency optimization with affordable complexity as suggested in [9]. Random Matrix theory and stochastic geometry appear as suitable statistical models for evaluating the randomness within the wireless networks, but a thorough research on energy efficiency needs to be conducted employing these tools.

Finally, the avenue of self-learning mechanisms is still less explored. Since local caching has been considered a potential answer for reducing the load on backhaul networks, novel approaches including this consideration need to be developed.

8. Conclusions

In this paper, we provide a survey of the state-of-the-art in energy-efficiency efforts in 5G. These new studies touch on several novel paradigms such as new radio, NOMA, ML-driven techniques and cache-enabled networks. Although there are several studies surveying the literature, our paper provides a clear classification of the proposed techniques with in-depth comparison. The paper is expected to be a road map for researchers in this field.

Funding: This work is supported by the U.S. National Science Foundation (NSF) under Grant CNS-1647135.

Conflicts of Interest: The authors declare no conflict of interest.

References

1. Ge, X.; Yang, J.; Gharavi, H.; Sun, Y. Energy Efficiency Challenges of 5G Small Cell Networks. *IEEE Commun. Mag.* **2017**, *55*, 184–191. [CrossRef] [PubMed]
2. Buzzi, S.; Li, C.; Klein, T.E.; Poor, H.V.; Yang, C.; Zappone, A. A Survey of Energy-Efficient Techniques for 5G Networks and Challenges Ahead. *IEEE J. Sel. Areas Commun.* **2016**, *34*, 697–709. [CrossRef]
3. Lorincz, J.; Matijevic, T. Energy-efficiency analyses of heterogeneous macro and micro base station sites. *Comput. Electr. Eng.* **2014**, *40*, 330–349. [CrossRef]
4. Zhang, Y.; Xu, Y.; Sun, Y.; Wu, Q.; Yao, K. Energy Efficiency of Small Cell Networks: Metrics, Methods and Market. *IEEE Access* **2017**, *5*, 5965–5971. [CrossRef]
5. Abrol, A.; Jha, R.K. Power Optimization in 5G Networks: A Step Towards GrEEn Communication. *IEEE Access* **2016**, *4*, 1355–1374. [CrossRef]
6. Ryoo, S.; Jung, J.; Ahn, R. Energy efficiency enhancement with RRC connection control for 5G new RAT. In Proceedings of the 2018 IEEE Wireless Communication and Networking Conference (WCNC), Barcelona, Spain, 15–18 April 2018; pp. 1–6.
7. Hajri, S.E.; Assaad, M. Energy Efficiency in Cache-Enabled Small Cell Networks with Adaptive User Clustering. *IEEE Trans. Wirel. Commun.* **2018**, *17*, 955–968. [CrossRef]
8. Zhang, J.; Zhang, X.; Imran, M.A.; Evans, B.; Wang, W. Energy Efficiency Analysis of Heterogeneous Cache-Enabled 5G Hyper Cellular Networks. In Proceedings of the 2016 IEEE Global Communications Conference (GLOBECOM), Washington, DC, USA, 4–8 December 2016; pp. 1–6.

9. Aydin, O.; Jorswieck, E.A.; Aziz, D.; Zappone, A. Energy-Spectral Efficiency Tradeoffs in 5G Multi-Operator Networks With Heterogeneous Constraints. *IEEE Trans. Wirel. Commun.* **2017**, *16*, 5869–5881. [CrossRef]
10. Rehan, S.; Grace, D. Efficient Joint Operation of Advanced Radio Resource and Topology Management in Energy-Aware 5G Networks. In Proceedings of the 2015 IEEE 82nd Vehicular Technology Conference (VTC2015-Fall), Boston, MA, USA, 6–9 September 2015; pp. 1–2.
11. Zhang, S.; Cai, X.; Zhou, W.; Wang, Y. Green 5G enabling technologies: An overview. *IET Commun.* **2019**, *13*, 135–143. [CrossRef]
12. Sun, Y.; Wang, Y.; Zhong, Y. Fronthaul Constrained Coordinated Transmission in Cloud-Based 5G Radio Access Network: Energy Efficiency Perspective. *IEICE Trans.* **2017**, *100-B*, 1343–1351. [CrossRef]
13. Kitanov, S.; Janevski, T. Energy efficiency of Fog Computing and Networking services in 5G networks. In Proceedings of the IEEE Eurocon 2017—17th International Conference on Smart Technologies, Ohrid, Macedonia, 6–8 July 2017; pp. 491–494.
14. Demestichas, K.; Adamopoulou, E.; Choraś, M. 5G Communications: Energy Efficiency. *Mob. Inf. Syst.* **2017**, *2017*, 5121302. [CrossRef]
15. Cavalcante, R.L.G.; Stanczak, S.; Schubert, M.; Eisenblaetter, A.; Tuerke, U. Toward Energy-Efficient 5G Wireless Communications Technologies: Tools for decoupling the scaling of networks from the growth of operating power. *IEEE Signal Process. Mag.* **2014**, *31*, 24–34. [CrossRef]
16. Isabona, J.; Srivastava, V.M. Downlink Massive MIMO Systems: Achievable Sum Rates and Energy Efficiency Perspective for Future 5G Systems. *Wirel. Pers. Commun.* **2017**, *96*, 2779–2796. [CrossRef]
17. Prasad K.S.V.; Hossain, E.; Bhargava, V.K. Energy Efficiency in Massive MIMO-Based 5G Networks: Opportunities and Challenges. *IEEE Wirel. Commun.* **2017**, *24*, 86–94. [CrossRef]
18. Nawawy, N.A.; Mohamed, N.; Dziyauddin, R.; Sam, S.M. Functional Split Architecture for Energy Efficiency in 5G Backhaul. In Proceedings of the 2018 2nd International Conference on Telematics and Future Generation Networks (TAFGEN), Kuching, Malaysia, 24–26 July 2018; pp. 98–102.
19. Rizvi, S.; Aziz, A.; Jilani, M.T.; Armi, N.; Muhammad, G.; Butt, S.H. An investigation of energy efficiency in 5G wireless networks. In Proceedings of the 2017 International Conference on Circuits, System and Simulation (ICCSS), London, UK, 14–17 July 2017; pp. 142–145.
20. Aligrudic, A.; Pejanovic-Djurisic, M. Energy efficiency metrics for heterogenous wireless cellular networks. In Proceedings of the 2014 Wireless Telecommunications Symposium, Washington, DC, USA, 9–11 April 2014; pp. 1–4.
21. Bouras, C.; Diles, G. Energy efficiency in sleep mode for 5G femtocells. In Proceedings of the 2017 Wireless Days, Porto, Portugal, 29–31 March 2017; pp. 143–145.
22. Beitelmal, T.; Szyszkowicz, S.S.; G, D.G.; Yanikomeroglu, H. Sector and Site Switch-Off Regular Patterns for Energy Saving in Cellular Networks. *IEEE Trans. Wirel. Commun.* **2018**, *17*, 2932–2945. [CrossRef]
23. Lorincz, J.; Matijevic, T.; Petrovic, G. On interdependence among transmit and consumed power of macro base station technologies. *Comput. Commun.* **2014**, *50*, 10–28. [CrossRef]
24. Yang, C.; Li, J.; Ni, Q.; Anpalagan, A.; Guizani, M. Interference-Aware Energy Efficiency Maximization in 5G Ultra-Dense Networks. *IEEE Trans. Commun.* **2017**, *65*, 728–739. [CrossRef]
25. Zappone, A.; Jorswieck, E.A. Energy Efficiency in Wireless Networks via Fractional Programming Theory. *Found. Trends Commun. Inf. Theory* **2015**, *11*, 3–4. [CrossRef]
26. Li, W.; Wang, J.; Yang, G.; Zuo, Y.; Shao, Q.; Li, S. Energy efficiency maximization oriented resource allocation in 5G ultra-dense network: Centralized and distributed algorithms. *Comput. Commun.* **2018**, *130*, 10–19. [CrossRef]
27. Boumard, S.; Harjula, I.; Kanstrén, T.; Rantala, S.J. Comparison of Spectral and Energy Efficiency Metrics Using Measurements in a LTE-A Network. In Proceedings of the 2018 Network Traffic Measurement and Analysis Conference (TMA), Vienna, Austria, 26–29 June 2018; pp. 1–8.
28. Kanwal, K.; Safdar, G.A.; Ur-Rehman, M.; Yang, X. Energy Management in LTE Networks. *IEEE Access* **2017**, *5*, 4264–4284. [CrossRef]
29. Yusoff, R.; Baba, M.D.; Ali, D. Energy-efficient resource allocation scheduler with QoS aware supports for green LTE network. In Proceedings of the 2015 IEEE 6th Control and System Graduate Research Colloquium (ICSGRC), Shah Alam, Malaysia, 10–11 August 2015; pp. 109–111.

30. Feng, Y.; Shen, X.; Zhang, R.; Zhou, P. Interference-area-based resource allocation for full-duplex communications. In Proceedings of the 2016 IEEE International Conference on Communication Systems (ICCS), Shenzhen, China, 14–16 December 2016; pp. 1–5.

31. Wen, K.; Chen, Y.; Hu, Y. A resource allocation method for D2D and small cellular users in HetNet. In Proceedings of the 2017 3rd IEEE International Conference on Computer and Communications (ICCC), Chengdu, China, 13–16 December 2017; pp. 628–632.

32. Qi, Z.; Fan, J.; Ji, P.; Xia, F.; Huang, X.; Zhao, S. Highly Energy-Efficient Resource Allocation in Power Telecommunication Network. In Proceedings of the 2017 International Conference on Computer Systems, Electronics and Control (ICCSEC), Dalian, China, 25–27 December 2017; pp. 488–492.

33. Yen, C.; Chien, F.; Chang, M. Cooperative Online Caching in Small Cell Networks with Limited Cache Size and Unknown Content Popularity. In Proceedings of the 2018 3rd International Conference on Computer and Communication Systems (ICCCS), Nagoya, Japan, 27–30 April 2018; pp. 173–177.

34. Yan, Z.; Peng, M.; Wang, C. Economical Energy Efficiency: An Advanced Performance Metric for 5G Systems. *IEEE Wirel. Commun.* **2017**, *24*, 32–37. [CrossRef]

35. Vu, T.X.; Chatzinotas, S.; Ottersten, B. Energy-efficient design for edge-caching wireless networks: When is coded-caching beneficial? In Proceedings of the 2017 IEEE 18th International Workshop on Signal Processing Advances in Wireless Communications (SPAWC), Sapporo, Japan, 3–6 July 2017; pp. 1–5.

36. Erol-Kantarci, M. Content caching in small cells with optimized uplink and caching power. In Proceedings of the 2015 IEEE Wireless Communications and Networking Conference (WCNC), New Orleans, LA, USA, 9–12 March 2015; pp. 2173–2178.

37. Zhou, X.; Lu, Z.; Gao, Y.; Yu, Z. An Effective Cooperative Caching Scheme for Mobile P2P Networks. In Proceedings of the 2014 International Conference on Computational Intelligence and Communication Networks, Bhopal, India, 14–16 November 2014; pp. 408–411.

38. Georgakopoulos, A.; Margaris, A.; Tsagkaris, K.; Demestichas, P. Resource Sharing in 5G Contexts: Achieving Sustainability with Energy and Resource Efficiency. *IEEE Veh. Technol. Mag.* **2016**, *11*, 40–49. [CrossRef]

39. Arbi, A.; O'Farrell, T. Energy efficiency in 5G access networks: Small cell densification and high order sectorisation. In Proceedings of the 2015 IEEE International Conference on Communication Workshop (ICCW), London, UK, 8–12 June 2015; pp. 2806–2811.

40. Dinh, T.H.L.; Kaneko, M.; Boukhatem, L. Energy-Efficient User Association and Beamforming for 5G Fog Radio Access Networks. In Proceedings of the 2019 16th IEEE Annual Consumer Communications & Networking Conference (CCNC), Las Vegas, NV, USA, 11–14 January 2019; pp. 1–6.

41. Xu, B.; Chen, Y.; Carrión, J.R.; Zhang, T. Resource Allocation in Energy-Cooperation Enabled Two-Tier NOMA HetNets Toward Green 5G. *IEEE J. Sel. Areas Commun.* **2017**, *35*, 2758–2770. [CrossRef]

42. Benjebbour, A.; Saito, Y.; Kishiyama, Y.; Li, A.; Harada, A.; Nakamura, T. Concept and practical considerations of non-orthogonal multiple access (NOMA) for future radio access. In Proceedings of the 2013 International Symposium on Intelligent Signal Processing and Communication Systems, Naha, Japan, 12–15 November 2013; pp. 770–774.

43. Rabee, F.A.; Davaslioglu, K.; Gitlin, R. The optimum received power levels of uplink non-orthogonal multiple access (NOMA) signals. In Proceedings of the 2017 IEEE 18th Wireless and Microwave Technology Conference (WAMICON), Cocoa Beach, FL, USA, 24–25 April 2017; pp. 1–4.

44. Choi, J. On the spectral efficient nonorthogonal multiple access schemes. In Proceedings of the 2016 European Conference on Networks and Communications (EuCNC), Athens, Greece, 27–30 June 2016; pp. 277–281.

45. Yang, K.; Yang, N.; Ye, N.; Jia, M.; Gao, Z.; Fan, R. Non-Orthogonal Multiple Access: Achieving Sustainable Future Radio Access. *IEEE Commun. Mag.* **2019**, *57*, 116–121. [CrossRef]

46. Fu, Y.; Liu, Y.; Wang, H.; Shi, Z.; Liu, Y. Mode Selection Between Index Coding and Superposition Coding in Cache-Based NOMA Networks. *IEEE Commun. Lett.* **2019**, *23*, 478—481 [CrossRef]

47. Ng, D.W.K.; Breiling, M.; Rohde, C.; Burkhardt, F.; Schober, R. Energy-Efficient 5G Outdoor-to-Indoor Communication: SUDAS over Licensed and Unlicensed Spectrum. *IEEE Trans. Wirel. Commun.* **2016**, *15*, 3170–3186. [CrossRef]

48. Al-Quzweeni, A.N.; Lawey, A.Q.; Elgorashi, T.E.H.; Elmirghani, J.M.H. Optimized Energy Aware 5G Network Function Virtualization. *IEEE Access* **2019**, *7*, 44939–44958. [CrossRef]

49. Al-Quzweeni, A.; El-Gorashi, T.E.; Nonde, L.; Elmirghani, J.M. Energy efficient network function virtualization in 5G networks. In Proceedings of the 2015 17th International Conference on Transparent Optical Networks (ICTON), Budapest, Hungary, 5–9 July 2015; pp. 1–4.

50. Abdelwahab, S.; Hamdaoui, B.; Guizani, M.; Znati, T. Network function virtualization in 5G. *IEEE Commun. Mag.* **2016**, *54*, 84–91. [CrossRef]

51. Al-Quzweeni, A.; Lawey, A.; El-Gorashi, T.; Elmirghani, J.M.H. A framework for energy efficient NFV in 5G networks. In Proceedings of the 2016 18th International Conference on Transparent Optical Networks (ICTON), Trento, Italy, 10–14 July 2016; pp. 1–4.

52. Xiao, Y.; Zhang, J.; Ji, Y. Energy Efficient Placement of Baseband Functions and Mobile Edge Computing in 5G Networks. In Proceedings of the 2018 Asia Communications and Photonics Conference (ACP), Hangzhou, China, 26–29 October 2018; pp. 1–3.

53. Sabella, D.; De Domenico, A.; Katranaras, E.; Imran, M.A.; Di Girolamo, M.; Salim, U.; Lalam, M.; Samdanis, K.; Maeder, A. Energy Efficiency Benefits of RAN-as-a-Service Concept for a Cloud-Based 5G Mobile Network Infrastructure. *IEEE Access* **2014**, *2*, 1586–1597. [CrossRef]

54. Ren, Y.; Phung-Duc, T.; Chen, J.; Yu, Z. Dynamic Auto Scaling Algorithm (DASA) for 5G Mobile Networks. In Proceedings of the 2016 IEEE Global Communications Conference (GLOBECOM), Washington, DC, USA, 4–8 December 2016; pp. 1–6.

55. Phung-Duc, T.; Ren, Y.; Chen, J.; Yu, Z. Design and Analysis of Deadline and Budget Constrained Autoscaling (DBCA) Algorithm for 5G Mobile Networks. In Proceedings of the 2016 IEEE International Conference on Cloud Computing Technology and Science (CloudCom), Luxembourg City, Luxembourg, 12–15 December 2016; pp. 94–101.

56. Abdullaziz, O.I.; Capitani, M.; Casetti, C.E.; Chiasserini, C.F.; Chundrigar, S.B.; Landi, G.; Talat, S.T. Energy monitoring and management in 5G integrated fronthaul and backhaul. In Proceedings of the 2017 European Conference on Networks and Communications (EuCNC), Oulu, Finland, 12–15 June 2017; pp. 1–6.

57. Klapez, M.; Grazia, C.A.; Casoni, M. Energy Savings of Sleep Modes Enabled by 5G Software-Defined Heterogeneous Networks. In Proceedings of the 2018 IEEE 4th International Forum on Research and Technology for Society and Industry (RTSI), Palermo, Italy, 10–13 September 2018; pp. 1–6.

58. Jiang, C.; Zhang, H.; Ren, Y.; Han, Z.; Chen, K.; Hanzo, L. Machine Learning Paradigms for Next-Generation Wireless Networks. *IEEE Wirel. Commun.* **2017**, *24*, 98–105. [CrossRef]

59. Miozzo, M.; Giupponi, L.; Rossi, M.; Dini, P. Switch-On/Off Policies for Energy Harvesting Small Cells through Distributed Q-Learning. In Proceedings of the 2017 IEEE Wireless Communication and Networking Conference Workshops (WCNCW), San Francisco, CA, USA, 19–22 March 2017; pp. 1–6.

60. Li, Y.; Chai, K.K.; Chen, Y.; Loo, J. Duty cycle control with joint optimisation of delay and energy efficiency for capillary machine-to-machine networks in 5G communication system. *Trans. Emerg. Telecommun. Technol.* **2015**, *26*, 56–69. [CrossRef]

61. Li, Z.; Lu, Z.; Wen, X.; Jing, W.; Zhang, Z.; Fu, F. Distributed Power Control for Two-Tier Femtocell Networks with QoS Provisioning Based on Q-Learning. In Proceedings of the 2015 IEEE 82nd Vehicular Technology Conference (VTC2015-Fall), Boston, MA, USA, 6–9 September 2015; pp. 1–6.

62. Park, H.; Hwang, T. Energy-Efficient Power Control of Cognitive Femto Users for 5G Communications. *IEEE J. Sel. Areas Commun.* **2016**, *34*, 772–785. [CrossRef]

63. AlQerm, I.; Shihada, B. Enhanced machine learning scheme for energy efficient resource allocation in 5G heterogeneous cloud radio access networks. In Proceedings of the 2017 IEEE 28th Annual International Symposium on Personal, Indoor, and Mobile Radio Communications (PIMRC), Montreal, QC, Canada, 8–13 October 2017; pp. 1–7.

64. Fernández-Fernández, A.; Cervelló-Pastor, C.; Ochoa-Aday, L. Energy Efficiency and Network Performance: A Reality Check in SDN-Based 5G Systems. *Energies* **2017**, *10*, 2132. [CrossRef]

65. Wu, G.; Yang, C.; Li, S.; Li, G.Y. Recent advances in energy-efficient networks and their application in 5G systems. *IEEE Wirel. Commun.* **2015**, *22*, 145–151. [CrossRef]

MDPI
St. Alban-Anlage 66
4052 Basel
Switzerland
Tel. +41 61 683 77 34
Fax +41 61 302 89 18
www.mdpi.com

Sensors Editorial Office
E-mail: sensors@mdpi.com
www.mdpi.com/journal/sensors

www.ingramcontent.com/pod-product-compliance
Lightning Source LLC
Chambersburg PA
CBHW051708210326
41597CB00032B/5413